PROFESSIONAL
CONSTRUCTION
MANAGEMENT

Including C.M., Design-Construct,
and General Contracting

Also Available from McGraw-Hill

Schaum's Outline Series in Civil Engineering

Most outlines include basic theory, definitions, and hundreds of solved problems and supplementary problems with answers.

Titles on the Current List Include:

Advanced Structural Analysis
Basic Equations of Engineering
Descriptive Geometry
Dynamic Structural Analysis
Engineering Mechanics, 4th edition
Fluid Dynamics
Introduction to Engineering Calculations
Introductory Surveying
Mathematical Handbook of Formulas & Tables
Mechanical Vibrations
Reinforced Concrete Design, 2d edition
Space Structural Analysis
State Space & Linear Systems
Statics and Strength of Materials
Strength of Materials, 2d edition
Structural Analysis
Structural Steel Design, LFRD Method
Theoretical Mechanics

Schaum's Solved Problems Books

Each title in this series is a complete and expert source of solved problems containing thousands of problems with worked out solutions.

Related Titles on the Current List Include:

3000 Solved Problems in Calculus
2500 Solved Problems in Differential Equations
2500 Solved Problems in Fluid Mechanics & Hydraulics
3000 Solved Problems in Linear Algebra
2000 Solved Problems in Numerical Analysis
800 Solved Problems in Vector Mechanics for Engineers: Dynamics
700 Solved Problems in Vector Mechanics for Engineers: Statics

Available at your College Bookstore. A complete list of Schaum titles may be obtained by writing to: Schaum Division
McGraw-Hill, Inc.
Princeton Road, S-1
Hightstown, NJ 08520

PROFESSIONAL CONSTRUCTION MANAGEMENT

Including C.M., Design-Construct, and General Contracting

Third Edition

Donald S. Barrie

President
CM Consultants

Boyd C. Paulson, Jr.

Professor of Civil Engineering
Stanford University

McGraw-Hill, Inc.

New York   St. Louis   San Francisco   Auckland   Bogotá   Caracas
Lisbon   London   Madrid   Mexico   Milan   Montreal   New Delhi
Paris   San Juan   Singapore   Sydney   Tokyo   Toronto

PROFESSIONAL CONSTRUCTION MANAGEMENT
Including C.M., Design-Construct, and General Contracting

International Edition 1992.

690.068 BAR

Exclusive rights by McGraw-Hill Book Co-Singapore for manufacture and export. This book cannot be re-exported from the country to which it is consigned by McGraw-Hill.

2 3 4 5 6 7 8 9 0 KHL SW 9 6 5 4 3 2

ISBN 0-07-003889-9

1064463.6

Library of Congress Cataloging-in-Publication Data

Barrie, Donald S.
  Professional construction management: including C.M. design.
construct and general contracting/Donald S. Barrie, Boyd C.
Paulson, Jr. — 3rd ed.
      p.   cm.
  Includes index.
  ISBN 0-07-003889-9
  1. Construction industry — Management.   2. Value analysis (Cost control) I. Paulson, Boyd C. II. Title.
TH438.B23   1992
624'.068 — dc20   91-27799

This book was set in Times Roman by Electronic Technical Publishing Services Company.
The editors were B. J. Clark and John M. Morriss;
the production supervisor was Richard A. Ausburn.
The cover was designed by Joe Gillians.
New drawings were done by ETP Services Company.
Project supervision was done by Electronic Technical Publishing Services Company.

When ordering this title, use ISBN No: 0-07-112917-0.

Printed in Singapore.

ABOUT THE AUTHORS

Donald S. Barrie holds a B.S. in civil engineering from the California Institute of Technology and has 40 years' experience managing construction in the field and the home office.

He held an appointment as consulting professor at Stanford University and has taught in the graduate programs at Stanford and the University of California. He is author or co-author of three books and is a registered civil engineer in Washington. He holds general engineering, building and electrical contractor's licenses in California.

From 1950 to 1984 he was with Kaiser Engineers, holding positions as resident engineer, general superintendent and project manager on a number of field and office assignments. Major field projects under his direction included the American River Project featuring 25 miles of tunnel, seven dams and four powerhouses. Other field assignments included industrial, power, building, and heavy construction projects in the United States and Canada. Home office assignments include estimator, project manager, department manager, and vice president, general construction division. Home office responsibilities included management of construction and construction management projects valued at over $1.5 billion. Construction projects included the shuttle assembly building and launch mount at Vandenberg AF Base, a refinery, an experimental power plant and electrical construction for a nuclear plant. Construction management projects included a major interstate highway program, coal mine development, and a number of food manufacturing and distribution facilities.

Currently he is president of CM Consultants, Incorporated, a construction-oriented management consulting firm providing broad services to owners, designers, and contractors.

Boyd C. Paulson, Jr., holds an endowed chair, the Ohbayashi Professorship of Engineering, in Stanford University's graduate program in construction engineering and management, where he has taught since 1974. He earned his doctorate in civil engineering from Stanford. He has also taught at the University of Illinois, and has been a visiting professor at the University of Tokyo in Japan, the Technical University of Munich in Germany, and at the University of Strathclyde in Scotland. He has extensive practical experience working in the construction industry, and has been a consultant to various companies and public agencies.

Professor Paulson's research interests are in computer applications in construction, including operations simulation and automated process control. His teaching focuses on the analysis and design of construction operations, computer applications, and construction equipment and methods. He is a member of ACM, ASCE, ASEE, and IEEE Computer Society. He was past chairman of the ASCE Construction Division Executive Committee. His honors and awards include ASCE's 1980 *Walter L. Huber Civil Engineering Research Prize* and 1984 *Construction Management Award*, Germany's *Humboldt Foundation Fellowship* in 1983, the Project Management Institute's 1986 *Distinguished Contributions Award*, and he was a 1990–91 scholar under both the Fulbright Foundation and The British Council.

CONTENTS

ix

Part 3   Methods in Project Management

Part 4 Business Methods in Managing Construction

PREFACE TO THE THIRD EDITION

In the years since this book's first and second editions were published, construction management (CM) has become a generally accepted alternative to traditional construction contracting procedures. It is by no means a panacea, but there are a variety of conditions where this is the best way of meeting an owner's project objectives as well as other situations where different management methods may be preferable. It is therefore important for owners, construction managers, contractors, and architect/engineers to understand the main available alternatives and to be able to select and apply the management alternative and type of contract that best fits the needs of each project.

Although the original emphasis of the first and second editions was on the professional construction management approach, it mainly used this perspective to explain almost all major alternatives and described the advantages and disadvantages of each. The majority of sections in the book dealt with methods and procedures that applied to any form of construction contracting.

In response to suggestions from readers and reviewers, several new sections and chapters have been included in the third edition. First, the entire test has been expanded to include detailed discussions, comparisons, and examples covering general contracting, design-build, and program management, as well as construction management. Chapter 2 adds a description of organizational options including functional, task force, line and staff, and matrix concepts. Chapters 4 through 9 have incorporated practical examples of general contracting and design-build concepts in addition to the construction management principles previously included. Chapter 11 has been modified extensively to incorporate discussions, comparisons and examples from building, heavy construction, and industrial estimating practice. A new Chapter 18 includes information on risk management, insurance, bonding, and liens and licensing. New Chapter 20 covers current practice in claims preparation, defense and alternate dispute resolution methods. Other chapters have been updated to conform to present practice, including a review of two decades of construction management nomenclature and practice set forth in Chapter 21.

Together with revisions and updates to the original material, this book now provides improved broad coverage and practical details for students and working professionals seeking to improve their management of future construction projects.

The following reviewers have been helpful in giving constructive suggestions for improved continuity and accuracy: Frederick Gould, Wentworth Institute of Technology; Charles Honigman, Pennsylvania State University; Oktay Ural, Florida International University; and Trefor P. Williams, Rutgers University.

Donald S. Barrie
Boyd C. Paulson, Jr.

PREFACE TO THE SECOND EDITION

In the years since this book's first edition was published, professional construction management has become a generally accepted alternative to traditional construction contracting procedures. It is by no means a panacea, but there are a variety of conditions where this is the best way of meeting an owner's project objectives. It is therefore important for owners, contractors, and architect/engineers to understand all of the main alternatives and to be able to select and apply the ones that best fit the needs of each project.

Although the original emphasis of the first edition was on the professional construction management approach, it mainly used this perspective to explain almost all major alternatives and described the advantages and disadvantages of each. The majority of sections in the book dealt with methods and procedures that applied to any form of construction contracting.

In response to suggestions from readers of the first edition, several new sections and chapters have been included in the second edition. First, Chapter 2 adds a description of program management which is another contract administration alternative gaining wider acceptance. Chapter 9 adds a new section on marketing. For readers needing a review of basic critical-path network-scheduling techniques, Appendix E has been added to supplement the applications-oriented topics in Chapter 12. Chapters 17 and 19 broaden project planning and control methods to include material on computer applications and industrial relations. Chapter 20 is also new and provides guidelines for implementing concepts covered in the book.

Together with revisions and updates to the original material, this book now provides both broad coverage and practical details for students and working professionals seeking to improve their management of future construction projects.

The following reviewers have been helpful in giving constructive suggestions for improved continuity and accuracy: George Blessis, North Carolina State University; Stephen Nunnally, North Carolina State University; and Jerald L. Rounds, Iowa State University.

Donald S. Barrie
Boyd C. Paulson, Jr.

PREFACE TO THE FIRST EDITION

In the years ahead, the construction industry will be challenged by increasingly difficult and complex problems in both engineering and management. This book offers its readers and the industry a challenge of another sort: to aspire to a still higher degree of professionalism in construction and, particularly in the management of construction, to overcome what otherwise might become insurmountable obstacles to the industry's continued prosperity.

In the past decade, professional construction management has in many applications emerged as a strong alternative to traditional construction contracting procedures. Toward this end, we address this subject on two levels. First, in a narrow sense, "professional construction management" means a three-party team, consisting of an owner, an architect/engineer, and a professional construction manager, united in a nonadversary contractual relationship to best serve the needs of the owner's project. In this case, considerable practical, experience-based guidance is given for the appropriate applications and limitations of this contractual arrangement as an alternative to more traditional approaches. Part 2, in particular, focuses on this alternative. Second, on a broader level, both the philosophy inherent in professional construction management and most of the methods and procedures that are discussed herein can enhance almost all the available contractual approaches, including design followed by competitively bid general contracts, "design-construct" or "turn-key" projects, "separate contracts," "design-manage," and even an owner's in-house "force-account" work, as well as professional construction management itself. Industry needs all these options to accomplish the challenging projects in its future. Part 3 of this book therefore goes into some detail on several of the basic management planning and control tools that apply in all these alternative contracting methods.

In both organization and content, this book has been designed to provide students and practitioners with a practical, in-depth introduction and orientation to this challenging subject and to further acquaint them with the major engineering and management techniques used in the professional construction management approach. A major feature of the book is the extensive use of examples related to a hypothetical project that is based on the senior author's own successful experience with professional construction management in the construction of several real projects.

As a college text, this book is written to be effective on several levels. First, it can serve as the text for a self-contained survey introduction to construction engineering and management. Such a course is often included in conventional undergraduate civil engineering and architecture programs. In this case, the main prerequisite is a major in such a related field. On a higher level, the student may gain even more from the text in a course that focuses more directly on construction administration or on construction planning and controls. In this case, it will be helpful, though not essential, for the student to have first had courses in either or both (1) construction specifications, contracts, and law; and (2) construction planning and scheduling. In either case, the example project that is introduced in Chapter 4 and amplified in Appendix A will provide realistic data for assignments designed to reinforce the student's understanding of new concepts as they are introduced. Appendix A also gives information on how to obtain reproducible copies of the original drawings and specifications.

Several other prople contributed to the preparation of this book. In particular, we would like to thank Leo Rosenthal, registered architect and engineer in Denver, Colorado, for preparing the drawings that are reproduced in Appendix A. We are grateful to the American Society of Civil Engineers for permission to draw extensively upon some of our own earlier papers published by the Society, and for permission to reproduce in Chapter 16 the quality assurance drawings referenced therein. Kaiser Engineers also granted us permission to use several of its figures and reports. We are also indebted to our families for their patience and support while we worked evenings, weekends, and holidays that would have been more enjoyable spent together. Finally, we acknowledge the following persons for their helpful coments regarding the manuscript: Professor Joseph E. Bowles, Bradley University; Professor Keith C. Crandall, University of California at Berkeley; Professor Ben C. Gerwick, Jr., University of California at Berkeley; Dr. Thomas C. Kavanaugh, Iffin, Kavanaugh, and Waterbury of New York City; Professor Walter L. Meyer, University of Colorado at Boulder; Mr. James J. O'Brien, Professional Engineer; and Professor C. H. Oglesby, Stanford University.

Donald S. Barrie
Boyd C. Paulson, Jr.

PROFESSIONAL CONSTRUCTION MANAGEMENT
Including C.M., Design-Construct, and General Contracting

PART **ONE**

CONSTRUCTION INDUSTRY AND PRACTICE

1

MANAGEMENT IN THE ENGINEERING AND CONSTRUCTION INDUSTRY

From architects' and engineers' dreams to the final coat of paint, construction is responsible for many of our noblest works—and for some of our most humble. Designers and constructors from history left us the Mayan and Egyptian pyramids, the Gothic cathedrals, the Great Wall of China, and, quite literally, the physical as well as the technological foundations upon which many of our modern structures are built. The scope of the industry today is immense: from suburban homes to 100-story skyscrapers; from city sidewalks to dams and tunnels for irrigation and hydroelectric power; from recreational marinas to complete harbors and even structures in the deep open sea; from bicycle shops to aircraft factories; thermal power plants, petroleum refineries, and mining developments; bridges, highways, and rapid transit systems that not only span physical spaces, but bring people together in their social, political, and economic endeavors. For better or worse, the constructed environment is among the most ubiquitous and pervasive factors in our lives.

CONSTRUCTION'S FUTURE

The most profound recent developments in construction are the increasing size of many of its projects and organizations, the increasing technological complexity of such projects, more complex interdependencies and variations in the relationships among its organizations and institutions, and proliferating regulations and demands from government. At the project level, management has just begun to integrate design, procurement, and construction into one total process. There are now, and will continue to be, shortages of resources, including materials, equipment, skilled workers, and technical and supervisory staff. There will be more and more governmental regulation of the safety of design and of field construction methods, environmental consequences of projects, and personnel policies at all levels. Management must also cope with new economic and cultural realities resulting from inflation, energy shortages, changing world development patterns, and new societal standards. These trends have been accelerating and will probably continue into the future. Figure 1-1 summarizes some of the elements that are involved.

Clearly, economic difficulties and increasing shortages of materials and other resources play a major role in the problems now facing today's projects. But this is not to say that engineers and managers must sit by hopelessly while urgently needed projects run out of control or are abandoned altogether. On the contrary, it is all the more critical that the skills of project engineers and managers improve, and that they have better tools with which to work, so they can optimize the planning and control of available resources and better cope with challenging realities imposed by new economic constraints. In spite of continuing economic problems, there is an ongoing need for the construction industry to expand and improve its capabilities and its scope of operations to meet changing and, in the long run, growing demands for its services.

The potential benefits from improved methods of accomplishing the management of future projects are worth seeking. Consider the experience of nuclear power plants in recent decades. A number of plants which were originally estimated to cost $500 million in the 1960s came on line at 2 billion or more in the mid 1980s with construction schedules as long as three times original estimates. Many plants were never completed or operated and some are in the process of being converted to fossil fuel. While environmental and anti-nuclear sentiment has played a major role in preventing any new plants being started for over a decade, the bottom line is that, due in part to an overall management and regulatory system failure, constructing new nuclear power plants in the United States is not economical even though such plants continue to be competitive in other developed countries. Currently proponents of nuclear power are concentrating in developing standard designs, smaller plants, greater prefabrication under manufacturing type conditions and other management innovations in an effort to once again experience the anticipated economies of nuclear fuel in the future. Similar management improvements will be required for urban rapid transit systems, refinery and chemical plants, pipelines, mineral resource developments, waste water plants and the design and construction of other projects to implement the advanced technologies that will be required even to maintain, let alone improve, our standard of living.

Time, money, equipment, technology, people, materials; these are resources. Organize them into activities, perform the activities in a logical sequence, and one has a project. Whether it is to construct a cottage at the beach or to design and construct a billion-dollar rapid transit system, the pattern is the same. Practice has been to assign total responsibility for all these factors to one person: a project manager. Over the years, this has proven to be a good approach. Intelligent, competent, experienced project managers have succeeded in "putting it all together." Can they continue? Why the past decades' failures? Now, more than ever, planning and control of the resources required to successfully accomplish today's increasingly complex projects remain among the most difficult and perplexing management responsibilities. Success will require the fullest understanding of all facets of the construction industry.

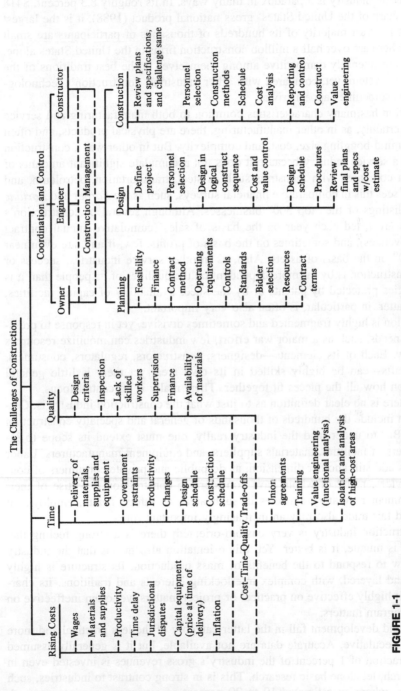

FIGURE 1-1
Challenges in construction's future. (From Boyd C. Paulson, Jr., *Goals for Basic Research in Construction, Technical Report No. 202,* Stanford University, Dept. of Civil Engineering, The Construction Institute, Stanford, Calif., July 1975.)

THE NATURE OF THE CONSTRUCTION INDUSTRY

The construction industry is a paradox in many ways. In its roughly 8.3 percent, $418 billion-plus share of the United States' gross national product (1988), it is the largest industry, but the vast majority of its hundreds of thousands of participants are small businesses. There are over half a million construction firms in the United States alone. These firms are intensely competitive among themselves in the best traditions of the free enterprise system, yet, compared with other industries, construction's technological advances sometimes appear trivial.

Construction has many characteristics common to both manufacturing and service industries. Certainly, as in other manufacturing, there are physical products, and often these are of mind-boggling size, cost, and complexity. But in other ways, construction is more like a service industry because it does not accumulate significant amounts of capital when compared with industries such as steel, transportation, petroleum, and mining. One sees this in comparative financial surveys, such as the *Forbes* and *Fortune* magazines' listings of the "top 500" businesses. Although several of construction's largest firms are listed each year on the basis of sales (cumulative annual contract awards or revenues), and sometimes on the basis of profits, few, if any, are even near the "top 500" on the basis of assets. Also, as in other service industries, success or failure in construction is by far more dependent on the qualities of its people than it is on technologies protected by patents or by the sheer availability of capital facilities, though the latter, in particular, is often also very important.

Construction is highly fragmented and sometimes divisive, yet in response to pressing national needs, such as a major war effort, few industries can mobilize resources more quickly. Each of its elements—designers, constructors, regulators, consumers, suppliers, crafts—can be highly skilled in its own area, yet there is little general perspective on how all the pieces fit together. There really is no central focus.

Indeed, there is no clear definition as to just what the construction industry is. Certainly it must include the hundreds of thousands of general and specialty construction contractors. But to understand the industry really, one must extend its scope to include designers of facilities, materials suppliers, and equipment manufacturers. Labor organizations add still another dimension, as do public and private consumers of construction services, many of whom have considerable construction expertise of their own. Government regulatory agencies in such areas as safety, health, employment practices, and fair trade also play an increasingly important role.

The construction industry is very custom-oriented; there is a strong feeling that if something is unique, it is better. Yet, this orientation also means that the industry has been slow to respond to the benefits of mass production. Its structure is highly specialized and layered, with complex interlocking interests and traditions. Its character makes it highly effective on practical or project matters, yet often ineffective on general or program matters.

Research and development fall in the latter category of the less practical and more general and speculative. Accurate data are not available, but it is generally assumed that only a fraction of 1 percent of the industry's gross revenues is invested even in applied research, let alone basic research. This is in strong contrast to industries, such as electronics, where an estimated 10 to 20 percent of revenues goes into research and development. This investment, in turn, at least partially accounts for the quantum leaps the high-technology industries have taken in recent years.

It has been observed that the construction industry is almost completely incentive-oriented. If there is little programmatic activity, it is likely that there is little incentive for investing in it. This reluctance to invest probably results in part because advances in construction tend to develop from innovations, or "better ideas." Most of these cannot be protected by either secrecy or patents, and therefore disseminate rapidly through the industry. Thus, there is little incentive for one firm to invest heavily in new developments that can soon be expected to benefit its competitors equally.

Owing to the comparatively large numbers and small sizes of its businesses, its fragmentation and divisiveness, and its service characteristics, the construction industry, as a whole, cannot significantly influence the demand for its output or control the supply. The consequent instability of demand thus dominates everything. For example, seasonality is chronic, and construction has an amplified reaction to basic business and economic cycles. Other economic problems in the industry relate to the lack of mobility of resources. Consequently, there is often too much work in some regions at the same time that others are suffering localized recessions. Major problems recur in funding both large and small projects, and these difficulties can be aggravated by government competition for and manipulation of the finite funds that are available. Construction also is often placed in the forefront of government fiscal and social policy.

TYPES OF CONSTRUCTION PROJECTS

Construction intersects almost all fields of human endeavor, and this diversity is reflected in its projects. Designers of hospitals interact closely with medical professionals best to serve the needs of patients. Educational philosophies and practices take shape in the architecture of schools and colleges, while governments and corporations express their "images" with structures that house their offices and production facilities. The design and construction of refineries, factories, and power plants generally require that the builders be more knowledgeable of the related industrial technologies than the manufacturers and utilities that operate them. Builders of dams, tunnels, bridges, and other civil works today must be geologists, ecologists, and sociologists as well as architects, engineers, and managers. And most of us, in our homes, recognize how intimately the design and quality of our constructed environment either enhance or frustrate our personal lives.

It is difficult, if not impossible, to categorize neatly so great a spectrum of projects. The exceptions, the ones that transcend the boundaries, often seem to outnumber those that are clearly recognizable. What follows, nevertheless, are four somewhat arbitrary but generally accepted major types of construction. In large measure, these categories parallel the general specialties into which designers and constructors tend to group themselves.

Residential Construction

Residential construction includes single-family homes, multiunit town houses, garden apartments, high-rise apartments, and condominiums. The latter, in particular, are technologically less closely related to residences than to the following description of non-residential building construction and are sometimes incorporated as part of multipurpose commercial developments. They are classified here from the user's point of view.

Residential construction accounts for about 30 to 35 percent of construction expenditures in an average year. Although largely financed by the private sector, the supply and demand for residential construction are heavily impacted by governmental regulation and fiscal policy. There are a few very large firms, but as a rule the low capital and technology requirements in this sector of the industry mean that it is characterized by large numbers of very small firms. Demand instability, among other things, causes a high rate of business failures among them. Designs are generally done by either architects, home designers, or the builders themselves, and construction is usually handled by either independent contractors or developer-builders. Whether in single units or in large developments, traditional construction has been field-labor-intensive, with on-site hand fabrication and installation of literally thousands of pieces per dwelling unit. In recent decades, however, there has been a small but growing trend toward industrialization and factory mass production of at least some major components, and even of complete modular homes.

Building Construction

Building construction produces structures ranging from small retail stores to urban redevelopment complexes, from grade schools to complete new universities, hospitals, churches, commercial office towers, theaters, government buildings, recreation centers, light manufacturing plants, and warehouses. For most of us, these structures form our nonresidential environment during our commercial, educational, institutional, governmental, social, religious, and recreational activities. Economically, this sector typically accounts for 35 to 40 percent of the construction market. Though labor- and materials-intensive like residential construction, the scope and technology of these buildings are generally much larger and more complex.

Most of these structures are financed and built by the private sector of the economy. Design is typically coordinated by architects working together with engineering specialists for the structural, mechanical, and electrical subsystems. Construction is usually coordinated by general contractors or construction managers, who, in turn, subcontract substantial portions of the work to specialty firms. In some cases, such as hospitals and schools, design requires a good working knowledge of the activities to take place within them. In others, such as commercial office space, an in-depth knowledge of the tenants' businesses is less important.

Heavy Engineering Construction

Though accounting only for some 20 to 25 percent of the market, heavy engineering construction includes many of the structures for which the industry is best known. Dams and tunnels provide hydroelectric power, flood control, and irrigation; bridges range from footpaths to internationally famous landmarks such as that spanning San Francisco's Golden Gate; other transportation structures include interstate railways, airports, highways, and urban rapid transit systems; ports and harbor structures fall into this category, as do many of those in the deep open sea. Pipelines are included here, as are some of our more utilitarian structures, such as water treatment and distribution systems, sewage and storm water collection, treatment and disposal systems, power lines, and communication networks.

Both the design and construction phases of heavy construction are primarily the domain of civil engineers, though almost all disciplines play important roles. The construction phase is much more equipment-intensive, characterized by fleets of large earthmovers, cranes, and trucks, working with massive quantities of basic materials such as earth, rock, steel, concrete, timber, and pipe. Another major distinction is that many, if not most, heavy construction projects are publicly financed, and this fact in turn limits the alternative contractual arrangements in this sector. Typically, design is done either by, or under contract with, a public agency, and construction is by competitive open bidding. Construction contractors here usually require much greater expertise in engineering and geology than do those in building and residential construction.

Industrial Construction

Industrial construction represents only about 5 to 10 percent of the market, but it has some of the largest projects and is dominated by some of the largest engineering and construction firms. These projects include petroleum refineries and petrochemical plants; synthetic fuel plants; fossil-fuel and nuclear power plants; mine developments, smelters, steel mills, and aluminum plants; large heavy-manufacturing plants; and other facilities essential to our utilities and basic industries.

Both design and construction require the highest levels of engineering expertise, from not only civil, but also chemical, electrical, mechanical, and other disciplines, and typically all phases of the project are handled by the same firm on a negotiated design-construct or "turnkey" contractual arrangement, with considerable overlap between design, procurement, and construction. The design-constructors must be intimately familiar with the technology and operations of the facility from the owner's point of view, and often they hold some of the key patents for advanced process technologies needed therein. In the Western free-enterprise countries, most of this work is privately financed.

In contrast with the basic materials characteristic of heavy engineering construction, the major factors in industrial construction generally consist of large amounts of highly complex mechanical, electrical, process piping, and instrumentation work. This work tends to be much more labor-intensive, though some of the largest hoisting and materials-handling equipment is also required.

EVOLUTION OF ALTERNATIVE APPROACHES TO THE MANAGEMENT OF CONSTRUCTION

In recent decades, traditional construction practices fell under increasing economic pressures from three separate, broad groups: (1) owners who wanted to achieve the best value for their expenditures; (2) contractors (and subcontractors) desiring to bid low enough to get the work but high enough to realize a fair profit on investment;

and (3) workers hoping to achieve increasingly better living standards and working conditions. Although a significant amount of negotiated work existed in various forms, the basic interactions in determining prices and in negotiating wage rates continued along classic economic lines. On the other hand, the architect and the consulting engineer were professionals who, to a significant degree, were dissociated from the economic interests of the three groups.

Owing primarily to the rapid economic growth that took place in the United States in the 1950s and 1960s, the demand for construction accelerated rapidly. The effect upon the traditional methods in each component of the construction industry is worth reviewing.

Modification of Traditional Concepts

In heavy construction, the demand took the form of multimillion-dollar projects of a complexity never before undertaken. Spurred by this growing market, contractors enlarged their scope of operations so that intense competition prevailed in spite of the demand. In fact, it was a profitless prosperity for a large number of construction firms which expanded rapidly to acquire major contracts through competitive bidding, only to find that profits on some jobs were difficult to achieve. The missile program of the United States government, encompassing Atlas, Titan, and Minuteman, is but one example where expanding demand for new technology, under extremely competitive conditions, brought disappointment to many contractors.

In building and light industrial work, private owners increasingly realized the importance of the time value of money in the construction process. The emergence of the design-construct engineer-contractor, and the negotiated contract with a selected general contractor for architect-designed projects, were two developments that fulfilled a need to shorten the overall elapsed time from concept to occupancy. Public and institutional work, until recently, clung to the traditional method, employing a separate designer and a lump-sum contractor. Today many of the largest rapid transit, wastewater and other public projects are utilizing program and construction managers to plan and help manage the work using a phased schedule with overlapping design and construction contracts.

However, it was in the industrial field that the design-construct concept, often on a cost-plus-fixed-fee contract, became particularly attractive to owners. Once demand for a new plant was established, the design-constructor's combination of process know-how and construction, and its ability to shorten the project duration through simultaneous design and construction, was clearly advantageous to industrial owners when compared with traditional methods. One year saved in overall time, compared with competitors, could mean the difference between economic survival, increased market penetration, increased profits—or bankruptcy. This design-construction ability helped increase the stature of the engineer-constructors constituting the National Constructors Association, which represents many of the larger firms that have captured a major share of this type of work.

Briefly summarized, influential trends included the following: heavy and highway construction continued with traditional methods; building and industrial work tended more and more toward design-construct or negotiated contracts to shorten the overall design-construct period; general contractors adopted or fought the inroads upon traditional methods; and owners tried to maximize investment returns. Add to these influences the increasing demands of labor for better working conditions and higher pay; and further, add the unfamiliarity of many architects, engineers, and owners with the basic conflicts and postures inherent within the industry. An erosion of traditional values and concepts was inevitable.

The Partial Breakdown

It is beyond the scope of this book to describe further the many problems facing the construction industry. Labor has blamed contractors; contractors have blamed labor and designers; and the owners have blamed them all. Declining productivity almost defied solution in certain sections of the country while in others a reasonable balance has been maintained or restored. Too often, however, labor has been just a scapegoat for poor designs and poor construction industry management.

Under traditional methods, there was a major economic incentive for contractors to protect their potential profits in collective bargaining with labor. Historically, the Associated General Contractors or other local general and specialty contractor associations bargained with unions for wage and working conditions largely without outside interference. Owners generally included "force majeure" (unforeseen event) or extension clauses in contracts to cover strikes and labor disputes. Many contracts between contractors and labor organizations included full protection against unanticipated wage increases in all work started during the contract period. Future wage increases would apply only to work started after the new collective bargaining agreement went into effect. Owners remained aloof from the economic bargaining. A reasonable balance was maintained, but it was better in some sections of the country than in others.

When owners discovered that design-construction and negotiated contracts in various forms could significantly reduce project durations, they intensified pressures on contractors to get facilities into production or occupancy at the earliest possible moment to maximize returns on invested capital. Construction was increasingly programmed to proceed simultaneously with design in the industrial and building fields.

Since substantial work was performed under some form of negotiated cost-reimbursable basis, owners began to exercise more control over the work. The sweeping acceptance of the **critical path method** (CPM) pointed out the importance of scheduling; through "crashing," segments of the work could be accelerated to make up delays or shorten completion times. Crashing could normally be performed in two ways: by working substantial overtime; or by adding more workers and equipment.

Contractors increasingly applied both methods to maintain schedules. As work volume increased, labor shortages on many projects became critical. Good craftworkers could be attracted from other areas only by offering scheduled overtime premiums, such as workweeks of six 10-hour days. Thus, the dominant goal of shortening overall design-construct periods became counterproductive on many individual projects.

The Long Road Back

The basic laws of economics continued to operate in construction as they do in other industries, although the results may not always be as readily apparent. For the reasons just given, industrial and building owners were primarily responsible for the emergence of the negotiated contract. As long as escalating costs could be offset by increased revenues from early utilization, the problems created by the growing power of the labor unions, and by the weakening stature of construction contractors and their local bargaining committees, were tolerated.

However, as the nation's economic climate deteriorated, excessive construction costs started to outweigh early returns from facilities. Demand for manufactured products began to suffer from skyrocketing prices, and owners began to see that the forecast market returns from some new facilities would not be realized. In other areas, such as electric power, new capacity was not to be required as soon as had been contemplated. This realization had a sweeping effect on the industry. Overtime was curtailed; site labor demand slackened; and contractor and owner organizations actively started to publicize the evils that had crept into the industry.

The steady growth of open shop—always strong in the south and southwest—began to spread to other areas of the country long considered union strongholds. Open-shop market share of the total market including home building went from about 24 percent in 1969 to 60 percent in 1979 and is estimated to be even higher in 1990 according to the Associated Builders and Constructors, the leading open-shop employer organization and the proponent of the Merit Shop. Open shop is currently competing with union work in most of the large metropolitan centers. Construction wage rates have remained reasonably stable during the mid to late 1980s, due in part to less demand and in part to growing non-union competition. All in all, the current labor climate seems ideally positioned to follow through to implement research and management breakthroughs in the 1990s if such innovations can in fact be developed.

Many owners have reappraised the advantages of negotiated contracts. A return to traditional fixed-price, competitively bid contracts, with a single lump-sum general contractor following a completed design, became more attractive, provided that sufficient lead time was available. In numerous other cases, shortening the design-construct period remained a paramount objective to achieve expected economic benefits. In the latter situation, utilization of professional construction management and program management utilizing phased construction and fixed-price contracts has gained increasing consideration.

Construction Organizations for the Future

The history that has brought us to the present time certainly has had its share of problems, but it has left us with several good alternative forms of project organizations and methods of contracting which, if applied knowledgeably and in the right circumstances, can lead to the effective management of construction in the future. The next two chapters will explain some of the most important alternatives and indicate the circumstances where they are most effective.

SUMMARY

Construction faces challenging projects in the future, including work of unprecedented scope and technical complexity. Problems accompanying these projects include the need for improved organization and management structures; increasing social, economic, and environmental constraints; and antiquated customs, interrelationships, and interests inherent in the underlying nature of the industry. One of the most basic obstacles is that fragmentation and divisiveness among its participants inhibits the type of programmatic efforts that can help improve the industry's long-term prospects.

Present and future construction projects have been divided into four main categories: (1) residential construction; (2) building construction; (3) heavy engineering construction; and (4) industrial construction. Each of these, and the many subdivisions within them, has its own special needs, characteristics, participants, and clientele; each will be important to the future development of our constructed environment.

While there have been many problems in construction in recent decades, particularly in the management of large and complex projects, new forms of organization, contracting and project control systems have been evolving that are helping to provide better management in the future. This book will present some of the best now available.

2

DEVELOPMENT AND ORGANIZATION OF PROJECTS

From concept to implementation, the stages in the development of construction projects fall into broadly consistent patterns, but in timing and degree of emphasis each project takes on its own unique character. Depending upon circumstances, the basic stages can occur sequentially in the traditional approach, or they can overlap to varying degrees as a part of a phased construction program. Based upon the degree of overlap desired or permitted, alternative contractual and organizational structures are available to provide the best means to achieve the owner's cost, time, and quality objectives.

This chapter first describes six major stages that compose the "life cycle" of a typical project. Then it briefly describes the management processes that organizations must perform to successfully complete a project. It next introduces the principal types of organizational and contractual arrangements employed on today's construction projects. Together, these subjects provide a broad perspective within which subsequent chapters can focus on the management of construction projects.

THE LIFE CYCLE OF A CONSTRUCTION PROJECT

Six basic phases contribute to developing a project from an idea to reality:

Concept and feasibility studies
Engineering and design
Procurement
Construction
Start-up and implementation
Operation or utilization

Figure 2-1 is a bar chart showing a typical chronology for these steps. In practice, of course, the degree of overlap among phases, in both time and operations performed, varies widely from one project to another, as does the distribution of responsibilities.

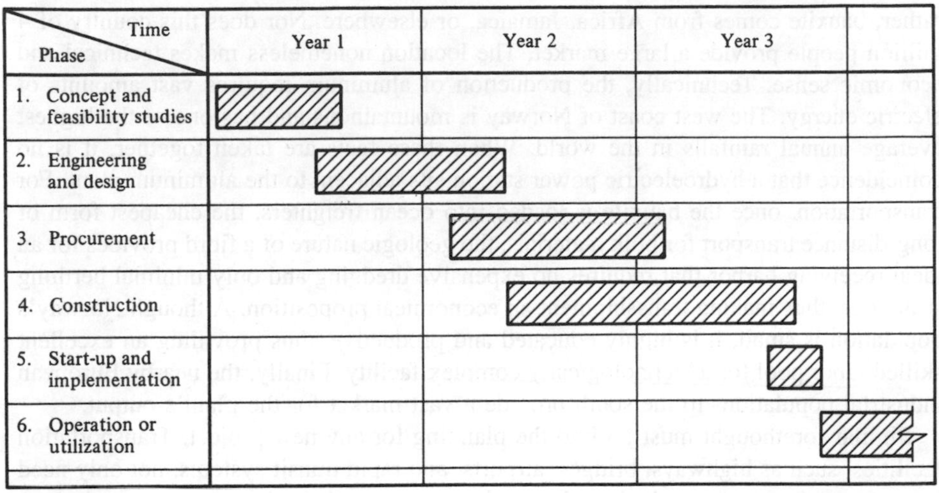

Phase \ Time	Year 1	Year 2	Year 3
1. Concept and feasibility studies	▨		
2. Engineering and design	▨		
3. Procurement		▨	
4. Construction		▨	
5. Start-up and implementation			▨
6. Operation or utilization			▨

FIGURE 2-1
The life cycle of a construction project.

This section describes each phase and shows how they all fit together. Functions, responsibilities, and interrelationships of the key parties involved are also briefly introduced. Part 2, Chapters 4 through 9, develop these subjects in much more detail.

Concept and Feasibility Studies

Most construction projects begin with recognition of a need for a new facility. Long before designers start preparing drawings, and certainly well before field construction can commence, considerable thought must go into broad-scale planning. Elements of this phase include conceptual analyses, technical and economic feasibility studies, and environmental impact reports.

For example, location is fundamental to planning for a new industrial plant. Where can the plant be located to provide desirable, nearby employment for an adequate supply of skilled, productive workers? What are the present and projected costs and customs associated with the labor force? Depending on the nature of its raw-materials input and its products, will the plant have access to the most appropriate and economical forms of transportation, be they air, water, highway, rail, or pipeline? Does the location provide access to raw materials and to markets? Are there adequate sources of energy, including gas, oil, and electricity; and are there convenient communication facilities? What political or institutional factors may ease or impede the development and operation of the facility? What will be the sociological and economic impact of this plant on the community? What will be the environmental impact? What do all these factors, taken as a whole, mean for the technical and economic feasibility of the project?

To illustrate, one might wonder why there is a large aluminum plant on the north shore of Norway's Hardanger Fjord. Norway does not produce the raw material;

rather, bauxite comes from Africa, Jamaica, or elsewhere. Nor does this country of 4 million people provide a large market. The location nonetheless makes technical and economic sense. Technically, the production of aluminum requires vast amounts of electric energy. The west coast of Norway is mountainous and has one of the highest average annual rainfalls in the world. When these facts are taken together, it is no coincidence that a hydroelectric power station sits adjacent to the aluminum plant. For transportation, once the bauxite is loaded into ocean freighters, the cheapest form of long-distance transport for bulk materials, the geologic nature of a fjord provides for an ideal receiving harbor that requires no expensive dredging and only minimal berthing structures, thus making transshipment an economical proposition. Although Norway's population is small, it is highly educated and productive, thus providing an excellent skilled labor pool for a technologically complex facility. Finally, the nearby European industrial populations to the south provide a vast market for the plant's output.

Similar forethought must go into the planning for any new project. Transportation facilities, such as highways, bridges, airports, and rapid transit systems, not only need forecasts of future demands, but also analyses of how the existence or nonexistence of these structures will actually affect social, economic, and demographic patterns and thus influence the demands the structures are intended to create or fulfill. The same applies to water supply systems, wastewater treatment plants, and new or more economical sources of energy.

Traditionally, these early stages are handled by the owner alone, or by the owner working with consultants knowledgeable of the most important factors affecting the situation. Considerable amounts of "free" information are available from, or offered by, public and private organizations that may benefit from, or be adversely impacted by, a new facility. To some extent, architect/engineer consultants, design-constructors, professional construction managers or program managers can become involved in this early activity, but normally they are not brought in at least until the latter stages, if at all.

Engineering and Design

Engineering and design have two main phases: (1) *preliminary* engineering and design; and (2) *detailed* engineering and design. These phases are traditionally the domain of architects and design-oriented engineers. Increasingly, however, the owner's operations and utilization knowledge and the field constructor's experience are being more strongly injected at this stage through both direct participation and stringent review procedures. This involvement should be the case especially with professional construction management, design-construct or program management, and it is one of the strong points of these approaches.

Preliminary Engineering and Design   Preliminary engineering and design stress architectural concepts, evaluation of technological process alternatives, size and capacity decisions, and comparative economic studies. To a great extent, these steps evolve directly from the concept and feasibility stage, and it is sometimes difficult to see where one leaves off and the other begins.

To illustrate, in a high-rise building the preliminary design determines the h̸ and spacing of the stories, the general layout of the service and occupied floor spa̸ general functional allocations (parking, retail, office space, etc.), and the overall desig̸ approach. The last-mentioned factor involves decisions such as the choice between a bolted structural-steel frame or a reinforced-concrete structure. Further refinements determine whether the structure will be precast or cast-in-place concrete. In building construction, the architect has the primary responsibility for preliminary design.

In heavy construction, engineers are responsible for the preliminary design, but they often need substantial input from geologists, hydrologists, and increasingly from ecologists and other professionals in the natural sciences. For example, in designing a dam for flood control, hydroelectric power, recreation, or water storage for agricultural, domestic, or industrial uses, or for regulating water quality, preliminary design requires analysis of the watershed's hydrologic characteristics as they relate to the purpose of the structure to determine the necessary reservoir storage characteristics; the geologic nature of the foundation and abutments determines the precise location of the dam on its site; the geology, size, shape, and availability of materials influence the choice among basic structural types, such as concrete, earth-fill, or earth-rock. A concrete structure might be further specified to be a gravity, arch, or buttress design, and an earth-fill might require decisions on the type of impermeable barrier, filters, and foundation cutoffs. These and succeeding decisions result in a set of preliminary plans and specifications that are first subject to review and refinement, and then serve as the departure point for the detailed engineering and design process.

Preliminary engineering and design in industrial construction involve input and output capacity decisions, choices between basic process alternatives, general site layout, and often the preparation of overall process flowsheets. In a mining and ore-processing operation, engineers and geologists work out the mine development scheme, choose between alternative ore beneficiation methods, and specify other related processes. In a thermal power plant, it is necessary to decide between the types of steam generation systems. An oil refinery or petrochemical plant often involves decisions between licensing several alternative patented processes. These decisions demand close cooperation among specialists from several engineering disciplines, and they require considerable interaction between the owner's staff and the design-constructor's personnel.

Once preliminary engineering and design are essentially complete, there is generally an extensive review process before detailed work is allowed to proceed. In private work, such as industrial construction and commercial building, the review focuses mainly on seeking approval from higher levels of management and from sources of external financing, where required. But increasingly this review involves regulatory bodies that look for compliance with zoning regulations, building codes, licensing procedures, safety standards, environmental impact, etc. In both public and private works, agencies are providing more and more opportunities for direct involvement of the general public. There are also complicated funding cycles in legislative and executive bodies, and most of the constraints from regulatory bodies and others also apply much as they do in private construction.

eering and Design Detailed engineering and design involve the ively breaking down, analyzing, and designing the structure and it complies with recognized standards of safety and performance design in the form of a set of explicit drawings and specifications nstructors exactly how to build the structure in the field.

This detailed phase is the traditional realm of design professionals, including architects, interior designers, landscape architects, and several engineering disciplines, including chemical, civil, electrical, mechanical, and other engineers as needed. The types of design professionals involved vary by type of work (building, heavy, or industrial) and are much the same as in the preliminary design phase, but the staffs become much larger and are generally augmented by various people at the technician and technology level, such as draftsmen and soils testers. In addition to designing the structure itself, the design professional often conducts detailed field studies to get good engineering information on foundation conditions, slope stability, and structural properties of natural materials. Such studies can require further input from experts in other disciplines, such as geologists, economists, and environmental scientists.

Again, it is becoming increasingly common for field construction methods and cost knowledge to be injected into the detailed engineering and design process. This is especially true in the design-construct, professional construction management, and program management approaches.

Procurement

Procurement involves two major types of activities. One is contracting and subcontracting for services of general and specialty construction contractors. The other is obtaining materials and equipment required to construct the project. Allocation of responsibilities for these two functions varies widely, and it is especially dependent on the contractual approach taken for a particular project.

The traditional form for procuring construction services as well as most of the materials and equipment required for a project is to solicit competitive bids for a single general contract. This takes place soon after the detailed engineering and design phase has produced a comprehensive set of plans and specifications. The general contractor then handles all subcontracting, plus the procurement of materials and equipment. In design-construct projects, the contractor also handles all these services, but awarding of subcontracts and procurement of major equipment and materials items can proceed incrementally and can considerably overlap the design phase. In professional construction management, the construction manager often coordinates all these functions, including the letting of several prime contracts instead of subcontracts, while acting as the agent of the owner. The program manager's role would be similar.

Construction

Construction is the process whereby designers' plans and specifications are converted into physical structures and facilities. It involves the organization and coordination of all the resources for the project—labor, construction equipment, permanent and temporary materials, supplies and utilities, money, technology and methods, and time—to complete the project on schedule, within the budget, and according to the standards of quality and performance specified by the designer.

The key roles at this stage are played by the contractors and subcontractors and their employees from the building trades. There is also considerable input for inspection and interpretation from the architect/engineer. Supporting roles are played by suppliers of materials and equipment, specialty consultants, shipping and transport organizations, etc.

Since much of this book focuses on construction, we will not discuss it further at this stage.

Start-up and Implementation

Most structures and facilities of any significance involve a start-up and implementation phase. In both simple and complex cases, much testing of components is done while the project is underway. Nevertheless, as the project nears completion, it is important to be sure that all components function well together as a total system. In some cases, this mainly involves testing, adjusting, and correcting the major electrical and mechanical systems so that they perform at their optimum level. Often this phase also involves a warranty period during which the designer and the contractors can be called back to correct problems that were not immediately evident upon initial testing and to make adjustments to better suit the facility to the owner's needs after he has had a chance to try it out.

In many projects, especially large industrial facilities such as power plants, refineries, and factories, start-up is a highly complex process that pushes the facility to its technological limits, as well as seeing that it operates efficiently under "normal" conditions. In this case, start-up is a project in its own right; it requires months of careful advance planning and, once underway, demands good coordination and supervision. Often, spares for critical components will be kept on hand in case something goes wrong.

Operation and Utilization

The functional value of the project will depend upon the decisions and implementation of the objectives developed during the preceding phases. With a projected operational life of 20 to 25 years or more, it is evident that the overall cost and value to the owner throughout the operating life are determined largely during the period from conception through start-up.

Parties involved at this stage range from homeowners doing weekend maintenance, through janitors and equipment specialists in buildings, to public works staffs maintaining highways and operating dams and bridges, and on to the skilled engineers and technicians who operate factories, refineries, power plants, and mines. In the case of major alterations or expansions, the operations phase can also involve recycling through the first five phases of a project mentioned above, whether the work is done in-house or by contract.

BASIC MANAGEMENT ACTIVITIES

Management theory identifies four essential management activities that must be accomplished in any successful organization. Organizations can be designed to best perform these according to the needs of a specific project:

Scoping—Clearly define desired project objectives.

Planning—Predetermine a course of action to achieve project objectives.

Organizing—Integrate individual, consultant and contractor efforts into an effective team.

Controlling—Monitor, influence, and direct achievement of project objectives throughout the performance phase.

Scoping

Scoping involves establishing realistic and specific objectives which establish in advance the desired results. Objectives must be stated in definite and measurable terms which cover costs, schedules and quality or performance requirements. Full and unequivocal communication of project objectives to project team members is essential if maximum performance is to be achieved. Objectives must be reasonable and achievable. Project team members will quickly become skeptical of unworkable or overly optimistic objectives.

Planning

Planning activities include programming, costing and scheduling. For most projects these activities are highly interrelated and are developed in overlapping phases rather than sequentially. Planning for most projects will evolve from a high-level plan in the early stages to a very detailed implementation plan during the performance phase. An integrated plan will involve a work breakdown structure of codes for estimating, scheduling, and costing direct and indirect work activities. Using man hours as the basis for estimating labor costs, for manloading schedules and for measuring earned value and productivity represents a common parameter interrelating both cost and schedule performance. These concepts will be discussed in detail in Chapter 8, *Application of Controls*.

Organizing

Organization is the process used by managers to relate tasks to people, other firms, regulatory agencies and other interested groups in order to achieve an economical and timely performance. In developing an efficient organization, the manager must deal with the design of the structure, delegation of responsibility, working relationships between individuals and groups, and creation of a communications program designed to keep everyone fully informed. In general, the number of managerial levels should be kept to a minimum, thus reducing management interfaces. On the other hand, too "flat" an organization may exceed the manager's effective span of control. The optimum number of persons reporting to a single manager will vary considerably depending upon the manager's effectiveness, employee skill and temperament, as well as the nature of the work.

Controlling

Control requires an awareness of the current status of cost, schedule and quality performance compared to project goals. Control can be achieved through frequent personal inspection of the operation by a knowledgeable person in order to judge whether or not the work is being properly performed. This type of control is normally called supervision as practiced by foremen and front line supervisors. As the manager's organizational span increases it is apparent that all of the work can no longer be supervised. Here, the manager must utilize a systematic process to identify deviations from the plan at an early stage when remedial action is still possible. Part Three sets forth the concepts and examples for an integrated project control system. Control systems can only keep the manager informed of variations from the plan. Effective remedial action is necessary if control is to be achieved.

ORGANIZATIONAL CONCEPTS

There are several possible organizational approaches to the design and construction of a project. The major concepts addressed here include (1) functional, (2) task force, (3) line and staff, and (4) matrix. Figure 2-2 provides simplified illustrations of these organizational types, and others are given in this section and later in the book.

Functional Organizations

Functional approaches have traditionally been used in the construction industry. The strengths of these approaches will include high stability; high professional standards, incorporation of the latest technology; and an excellent corporate memory.

In a functional organization, everybody knows where they stand compared to others, understands their tasks, and has a permanent home base. Weaknesses of functional organizations can include low adaptability, minimum appreciation of overall project objectives, overly rigid operating rules, resistance to change, and difficulty in developing well-rounded project managers. Many individual managers may tend to over optimize their particular specialty to the possible detriment of the overall project.

Notable success has been achieved by functional organizations when design and construction do not fully overlap, thus minimizing interaction. In this case, the overall project management plan can be said to be on a functional basis, using a separate designer and a separate general contractor or construction manager. Most fixed- or unit-price types of construction contracts fall in this category. An overall functional organization can be helpful when the owner acts as project manager with a minimum staff, depending upon others for the functional expertise. Functional organizations work best when overall managers are skillful, people-oriented, and can help avoid internal conflict with other functional groups. The functional organization permits the tightest discipline control of any organizational concept. Figure 4-7 shows a functional

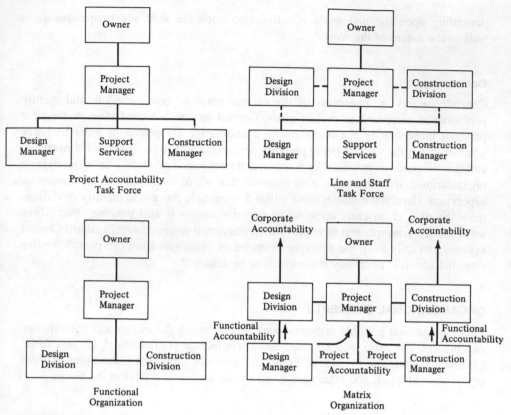

FIGURE 2-2
Organizational concepts.

organization for a construction management project for an owner with a minimum staff.

The Project Task Force

The task force has been notably successful when a self-sufficient organization is required. The strengths of the task force include high adaptability and high understanding of the overall task and can foster an excellent team spirit if given the proper leadership. The task force features close personal relationships and can be responsive to new ideas and methods. Figure 2-2 shows a task force for project management covering all phases of a project. Figures 7-4 and 7-5 illustrate task forces for major construction projects covering only the construction phase.

Autonomous project task forces (sometimes called integrated organizations) are increasingly being utilized by public and private owners where full-time task forces are staffed with individuals who are employees of both the owner and outside firms. Some very successful construction companies have been organized on a task-force basis, taking one major project at a time while devoting essentially the entire corporate resources to one individual project. Interestingly enough, when financial success has permitted branching out into several projects constructed simultaneously, profit margins and performance have often shown considerable erosion.

The weaknesses of a task force feature poor stability. Everyone may not have a corporate home for long-term career development and for continuity of employment between assignment periods. The task force by itself has no corporate memory, the "memory" being that of the individuals assigned to the project. Everyone may not understand his or her own task, and there are no functional checks and balances to preserve workmanship quality and accepted standards. The task force may be preoccupied in inventing the wheel because of the unavailability of the corporate memory personified by experienced senior functional managers.

Task forces work best when all team members can be located physically in the same area to foster closer personal relationships, when attempting something new without recognized standards and when the team is made up of very experienced members. Large remote overseas construction projects are generally organized on a task-force basis, as are numerous large projects and programs in the United States and Canada.

Line-and-Staff Organizations

The military line and general staff organizations and the General Motors organization as developed by Alfred P. Sloan form outstanding examples of this method of organization. The line-and-staff organization has worked well in the manufacturing industries. Many construction companies evolved from a functional organization to a line-and-staff organization as growth required additional management strengths. Line responsibility is often indicated by a solid line on organization charts while staff responsibility is shown as a dotted line. Figure 2-3 illustrates a corporate organization for a construction company based upon principles of line and functional staff.

Strengths of the line-and-staff organization include the combination of functional strengths and expertise with the project-oriented team. This form of organization strikes a balance between overall control of both craftsmanship and project objectives. The functional staff provides a high corporate memory and, if managers are given diversified assignments, offers an excellent climate for developing project managers.

Weaknesses are sometimes evident in the conflict between the operating organization and the functional staff. Organizational structures tend to be somewhat top heavy, and overall costs may exceed more simplified operating concepts.

The Matrix Organization

Most design-construct organizations and a number of public owners have developed some form of the matrix organization. This organizational form is something of a cross between a task force and a functional organization and represents an attempt to preserve the strong points of each. Figure 2-4 illustrates strong, balanced, and weak matrix organizations. A strong matrix assigns maximum power to the project manager; a weak matrix assigns maximum power to functional managers; while a balanced matrix divides the power into functional and project responsibilities. The matrix or-

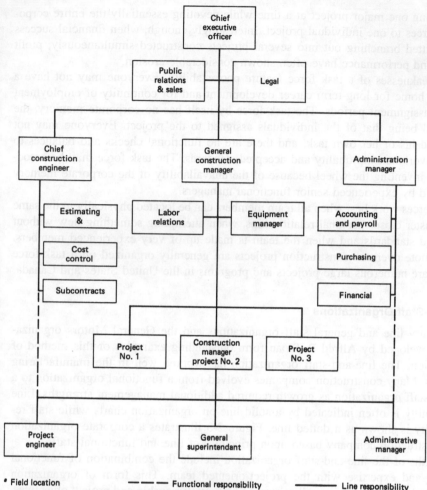

FIGURE 2-3
Construction company line and staff organization.

ganization endeavors to solve the conflicts between the operating line organization as represented by the project manager and the functional staff by opening up lines of communications at all levels and through assigning subordinate managers dual reporting responsibility. Project responsibilities such as scope, cost, and schedules are the responsibility of the project manager. Functional objectives such as quality assurance, design standards, and internal company policies are the responsibility of the functional staff. Conflicts are normally worked out between the project manager and the functional staff manager or settled at the executive level.

The matrix organization combines functional strengths with the advantages of a project-oriented team. It fosters an excellent climate for developing project managers,

MATRIX ORGANIZATIONS

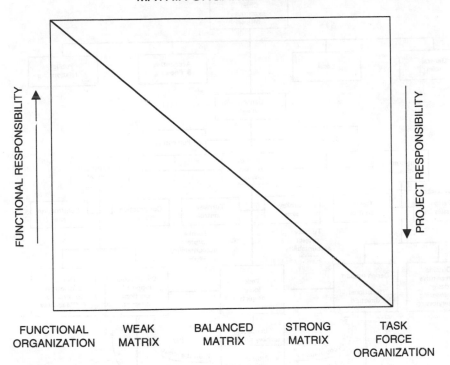

FIGURE 2-4
Project matrix organizations.

retains access to the corporate memory and allows control of both craftsmanship and project objectives. Disadvantages include difficulties in precisely defining accountability to both functional and project managers. Observation has shown that in many organizations the strongest managers often dominate and that individual team members may be unsure of their position in the pecking order. This organizational form generally has the highest overall management and administrative cost.

The matrix organization is appropriate when both project accountability and functional expertise are required. It works best when experienced and people-oriented managers are involved and when functional and project authority can be explicitly defined and divided. Figure 2-5 illustrates a matrix organization applicable to a design-construct company.

CONTRACTUAL RELATIONSHIPS

There are numerous alternative contractual approaches to bringing together a team for the design and construction of a project. The principal categories addressed in this section include the traditional approach, the owner-builder, turnkey (design-construct or design-manage), professional construction management, and program management.

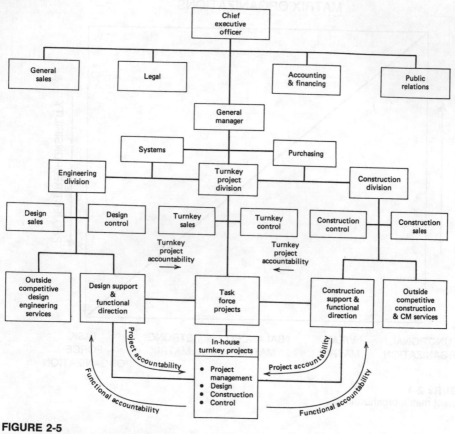

FIGURE 2-5
Company matrix organization.

Each has its advantages and disadvantages for a particular application, and each has developed a certain degree of flexibility so that, in reality, many of the individual alternatives overlap one another; in practice, it is sometimes difficult to categorize any one particular project's contractual arrangement precisely.

The following descriptions outline major differences among approaches and explore some of their variations, advantages, and disadvantages, as well as their similarities. Figure 2-6 gives a simplified chart of most of the approaches. Program management is described in Figure 2-7.

The Traditional Approach

Members of the **Associated General Contractors of America** (AGC) have generally advocated and operated under the traditional method. Here the owner employs a designer (architect, architect/engineer, or engineer) who first prepares the plans and specifications, then exercises some degree of inspection, monitoring, or control during

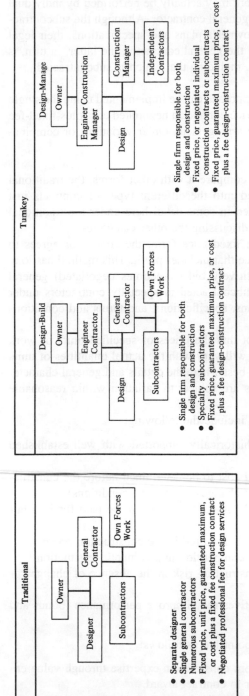

Traditional

- Separate designer
- Single general contractor
- Numerous subcontractors
- Fixed price, unit price, guaranteed maximum, or cost plus a fixed fee construction contract
- Negotiated professional fee for design services

Owner-Builder

- Owner responsible for design and construction
- Optional own forces work contractors and subcontractors
- Fixed price, unit price, or negotiated construction contracts

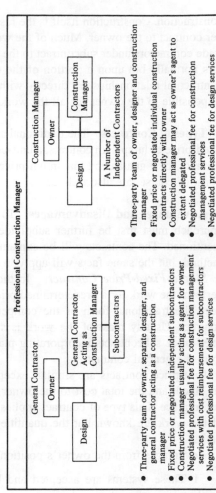

Turnkey

Design-Build

- Single firm responsible for both design and construction
- Specialty subcontractors
- Fixed price, guaranteed maximum price, or cost plus a fee design-construction contract

Design-Manage

- Single firm responsible for both design and construction
- Fixed price, or negotiated individual construction contracts or subcontracts
- Fixed price, guaranteed maximum price, or cost plus a fee design-construction contract

Professional Construction Manager

General Contractor

- Three-party team of owner, separate designer, and general contractor acting as a construction manager
- Fixed price or negotiated independent subcontractors
- Construction manager usually acting as agent for owner
- Negotiated professional fee for construction management services with cost reimbursement for subcontractors
- Negotiated professional fee for design services

Construction Manager

- Three-party team of owner, designer and construction manager
- Fixed price or negotiated individual construction contracts directly with owner
- Construction manager may act as owner's agent to extent delegated
- Negotiated professional fee for construction management services
- Negotiated professional fee for design services

FIGURE 2-6
Alternate contractual approaches.

construction. Construction itself is the responsibility of a single general contractor under contract to the owner. Much of the work may actually be performed by individual trade contractors under subcontract to the general contractor. Although the subcontractors normally bid upon a portion of the owner's plans and specifications, their legal contractual relationships are directly with the general contractor; the latter, in turn, is responsible to the owner for all the work, including that which is subcontracted.

Types of Contracts   The traditional approach can be implemented using the single fixed-price or lump-sum contract, a unit-price contract, a negotiated cost-plus-fixed-fee contract, or a guaranteed maximum-price arrangement. Other variations or combinations, of course, are also utilized.

Advantages and Disadvantages   To compare it with other forms, the traditional approach must first be further subdivided into the different types of contracts just mentioned. The contracts will be explained in some detail here under the traditional method, but the same facts will apply in discussing the other categories.

Single Fixed-Price Contracts   In the fixed-price form, the contractor agrees to perform the work for a predetermined price that includes profit. This method has long been the traditional form of the competitively bid (sometimes negotiated) general contract. Usually most of the work is subcontracted to specialty contractors under fixed-price subcontracts incorporating plans, specifications, and terms and conditions from the general contract.

Unit-price contracts are similar, except that the prices of specified units of work are fixed, and the total cost to the owner will vary with the actual quantities of units put in place. This type of contract applies best where the details and general character of the work are known but the quantities are subject to variation within reasonable limits.

Advantages from the owner's position include the following:

1 These systems are accepted and historically supported with well-established legal and contractual precedents.

2 The lump-sum type permits overall cost to be determined before the construction contract is awarded (except for scope changes or changed conditions).

3 The unit price type permits variable amounts of work to be paid for in a fair and equitable manner.

4 Minimal involvement of the owner is required in the construction process.

5 The owner may benefit from price competition in a competitive situation.

6 The contractor takes all of the construction risk in the absence of changes or impacts unforeseen by either party.

7 Contractor incentives and disincentives may improve performance (bonus and liquidated damages).

Disadvantages from the owner's viewpoint are as follows:

1 Design usually does not benefit from construction expertise through value engineering or constructibility analysis prior to contract award.

2 Overall design-construct time is usually the longest.

3 The owner is often in an adversary position with the general contractor.

4 The designer is often in an adversary position with the general contractor, and the owner may be required to be the referee.

5 Changes to the work or unforeseen difficulties will often end in disputes and litigation that can drive up costs in spite of the fixed-price concept.

6 The owner has minimal control over performance of the work.

7 The unit price type requires a substantial expense to measure pay quantities. Initial estimates may be exceeded if measured quantities are greater than the initial estimates.

8 Contractor pressures to submit the lowest bid may result in use of marginal subcontractors.

Advantages from the contractor's standpoint include:

1 The contractor can name his own price for the work as well as his profit objective in his bid.

2 There is minimum involvement of the owner or architect/engineer in the details of the building process other than for quality, schedule and change control.

3 The innovative contractor may obtain an opportunity to maximize profits through innovation.

4 Administrative requirements are based upon applicable law, the contract and the contractor's own determination.

5 The contractor may pass on much of the risk to lower-tier specialty contractors when feasible.

Disadvantages from the contractor's standpoint include:

1 To be competitive the builder must often use marginal subcontractors who may have problems performing the work.

2 On many contracts too many bidders may make it difficult to obtain the work for a fair price and in weak markets the cumulative cost of preparing the proposal by all contractors may exceed the profit potential of the successful bidder.

3 The owner controls the funding on disputed extra work or changed conditions, and the contractor must often resort to expensive arbitration or litigation with no assurance that it will recover for the additional costs.

4 The contractor usually bears the economic risk of unusual weather conditions, strikes, or other external factors that influence a contractor's cost but which may not be directly under its control.

5 Last minute telephone quotations may contribute to misunderstandings with material suppliers and subcontractors.

Negotiated Cost-Plus-A-Fee Contract   In negotiated cost-plus contracts, the contractor agrees to perform the work for a fixed or variable fee covering profit and home-office costs, with all field costs being reimbursable at actual cost. Common variations will include a fixed fee to cover profit and general and administrative costs only, with both direct home-office costs and field costs being reimbursable. Incentive fees, where some or all of the fee is dependent upon achieving certain cost or schedule goals, are becoming increasingly popular. Cost-plus-percentage-fee contracts are generally not favored, except for extra work or for minor work where the scope is indeterminate.

Fees are usually based upon the size and complexity of the project; they may be expressed as a flat dollar sum, as a percentage of a "definitive estimate" prepared at a specific point, or as a sliding scale tied to cost or estimated cost. Fees will usually involve a detailed understanding and agreement on cost-accounting methods, reimbursable items, and non-reimbursable items.

A variation on negotiated contracts, the guaranteed maximum price, has some of the features of competitively bid lump-sum work. In this form, the contractor agrees, for a fixed fee for profit and sometimes home office overhead, to complete a project at a cost not to exceed a pre-established maximum or upset price. Costs above the guarantee are absorbed by the contractor. Savings where the final cost is under the guarantee may revert entirely to the owner or more often be divided in a previously agreed-upon proportion with the contractor. Incentive or dis-incentive fees may also reward or penalize contractors based upon schedule, cost , safety performance or other criteria established by the owner. Award fees are also used where the contractor is guaranteed a minimum fee and receives additional compensation up to a set maximum depending upon the owner's evaluation of periodic objectives agreed to in advance.

Advantages from the owner's point of view include the following:

1 These contract systems are also accepted and historically supported.

2 They permit reduction of design-construct time by utilizing phased construction.

3 This approach enables the contractor to react quickly to major design changes and unforeseen conditions, and, in part, minimizes the adversary position.

4 There is an opportunity to utilize contractor expertise during the design phase to help minimize overall costs.

5 The owner pays for the actual cost of the work plus a negotiated fee which may result in savings in periods of high demand for construction work.

6 The owner and the general contractor can still subcontract a substantial portion of the work to prequalified subcontractors, thus achieving the advantage of eliminating marginal subcontractors as well as achieving fixed prices for specialty work.

7 The owner may participate fully in the management and control of the project to the extent that he has qualified personnel and may exercise control of expenditures in advance, may participate in major decisions, or actually supply certain services or materials to the contractor.

8 Under the guaranteed maximum price option, the owner may pass on some portion of the construction risk to the contractor.

Disadvantages from the owner's position are the following:

1 Cost plus a fixed fee may not be the most economical alternative in a competitive market.

2 Disreputable, unskilled, or unknowledgeable contractors can abuse this arrangement if the owner is not careful in selection.

3 The guaranteed maximum, while theoretically setting a ceiling, may not stand up in the event of poor initial scope, changed conditions, price increases, design delays, or major design changes which may result in numerous change orders.

4 Owner involvement (or that of the designer) is increased over the lump-sum method in view of the necessity for controls on expenditures, audits, approvals, and other administrative requirements that are considered good practice when this form of contract is employed.

5 Definition of reimbursable items of cost, particularly those for contractor tools and equipment, contractor home office expense and other items that directly affect contractor profit, may be the subject of overruns, disputes and foster adversarial relationships.

6 Under the guaranteed-maximum-cost option, design and scope changes may negate the advantages of the guarantee.

Advantages to the contractor will include:

1 The contractor has eliminated the risk inherent in fixed price contracting as a trade-off for a lower guaranteed fee.

2 The contractor is generally paid for the preparation of his initial planning, including development of cost estimates, schedules and other work plan items which he must absorb in the preparation of a fixed-price bid.

3 The contractor has an opportunity to obtain future work from the owner with minimal or no competition if the program objectives can be achieved in a harmonious relationship.

4 The jobsite can be staffed to provide most of the required services on a reimbursable basis, thus minimizing the contractor's home office overhead.

5 The contractor can profit from equipment and tool rental in excess of amounts usually included in fixed-price quotations.

Disadvantages from the contractor's standpoint include:

1 Fees may be minimal in comparison to profit potential in areas of known performance with a favorable risk/reward ratio.

2 Contractor supervision and management may resent major decisions being made or questioned by the client in areas where they would normally have full responsibility.

3 Planning and control functions and assignment of personnel are made increasingly difficult because of the simultaneous nature of construction and design. Numerous changes may foster a hip-shooting response that may not be the best training ground for young managers.

4 The contractor's reputation may suffer in the event of significant delays, cost overruns, or compatibility or personal clashes with owner personnel.

5 Under the guaranteed-maximum-price option, the contractor may bear risks for items not under his control. The owner may resent amounts paid to contractor in event of a large underrun.

The Owner-Builder

Historically, many city, county, and state public works departments, federal government agencies such as the Tennessee Valley Authority, and private companies such as DuPont and Coors have performed both their own design work and some or all of the actual construction with their own forces. This approach is often referred to as "force account." Other owners (or owners' representatives), such as the Army Corps of Engineers, the Bureau of Reclamation, the Public Building Service of the General Services Administration, and Proctor & Gamble in the private sector, while retaining many of the management and conceptual design responsibilities, have utilized consultants for some or all of the detail design, and have depended upon construction contractors for the actual hiring and supervision of the labor force.

Types of Contracts   Owner-builders have utilized many of the contractual forms discussed above for the traditional approach, and they are increasingly moving to professional construction management methods. Actually, the owner-builder can be likened to the design-constructor, except that the ultimate product is utilized in-house rather than developed for an outside owner. Many of the owner-builders have developed design-construct divisions that are of a size comparable with those of many of the larger turnkey builders.

Advantages and Disadvantages   The circumstances of each individual owner determine the advantages and disadvantages of the owner-builder approach. However, it appears that this method of performing design and construction can be best justified when the volume of work is relatively large and relatively constant over a long period of time, and where project management can be separated from operational management.

The owner-builder can employ all the techniques of the design-constructor, the professional construction manager, and the traditional approach. However, the advantages of this type of approach are best suited to a relatively few, favorably situated companies or agencies. Further discussion is beyond the scope of this book, but would generally include components of all the other concepts, altered as necessary to fit the owner-builder's objectives.

As a sideline to their basic business, some successful construction companies have themselves turned into owner-builders, constructing apartment houses, office buildings, and other rental or lease-back facilities. In a few instances their success as developers has minimized or eliminated the general contracting or design-construct parent company.

Design-Construct or Design-Manage (Turnkey)

Some authorities differentiate between "design-construct" and "turnkey." General usage, however, treats them interchangeably. In this method, all phases of a project, from concept through design and construction, are handled by the same organization.

In the case of design-construct, the constructor acts as a general contractor with single-firm control of all subcontractors. Usually, but not always, there is some form of negotiated contract between design-constructor and owner. In the case of design-manage, construction is performed by a number of independent contractors in a manner similar to the professional construction management concept. Under either design-construct or design-manage, construction can readily be performed under a phased construction program to minimize project duration. This form of completing projects has been used for the majority of process-oriented heavy industrial projects constructed in the United States in the last few decades. Reference to *Engineering News-Record*'s annual list of the 500 largest designers shows that the design-constructors are heavily represented in the top 20.

Background and Evolution   In the 1950s, the typical design-construct or engineer-contractor organization was organized along functional lines. Design, procurement and construction departments were coordinated by a project engineer who was responsible for the overall project with little or no authority to direct the functional groups. The real responsibility initially rested with the design group and shifted to the construction group as the design neared completion. As projects became more complicated and early completion became increasingly important, clients began to demand that one person be placed in charge who could speak for all departments and be accountable to the owner for the overall performance.

A number of design-construct firms created the project manager as the leader of a task force type of organization in order to be more responsive to owner requirements. Here the project manager was normally placed in full charge of engineering, procurement, construction and support personnel. The task force operated as a separate organizational unit made up of functional personnel loaned from their respective departments on a full-time basis for the duration of the project assignment.

The task force headed by the project manager proved popular with both the client and top management, and the use of the task force, which had previously been utilized mainly on remote foreign projects, became the prevailing practice domestically as well. As the task forces became increasingly self-sufficient, relationships with the functional departments became more and more remote. The responsibility shift to the project manager was not readily accepted by the functional managers and in the early years considerable friction developed. However, with the strong backing of the client and company management, the project manager's authority continued to be enhanced.

The task forces worked well during the short range when informal ties to the functional departments remained in force to enhance communications and exchange of engineering and construction techniques. However, in many organizations the projects became increasingly isolated and ties with the functional departments were often almost nonexistent. Feedback data to the engineering and estimating departments in a form necessary to maintain corporate records was often minimized or eliminated by clients or project managers who were more concerned with project performance and short-range economy than with enhancing the parent organization's long-range knowledge base. Engineering design, estimating and construction capabilities in the functional departments no longer continued to be improved based upon the evaluation of the effectiveness of project discipline performance compared to original plans. As work continued to expand, project teams increasingly were made up of less senior engineers and managers who in turn were unaware of successes and failures in other similar projects which had previously helped to avoid "reinventing the wheel." Upon completion of a project, many participants found that they had no real acquaintance or credibility with the functional department managers who gave priority in staffing future projects to more well known individuals. Specific projects sometimes modified or changed standard company design specifications and criteria, and/or cost and schedule control systems to better meet short-term project or client needs. Overall consistency began to deteriorate. Top management began to perceive that the deterioration in engineering and performance quality and in the ability to plan and estimate new projects were often the long-range consequences of the task-force method of operation.

The matrix organization was designed to preserve the advantages of the project manager led task force while regaining the optimization of the discipline excellence of the initial functional organization. Sometimes called a task force within a matrix organization, the new concept featured a dual reporting role for key members of the project team. The project manager continued to be fully responsible for scope, costs and schedule objectives as developed for a particular client. However, the functional department managers regained the responsibility for compliance with approved company-wide design, administrative and construction standards and procedures. Functional discipline managers were often given back responsibility for discipline training, dissemination of information on new technology, and for supplying a permanent home and career path progression to team members.

Types of Contract   The turnkey approach can be utilized under just about any form of contract, including lump sum, cost plus a fixed fee, cost plus an incentive fee, and guaranteed maximum price. Unit-price contracts do not generally lend themselves to the turnkey approach. However, most of the turnkey work performed in recent years has been through some form of negotiated cost-plus-a-fee type of arrangement.

Advantages and Disadvantages   Advantages and disadvantages of the turnkey approach depend upon the individual owner, the importance of the process in an industrial project, and the skill of the design-construct firm. The choice between design-construct and design-manage is usually one of minimizing the overall project cost (including benefits of shortening the schedule); it is dependent upon the location of the project, the skill and availability of local contractors, the contemplated number of potential changes, factors pertinent to the economic climate, availability of competition, and other considerations.

Advantages to the owner include the following:

1 There is but one overall contract for the owner, with design, construction, and often process know-how furnished by a single organization.

2 Minimal owner coordination is needed between construction, design, and other project elements. This can be of great benefit to an unknowledgeable owner.

3 Design-construct time can be reduced through phased construction.

4 There is considerable opportunity for construction expertise to be incorporated during the design phase.

5 Implementation of changes is simplified throughout the construction program.

Disadvantages from the owner's viewpoint are these:

1 Usually no firm project cost is established until construction is well underway.

2 If performed under a lump-sum or guaranteed maximum-price contract, overall quality and performance may be subordinated to ensure profitable operation by the design-constructor.

3 There are few checks and balances, and the owner is sometimes not advised or aware of design or construction problems that may greatly affect cost or schedule.

4 Because of the minimum involvement of the owner, the final result may not fully comply with expectations.

5 Successful integration of design and construction functions and avoidance of changes are largely left up to the design-construct firm; the owner may not be aware of weaknesses that interfere with economical and timely project completion.

6 Insistence by the owner's personnel on making major decisions, the consequences of which (such as working unreasonable overtime) may not be understood, can prejudice the overall economic result.

7 Other disadvantages are similar to those listed for the negotiated contract under the traditional method.

Professional Construction Management[1]

Professional construction management treats the project planning, design, and construction phases as integrated tasks. This approach unites a three-party team consisting of owner, designer, and construction manager in a non-adversary relationship, and it provides the owner with an opportunity to participate fully in the construction process. A prime construction contractor or funding agency may also be part of the team. The team works together from the beginning of design to project completion, with the common objective of best serving the owner's interests. Contractual relationships among members of the team are intended to minimize adversary relationships and contribute to greater responsiveness within the management group. Interactions relating to construction cost, environmental impact, quality, and completion schedule are carefully examined by the team so that a project of maximum value to the owner is realized in the most economical time frame.

Professional construction management is competitive in overall design-construct time with a negotiated contract under the traditional method and with the turnkey approach. It usually features a number of separate lump-sum or unit-price construction contracts which, under certain circumstances, may prove more competitive than either the general contract or the cost-plus-a-fee approach. If phased construction is used, it, like phased construction under other methods, involves the owner in some degree of risk in overrunning budgets.

A professional construction manager is a firm or an organization specializing in the practice of professional construction management, or practicing it on a particular project, as a part of a project management team. In keeping with the non-adversary

[1]The definitions for "professional construction management," and "professional construction manager," and the rationale that follows are based on the work of the American Society of Civil Engineers' (ASCE's) Task Committee on the Management of Construction Projects. They were published in D. S. Barrie and B. C. Paulson Jr., "Professional Construction Management," *Journal of the Construction Division*, ASCE, vol. 102, no. CO3, September 1976, pp. 425–436.

relationship of the team members, the professional construction manager does not normally perform significant design or construction work with its own forces, although it may provide the general conditions of the site.

The term "professional construction manager" is certainly not intended to exclude qualified general contractors or any other qualified organization. The definition could well include a qualified general contracting organization, a qualified design-construct or design firm, or a qualified construction management firm. Many general contractors have been providing construction management services for years under a negotiated contract with the owner in which they perform certain portions of the work with their own forces. Many design-construct firms have also prepared the design and performed construction management services utilizing a number of independent contractors to do the actual construction. The terms "professional construction management" and "professional construction manager" are intended to define both another alternative for the procurement of constructed facilities and the interrelationships among the three parties while practicing this form of construction management. In practice, firms proposing to manage construction under a negotiated contract will compete with general contractors acting as professional construction managers, as well as with construction management consultants, designers, and design-constructors who have obtained the necessary construction skills—skills that can probably best be learned by working as a general and specialty contractor.

Type of Contracts   Both the AGC and the **American Institute of Architects** (AIA) have developed model contracts along the lines of a negotiated construction contract; the principal difference is the guaranteed maximum-price alternative included in the AGC document. The AGC document is reproduced in Appendix C.

Usually, professional construction management contracts with the owner will provide for full reimbursement of field costs, plus a fixed fee to cover home-office costs and profit. An alternative form preferred in certain situations is to provide for reimbursement of home-office costs plus a fixed fee for profit only. This arrangement can also include a guaranteed maximum price for home-office costs, and sometimes also for the field costs incurred by the manager's forces.

Advantages and Disadvantages   Advantages from the owner's position are as follows:

1 Special construction skills may be utilized at all stages of the project with no conflicts of interest between the owner and the designer.

2 Independent evaluation of costs, schedules, and overall construction performance, including similar evaluation for changes or modifications, helps assure decisions in the best interest of the owner.

3 Full-time coordination between design and the construction contractors is available.

4 Minimum design-construction time can be achieved through use of phased construction.

5 The professional construction manager approach allows price competition from local contractors akin to the traditional lump-sum or unit-price methods.

6 Significant opportunities are provided for value engineering in the design, bidding, and award phases.

Disadvantages from the owner's position are as follows:

1 If the professional construction manager recommends phased construction, the owner begins the project before the total price is established. Early completion may not provide a sufficient trade-off for this risk.

2 If the owner has only a fixed amount to spend, and would not build the project if its cost would exceed this amount, the traditional method would be preferable in such a go/no go situation.

3 The owner has certain responsibilities and obligations that must be fulfilled in a timely manner.

4 Success of the program depends greatly upon the planning, scheduling, estimating, and management skills of the professional construction manager.

5 The professional construction manager does not usually guarantee either the overall price or the quality of the work; this situation contrasts with that of the general contractor in the traditional lump-sum approach.

Program Management

Program management (sometimes called *project management*) is an emerging concept being used on some of the very largest projects. Program management services may include no design or direct construction but the program manager could handle overall management of a number of individual projects related to an overall program. Program management has been applied on several major projects where the owner has participated heavily in managing the program in an integrated program management team utilizing owner top management personnel and management and specialized personnel from a **Construction Management** (CM) firm, architect/engineer, or other consultant.

Background The roots of modern program management began with the major design-construct companies. Domestically, the companies initially performed direct-hire construction work, generally operating under national agreements with the AFL-CIO craft labor unions throughout the United States and portions of Canada. On overseas projects, as the American worker and middle manager became increasingly costly, many of the projects contracted out the actual construction work to local or international contractors while continuing to manage the overall project.

Stephen D. Bechtel, Jr., in the keynote address to the 1988 **Project Management Institute** (PMI) Seminar/Symposium held in San Francisco, stated, "In 1951, Trans-Mountain (pipeline project) was the first project where we, as an organization, actually functioned as the Project Manager. We were the engineer and the engineering management organization, performed the engineering, did the procurement , and managed the construction, with the actual construction work being done by other contractors. Thus, although we didn't call it Project Management then, the approach and organization was a forerunner of things to come."

Evolution   In many ways, Mr. Bechtel's description captures the essence of program management, although his description did include design services within his firm. The program management concept as now emerging is in some ways similar to the project management concept generally favored by the design-construction companies, where the project manager is in overall charge of the firm's design, administrative, and construction functions, and the task force is a favorite operational method. In the more narrow sense today, the term implies that other parties do most of the engineering and design as well construction. The program manager serves as the single point of contact to the owner to coordinate and manage the various other parties involved in planning, design, procurement and construction. This differs significantly from the professional construction manager, who is but one of two or three parties reporting more directly to the owner, and the owner is usually more closely a part of the team. The program management concept utilizes an overall management organization that may manage a number of design firms, construction contractors, material and equipment suppliers, and other participants in the building program. Figure 2-7 shows two alternate program management concepts.

Program management organizations can be organized using functional, task force, and matrix organizations. Each type of organization has its own advantages and disadvantages when applied to a particular project. Today, program management has been adopted by public and private owners, contractors, design firms, developers and other organizations as well as by the design-construct firms who popularized the concepts. Program management concepts have been applied to several rapid transit systems, to multi-billion-dollar municipal sewerage projects, to power projects, and to major private undertakings. Its use is also spreading to many non-construction projects such as the development of a major software program, a relocation of company headquarters, a research program for a new drug, and other non-construction oriented projects.

One interesting innovation being discussed for a billion dollar manufacturing upgrading and renovation program is for the owner to employ a program manager who will assist in developing common programs and standards for the execution of the individual projects. This particular program consists of four major multimillion dollar projects, each contemplated to be managed by a construction management firm reporting to the owner's program management organization, which is assisted by a program management consultant. This illustrates how, in practice, real projects draw from several of the basic concepts described in this chapter to form their own unique organizational and contractual approach to best suit their needs.

Advantages and Disadvantages   The advantages and disadvantages of program management are something of a synthesis of those of the design-construction approach (with the single point of contact for the owner for design and construction) and professional construction management approach (with the more detached, impartial view resulting from the manager's lack of vested interest in the actual design and construction that is handled by others). Rather than duplicate and rehash those points here, the reader is encouraged to critically reexamine the pertinent lists in those earlier sections to decide which are advantages and which are disadvantages in the case of program management.

PROGRAM MANAGEMENT

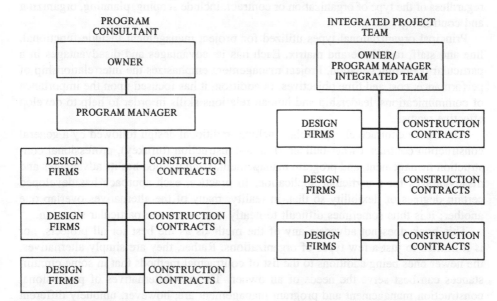

- SEVERAL DESIGN FIRMS
- PROGRAM MANAGER RESPONSIBLE FOR OVERALL PROJECT
- PROGRAM MANAGER USUALLY ACTS AS CONSTRUCTION MANAGER
- FIXED, UNIT PRICE OR NEGOTIATED CONSTRUCTION CONTRACTS

- SEVERAL DESIGN FIRMS
- INTEGRATED TEAM RESPONSIBLE FOR OVERALL PROJECT
- INTEGRATED TEAM USUALLY ACTS AS CONSTRUCTION MANAGER
- FIXED, UNIT PRICE OR NEGOTIATED CONSTRUCTION CONTRACTS

FIGURE 2-7
Program management.

SUMMARY

The life cycle of a construction project can be classified into the following six major identifiable phases:

- Concept and feasibility studies
- Engineering and design
- Procurement
- Construction
- Start-up and implementation
- Operation or utilization

Owners often handle the concept and feasibility studies themselves, with assistance from specialized consultants as required. This phase includes broad-scale planning, conceptual analysis, and technical and economic feasibility studies; it culminates in defining the general location and components of the project. The other construction phases can usually be performed sequentially, as they are in traditional methods, or be subject to overlapping in a phased construction or "fast-track" program.

Basic management activities that must be accomplished within each phase, and regardless of the type of organization or contract, include scoping, planning, organizing and controlling.

Principal organizational types utilized for project management include functional, line and staff, task force and matrix. Each has its advantages and disadvantages in a particular operating situation. Project management emphasizes the interrelationship of performance, cost and time objectives. In addition, it has focused upon the importance of communications, leadership and human relations skills in order to help to develop effective teams.

Alternative contractual approaches include traditional design followed by a general construction contract, owner-builder, design-construction (turnkey), professional construction management, and program management. Each method has its advantages and disadvantages for a particular application. In practice, each approach has developed certain degrees of flexibility so that, in reality, many of the alternatives overlap one another; it is thus sometimes difficult to neatly categorize one particular situation.

This book does not advocate any of the methods as the best for all projects, nor is it limited to just a few types of organizations. Rather, they are simply alternatives, the newer ones being additions to the list of contractual methods that in some circumstances can best serve the needs of an owner. The new alternatives of professional construction management and program management are, however, uniquely different approaches that have justifiably established their place on the construction scene.

CHAPTER 3

APPLICATIONS AND REQUIREMENTS FOR MANAGEMENT ORGANIZATIONS

Professional construction management and program management have emerged as important alternatives among several time-tested methods for organizing a construction project. Like other approaches, however, their advantages and limitations depend upon their specific applications. Each of the parties—owner, designer, and manager—has relationships and responsibilities to each of the other team members, to others in the construction industry, and to the public. To succeed, construction management organizations must create a non-adversary team relationship involving all parties. Just as the designer must have a sound organization with proven qualifications and skills, the professional construction manager or program manager must have the personnel, qualifications, and applicable construction experience to develop and implement a well-thought program for successful accomplishment of the project.

This chapter will first explore some of the applications as well as the limitations of professional construction management, general contracting, design-construct, program management and other alternatives. It will then examine key relationships among, and responsibilities of, the team members. Finally, it will specify some of the important planning and execution requirements that are assumed by a qualified management firm.

APPLICATIONS AND LIMITATIONS

Most construction projects are unique, with plans and specifications developed on an individual basis. However, one can at least perceive certain broad classifications of owner objectives that in turn influence the most appropriate contractual and organizational arrangements for a particular job. Although professional construction management or program management can operate successfully in almost all these areas, certain situations are generally more suitable than others. Factors that should be considered in selecting the best approach include budget, geographic location, technical and managerial characteristics of the project itself, and the needs and capabilities of the owner. Each of these will be discussed in this section.

Budget Considerations

The knowledgeable owner is interested in achieving lowest overall project costs commensurate with maximum return on investment while maintaining required value. Capital costs for a project with a typical distribution would include, but not be limited to, the following:

Property acquisition	5%
Preliminary planning	3%
Design	7%
Construction	54%
Construction administration	6%
Financing	20%
Owner's internal costs	5%
Total capital cost	100%

When shortest elapsed design-construction time is of sizable economic importance, the professional construction management or program management approaches, combined with a phased construction program, should receive serious consideration. However, an owner might find other ways of organizing the project to be more advantageous when other influences are more important. We now discuss a few of these.

Little- or No-Risk Situations   One situation where professional construction management or program management, combined with phased construction, may not be suitable occurs when the project definitely must be constructed within a fixed sum of money; some form of construction cost guarantee is then required. The tightest guarantee, of course, is the traditional single-contract method, where no award is made if the low bid is over budget. Other possible answers to this situation include negotiating a lump-sum or guaranteed maximum-price contract with a design-constructor to produce a structure with the desired design and performance standards, or negotiating a similar contract with a qualified developer who would obtain its own design services to achieve the same result. In fact, a construction manager who is indeed a professional would recommend these alternatives in a go/no-go situation of this type.

Early Completion Economically Significant   When the owner can assume some risk regarding the project's ultimate construction cost and when earliest completion is required, it is still both feasible and prudent for the owner to approach choosing the method of construction and the individual companies involved with an underlying goal of minimizing not only time, but also risk and overall project costs. Here, the owner has a number of time-tested general methods for consideration, all featuring simultaneous design and construction. It may:

Employ a design-construct firm as both designer and general contractor on the basis of cost-plus-a-fee, or cost-plus-an-incentive-fee, utilizing lump-sum contractors.

Employ a separate architect, negotiate a contract with a qualified general contractor on a cost-plus-a-fee or incentive-fee basis, and use lump-sum subcontractors.

Employ a design-manage firm to prepare the design and to administer and manage a phased construction program with lump-sum bid packages.

Employ an overall program management firm that will coordinate independent design firms and use lump-sum contracts for bid packages let to the construction subcontractors.

Employ a separate designer coupled with a professional construction manager to develop, administer, and manage a phased construction program with lump-sum bid packages.

Figure 3-1 shows representative overall design-construct schedules for the above methods; it further compares these methods' phased-construction program with a single fixed-price contract based upon complete documents. Figure 3-2, developed by the General Services Administration, also compares phased construction with traditional methods.

Long-lead Planning with Minimum Risk   When advance planning and/or project completion requirements permit the design to be completed before beginning construction, and without significant economic penalty, the traditional approach, with a separate designer and a lump-sum general construction contract, remains attractive. However, in special situations, each of the methods previously outlined may also prove advantageous, depending upon the special qualifications of the firms involved and upon the geographic location of the project.

Owner Evaluation   Each of the mentioned classifications of firms is actively promoting its concepts as the best solution to the owner's problems. The owner must

FIGURE 3-1
Phased construction program comparison.

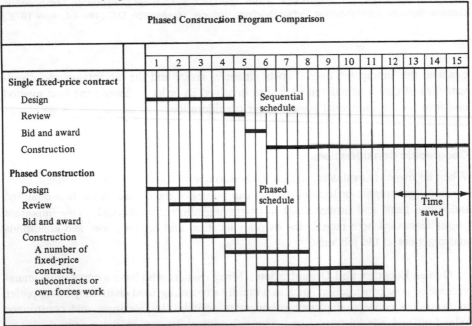

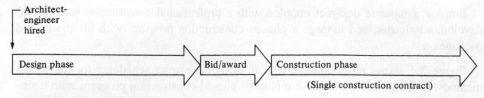

Traditional Construction Method

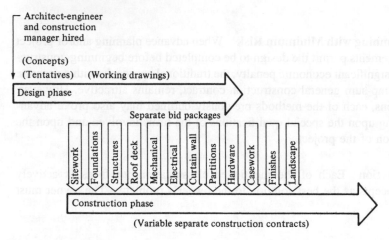

Phased Construction Method

FIGURE 3-2
Traditional versus phased construction. (From *The GSA System for Construction Management*, General Services Administration, Public Buildings Service, Washington, D.C., rev. ed., April 1975.)

evaluate the anticipated overall costs, financial risks, and the advantages and disadvantages of each method and each firm involved in order to choose the best method and firm for the individual project under consideration.

Geographic Considerations

When an owner is evaluating which method to use in today's complex business environment, geographic or location factors can influence the choice of the best-qualified individual firms and further assist in comparing alternative methods. Two important points discussed here include the owner's location and organization, and conditions and practices at the job site.

Owner Location and Organization   Many owners who have a continuing annual volume of new construction (such as a rapidly expanding food chain may have) prefer to centralize conceptual planning, budget responsibilities, designer and constructor selection, and administration and financial control in one or more central offices.

Project sites all over North America and even in foreign countries will be controlled and administered from the home or branch office.

In this situation, a typical project may include the owner's project manager located in one city, an architect in another, a general contractor in still a third, a construction site in yet another, and possibly an associate architect in the city where the project is being constructed. Consider the vastly different communication problems present in this situation when compared with a project constructed in the owner's home city with a local architect and a local general contractor.

Even with the assistance of a program management or a professional construction management firm, similar geographic separation will introduce additional complexities and communication difficulties. Again, each construction project is unique. The intelligent and economical construction management plan will take full advantage of area location and familiarity when significant, and will develop individual methods consistent with overall project economy.

For example, an owner who has developed a strong central construction and design control group over the years, and who has a reasonably level or accelerating volume, may well find it advantageous to use a professional construction manager with offices located near the same central location. In this way, face-to-face analysis of recommendations and problems, and joint determination of solutions affecting the remote site are more easily handled.

On the other hand, a one-time or infrequent owner who delegates procurement action to a program manager may well prefer that such a firm be located in the same city as the job site to facilitate on-the-spot decisions, especially if the architect is also located in the same area. Or, alternatively, the project office can perform all work with minimal reliance upon a remote home office.

Job-Site Construction Practice   Any method of construction management must take into account existing conditions at the job site. This factor is increasingly important when the designer is located some distance away, since practices differ in various sections of the country. Specifications may require particular brand-name products that are not readily available, or other peculiar methods may be called for that will result only in excessive costs. Designers cannot be expected to understand construction trade jurisdictions in a multiplicity of locations, or to analyze construction craft productivity, or to be knowledgeable about continuing local jurisdictional or other labor disputes among different crafts.

One of the major contributions that a professional construction manager or program manager can make in minimizing construction costs under any method of construction is to influence the design and specifications to take full advantage of proven local methods and materials understood by all potential bidders. Equally important is the development of work packages, under a phased- or multiple-contract program, that generally fit into the prevailing method of doing business in the area. Conforming to trade jurisdiction in work packages, avoiding continuing jurisdictional squabbles, using locally favored materials and methods, and planning construction packages best fitted to local weather conditions and labor availability and contractor workload in the area can materially reduce the construction cost as well as expedite earlier completion.

It is, of course, helpful—although not necessary—for key personnel in the manager's organization to have worked in a responsible capacity in the job-site area. Although the cited factors may be unique in a particular area, their general existence is known to any construction professional qualified to develop a plan for managing the program. Pure logic, coupled with design and estimating experience, no matter how well performed, cannot discharge the coordinating manager's planning and value-engineering functions unless accompanied by a sound knowledge of job-site conditions.

An apprenticeship in "hard-dollar" estimating, and in profit-and-loss responsibility for lump-sum general and specialty contractors, is by far the best way to gain an appreciation for the importance of this phase of planning and value-engineering review. This kind of background will enable a team of qualified construction professionals quickly to grasp essential project needs and constraints while visiting the area and meeting with representatives of the local building trades council, individual labor unions, contractor associations, individual local contractors, subcontractors, and other local contacts. By comparing answers to questions and listening to the knowledgeable local representatives, a competent construction professional can develop a reasonably accurate picture of local conditions affecting design and contract package preparations in the area in a surprisingly short time. The importance of a job-site visit by qualified personnel prior to design review and bid package development cannot be overemphasized.

Type of Project and Approach

Broad categories in the construction industry were listed in Chapter 1 as follows:

1 Heavy engineering construction
2 Building construction
3 Industrial construction
4 Residential construction

Each category has its own special demands that influence the advantages and disadvantages of alternative organizational and contractual approaches.

Heavy Engineering Construction   Is there a future for construction management and the emerging program management on heavy construction projects? To date, the approach, or variations on it, has been applied frequently to airports, including those at Dallas-Fort Worth, Newark, Salt Lake City, and Cincinnati. A modified form of construction management has also been practiced for many years by certain design-construct firms. One of the early major projects was the Bay Area Rapid Transit facility in San Francisco. Here, a joint venture of Parsons-Brinckerhoff-Tudor-Bechtel supplied construction consultation during design; packaged the work into a number of phased lump-sum and unit-price contracts; provided cost and management control, planning, and programming; supplied field coordinating and inspection services; and provided other functions supplied by a program manager in addition to design. Variations of this approach have also been applied on new rapid transit projects in Washington, D.C., Atlanta, Baltimore, Miami, Montreal, and other cities. The Alaska pipeline is of course one of the largest projects which utilized construction management principles as well as program management concepts.

A smaller but similar earlier example was a new water supply and treatment project constructed for the City of Vallejo, California, in the early 1950s. On this project, Kaiser Engineers acted as consulting engineer for the City of Vallejo, The project included phased-construction packages including a lump-sum water treatment plant, a 30-mile pipeline under a unit-price contract, a reservoir under a unit-price contract, and a pumping station constructed on a lump-sum contract. While the on-site representative was called a "resident engineer" on this project, all the functions usually supplied by a professional construction manager, plus design and feasibility studies, were provided by Kaiser.

It is believed that professional construction management and program management have bright futures on sizable heavy construction projects. Through phased construction, separate lump-sum and unit-price bid packages can be developed using a number of individual designers and can be bid by several general or trade contractors, all managed and coordinated by a professional construction or program manager responsible to the owner. Although the preponderance of projects will undoubtedly continue in the traditional manner, in certain circumstances a carefully planned professional construction management or program management approach can be of substantial benefit to an owner in achieving earliest completion as well as additional economies of increased competition on separately bid packages.

Building Construction   Professional construction management initially received wide acceptance by private owners in building construction, but in recent years has received increased competition from both developers and general contractors. Program managers practicing construction management have also handled a number of the largest projects in both the public and private sectors. As in other types of construction, the value of professional construction management will depend upon individual project and owner considerations. Some of the situations for which these newer methods should receive consideration, with or without a phased construction program, include the following:

Unknowledgeable Owner   Where the owner is neither capable nor desirous of performing value engineering and of coordinating and adjudicating differences between the designer and contractor, it may wish to consider delegation of this responsibility to a program manager.

Overloaded Owner   Where an owner is staffed to handle value engineering and overall coordination and management for a level workload, it may wish to engage a program manager or a professional construction manager to perform this function for individual major projects or peaks in the total program. In this case, the owner uses the manager as an extension to its own way of working.

Knowledgeable Owner   In some cases, an owner may recognize the potential savings possible through fresh design review and overall program planning, even though it may be capable of performing these functions. Here, the program manager or professional construction manager can prove useful for a wide variety of spot consulting services at all stages of the program.

Industrial Construction Much of the previous discussion is also applicable to industrial construction, but there are certain additional considerations that are largely due to the importance of the project's industrial process. First, when drawings are completely finished because of far-sighted planning, the lump-sum single contract, with or without separate bidding for mechanical, electrical, or other significant specialties, is often preferable. In this case, the process itself is the significant contribution of the designer. The opportunity for outside professional construction manager consultation during the design phase is considerably minimized. The typical solution is to employ a design-constructor or program manager with process skills for the design work, assign to it construction control or resident engineering services during the construction phase, using the single lump-sum contract, or a modified version of it with separate electrical, mechanical, or other specialties. In this situation, contracts can either be assigned to the prime contractor for coordination and control or be handled by the resident engineer.

On the other hand, where drawings and specifications cannot be completed before construction begins, the owner generally has the choice of adopting a phased construction program utilizing a design-constructor or perhaps a program manager skilled in process know-how, or of proceeding on a negotiated construction contract either with the design-constructor or a separate general contractor. When earliest completion is economically significant, and where the process is complicated and forms the basis of the success or failure of the entire project, justification of a separate professional construction manager becomes more difficult. It appears that the use of a three-party professional construction management team in this situation may have the least potential advantage to the owner.

Residential Construction Again, much of the discussion of building construction should apply also in residential work, especially on larger developments. Many developers are in fact skilled construction managers performing all the basic functions and duties while hiring a separate designer and completing construction using lump-sum, phased packages, or in the conventional manner as a general contractor.

Other Considerations

Other considerations would include the different objectives, policies, and organizations of owners in the following three basic categories:

One-time owner with a single project

Owner with a continuing level of workload with a reasonably predictable number of projects over a given time period

Owner with a fluctuating workload with major peaks and valleys over a given time period

Single Project Most owners with a single project to build are not equipped with internal staffing or internal controls geared to performance of construction work. The program manager or professional construction manager can serve as a knowledgeable representative for such an owner while being free to recommend the method as well as prequalification of individual firms most qualified to achieve the objectives of a particular project. However, the design-constructor, the separate designer–general contractor combination, and the developer all perform work for this class of owner with varying degrees of success. The principal advantage of the qualified professional construction manager or program manager is the freedom from economic conflict with the owner that is present to some degree in other forms. However, the single most important factor is choosing a firm, whichever method is employed, that is thoroughly qualified and of unquestionable integrity.

Level Continuing Workload   The owner with a level continuing workload has great flexibility in choosing the methods to be employed for a particular project. It can, by developing its own organization, achieve a significant degree of in-house construction management or even construction capability. Its benefits from use of a professional construction manager will depend upon its realistic appraisal of the strengths and weaknesses of its own organization. This class of owner has the opportunity to develop a staff of knowledgeable people who can analyze each project for its unique objectives, then choose the performance method as well as the individual firms that offer the greatest potential. It can develop technical skills and resources that will enhance the handling of construction management functions ranging from entire projects to specialization in one or more technical or process-oriented major portions of typical installations. The outside program manager or professional construction manager, performing on selected projects, can supplement its own forces under relatively set procedures, or inject new ideas, techniques, or knowledge of local area practices that can prove advantageous.

Fluctuating Workload   Major peaks and valleys superimposed on a continuing workload create a greater need for outside planning and control services for certain individual projects, coupled with parallel development of in-house capabilities for the level portion of the workload. Assignment of management and control responsibilities for major peaks, as well as the choice of the particular methods, can again upgrade the basic organization through cross-fertilization from knowledgeable outsiders. Furthermore, this type of owner must be able to develop the in-house capacity to analyze and choose the proper course for the major peaks, consistent with its own capabilities and requirements.

An Added Thought   Many owner organizations become specialized. Their construction management personnel are constantly being exposed to only one side—that of the owner—of the complex problems and interactions within the industry. Use of a professional construction manager or program manager, one who is up to date and knowledgeable about the problems and latest techniques of contractors, subcontractors, labor unions, and other groups, helps the owner's forces to adapt to a sometimes chaotic and constantly changing construction environment.

RELATIONSHIPS AND RESPONSIBILITIES

The functions of the program manager or professional construction manager are to plan, administer, and control an overall construction program objectively and conscientiously, in a manner best suited to the individual project objectives of the owner, while maintaining a fair and businesslike relationship with others involved in the program. Objectives of the owner will include minimum overall project cost, involving the economic benefits of minimum design-construction time, compliance with recognized owner administrative and control requirements, and assurance of specified quality and utility in the finished product.

Responsibilities to the Owner

The coordinating manager's duties and responsibilities toward the owner include faithful and professional representation and advice, free from economic conflict. This advice and representation will be objectively handled within the framework of the delegations of responsibilities that the owner may elect to assign the manager. The manager should keep the owner fully informed at all times regarding the current status of the project in comparison with the overall plan.

Responsibilities to the Designer

The manager's relationship with the designer must be thoroughly professional; to succeed, he must obtain full cooperation from the architect or engineer designing the project. Only by working together can the full benefit of a design-phase value-engineering program be achieved; credit for successful reduction of project cost while preserving value must be equally shared with the designer if the relationship is to survive and prosper. The manager provides his economic knowledge of the construction industry as a resource for the designer in furthering the overall objectives of the owner. If, in turn, the continuing design responsibilities of the designer to the owner are preserved and acknowledged by the manager, he will then have a valuable partner. By working together to attain their mutual economic interests, the designer, owner, and manager will aid in achieving the owner's objectives.

Responsibilities to the Contractors

The manager's relationship with the project contractors must be equally professional. He must accurately interpret plans and specifications and promptly request clarification from the designer when necessary. Many years ago, resident engineers interpreted plans and specifications in an impartial manner and decided questions of fact and conflicts between owner and contractor. Today, the program manager or professional construction manager must fully discharge this same responsibility to the contractors. The manager must insist upon compliance with plans and specifications to ensure achieving the owner's objectives, but must equally insist upon fair compensation to contractors for changes and modifications initiated by the owner or designer, or caused by managerial omissions.

Responsibilities to Others

The manager also has a duty toward labor through recognition of the collective bargaining agreements under which contractors and labor unions operate, and must have a reasonable knowledge of craft jurisdiction as practiced in the project area. In open-shop areas the manager has similar, though different, responsibilities.

The professional construction manager or program manager has a duty to the industry and to the general public. If qualified, this person is aware of the many problems facing the industry. The manager should act as a knowledgeable professional in advising the owner and in fulfilling the required responsibilities to assist in solving the underlying problems and economic conflicts that are always present to some degree in a particular project area.

REQUIREMENTS OF THE PROFESSIONAL CONSTRUCTION MANAGER OR PROGRAM MANAGER

The professional construction manager or program manager must first obtain the facts, then develop a sound plan and implement the plan during construction.

Planning Requirements

The success of a management program depends on sound planning. The plan forms the standard upon which the project control system is based and by which future performance is judged.

For best results, the manager should be appointed prior to the beginning of detailed design. Sufficient preliminary planning by the owner and its designer should be available so that the general scope of the project is apparent at that time.

Depending upon the owner's method of selection, some preliminary planning may have been accomplished in a proposal submitted by the prospective manager. In any event, the manager's preliminary and final planning is performed during the design stage of the project.

The program manager's or professional construction manager's initial planning is divided into several major stages:

Fact Finding This step is often neglected but represents a major key to unlock the essential facts and information necessary to construct a successful project. Each construction project is unique in terms of both structures and geographic and economic factors prevalent at the work site. If, by virtue of long-time associations, the manager is familiar with the project area and its local economic conditions, fact finding for a particular project can be confined to visiting the work site itself and becoming familiar with the planned structures, and understanding the objectives, needs, and requirements of the owner and designer. Fact finding will include the following considerations:

Owner's Objectives and Requirements The manager will meet with the owner's representatives to understand its objectives and requirements. The manager, among other things, will:

Determine project duration, completion priorities, and other scheduling information
Obtain preliminary cost estimates, cost criteria, appropriations, and other budget considerations
Obtain owner's drawings, specifications, and preferred construction techniques

Obtain owner's operating procedures, including contractual requirements, bidder qualifications, bonding requirements, and other internal procedures required or preferred by the owner

Define responsibilities of owner, designer, and construction manager, as well as the extent of delegation to each

Determine specific functions the owner intends to perform for himself, and the extent to which supplementary assistance may be required

Define responsibilities of key individuals on the staffs of both the owner and the professional construction manager

Designer Objectives and Requirements The manager will meet with the designer's representatives in order to understand its objectives and requirements and to establish ground rules for a mutually rewarding professional relationship. The purposes of these meetings will be to:

Review design criteria, conceptual planning, and detail design to date

Review or develop a preliminary design schedule; this will be significant in developing a phased-construction program

Develop the basic understandings necessary to commence a partnership value-engineering program utilizing the manager, designer, and owner

Determine designer experience in the area and its understanding of job-site economic factors relating to the construction work

Review overall completion requirements and agree on preliminary scheduling

Establish the type of professional relationship that will enhance the standing of the designer with the owner by facilitating a new resource for planning, implementing, and controlling the program

Establish a meeting of the minds with the designer on its construction responsibilities to both the owner and the manager, and establish the manager's responsibilities to the designer

Review the delegation of authority to each party by the owner

Define responsibilities of key individuals in both organizations

Area and Site Visit The professional construction manager or program manager will spend sufficient time at the job site and surrounding area to appreciate local factors and requirements. The manager will:

Review local work practices and jurisdiction

Ascertain local craft productivity and availability

Obtain collective bargaining agreements

Determine locally favored methods and materials

Ascertain key local prices for standard items

Obtain climate information for use in developing weather constraints

Screen local contractor capabilities, workload, and interests

Visit key local contractors, trade associations, labor union representatives, and other knowledgeable local industry representatives

Determine building-permit requirements and local agency jurisdiction and permit requirements

Development of Preliminary Program After determination of own... signer objectives and requirements, and on the basis of a thorough knowle... job site and its surrounding area, a "Preliminary Program" for the proje.. can be developed. This program will include some or all of the following items, depending upon information available:

Development of criteria and conclusions from the site and surrounding area investigation

Development of a proposed "Work Plan," setting forth in detail the recommended approach to the project, including:

- Overall approach
- Home-office services
- Field management services
- List of proposed work packages
- List of proposed contractors for further screening
- Preliminary design schedule and package procurement schedule
- Value-engineering program
- Preliminary construction schedule
- Preliminary CPM precedence diagram
- Bar-chart schedule

Submission of preliminary magnitude estimate of project cost for preliminary control purposes and for prequalifying selective bidders

Submission of a detailed estimate of professional construction management or program manager costs, if not already presented as a part of the proposal

Assignment of key personnel and submission of schedules for contemplated future personnel to be assigned

Development of Final Program After review of the preliminary plans with the owner and with the designer, the *Project Plan* is completed and issued to all key parties in the overall project for use in controlling the actual progress. Final planning will include the following:

Breakdown of design schedule by contemplated construction contract packages

List of proposed contract packages, including detailed scope of each

Completion and issuance of overall project CPM schedule

Establishment of project control systems

Beginning of value-engineering program

Issuance of a procedures manual for the project, setting forth key duties and responsibilities

Implementation Requirements

Execution of the plan is divided into two equally important objectives. The job must be "bought out" at a price within the budget, that is, firm prices must be obtained for materials and construction contracts; and it must be completed as designed and on schedule.

Bidding and Award Phase In a phased construction program, the first packages must be developed shortly after commencement of detail design, and will proceed simultaneously with the detail design work. Items to be considered during this phase will include the steps necessary to:

Finalize preliminary contractor bid lists by contract package
Prequalify selected contractors based upon qualifications criteria
Issue final invited-bidder list by contract package
Prepare bid packages
Review bid packages
Issue "Requests for Quotation"
Prepare detailed fair-cost estimate for each bid package
Review and analyze bids
Recommend contract awards
Issue "Notices to Proceed with Field Work"

Construction Phase The field construction phase will begin prior to the award of the first contract. Home-office management and control will parallel the field effort. On certain large projects, the two may be combined in the field location through choice of proper personnel. The manager's function during the construction phase will include responsibilities to:

Establish the field office
Hire testing laboratory and surveyor
Obtain necessary permits
Manage, coordinate, and inspect the work of individual contractors
Maintain job diaries, drawing register, and other records
Prepare and approve progress-payment requests
Maintain progress records and photographs
Prepare input for project control system
Prepare field reports and schedules
Prepare contract closeout and acceptance documents

Controls The development and implementation of a comprehensive project control system are essential if the full potential of a program management or professional construction management approach is to be realized. The controls must be based upon realistic goals developed during the planning and design phase. The control system, by itself, will not manage the project, but it will, if properly designed, measure the current status against programmed goals, so that corrective management action can be applied when warranted. Features of a comprehensive project control system will include:

An updated and current CPM network
A design and procurement schedule showing actual progress compared with that scheduled
CPM summary schedule showing actual contract progress compared with scheduled progress for each contract

Cost report comparing forecast-at-completion costs, including committed and estimated contract costs to complete, compared with budget estimates

Value-engineering summary showing results of program to date

Weekly progress reports listing significant progress, lack of progress, current problems, proposed solutions, and other pertinent information

Monthly progress reports summarizing pertinent information developed from the above control information

Special studies developing recommended solutions or alternate solutions to current or anticipated problems

SUMMARY

Professional construction management and program management are among many methods for organizing a construction program. Each of these time-tested methods has certain advantages and disadvantages in individual cases, depending upon the owner's objectives. The qualifications of an individual firm, represented by its management and technical personnel actively assigned to the project, may in many cases be more important than the particular method chosen.

The three-party professional construction management team has the advantage of freedom from the adversary relationship prevalent in older, more traditional methods. If all parties of the team are knowledgeable, technically proficient, and professional, this combination of efforts can often result in achieving minimum costs while preserving phased construction's reduced design-construction time schedules.

The program management approach gives the owner a single point of contact and accountability, where the program manager procures, contracts and coordinates the services of both design firms and construction contractors. It can also minimize adversary relationships while allowing competitive procurement of construction contracts.

PROFESSIONAL CONSTRUCTION MANAGEMENT IN PRACTICE

PART TWO

CHAPTER 4

INTRODUCTION TO AN EXAMPLE PROJECT[1]

This chapter introduces an example project that explains practical implications of concepts and procedures described in the chapters that follow. The project is large enough to illustrate all the main procedures and tools associated with construction management, yet small enough to emphasize overall concepts without obscuring them in the maze of detail associated with large projects. Many of the schedules, estimates, reports, and other exhibits in this and later chapters are also based on the example project to illustrate better an integrated approach to project planning and control.

To describe the example in sufficient detail, this chapter begins with a technical description of the project and an overview of its life cycle. To illustrate more realistically the project's objectives, organization, and administration, the chapter's format then shifts to a summary of the actual professional construction management proposal, followed by a copy of the procedure outline.

The chapter then reviews other ways in which the owner may construct the project, including use of a single general contractor, a design-construct firm, or a developer. Alternate ways to provide for overall management, contract administration and quality control are also discussed.

THE EASYWAY WAREHOUSE PROJECT

With some 150,000 square feet under roof, the Easyway Food Company's new warehouse in Mountaintown, Westamerica, will provide storage for grocery and nonfood items. Although primarily equipped with pallet racks for storage of items handled by forklift trucks, it also includes flow racks that permit individual items to be packed in assortments for shipment to retail stores.

---

[1] This example is based on a real project, but names, dates, and places have been changed for confidentiality.

Dry Storage Warehouse,
Mountaintown, WestAmerica

Sheet   of
Date
By E.F. McManus

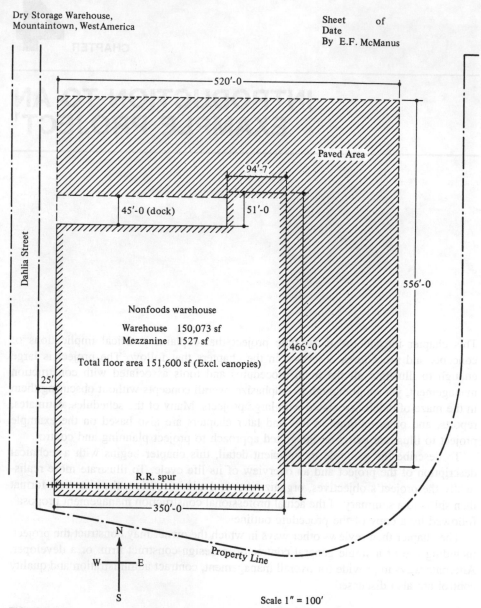

FIGURE 4-1
Plan view of example project.

Figure 4-1 gives a simplified plan view of the new warehouse. Figure 4-2 shows photographs from a similar project.

Other principal features are:

Heavy-duty concrete yard paving
Loading-dock-high floor slab

Initial nonfoods storage is shown in flow racks (lower tiers) and pallet racks above.

Dry storage warehouse has a nominal floor area of 150,000 square feet and is used to warehouse a large variety of nonperishable items.

FIGURE 4-2
Photographs from a similar project.

Special hardened topping on floor slab
Double-tee precast concrete walls
Structural steel framing with metal roof deck
Sprinklered throughout
Built-up roof
Required utilities, including domestic water, fire water, storm sewer, sanitary sewer, and natural gas
Natural gas unit heaters
Mercury vapor lighting
Rail spur for incoming items
Usual appurtenances

Six working drawings were prepared to describe the new warehouse. A comprehensive set of technical specifications was also required. The following drawings were prepared by the architect.[2]

Drawing SI-1: Plot Plan and Utility Plan
Drawing A-1: Foundation and Floor Plans, and Finish Schedule
Drawing A-2: Roof Framing, Plan and Details
Drawing A-3: Elevations and Roof Plan
Drawing PH-1: Plumbing and Heating Plan
Drawing E-1: Electrical Plan

Included in Appendixes A and B are a simplified specification, a CPM precedence diagram, fair-cost estimates, and a sample-bid package for one of the construction contracts.

PROJECT LIFE CYCLE

Chapter 2 separated the life cycle for a new project into the following key phases:

Concept and feasibility studies
Engineering and design
Procurement
Construction
Start-up and implementation

This section will summarize what each involved for the Easyway Warehouse.

Concept and Feasibility Studies

The Design and Construction Department of the Easyway Food Company, an expanding grocery chain with regional distribution and retail outlets, did the initial feasibility studies, then prepared a conceptual drawing based upon overall requirements obtained from the Operating Division. The Operating Division specified that the new warehouse cover approximately 150,000 square feet, that it provide for incoming shipments by

---

[2]Instructions for obtaining full-size reproducible copies of these drawings are also given in Appendix A.

both truck and rail, and that the storage-rack layout conform to existing standards. Other standards were set by Easyway's insurance underwriter, the Mountaintown local building department, and other agencies. The Operating Division also required that the occupancy date coincide with the anticipated need for the new facilities. The conceptual drawing, accompanied by company-standard details and specifications, gave sufficient detail to enable an architect to prepare detailed drawings and specifications.

Easyway's Design and Construction Department had to choose the architect, develop a preliminary schedule for achieving the required completion date, and select the method of designing and constructing the project.

In the case of this new warehouse, Easyway prepared a comparison between a sequential schedule (traditional lump-sum single contract) and a phased schedule. Figure 4-3 shows this comparison.

Sequential Schedule If time available for completion was 12 months or more, the traditional approach would probably be chosen. This method would include plans and specifications prepared by a local architect, followed by construction performed under a single lump-sum contract. Design would take 3 months; 1 month would be allowed for bidding and award, and 8 months for construction. Barring changes after award, the cost of the project would be fully known before construction begins.

FIGURE 4-3
Alternate design-construct schedules.

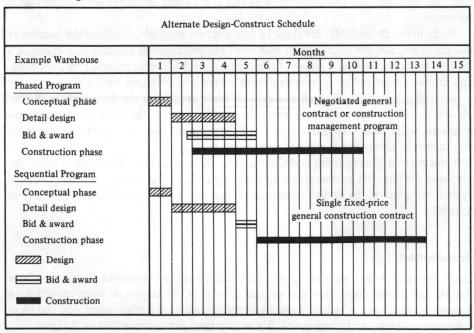

Phased Schedule   In this example, only 10 months were available for design and construction if required completion dates were to be met. Therefore, the owner's choice narrowed to turnkey, traditional with a negotiated construction contract, and professional construction management. All these methods could phase the construction program and, if successful, would complete the facility by the required occupancy date.

The owner reviewed the three methods, the capabilities of firms in the area, and Easyway's own internal qualifications and objectives. The company determined that, for this particular project, professional construction management offered advantages of lump-sum contracts coupled with completion by the required date.

Because of the tight schedule, Easyway immediately negotiated a contract with a local architect and selected a professional construction manager in accordance with established procedures.

For the new warehouse it took about 2 weeks for achievement of conceptual agreement with the architect, Rushmore, Owens and Peril (ROP), and another 2 weeks for the selection of Construction Management and Control, Inc. (CMC) as the professional construction manager. With the manager thus selected, the project moved to the engineering and design phase.

Engineering and Design

The professional construction manager and the architect met and reviewed the proposal previously submitted to the owner. After making required modifications, they reached a basic understanding, and CMC prepared and issued for approval a "Procedure Outline" spelling out the relationships between the owner, architect, and construction manager.

Early in this phase CMC developed a procurement schedule, finalized the number of bid packages, and the team agreed that, with the professional construction manager's review, the architect would put out the bid packages. Bids were received by the owner, and an evaluation and recommendation for award was prepared by CMC.

Several value-engineering alternates were reviewed during the design phase. Some, chosen as being clearly most economical, were included in the final plans and specifications, while others were included as alternates in the bidding documents to enable contractors to determine the least costly solution.

Construction Management and Control, Inc. developed the CPM control schedule, prepared the fair-cost estimates during the design phase, and set up the project control system to monitor the actual performance against the budget estimate and project schedule.

Procurement

Since the project was built using phased construction, procurement overlapped both design and construction. For this project a total of 10 contracts were chosen as a compromise between scheduling requirements, engineering design schedules, and local contractor interests and capabilities. These contracts are summarized as follows:

1 *Site earthwork*   Includes cut and fill required for site grading, imported fill for the dock-high warehouse pad, fencing, and other work shown on the drawings.

2 *Foundation and slab concrete*   Includes concrete foundations complete with structural excavation and backfill. Also includes exterior concrete slabs to make the overall package more attractive to better-qualified contractors. Includes installation of anchor bolts, frames, and other embedded items furnished by the structural steel contractor.

3 *Structural steel*   Includes furnishing and erection of all structural steel, including the metal roof deck. Does not include miscellaneous steel that is not fully designed at this time.

4 *Precast concrete walls*   Includes furnishing and erection of double-tee wall system.

5 *Fire protection*   Includes design and construction of an overhead dry-type sprinkler system to performance specifications. Includes interior sprinkler system and exterior fire mains as indicated on the drawings.

6 *Plumbing, heating, ventilating, and air conditioning*   The mechanical contract includes rough and finish plumbing, gas-fired unit heaters, gravity roof ventilation, and office air conditioning. Also included are sheet metal flashings required for the roofing in conformance to trade practices in the area, and site underground utilities, such as domestic water, storm sewer, and gas.

7 *Electrical*   Includes all power wiring, lighting, and electrical controls.

8 *Roofing*   Includes roof insulation, built-up roof, and required accessories.

9 *Special floors*   Includes the base slab and $\frac{3}{4}$-inch-topping wear surface for the warehouse interior.

10 *Building finish*   This contract includes balance of specialty items, such as remaining concrete, concrete block, drywall, painting, tile, glazing, finish carpentry, dock hardware, and items not otherwise covered.

Construction

On-site work began with award of the site earthwork contract. CMC appointed a field construction manager and hired a clerk typist. Initial layout was handled by a local professional surveyor, and a testing laboratory was chosen for soils, concrete, and other specialized inspection and testing. The local architect visited the project frequently in accordance with his overall responsibilities during the construction phase. All contractors were directed through the job-site office. The "Procedure Outline," given later in this chapter, describes significant features during the construction phase, along with key forms and reports utilized in managing the project.

Start-up and Implementation

For the new warehouse the owner handled moving in of the stock and initial operations. Since the owner performed such functions with the company's own forces, start-up responsibilities of the professional construction manager were limited to initial owner familiarization with operating equipment, and to troubleshooting throughout the start-up phase.

THE SUCCESSFUL PROPOSAL

The successful proposal was submitted generally in accordance with established methods. In this example, the owner also requested proposals from two local general contractors. After evaluating the proposals, the owner chose the professional construction management approach as being best suited to the requirements of this particular project.

A summary of the successful CM proposal is set forth on the following pages. The proposal consists of a letter that outlines proposed services and quotes a fixed fee for home office services, general overhead, and profit. The following key exhibits, illustrating preliminary planning which has been performed as a part of the proposal, are also included at the end of the proposal.

Preliminary Schedule (Figure 4-4)
Preliminary Cost Estimate (Figure 4-5)
Estimate of Construction Management Costs (Figure 4-6)
Proposed Organization Chart (Figure 4-7)
Preliminary CPM Diagram (Figure 4-8)

CM PROPOSAL
Easyway Dry Storage Warehouse
CONSTRUCTION MANAGEMENT & CONTROL, INC.
September 1, 1991

Proposal No. 91-17

Mr. Peter J. Cleaveland
Manager, Design & Construction Department
Easyway Food Company
200 Madison Street
Mountaintown, WestAmerica 99999

Subject: Professional Construction Management Proposal
         Mountaintown Dry Storage Warehouse

Dear Mr. Cleaveland:

In accordance with your request we are pleased to submit this proposal to furnish Professional Construction Management services for construction of the Mountaintown Dry Storage Warehouse.

Construction Management & Control, Inc., proposes to provide a management services program structured to meet the objective of achieving warehouse completion in ten calendar months from this date while preserving the benefits of fixed-price construction contracts. Through use of the "Fast Track" (or phased construction) approach, fixed-price construction contracts will be developed, bid, and awarded to permit building closure at the earliest possible date. This will enable interior work to continue during the winter months in order to permit owner occupancy on schedule next spring.

We propose herein that Construction Management & Control, Inc. will provide the following services:

1 *Prepare Control Schedule* Prepare a master control schedule showing the contemplated bid packages and the required construction duration for achieving project objectives.

2 *Develop Bid Packages* With the assistance of owner and architect, develop a detailed scope of the separate bid packages applicable for lump-sum bidding.

3 *Prepare Bidders List* Handle prequalification of prospective bidders having the specialized skills necessary for accomplishing the work. A bid list will be prepared in consultation with architect and owner.

4 *Prepare Fair-Cost Estimates* A fair-cost estimate for each bid package will be prepared for use in evaluating bids.

5 *Receive, Review, and Evaluate Bids* Bid openings will be conducted, bids evaluated, and recommendations prepared for contract award by Easyway Food Company, Inc.

6 *Manage, Coordinate, and Inspect the Work* It is our understanding that a representative of Easyway Food Company, Inc. will visit the work periodically, and that the architect will also make periodic visits as required. Construction Management & Control, Inc. will provide a full-time Field Construction Manager who will be assigned to the site for managing, coordinating, and inspecting all work performed on the project. This person's duties will include coordination of contracts; monitoring the schedule of individual phases of the work; making recommendations for adjusting the work to accommodate changing and unforeseen conditions if applicable; preparation of reports on the progress of the work; reviewing and recommending progress payments; obtaining required shop drawings and forwarding them to the architect for approval; obtaining testing laboratory services as required; inspecting the quality of materials and workmanship; maintaining daily logs and records; and such other services as are customarily required in order to manage the work in accordance with the owner's objectives.

7 *Provide Home-Office Support Services* Construction Management & Control, Inc., will designate one individual in the office who will be responsible to the owner's Project Manager for all work, and who will be utilized on an "as required" basis to the extent necessary for this purpose. In addition the resources of Construction Management & Control, Inc.'s management and technical personnel will be available for assistance to the owner throughout the project in the event of special need.

Attached for your review and consideration are the following exhibits illustrating the preliminary program that we have developed in an effort to achieve your overall objectives:

Preliminary Schedule
Preliminary Construction Cost Estimate
Estimate of Construction Management Costs (General Conditions)
Proposed Organization Chart
Preliminary CPM Diagram

We propose to provide the Professional Construction Management services for a fixed fee to cover home office services, with all field costs to be reimbursable. Our fixed fee will be Two Hundred Thousand Dollars ($200,000.00). Reimbursable costs are set forth in exhibit A.

Our proposal is subject to the negotiation of a mutually satisfactory agreement.

We greatly appreciate the opportunity to provide our proposal for Professional Construction Management services for your Mountaintown project. Since our proposal has been developed on the basis of preliminary information, we would be pleased to discuss the program further, and incorporate modifications if required, in order to more fully comply with your overall objectives.

Very truly yours,

CONSTRUCTION MANAGEMENT & CONTROL, INC.

JWH:mp J. Walter Harrington

Attach. President

EXHIBIT A
REIMBURSABLE COSTS

In addition to the fixed fee covering home office costs, general corporate overhead, and profit, the costs and expenses to be reimbursed to Construction Management & Control, Inc. shall consist generally of the following:

1 Field Costs
   A All salary and wage costs for Construction Management & Control, Inc. personnel assigned to the project field office for time expended in the performance of services, including salaries and wages, payroll taxes, vacation allowances, sick leave, welfare benefits, pensions, retirement benefits, and all other benefits pursuant to employee benefit programs.
   B The cost of any outside consultants retained for performance of necessary services.
   C All amounts paid or incurred by Construction Management & Control, Inc. under or in connection with any contracts, subcontracts, or purchase orders let or issued for the project.
   D Sales, use, turnover, gross receipts, or other taxes paid in connection with the project which are not measured by net corporate income.
   E All costs of the project field office. These costs include but are not limited to rental costs, furniture, fixtures, and utilities and telephone services.
   F All insurance and bond premiums paid or incurred in connection with field services.
   G All costs of incidental purchased labor and services.
   H All costs of long distance calls, telegrams, cables, postage, duplicating, photostating, fax, teletype, and other similar items of expense directly connected with the services.

I The cost of travel, transportation, moving, and living expenses (including relocation subsistence) incurred by construction management and supervisory staff personnel directly connected with the services.

J All other direct costs and expenses incurred in connection with the performance of the services.

2 Special Services

A The fixed fee for home-office costs, general overhead, and profit will include all normal home-office expenses associated with the anticipated program. In the event that changes to the work initiated by owner or architect result in additional planning or estimating costs in the home office, such work will be performed at Construction Management & Control, Inc.'s standard rates for such services, subject to prior approval of the owner.

B Should expediting services for owner-furnished equipment be required, Construction Management & Control, Inc. will perform such services in an effort to maintain project schedules at Construction Management & Control, Inc.'s standard rates for such services, subject to prior approval by the owner.

FIGURE 4-4
Preliminary schedule for example project.

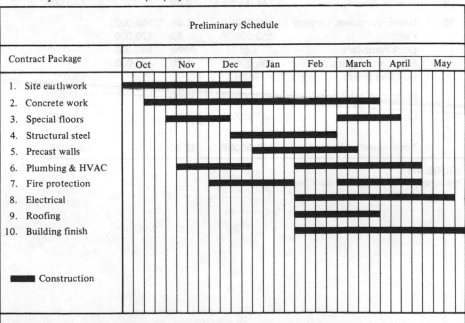

Preliminary Cost Estimate
Easyway Food Co.   Location Mtntown
Dry Storage Warehouse
(From Preliminary Drawings)

Job No.
Date   9-15
By   DSB
Sheet   1   Of 1

Code	Description	Quantity	Unit Cost	Amount	Total
1	Site earthwork	46,000CY	4.50		208,000
2	Structural concrete	450CY	350.00	158,000	
	Yard paving	3,500CY	113.00	396,000	
	Fencing	4,800LF	13.00	62,000	
	Struct. & yard concrete	Total			616,000
3	Structural steel	500T	1700.00	850,000	
	Metal roof deck	165,000SF	1.20	198,000	
	Structural steel	Total			1,048,000
4	Double tee walls	60,000SF	8.00		480,000
5	Fire protection	150,000SF	2.00		300,000
6	Plumbing	150,000SF	.60	90,000	
	Sheet metal	2,000LF	20.00	40,000	
	Heat—vent.—air cond.	150,000SF	1.00	150,000	
	Yard utilities	Lot	Allow	100,000	
	Mechanical	Total			380,000
7	Electrical	150,000SF	2.20		330,000
8	Roofing & insulation	165,000SF	1.80		298,000
9	Special floors	144,000SF	3.80		548,000
10	Doors, windows, carpets	150,000SF	1.00	150,000	
	Painting	150,000SF	.80	120,000	
	Dock hardware	Lot	Allow	140,000	
	Block, part, glass & misc.	150,000SF	1.20	180,000	
	Building finish	Total			590,000
	Estimated total cost	150,000SF	32.00		4,798,000
	Contingency	4,798,000	10%		482,000
	Total construction cost	150,000SF	35.20		5,280,000

FIGURE 4-5
Preliminary cost estimate—work items.

Preliminary Cost Estimate
Easyway Food Co.   Location Mtntown
CM & General Conditions

Job No.
Date   9-15
By   DSB
Sheet   1   Of 1

Code	Description	Quantity	Unit Cost	Amount	Total
	Field Construction Mgr	8 mos	4600	36,800	
	Clerk	7 mos	1600	11,200	
	Subtotal			48,000	
	Payroll taxes, ins., benefits		30%	14,400	
	Subtotal field labor				62,400
	Office trailer rent	8 mos	500	4,000	
	Telephone	8 mos	600	4,800	
	Reproduction	8 mos	200	1,600	
	Office equipment	Lot		4,000	
	Safety equip. & supp.	Lot		1,000	
	Utility bills—office	8 mos	400	3,200	
	Move in expense	Allow		10,000	
	Job pickup truck	8 mos	600	4,800	
	Travel & misc. exp.	Allow		2,200	
	Subtotal field expense				35,600
	Sanitary toilets	8 mos	250	2,000	
	Trash removal	8 mos	500	4,000	
	Guards (by owner)				
	Testing lab (by owner)				
	Surveyor (initial layout)	Allow		14,000	
	Temporary lighting	Allow		12,000	
	Final cleanup	Allow		16,000	
	Utility bills—field	8 mos	1500	12,000	
	Subtotal general cond.				54,000
	Estimated total				152,000
	Escalation	152,000	3%		4,000
	Contingency	156,000	10%		16,000
	Est. reimb. cost	5,280,000	3.25%		172,000
	Fixed fee (home office, profit, general o'head)	5,280,000	3.80%		200,000
	Total CM & general cond.	5,280,000	7.05%		372,000

FIGURE 4-6
Preliminary cost estimate—administration and general conditions.

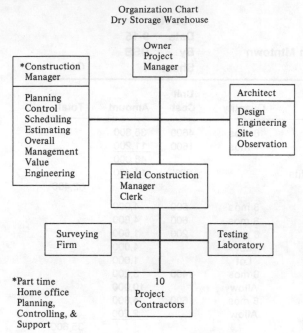

FIGURE 4-7
Organization chart for example project.

THE PROCEDURE OUTLINE

In a three-party professional construction management program, it is extremely important that the duties and responsibilities of each of the three team members be clearly set forth. The Procedure Outline, however, is not intended to define fully the design relationships between the owner and the architect. Rather, it illustrates the duties and responsibilities during the planning, procurement, and construction phases. The document serves as a general guide, and can be easily modified or added to throughout the course of the project.

For Easyway's Mountaintown dry storage warehouse, the professional construction manager met with both the architect and the owner immediately upon notification of his selection. After this meeting, a draft of the Procedure Outline was prepared and submitted to both the architect and the owner for review, modification, and approval. A summary of the Procedure Outline prepared for the Mountaintown project follows.

Mountaintown Warehouse Milestone CPM

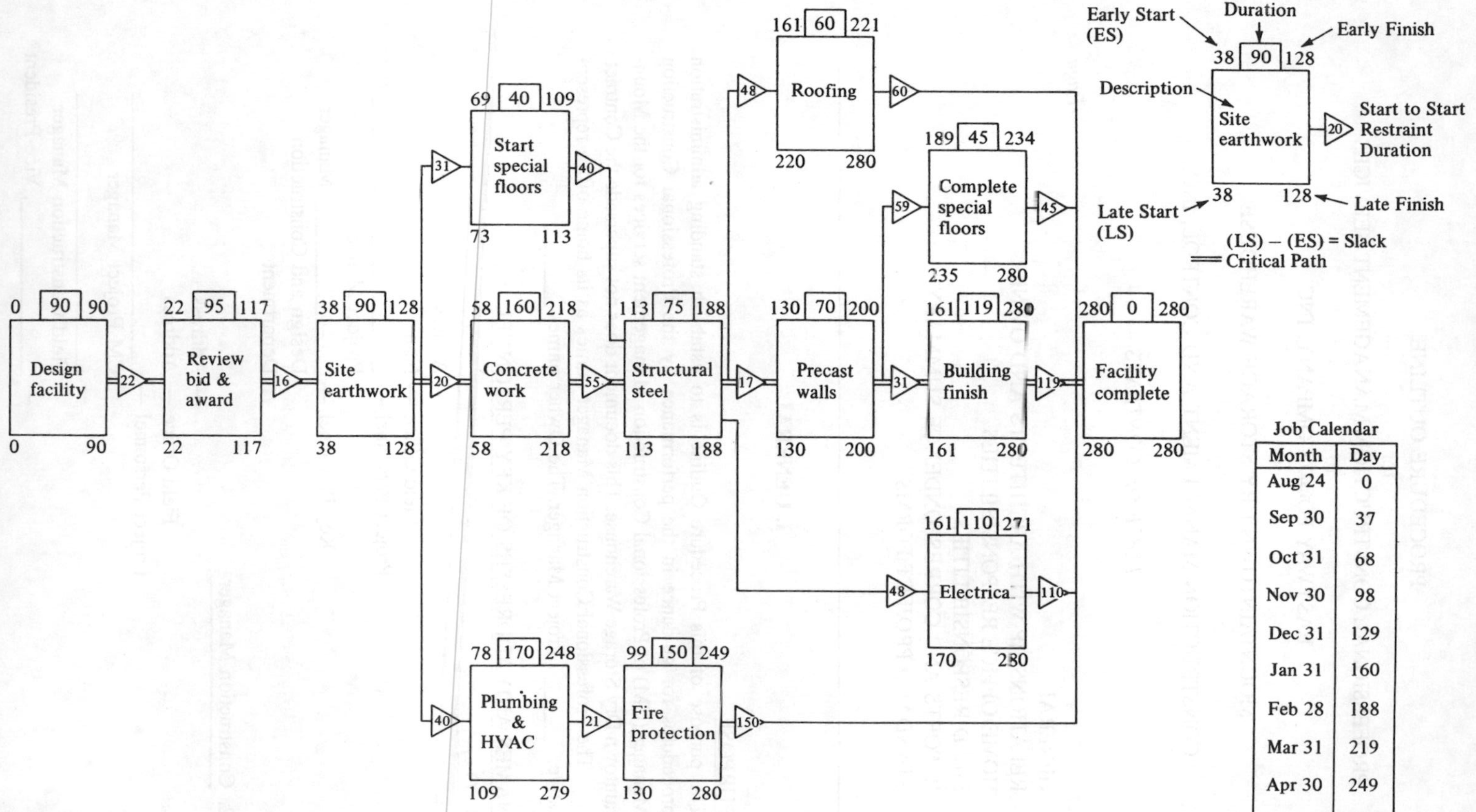

Legend

| Early Start (ES) | Duration | Early Finish |
| Late Start (LS) | Description | Late Finish |

Start to Start Restraint Duration

(LS) − (ES) = Slack
═══ Critical Path

Site earthwork: 38 | 90 | 128 / 38 | | 128

Job Calendar	
Month	Day
Aug 24	0
Sep 30	37
Oct 31	68
Nov 30	98
Dec 31	129
Jan 31	160
Feb 28	188
Mar 31	219
Apr 30	249
May 31	280

FIGURE 4-8
CPM milestone schedule.

PROCEDURE OUTLINE

for

PROFESSIONAL CONSTRUCTION MANAGEMENT SERVICES

to

EASYWAY FOOD COMPANY, INC.

on

MOUNTAINTOWN DRY STORAGE WAREHOUSE

by

CONSTRUCTION MANAGEMENT AND CONTROL, INC.

TABLE OF CONTENTS

I. GENERAL

A. <u>PURPOSE</u>

The purpose of this Procedure Outline is to establish standing administration procedures for guidance in the performance by the Professional Construction Manager (CM) of Professional Construction Management services for the Mountaintown Dry Storage Warehouse. This document does not supersede the Contract.

The Professional Construction Manager names as its home-office representative, _____, Project Manager. The Owner names _____

B. <u>NAMES AND ADDRESSES OF KEY PERSONNEL</u>

1. <u>Owner</u>

— Address

Field Office — None

Project Personnel — _____

Project Manager

Key Personnel — _____, Manager

Design and Construction

Department

2. <u>Construction Manager</u>

— Address

Field Office — Address

Project Personnel — _____

CM Project Manager

— _____

Field Construction Manager

— _____, Vice-President

3. Architect

— Address

Project Personnel — _____

_____

Mechanical Consultant — _____

Electrical Consultant — _____

C. CORRESPONDENCE

Official correspondence on all subjects between Owner and Professional Construction Manager shall be between _____ and _____.

All correspondence between Field Construction Manager and the Architects or Job-site Contractors shall be by Field Transmittal Memorandum (FTM). One copy of all FTM's shall be sent to Owner's Project Manager and to _____, Project Manager.

Job-site transmittal and other information requested by Owner shall be sent to FTM to _____ with a copy to _____.

Information or correspondence originating from Owner pertaining to field activity shall be sent directly to the Field Construction Manager with a copy to _____, Project Manager.

Important instructions received by telephone or determined in meetings shall be confirmed in writing by the Professional Construction Manager.

D. REPORTS TO OWNER

1. Monthly Progress Report

The CM Project Manager shall prepare and issue by the 15th calendar day of the following month a monthly progress report covering significant phases of the monthly activity.

A. The report shall include the following:

1 A short narrative covering the professional construction management activities during the period.

2 A summary of requests for quotations received, bids evaluated, recommendations for procurement by Owner, and contracts awarded to date.

3 A graphic (bar chart) schedule showing scheduled and actual progress during the month, and cumulative progress to date.

4 A summary cost report showing status to date, covering both contract awards status, and recorded and estimated costs at completion.

B. Distribution of the report will be as follows:

The Professional Construction Manager will distribute three copies to Owner's Project Manager, and one copy to the Architect.

Internal copies will also be distributed by the Professional Construction Manager as follows:

One copy CM Project Manager
One copy CM Field Construction Manager
One copy CM Vice President

2. Weekly Field Activity Report

The Field Construction Manager will prepare and issue a Weekly Field Activity Report listing significant items accomplished during the week.

A. The report will be typed on interoffice-memo forms, and shall include the following:

      1 Significant progress or lack of progress achieved during the week

      2 General description of weather conditions and their effect on progress

      3 Job visitors, including inspections by City and County personnel

      4 Status of schedule

      5 Construction problems

      6 Summary Force Report

  B. Distribution of the report will be as follows: Same as Monthly Progress Report.

3. Weekly Force Report

The Field Construction Manager will prepare and issue a Weekly Force Report listing number of CM and contractor personnel on the job.

Distribution will be the same as above.

4. Contract Status Report

The Field Construction Manager will prepare and issue a Monthly Contract Status Report.

  A. The report shall include the following:

      1 Contractor and description

      2 Original contract price

      3 Change orders

      4 Revised estimated contract price

      5 Percent complete

      6 Contract completion date

      7 Estimated completion date

  B. Distribution of the report will be as follows:

  Owner Project Manager

  CM Project Manager

  Architect

5. Other Reports

Other reporting requirements are discussed in Section IV, Field Responsibilities. No other monthly or weekly reports will be issued except at the request of Owner, unless approved in advance by the CM Project Manager.

II. RELATIONSHIP WITH ARCHITECTS AND OWNER

The following items are listed in order to clarify relationships and responsibilities:

1. Special Conditions and General Conditions Typical copies of general and special conditions for contracts, notice to bidders, and other standard documents were given to the Professional Construction Manager by Owner. The Professional Construction Manager made certain modifications to fit the professional construction management concept; these have been reviewed and approved by Owner's Project Manager and by the Architect. This package will be used as a model for remaining packages.

2. Drawings and Specifications Drawings and Specifications for all work will be developed by the Architect, who will furnish the required number of plans and specifications for all bid packages as per a predetermined bid list, and for contract revisions. All revisions prepared after contract packages have been prepared shall be accompanied by a description of the change and shall show revision numbers and dates.

3. Substitution Policy   In general, "or equals" are not specified. Value Engineering alternates are encouraged to be set forth in the original proposal. Alternates are generally given consideration for 30 days after award. All proposed modifications or alternates should be reviewed by the Professional Construction Manager and then be submitted directly to the Architect. A joint recommendation will be made to the Owner if found desirable.

4. Shop Drawings   All shop drawings shall be submitted directly to the Architect by the CM Field Construction Manager. After approval, four copies will be required for distribution as follows:

One copy Owner
One copy Architect
One copy Field Construction Manager
One copy Contractor

Additional copies may be requested as required. One additional copy will be held at field office during the approval period.

5. Modification and Changes   The Professional Construction Manager is not authorized to commit the Owner. In the case of changes as well as in original contracts, the Professional Construction Manager will make recommendation to the Owner's Project Manager. All commitments will be made by the Owner unless specifically authorized in advance. The Professional Construction Manager will prepare fair-cost estimates as required in order to review and evaluate changes.

6. Building Permit   The Architect (or Professional Construction Manager) will contact the proper authorities to obtain the Building Permit. The Architect has confirmed that earthwork can proceed in advance of obtaining the overall permit, and that a phased construction schedule is satisfactory.

7. Survey and Testing Laboratories   The Architect will request proposals from qualified firms. A joint recommendation for award will be submitted to the Owner by the Architect and Professional Construction Manager.

8. Proposed Bid Packages   One week's review period has been allowed in the Procurement Schedule. Copies of prepared bid packages shall be submitted for approval by the Architect as follows:

Field Construction Manager   one copy
CM Project Manager   one copy
Owner Project Manager   two copies

9. Approved Bid List   Plans and specifications will not be issued to any prospective bidder unless prior approval of the Owner is obtained. Proposed bid lists shall be submitted to Owner by the CM Project Manager in sufficient time to permit full evaluation of bidders financial qualifications by the Owner prior to issuing requests for quotations.

In general, all potential bidders will initially be screened by the Field Construction Manager. Potential bidders will be requested to supply the following information for use in the prequalification:

(a) Filled out contractor's reference form, setting forth general experience, annual volume, and other pertinent information.

(b) Current financial statement

(c) Certification, signed by a responsible officer or accounting firm, that financial statement is true.

10. Approved Bid Documents   The Professional Construction Manager will submit recommended bid documents to the Owner's Project Manager. After approval, forms will be printed by the Architect. Deviation from the approved form is not authorized unless prior approval of both Owner and Professional Construction Manager is obtained.

11. On-Site Inspection   The Professional Construction Manager will perform day-to-day construction inspection as required, under the direction of the Architect, and with the assistance of the Testing Laboratory and Surveyor. In addition, periodic visits by the Architect will be required. The Architect shall be kept fully informed at all times of the status of job and of the status of problems or required action.

12. Plan Interpretation   All design questions or determination of "or equals" shall be submitted to the Architect by the Professional Construction Manager.

III. HOME-OFFICE RESPONSIBILITIES

Home-office responsibilities will be directed and coordinated by the CM Project Manager. These responsibilities will include the following:

A.   OVERALL JOB SCHEDULE

   An overall job CPM network similar to the one included in Proposal No. 17 will be developed and updated as required. From this network, individual contract summary schedules will be developed for inclusion in bidding documents and for use in measuring actual progress. An overall summary schedule suitable for measuring job progress against scheduled progress as described in the Monthly Progress Report will be prepared and updated monthly.

B.   FAIR-COST ESTIMATES

   Fair-Cost or Engineer's Estimates will be prepared for each contract package, then will be transmitted to the Owner's Project Manager in advance of the bid due date for each contract.

   These estimates will be prepared in a format similar to normal fair-cost estimates. In the event of unusual discrepancies between the fair-cost estimate and bids, the estimate may be useful in negotiations with the low bidder or bidders.

C.   GENERAL HOME-OFFICE SUPERVISION

   Home-office supervision and support services will be kept to a minimum consistent with the needs of the project, unless other services are requested by Owner's Project Manager. Home-office supervision and support services will include duties set forth in the Agreement and summarized as follows:

   1.   Develop, with the assistance of the Owner and Architect, a detailed scope and bidding document for each proposed bid package, suitable for lump-sum bidding. Obtain the Owner's approval of above, and prepare necessary number of copies of documents not furnished by the Architect.

   2.   Assist the Owner in the Prequalification of proposed bidders, and present proposed bid lists for the Owner's approval or modification.

   3.   Evaluate bids, prepare recommendations for award by the Owner, and develop recommended Contract Format. All contracts will be prepared using Standard Owner Forms as approved for the Project.

   4.   Issue Monthly Progress Report.

5. Provide general home-office supervision of field activity, and maintain liaison with the Owner and Architect.

IV. FIELD RESPONSIBILITIES

Responsibility for the performance of all work at the job site will be delegated to the Field Construction Manager. He will establish a local trailer office and will hire a field clerk.

A. DUTIES AND RESPONSIBILITIES WILL INCLUDE:

1. Assist in the prequalification of prospective bidders and in the preparation of bid lists.
2. Assist in evaluation of bids and recommendations for award.
3. Manage, coordinate, and inspect all work performed by contractors on the project as set forth in the Agreement.
4. Administer and direct the Testing Lab and Surveying contracts jointly with the Architect.

B. SUPPLEMENTARY REPORTING AND RECORDS REQUIREMENTS

In addition to reporting requirements set forth in I-D, the following additional reporting and recording requirements will be followed when indicated:

1. Maintain job diaries for each contract package on a daily basis. These diaries shall be open to inspection by the Owner Manager and the Architect (or his representatives) at all times.
2. Maintain drawing register and records.
3. Document all correspondence by FTM (Field Transmittal Memorandum).
4. Initiate *Force Majeure* Delay Reports when required, reflecting effect on each contractor. Distribution will be as follows:

Owner Project Manager
CM Project Manager

5. Review, document, and recommend for payment or disapproval all Contractor Proposals and Requests for Change Order using Owner forms. All Change Order Requests shall be forwarded to the CM Project Manager with full documentation attached. A copy of the transmittal only shall go to the Owner's Project Manager. Recommendation regarding acceptance will be prepared by the CM Project Manager.
6. Review, modify, or approve all Contractor Monthly Progress Payment Reports, and forward them to the Owner's Project Manager for approval and payment using specified forms. Distribution will be as follows:

Owner Project Manager
CM Project Manager

7. Initiate completion and final acceptance procedure per Owner requirements.
8. Maintain progress and record photographs at the job site. No distribution will be made unless requested by the Owner's Project Manager.
9. Maintain liaison with the Architect in order to document and expedite approval of shop drawings, drawing preparation, clarify construction or materials requirements, or handle other significant items.

10. Maintain one record set of "as-built" drawings at the job site, marked up to show all field changes, locations of buried utilities, and other significant items. This information shall be turned over to the Architect for preparation of final record drawings.
11. Request architectural clarification or design interpretation from the Architect.

V. REPORTS AND CORRESPONDENCE SUMMARY

Attachment 1 summarizes the above-referenced reporting and correspondence procedures.

INDEX OF STANDARD FORMS

1. Notice to Proceed
2. Progress-Payment Report
3. Proposal and/or Request for Change Order
4. Sample Force Report
5. Contract Status Report
6. Contract Closeout Procedure
    (a) Final Progress Payment Report
    (b) Release and Waiver of Lien
    (c) Contract Completion and Acceptance Certificate

NOTE: Examples of the standard forms referenced above are further explained or set forth as exhibits in the following chapters.

SINGLE GENERAL CONTRACTOR

Easyway can choose between a cost-plus-a-fee, fixed price, or guaranteed maximum price contract depending upon objectives as discussed in Chapter 2.

Fixed Price Bids

When Easyway has sufficient lead-time, top management normally prefers a fixed price contract bid to pre-qualified general contractors. In this alternate, a separate contract would be negotiated with the architect for completion of the conceptual and detail design in four months as shown in Figure 4-3.

Easyway has developed experience and financial qualification criteria for selection of bidders. Easyway's policy is to select five or six of the most qualified firms although special circumstances may alter this number either up or down. Prequalification information is based upon experience in like types of work, past experience with Easyway and financial soundness. Easyway requires that proposed bidders submit a certified financial statement. Selection requirements call for the bidder to show net quick assets of 10 percent of the estimated contract cost. Easyway requires an optional 100 percent payment and performance bond and will reimburse the contractor for this expense if required. The bond requirement will often be waived for financially strong contractors.

TABLE 4-1
DOCUMENT DISTRIBUTION LIST (NOT COMPLETE; FOR EXAMPLE ONLY)

Codes
O – Originator
R – Review
I – Information
A – Action

	Owner	Arch./Engineer	Contractors	Others	Professional Const. Mgr. Field Office	Professional Const. Mgr. Home Office
Correspondence, CM to:						
Owner	A	–			–	O
Arch/Engineer (FTM)		A			–	O
Contractors		–			O	–
Others		–		A	O	–
Reports						
Monthly Progress					O	O
Weekly Field Activity					O	–
Weekly Force					O	–
Contract Status					O	–
Force Majeure			A		O	–
Progress Payment Request	A		A		O	–
Contract Documents						
Bid Documents	R	O	A		R	R
Bid Packages	R	R	A		R	O
A/E Drawings, Specs.	R	O	O		A	R
A/E Revisions	R	O	O		–	–
Shop Drawings	–	R	O		R	–
Change Requests	A	–	A		A	–
Change Approvals	O					

The successful general contractor will usually plan to perform most of the work using subcontractors. A typical contractor might perform the structural concrete work with its own forces along with any other work for which it is well qualified. It will accept prices from subcontractors for the balance of the work. Easyway requires that bids for major projects be submitted by mail to the home office for evaluation and award. Other owners may receive bids in person at a designated time and place. Most private owners do not hold public bid openings. However, Easyway makes a practice of notifying unsuccessful bidders shortly after contract award.

Guaranteed Maximum Price

A competitive alternate to professional construction management when shortest schedules are economically important utilizes the **guaranteed maximum price** (GMP). Here the owner supplies the contractor with conceptual plans and specifications along with a scope of work. A negotiated GMP is developed which often includes a fixed fee for home-office services and profit. Field costs and subcontracts are reimbursable up to the limit of the guaranteed maximum. Savings under the GMP are often split between owner and contractor under a predetermined formula. If the GMP is developed upon inadequate information, considerable friction can develop between owner, contractor and architect over the responsibility for overruns. This type of contract works best when both the contractor and owner are experienced in this type of contract, are familiar with the type of work, and have had a successful ongoing relationship on prior work. The architect must also be sensitive to both owner requirements and contractor costs. A major incentive to both contractor and architect can be the opportunity for future negotiated projects from repeat owners.

Knowledgeable owners often participate fully with the contractor in evaluating major material and equipment and subcontract proposals, and in other major decisions with economic or schedule considerations. The general contractor can usually obtain fixed price subbids upon completed portions of the design. If the contractor performs no work with its own forces, the method is equivalent to a construction management contract with a guaranteed maximum price as described in Appendix C.

Cost-Plus-Fixed-Fee

General contractors can also perform work on a phased program where all field costs are reimbursable and a fixed fee for home-office costs and profit is negotiated. Here the owner has maximum opportunity to participate in the management of the project. If the scope of the project has not been fully developed and overall minimum schedule is very important, this type of contract can prove to be advantageous to both parties. However, the owner must understand that it has assumed all of the price risk on the project.

DESIGN-CONSTRUCT

A number of firms specialize in a design-build or design-manage, or turnkey, approach. Easyway could contract with a single firm to provide both design and construction work on a phased-program aimed at earliest completion. All three contract types can also be utilized dependent upon schedule objectives and the price risk that the owner is prepared to accept. Here the initial scope must be sufficiently clear so that both parties understand their responsibilities and risks contemplated by the agreement.

DEVELOPER

In many ways, the rise of the developers during the 1980s is similar to the emergence of construction managers in the 1970s. Developers combine the skills of a construction manager, design constructor and real estate investor. The developer will normally own or control a parcel of land that he believes can be developed to obtain the maximum potential of the site. Easyway could also contract with a developer to build a warehouse based upon their design criteria. It would then lease the facility for an extended period of time with possible options to extend or to purchase. In recent years developers have proven to be extremely marketing and cost conscious and have reached domination in selected markets. Operationally many developers will hire general contractors to manage the construction while others will act as contractor or construction manager. In either event, the developer will be very active in performing value engineering studies, managing the design work and in the award of general and/or subcontracts in order to achieve cost objectives.

MANAGEMENT, FIELD ADMINISTRATION, AND QUALITY CONTROL

The owner, under any of the above construction programs, has several alternatives to provide for contract administration, field inspection and quality control. Many repeat owners have their own staff to provide these services. Others will delegate the responsibility to the architect, engineer, or to a construction manager. Repeat owners often supply a project manager to oversee all aspects of the project including dealing with in-house clients. Easyway usually supplies a project manager who has responsibility for several projects. Field administration, inspection and other responsibilities are delegated to the architect/engineer or construction manager as appropriate.

SUMMARY

The example project described in this chapter establishes a common data base for estimates, schedules, reports, procedures, and other tools which in later chapters will illustrate interrelationships among the methods utilized in professional construction management and in other management alternatives. The project management and control examples will illustrate the degree of accuracy achieved in similar real projects managed with the concepts described in this book. The example project, which shows the interrelationship among the various exhibits, will help convey the importance of integrating all the specialty components into the overall project plan and management control system.

The example project also has one other major objective. Its exhibits, reports, and other illustrations are indicative of the level of detail needed in a real-world compromise between the theoretical optimization of the component parts covered in Part 3, Methods in Professional Construction Management, and the time, financial, and resource limitations imposed by the profit-oriented marketplace. This device will thus enable the reader to understand the simplified approaches that must be adopted by successful firms in order to preserve the overall relationships among all components of the management system. Over-optimization of a few components, accompanied by neglect of the integrated relationship of all components to the overall project, generally results in a lack of overall control, resulting in cost overruns, delays, and owner dissatisfaction with the construction project.

5

PRECONSTRUCTION SITE INVESTIGATION, PLANNING, SCHEDULING, ESTIMATING, AND DESIGN

This chapter emphasizes the role of the professional construction manager in preconstruction planning. On design-construct or general contractor projects the fundamental requirements remain the same and must be covered by a member of the project team in advance of actual construction. Site investigations, planning, scheduling and estimating will later be carried out in greater detail by general and subcontractors who are bidding the work on fixed price projects.

Planning aims at a workable program that will achieve project goals and serve as a standard against which actual progress can be measured. The importance of fact finding at this stage of the project cannot be overemphasized. The manager must first understand the designer's objectives and operating methods, but, above all, he must thoroughly investigate, and become expert on, the local job-site conditions and area construction practices important to developing proposed contract packages, fair-cost estimates, realistic schedules, and the value-engineering program.

After the professional construction manager has obtained a thorough knowledge of job-site conditions that will affect performance of the work, preparation of the *work plan* for the project can begin. An early work plan for overall project execution is important in creating a team effort among the designer, owner, and professional construction manager, and it forms the basis for planning that will continue throughout the project as additional information becomes available. Approaches to initial planning will vary with project objectives, but the component parts of a project work plan will generally include the following items or their equivalents:

Preliminary estimate
Summary schedules
Work packages

Value-engineering program
Construction planning

Each of these will be discussed in the following sections.

CONSTRUCTION SITE CONDITIONS

Successful contractors and subcontractors native to the area are fully cognizant of factors affecting performance of construction work at the job site; those who are not soon fail. The professional construction manager must also become knowledgeable of these factors if he is to offer his services to the owner and the designer. Programs and bid packages that have worked well in one section of the country will not necessarily work as well in another. This section examines information that must be obtained before a meaningful and realistic program for completing the project can be developed.

Representatives of the professional construction manager must visit the work site. Their investigation is similar to that of a contractor planning to bid a project or a portion of a project, and likewise must be conducted by experienced construction professionals who can translate information obtained into the best way to minimize the construction costs which will later be evaluated by the bidders. The professional construction manager must develop the program in this manner to fulfill the obligation to the owner.

The items to investigate on the site are many and varied. A knowledgeable general contractor, design-constructor, developer or specialty contractor will have developed a particular method of appraising site conditions. The selection of items for investigation and the conclusions drawn are the result of many years of experience in managing and estimating construction work. Individuals may approach the investigation from different directions, but the overall conclusions must be similar. The items in the subsection below have been chosen to illustrate the importance of the site visit. The visit itself will turn up numerous other factors that must play a significant part in the overall project plan. As a guide to illustrate some of the items that should be investigated, Table 5-1 shows a checklist for the site investigation for the example project in Chapter 4. Many general contractors, design-constructors, developers, and subcontractors have considerably more detailed similar checklists.

Foundation and Earthwork Conditions

The knowledgeable construction manager first obtains an overall plot plan of the new facility, along with a copy of the soil's report and other information that may be available. Noting the soils engineer's proposed foundation recommendations, the earthwork phase can then be approached to determine the desirability of a separate earthwork contract and the broad scope and extent of the contract, and a preliminary judgment can be made as to methods that should prove most economical. In addition, the extent of stripping or removal of top soil or other unsuitable materials shown in the soils report should be determined. This information should be confirmed by surface inspection and hand excavation. The construction manager should also observe unusual conditions, such as the presence of surface rocks or water, peculiar drainage

TABLE 5-1
AREA INVESTIGATION GUIDELINES

1. Site Description
   (Vegetation, trees, terrain, depth of topsoil, drainage, existing structures, existing utilities, access, etc.)
2. Utilities Serving Site
   (Electricity, gas, water, sanitary sewer, storm sewer, railroad, highway, railroad siding, etc.)
3. Building Department
   (Contact, telephone number, building code, plan check time, fees, zoning, licensing, etc.)
4. Labor Unions
   (Membership, manpower shortages, manpower surplus, current agreements, wage rates, expiration dates, etc.)
5. Recommended Contractors
   (List recommended general and trade contractors for further consideration.)
6. Materials and Methods
   (List favored local materials including current quoted price for ready-mix concrete, lumber, imported granular base, plywood, masonry, and other key items.)
7. Equipment Rental
   (List local prices or key local quotations.)
8. Climatological Data
   (List average maximum and average minimum temperature, precipitation, and other significant data by months.)
9. Other Projects
   (Visit other projects noting productivity, favored methods, favored materials, subcontractors, etc.)
10. General Appraisal
    (Summarize results of site and area visit and recommend significant conclusions to be taken into account during the planning of the program.)

patterns, and other factors. In short, this approach mirrors that of a potential bidder. If the construction manager is not aware of the potential problems and difficulties, a meaningful estimate for the work cannot be prepared, nor can a bid package be framed that will minimize uncertainties, unneeded risks, and hence the contingencies that would be included by bidders in their prices.

On one professional construction management project involving substantial fill, the earthwork bid package was developed by a local architect with a standard specification requiring the bidders to strip top soil and unsuitable material up to 18 inches. Because of the rush nature of the project, the professional construction manager was not able to visit the job site or review the specifications until the bids were received. Upon visiting the site later, the manager determined that the indicated stripping depths were more probably of the order of 6 inches. Bidders were requested to reappraise their positions assuming an average stripping depth of 6 inches, with additional stripping to be paid by the owner at quoted unit prices. Every bidder reduced his quotation substantially, reflecting the decreased risk. As it turned out, the 6-inch stripping estimated by the professional construction manager proved sufficient, and the owner received a 10 percent decrease in the originally estimated contract price for this item. Had the manager been able to visit the site prior to requesting bids, the specifications could have been amended in the first place to achieve identical results. The saving would not have been quite so obvious to the owner, but it would have been just as significant.

General Planning

By visiting the site, the construction manager can see access roads, railroads, and other factors firsthand. He can then choose areas for locating temporary facilities, develop a preliminary plan for contractor storage areas, and later allow for existing electrical, water, or other service utilities in developing or evaluating bid packages and in reviewing owner-furnished items. He can observe interferences with existing facilities and develop a plan for site security. The investigator should also be alert for conditions on the site that may necessitate changes from preliminary design information that he may have. Again, the professional construction manager will approach the investigation of general conditions exactly as would a general contractor planning to bid the work.

On one professional construction management project, the manager visited the site a second time after preliminary earthwork drawings were received. These drawings had been prepared by a local architect on the basis of a contour survey prepared as a part of the property acquisition several years before. It immediately became obvious that someone had dumped a significant amount of loose fill on the site, completely changing the site conditions from those shown on the previous survey. All this material had to be removed so that unsuitable top soil could be stripped. Through a resurvey and modification of the plans and specifications prior to bidding, a lump-sum bid for the actual conditions was awarded. In comparing unit prices for additional work as actually bid by the low bidder, it was clear that the owner received a substantial saving over performance of the added work by the unit prices originally contemplated.

Site visits are generally the only way that items of the type described here can be taken into account in the overall program. As the project evolves, the professional construction manager must continue to be fully informed of new developments peculiar to the site, and he must be able to communicate his on-site knowledge to the designer, the owner, and his own personnel.

AREA CONSTRUCTION PRACTICE

Equally important to the job-site investigations, or even more important, is the investigation of the normal method of doing business in the project locale. Even if the manager is familiar with the area, he should systematically review the local conditions and practices. If he is operating in a new area, the investigation is of paramount importance. Some of the significant items that must be investigated in order to develop a suitable program are outlined in the following subsections.

Local Work Practices and Jurisdiction

Each area is unique in the local practices and jurisdictions that have evolved over the years. The professional construction manager handling a phased construction program is constantly faced both with fitting his construction packages to the design schedule and with tailoring them to the optimum size that will attract qualified contractors. In order to achieve this objective while making bid packages attractive to potential bidders, he must know the prevailing practices in the area.

For example, in some parts of the country, storm and sanitary sewer pipes outside the building lines are installed by utility contractors using laborers. In other sections, this work is customarily performed by plumbers. In still others, work outside the building lines can be performed by laborers if it is a part of the sitework, but must be performed by plumbers if it is included in the overall building plumbing. In certain areas, sheet-metal flashing is customarily furnished and installed by the roofing contractor. In other locations, flashings are furnished by a separate sheet-metal contractor, with the roofing contractor supplying only the built-up roof. Many open shop areas which commonly utilize a mixture of union and nonunion contractors have developed similar understandings.

The general and specialty contractors operating in the area are fully familiar with these types of area practices. The professional construction manager or out of town general contractor or design-constructor must become equally well informed to be able to define contract or subcontract packages in the most expedient and economical manner.

Labor Costs, Productivity, and Availability

Determination of craft productivity is difficult since it varies considerably from contractor to contractor and from trade to trade. Any conclusions by necessity must be subjective. Nevertheless, in discussions with local contractors and labor union personnel, numerous relevant facts will be obtained. Inspection of several projects in the area can also be helpful.

In some areas, key contractors and subcontractors employ a substantial number of workers on a year-round basis; depending on their business volume, they will obtain additional workers for seasonal peaks or for increased workload. Framing work packages to utilize the relatively permanent craftworkers can thus sometimes have a significant economic benefit to the owner.

Other large projects will have a completely different outlook, since substantial manpower must be recruited by all contractors involved. Productivity will also generally follow the extent to which contractor management has retained control over the labor force. Estimating construction costs in areas where extensive coffee breaks, long lunch hours, early quitting, and similar practices are uniformly tolerated is completely different from estimating similar work in areas where contractor management has good control over the work force.

Determination of labor availability is also significant. One simple way for an outsider to develop his own conclusions is to meet with key business agents of each trade, ascertain the size of the local union, and observe the number of workers on the bench. Key design decisions can sometimes be influenced to change specifications from those requiring chronically scarce craftsmen to alternate methods for which sufficient manpower is available.

Collective bargaining agreements for all crafts should always be obtained by the manager as early as possible. The agreements, of course, will be used in determining wage rates to be used in the estimate, and they can also help in forecasting productivity by revealing the presence or absence of restrictive work practices. Agreement expiration dates, anticipated strikes, and anticipated wage increases not yet negotiated are also significant.

Locally Favored Methods and Materials

Many designers habitually specify methods or name-brand items, with equals to be proposed by the bidder only as an alternate. If the designer is unfamiliar with local conditions, investigation of locally favored methods and materials, and their subsequent utilization where possible in the specifications, can significantly reduce project costs. It is unreasonable to require bidders to specify and price numerous alternates when a thorough investigation of locally favored and more economical items can result in specifications tailored to the use of more economical local materials and methods.

Key Local Prices

Local prices for standard items can be readily obtained, and they are of significant value in comparing alternative methods as well as in making the fair-cost estimates. Such local prices can include ready-mix concrete, sand and gravel, lumber, reinforcing steel, concrete blocks, precast concrete, pipe and fittings, cement, and other items.

In certain areas, precast concrete plants have developed standard sections that are very economical when compared with other methods. In other areas, no precast plant is readily available.

Local Contractors

The professional construction manager must develop a representative list of qualified, interested contractors for each proposed bid package. The list should be large enough to ensure competition, yet small enough to create significant interest in all bidders. By far the best procedure is to invite only fully qualified bidders to submit proposals, so that the award can be made to the lowest responsive bidder.

Some owners will require strict—sometimes overly strict—financial qualifications. One major owner, with a continuing, reasonably level, repetitive workload of upward of $150 million per year, insists upon a certified financial statement. This owner requires net quick assets of 20 percent of the estimated cost of the contract, or the contractor will not be allowed to submit a proposal. In addition, this owner requires a 100 percent payment and performance bond for every contract.

Other owners will depend upon the professional construction manager to screen potential bidders. Here the question of whether or not bonds shall be required is always present.

Preliminary lists of prospective bidders should generally be developed prior to financial screening. A knowledgeable professional construction manager, even if initially unfamiliar with the project area, will have developed local contacts who can give valuable information. Union representatives, contractor associations, local architects, and engineers, and many others can give valuable assistance in prescreening available contractors.

One method that offers significant fringe benefits to a professional construction manager relatively unfamiliar in the project area is initially to develop a preliminary list for further screening, as discussed above. Then, several days early in the planning stage can be allowed for scheduling short individual meetings with all interested contractors. These meetings can be preplanned and conducted in 20 to 30 minutes to accomplish the following:

1 Explain the overall program for the project, including the approximate scope and number of individual bid packages.

2 Review the preliminary procurement schedule, discussing the approximate periods when packages of possible mutual interest will be put out to bid.

3 Determine the interests of the potential bidder, and specifically the package for which he is interested in being considered for the invited list.

4 Discuss trade jurisdiction as applied to the particular contractor. Request his advice regarding inclusion of certain items in various work packages.

5 Obtain required information on previous jobs completed by the contractor; if possible, determine his present workload; and, if necessary, obtain required financial information.

A skilled professional construction manager will stimulate local interest with the above technique. By comparing answers to questions from similar contractors, he will also obtain in a surprisingly short time a reasonably accurate picture of how the construction business is conducted in the area.

Particular emphasis should be placed on meeting with reputable and leading general contractors. Many professional construction management jobs cannot be successfully managed unless certain phases of the work, such as a building finish package, are handled by a knowledgeable and skilled general contractor. If the program is properly explained, many leading general contractors, who may or may not be interested themselves, can and do offer suggestions for scoping work packages or for using local methods or materials that can be worth considerable savings to the owner upon implementation.

Operating from a hotel room, on one project the senior author met with over 35 prospective contractors for early bid packages for a phased construction program; appointments were scheduled every 20 minutes over a day and a half. Almost all the contractors arrived on time, most offered valuable suggestions, and all were very appreciative of the opportunity to hear about the project.

Other Key Local Contacts

In most areas, the local chamber of commerce can furnish economic data, discuss weather and climate conditions, confirm local business licenses, assist with tax information, and offer considerable other assistance.

The local building department is in many areas a key factor to a successful, early start for a phased construction program. Some areas require all plans and specifications to be approved before construction can begin. Others require special licenses for the professional construction manager's field construction manager who is in direct charge of the job-site work. All areas have special permits and fees required at various stages in the program, such as sewer and water connections. In some areas, these contacts are best handled by the designer, especially if he represents a local firm; but in others, the designer needs input from the professional construction manager.

Local utilities should be contacted so that an early determination of the method of supplying construction power, water, and other required temporary utilities can be made.

A large amount of local business information is often available. In any relatively unfamiliar area, the ingenuity of the professional construction manager is challenged by the need quickly to gain an understanding that will serve as a base for the planning phase.

Establishment of Project Field Office

The information developed in the early site visits is by nature preliminary. It is important to build upon this base continuously throughout the planning, design, and procurement phases so that new or revised information may be incorporated into the program.

Ideally, the field office should be established in advance of the award of the first contract so that potential bidders can be shown the work site and so that a local contact with other potential bidders, agencies, and others is maintained. The field construction manager will be the key representative in all dealings with local people; the earlier this position is assumed, the better for all concerned.

PRELIMINARY ESTIMATE

When the overall scope and conceptual design have evolved to the point where the manager has a reasonable idea of the requirements of the owner and the implementation program of the designer, preparation of a preliminary estimate can proceed. Figure 4-5 in Chapter 4 gives an example of the document prepared in this vital step.

The preliminary estimate initially serves to check the design against the owner's original budget or appropriation estimate. If costs appear to be over budget, alternative concepts can be explored through the value-engineering program before anyone launches into a significant amount of detail design.

The preliminary estimate is also necessary for preparing a realistic overall project schedule that forecasts occupancy dates and specifies completion schedules for individual construction contracts. The preliminary estimate forms the basis for cost control during design and procurement and is extremely useful in determining the proper size of individual contract packages that will stimulate maximum competition and interest among selected bidders.

SUMMARY SCHEDULES

Three separate but distinct summary schedules are important for effective control on most multiple-facility projects. These include a design and procurement schedule, a

construction schedule summarized by individual contracts, and a construction sche___
summarized by individual facilities.

Design and Procurement Schedules

For best results, a schedule for each proposed bid package must be prepared showing
the detailed design and specification period, package review and approval period,
bidding period, and evaluation and award period. This schedule must be developed
early and must be used by the designer, owner, and manager in performing their
assigned tasks. The schedule will form the control standard for monitoring actual
performance during the planning and design and the procurement phases, since the
construction schedule is wholly dependent upon award contracts by the required dates.

In general, most designers will prepare an overall design schedule. The manager
must take the proposed bid packages and, with the designer, develop a control design schedule by bid package. Depending upon construction schedule requirements,
adjustments can be made with the designer to schedule an orderly design completion
that fits the needs of the critical path.

A period for owner and manager review of the preliminary bid packages is a
necessity if the designer is preparing bidding documents under the manager's general
instruction regarding scope. If the manager prepares the bid packages from plans and
specifications furnished by the designer, a review period by owner and designer is
equally important. This review period is generally the last chance to avoid errors,
take advantage of recent knowledge, and avoid later plan changes which will result
in additional costs if made after contract award.

Reasonable bid periods should be scheduled by the manager, taking into account
his knowledge of the present bidding volume in the area. If sufficient time is planned
from the beginning, schedules can be more easily met, and more competitive bids will
normally be received.

The professional construction manager has a unique opportunity to solicit alternate quotations, either by specifying clear choices in the contract document or by
encouraging the ingenuity of the bidders. Evaluation of alternates, whether requested
or volunteered, takes time; a reasonable period for evaluation and award of each bid
package should therefore be included in the schedule. See Figure 5-1 for a design and
procurement schedule summary.

Summary Construction Schedules

When a preliminary estimate and a design schedule by contract package have been finalized, a CPM precedence diagram (or arrow diagram)[1] can be prepared setting forth
the logic of the contemplated program in sufficient detail to determine the critical path
and to develop key contract milestones. This diagram will enable adjustments to be
made to the design and procurement schedule so that critical items are taken into account by the designer, owner, and manager. See Figure 4-8 in Chapter 4 for an example
of a preliminary CPM diagram. A more detailed diagram is included in Appendix A.

---

[1]Chapter 12 provides more detail on these and other scheduling tools.

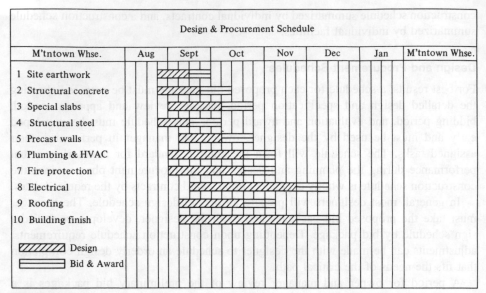

FIGURE 5-1
Design and procurement schedule.

After the planning is complete and the CPM logic is developed and reviewed, working summary bar-chart schedules can be prepared showing early- and late-start dates, early and late completion dates, the anticipated duration of each contract package, and also the interrelationships between the separate packages. Monitoring of actual performance when compared to early- and late-start scheduled performance will show status of schedule at all times, and is an integral part of the project control system. Figure 8-8 in Chapter 8 shows a project schedule summarized by contract.

On a multiple-feature project, a similar bar chart can be prepared, fully consistent with early- and late-start schedules, showing relationships of the separate facilities, and with provision for monitoring actual performance by facility in a similar manner.

WORK PACKAGES

After the professional construction manager has become thoroughly familiar with the project locale, after the preliminary design schedule is developed, and after the preliminary estimate is complete, the individual can define proper work packages and develop a reasonably detailed scope. Two of many important factors that should influence this process are construction economy and design constraints.

Construction Economy

Bid package development is one of the most significant contributions of the professional construction manager. The scope of packages should be designed to be of a size that will prove most economical by stimulating competition, that will minimize overall costs by avoiding unnecessary tiers of contractors and subcontractors, and yet that will take advantage of the coordination skills of the various general and trade contractors in the area.

Design Constraints

The packages must be scoped to fit a reasonable design schedule when earliest completion is important. Design constraints will modify the content of bid packages in balance with overall objectives. A successful phased-construction program is wholly dependent upon the care and skill that go into defining work packages in order to balance economic considerations with completion requirements to achieve maximum overall benefit to the owner.

VALUE-ENGINEERING PROGRAM

Value-engineering as used in this book does not necessarily comply with the requirements of the Society of American Value Engineers (SAVE), who have developed a certification program. Many practitioners use the term *constructibility analysis* or *value analysis* to connotate a similar program. Practical value engineering as defined and practiced by the successful professional construction manager sheds much of the methodology that has, to a degree, obscured the significant contributions that an organized program can make. The value-engineering program must be enthusiastically accepted and practiced by all members of the project team—owner, designer, and manager. All savings result from the team effort, and should be so acknowledged. Approaching value-engineering in this way can eliminate most or all of the natural resistance many designers display when design review and alternative suggestions are proposed. The manager's function is to provide an organized program and to stimulate creative analysis during all phases of the program, but especially in the planning and design, bidding, and award phases. Value-engineering or constructibility analysis considerations in three of these phases is now discussed.

Conceptual Phase

Early in the program, the manager and the designer can explore basic concepts and list possible alternative solutions. If the professional construction manager's historical cost data are properly tabulated, alternative comparable cost estimates can be prepared quickly and economically. If suggested alternates are indicated to be less expensive while preserving basic value, a refined, detailed estimate can be developed. Proposed savings can be presented to the owner for approval, or for consideration as a joint effort between professional construction manager and designer. For example, alternative evaluations of basic wall specifications for industrial and warehouse buildings can be developed easily and cheaply, using basic materials, prices, and labor estimates for the particular area in question. On one project, a simple low-cost study of precast double-tee wall sections, tilt-up concrete slabs, and concrete-block walls showed decidedly that the precast double-tee walls were less expensive, structurally equivalent, and architecturally more pleasing. See Figure 5-2 for a value-engineering study of these alternative wall systems. Without the study, a different wall system would have been specified. Significant opportunities of this type are available for an alert owner-designer-manager team to produce substantial cost savings at the conceptual stage.

Detail Design Phase

Similar opportunities are available during the detail design phase. A well-organized professional construction manager will keep an up-to-date book, indexed by standard architectural specifications sections, listing numerous alternative materials and methods, together with cost comparisons from previous jobs. As each new project is value-engineered, the book becomes more valuable. The professional construction manager can review proposed items and, when alternatives look promising, recost the alternatives for the individual project area. This kind of construction-cost-oriented advice greatly benefits designers by allowing them to stay within budgets. If properly handled, such recommendations are usually well accepted by the designers, and the cost advantages are realized by the owner.

Procurement Phase

Within limits, alternate quotations can be received in a phased construction program much more readily than under a lump-sum, single contract let under competitively bid conditions. If bidders are requested to price two alternates on a 15-contract phased construction management project, no unreasonable effort is required from any one bidder, and the owner receives the opportunity to achieve significant savings. Imagine the impracticality of asking for the same alternates for a single lump-sum bid assembled by general contractors accepting last-minute telephone quotations just prior to bid time.

An often overlooked but important resource for value-engineering savings consists of the bidding contractors themselves. To tap these ideas, specifications can encourage bidders to submit original cost-saving modifications as an alternate. If bidding contractors are aware of the possibilities of receiving an award through a voluntary submission of an equally desirable alternate, many will develop additional options for consideration.

A helpful technique is to list each agreed value-engineering saving by number and to make a complete record of the overall result of the program for an individual project. The record is then available to the manager, architect, and owner for use on future similar projects where applicable. Figure 8-7 in Chapter 8 tabulates value-engineering savings for the example warehouse project in Chapter 4.

CONSTRUCTION PLANNING

Basic construction planning during or before the detail design phase will include an organization chart, project staffing schedule, temporary facility requirements of the construction manager, selection of the particular individuals to be assigned, and

Title Value-Engineering Study   **Job No.**
Client Easyway Food Co.   Location Mountaintown Date 9-15
Subject Alternate Wall Systems   **By DSB**
                    Comparative Costs   **Sheet 1   Of 1**

Code	Description	Quantity	Unit Cost	Amount	Total
1	7½ in. concrete tilt up				
	Panel size 20 × 36 × 7.5″	720SF			
	Casting slab	240SF	1.00	240	
	Panel concrete	17CY	100.00	1,700	
	Reinforcing steel	1,440LB	.60	864	
	Embedded steel	360LB	1.20	432	
	Bond breaker	480SF	.10	48	
	Edge forms	112LF	3.00	336	
	Bracing & lifting inserts	720SF	.20	144	
	Finish trowel	720SF	.50	360	
	Winter protection & cure	17CY	20.00	340	
	Erect panels	720SF	1.20	864	
	Caulking	36LF	3.00	108	
	Subtotal	720SF		5,436	
	Overhead & profit	12.5%		680	
	Estimated total cost	720SF	8.50	6,116	8.50/SF
2	12 in. concrete block				
	Est. cost 720 SF				
	12 in. block—quote	814EA	4.00	3,256	
	Mortar	36CF	2.50	90	
	Grout	108CF	2.50	270	
	Scaffold	720SF	.40	288	
	Clean	1,440SF	.20	288	
	Waterproofing	720SF	.40	288	
	Wall truss reinf.	550LF	.20	110	
	Reinforcing steel	500	.50	250	
	Winter protection & cure	720SF	.80	576	
	Subtotal			5,416	
	Overhead & profit	12.5%		678	
	Estimated total cost	720SF	8.46	6,094	8.50/SF
3	Precast double tees				
	Local quote—furnish	720SF	6.00	4,320	
	Erection	720SF	1.20	864	
	Caulking	90LF	3.00	270	
	Overhead & profit	6.125%		344	
	Estimated total cost	720SF	8.06	5,798	8.00/SF

FIGURE 5-2
Value-engineering study.

delineation of their responsibilities. A complete cost estimate to serve as the manager's budget can be readily prepared if initial planning is sound.

Temporary Facilities

An important phase of construction planning is the analyzing of temporary utility and general conditions requirements for the project; this analysis is similar to a general contractor's appraisal. Temporary utilities can be furnished by the owner, be built into individual contract packages, or be obtained from others based upon local practice and job-site conditions. Utility bills can be paid for by the owner, or individual contractors can be billed or required to furnish their own utilities. Again, the best solution depends upon the professional construction manager's knowledge of the area.

Much of the detailed construction planning can be best accomplished from the job site. Sending in the field construction manager at an early date to develop construction planning details under job-site conditions is usually most productive.

Successful general contractors have developed the knowledge and skills necessary to plan temporary facility requirements and perform general conditions items in a manner most economical for the project. A qualified professional construction manager must have similar knowledge and skills. Refer to Figure 4-6 in Chapter 4 for an estimate of general conditions costs.

Procedure Outline

Each project is unique. One of the greatest advantages of a professional construction management program is its flexibility. The manager must be able to assess the conditions and problems as they develop and to react without delay to further the interests of the project.

However, with three parties involved in the management of the project, it is important that each understand the responsibilities and duties of the others. This was one of the main purposes of the procedure outline given in Chapter 4, and it is very important to construction planning.

Cash-Flow Requirements

An estimate of cash-flow requirements for the project can be readily prepared from the preliminary estimate and from the summary schedule. Some owners require more accurate cash-flow projections than others. A simple cash-flow projection based upon prior planning can be prepared as a part of the control package. If warranted, actual requirements can be tabulated monthly and compared with earlier forecast requirements. See Figure 5-3 for a cash-flow projection.

The cash-flow schedule for the Mountaintown Warehouse gives the owner estimated gross cash requirements for expenses throughout the program. Net cash estimates can be prepared taking into account investment income, retained funds, and the lag between billings and payment requirements.

Public works are often funded by bond issues. Here the proceeds from the bonds will normally be received by the issuing agency prior to commencement of the work.

Cash-Flow Schedule Mountaintown Warehouse

	Estimated cost	% Complete Period	% Complete Cumul	Payments period	Payments cumulative
Design & observation	$ 260,000				
October		15	15	39,000	39,000
November		15	30	39,000	78,000
December		20	50	52,000	130,000
January		20	70	52,000	182,000
February		10	80	26,000	208,000
March		10	90	26,000	234,000
April		10	100	26,000	260,000
				260,000	
Construction (all costs)	$5,600,000				
October		2	2	112,000	112,000
November		9	11	504,000	616,000
December		9	20	504,000	1,120,000
January		12	32	672,000	1,792,000
February		21	53	1,176,000	2,968,000
March		27	80	1,512,000	4,480,000
April		16	96	896,000	5,376,000
May		4	100	224,000	5,600,000
				5,600,000	
Equipment & owners	$540,000				
October		5	5	27,000	27,000
November		5	10	27,000	54,000
December		5	15	27,000	81,000
January		5	20	27,000	108,000
February		5	25	27,000	135,000
March		25	50	135,000	270,000
April		25	75	135,000	405,000
May		25	100	135,000	540,000
				540,000	
Total project cost	$6,400,000				
October		3	3	178,000	178,000
November		9	12	570,000	748,000
December		9	21	583,000	1,331,000
January		12	33	751,000	2,082,000
February		19	52	1,229,000	3,310,000
March		26	78	1,673,000	4,958,000
April		17	95	1,057,000	6,041,000
May		5	100	359,000	6,400,000
				6,400,000	

FIGURE 5-3
Cash-flow schedule (Mountaintown Warehouse).

Since payments will be spread throughout the life of the project, a reinvestment program can be a sizeable source of additional funds to the agency. Many public agencies and some private owners are now allowing contractors to post securities in lieu of retained funds throughout the contract with interest or dividends distributed to the contractor. If 10 percent is retained from all progress payments until 30 days after completion of the contract, it is evident that positive cash flow will be difficult to achieve and that a substantial amount of working capital will be required. Public agencies and most large owners can obtain funds at a lower interest rate than most contractors. Minimizing retained funds can often benefit owners in receiving lower bid prices by minimizing contractor interest expense.

General contractors, design-constructors, developers and subcontractors also use cash-flow schedules to determine the capital requirements for each project and the overall company cash flow projection is important in determining working capital requirements for the firm and in setting up lines of credit to borrow funds at favorable rates.

SUMMARY

Planning must be based upon facts if project goals are to be achieved. To start, the professional construction manager must understand the overall objectives of the owner, and must know the overall concepts and operating methods of the designer. However, the implementation of the owner's objectives and the designer's concepts must be done in the particular locality where the project is to be built.

Each area and locality has its own construction peculiarities. Programs, methods, and materials developed for certain sections of the country are totally inappropriate for other localities because of the vastly different conditions that affect the work. The professional construction manager must become fully acquainted with local conditions as well as with owner objectives, design concepts, and operating methods, to develop a planned program which will be both realistic and capable of serving as a standard against which actual project accomplishment can be measured.

In the planning stage, all portions of the project are developed in a straightforward and logical manner consistent with project knowledge and anticipated conditions. This development of a logical program will form the basis of the project control system that will assist in managing the project throughout the construction period.

During the planning and design phase of a construction project, planning will include the preparation of preliminary estimates and design, procurement, and construction schedules. From this information, construction contract packages can be chosen in a logical manner. A formal plan for a value-engineering program and initial construction phase planning complete the overall work plan to be used as a standard in measuring project performance at all stages.

As unanticipated events impact upon the project, the plan must be changed to fulfill its primary function. In this event, replanning will occur throughout the construction phase of the project. If the management control system is to have any effect at the project level, such replanning must be reflected in the control system so that actual results are being compared with a realistic plan.

CHAPTER

6

BIDDING AND AWARD

Basic requirements that must be met for any construction project to be successful from the owner's standpoint include: The project must be "bought out" within budget; it must be completed on schedule and be constructed in accordance with the plans and specifications.

The bidding and award phase explained in this chapter establishes the foundation for the project and offers an excellent opportunity for designer, owner, and professional construction manager to learn to work together, while enhancing their mutual respect and confidence. The initial planning process described in the previous chapter continues into the bidding and award phase. Here again, the importance of owner and designer review, input, and acceptance of the plan cannot be overemphasized. All members of the construction management team must carry out their assignments in keeping with project schedules. As new problems are encountered, new solutions must be fed into the overall plan if the project is to achieve its initial objectives.

This chapter will describe an approach typical of a professional construction management program in detail. With minor changes it is also applicable for use in bidding and award to a general contractor. In fact, the CM approach has evolved from a typical approach to general contractor selection through the addition of individual work packages under construction manager control rather than leaving these duties to a general contractor acting in the traditional manner. Bidding and award to general contractors is also discussed with particular emphasis upon the differences from construction management practices.

DEVELOPING CM CONSTRUCTION PACKAGES

Proposed contract packages outlined in the overall plan can be developed in detail by either the professional construction manager with designer review, or by the designer with manager review. The choice will depend in part upon the requirements of

101

the owner and the location and procedures of both the designer and the professional construction manager. Each method will have advantages and disadvantages for a particular project. Setting forth the required scope in a straightforward manner, so that the bidder understands what is required, is of paramount importance. Ambiguities or alternative interpretations in the scope of work are the forerunners of misunderstandings, claims, and litigation, so the qualified professional construction manager will pay particular attention to avoiding such problems. The manager's close communication and cooperation with the designer are essential; each must contribute in his own area of primary responsibility and act as a helpful advisor and counselor in the other's area.

For purposes of scoping construction contract packages, it is assumed here that all plans and specifications are prepared by, and are the responsibility of, the designer. No changes to plans and specifications can be made by the professional construction manager without the prior consent and approval of the designer. Within these guidelines, two different and distinct methods of preparing construction contract packages are commonly applied by an experienced manager. He may:

1 Prepare individual contract packages from relatively standard specifications and drawings similar to those prepared for a single general-contract project.

2 Prepare individual contract packages from individual specifications and drawings prepared with the overall scope of the desired contract packages clearly in mind.

Each approach is discussed in more detail in the following sections.

Standard Drawings and Specifications

Many design firms update or modify standard specifications they have developed over many years; these firms have geared their individual designers and specification writers to conform to the standard specifications. Many design firms also standardize construction details in a similar manner. The routine and smooth working methods of the design office in many instances depend upon use of such established standards.

When this situation is encountered, it is often preferable that the professional construction manager prepare the detailed scope sections of the construction contract package, and designate those drawings to be included as contract drawings and those to be included for reference and general information. Properly handled, assumption of this responsibility by the manager will avoid additional difficulties and possible added costs for the designer, and will contribute greatly to the partnership philosophy.

Neglecting for the moment the standard documents common to each work package, one method found to be successful in outlining the detailed scope is summarized thus:

1 The manager adopts all the specification sections pertinent to the work package in question and includes them in that package.

2 The manager writes a general scope of the work to be included, describing in general terms the work to be performed under the contract.

3 The manager prepares two summary schedules, setting forth each item by specification number and subnumber under "Work Included in Contract" or "Work Not

Included in Contract." All included standard specification sections and subsections are covered in one or the other of the two schedules.

4 In the event of conflict between the standard specification provisions and the requirements of the work package, an addendum, written jointly with the designer, makes the required modifications. Thus, the standard specifications as developed by the designer are never changed, and any modifications are clearly specified either in the schedules or in the addendum.

5 The manager lists by number all drawings to be included as contract drawings and identifies drawings to be included as reference drawings.

6 Before the manager incorporates the package into the bidding documents, the designer reviews the entire scope of the work package as just described. Owner review is also performed at this time if desired.

Individually Developed Drawings and Specifications

Other designers commonly develop plans and specifications to fit phased construction programs or multiple-contract projects. In this case, it is often more expedient and more economical for the designer to prepare the plans and specifications to fit the requirements of the program. One method found to be successful is outlined here:

1 In addition to the preliminary scope for each contract package, the manager prepares a detailed scope of individual items and describes the overall intent covering the scope of each work package.

2 The manager carefully reviews this overall intent and detailed scope with the designer and modifies it as required. The designer is encouraged to offer additional suggestions or modifications to fit his design schedules better as the design work progresses.

3 Prior to requesting bids, the manager blocks out a definite period for a detailed review of the completed packages. Owner review is accomplished simultaneously.

PREPARATION OF BIDDING DOCUMENTS

Bidding documents for a professional construction management project must be developed as a joint effort of the designer, owner, and manager. Bid package makeups will vary depending on owner requirements and on designer and manager procedures. A typical bid package might consist of the following items:

Invitation to bid
Bid form
Bid breakdown
Construction contract
General conditions
Special conditions
Work included in contract (optional)
Work not included in contract (optional)
Specifications, addenda, and drawings

Supplemental provisions
Owner-furnished items
Construction schedule

Each will be discussed in more detail in the following paragraphs. An example bid package is also included in Appendix B.

Invitation to Bid   Generally, the invitation to bid states the requirements and procedures for a responsive bid and gives additional information pertaining to the contract itself.

Bid Form   The bid form is completed and signed by the bidder and states the terms of his offer. Information commonly submitted in the bid form may include:

A statement that the bidder has examined plans, specifications, and the job-site location
The amount of compensation to be received for the work performed or offered by the bidder
The amount of liquidated damages if applicable
A statement that the bidder agrees to execute a contract if his bid is accepted
An agreement to submit a performance and payment bond, if required
Overhead and profit percentages applicable to extra work
Bid alternates, if applicable
Contract completion requirements, or number of days to complete the work
Special provisions that may be applicable

Bid Breakdown   The bid breakdown is filled in by the bidder, and it gives the individual price components that sum to the total contract price. This breakdown may later guide progress payments.

Construction Contract   A sample contract is included to inform prospective bidders of the type of contract each will be expected to sign if his executed proposal form is accepted by the owner.

General Conditions   The general conditions usually are part of the specifications; they state conditions applicable to all contracts to be awarded.

Special Conditions   The special conditions generally are a part of the specifications, and they set forth specific conditions applicable to the particular contract or group of contracts to be awarded.

Work Included in Contract (Optional)   This section may designate provisions of a standard specification applicable to the particular contract to be awarded.

Work Not Included in Contract (Optional)   This section may exclude provisions of a standard specification not applicable to the particular contract to be awarded.

Specifications, Addenda, and Drawings These items provide technical requirements of the contract; taken together, they fully define the scope, extent, and quality of the work.

Supplemental Provisions Supplemental provisions may include additional items not suited for inclusion in the special conditions, such as a definition of the status of the professional construction manager, and prevailing wage rates, if applicable.

Owner-Furnished Items This section describes all items to be furnished to the contractor by others. It may include varied items such as materials and equipment, temporary utilities, storage areas, water and sanitation facilities, and survey controls.

Construction Schedule This section shows scheduled milestones and overall completion requirements for the particular contract being bid, and provides an overall schedule showing general relationships between work packages and design activities.

CONTRACTOR QUALIFICATION, BIDDING, AND AWARD

Development of Bidders List

Prescreening techniques were discussed in Chapter 5. The success of any construction program depends upon utilization of reputable, skilled, and financially sound contractors. Prequalification of all contractors prior to issuing bidding documents is by far the best way to achieve this objective. Methods and techniques used to prequalify prospective bidders will vary, but they should include obtaining evidence of capability from previous projects and of financial strength sufficient to handle the project. Owner approval of the bidders list is often required, and many owners are quite specific about the financial qualifications they expect potential contractors to have.

The best way to ensure fair prices as well as performance for the owner is to receive competitive bids from a reasonable number of prescreened, prequalified bidders. The inclusion of too many bidders, while promoting competition, will reduce the attraction for many of the best-qualified contractors, and the use of too few bidders will generally increase prices because of lack of competition. The owner or his representative must strike a reasonable balance to stimulate high interest while assuring reasonable competition. Six to eight bidders is normally about right. With prequalification, the award can almost always be made to the one evaluated as the lowest responsive bidder. See Figure 6-1 for a sample bid tabulation.

Fair-Cost Estimates

There are many advantages to preparing a fair-cost estimate from the same bid package submitted to the bidders. It shows within reasonable limits a fair price for the work, and gives the professional construction manager or other owner's representative a chance to spot inconsistencies or conflicts in the detailed drawings or specifications that may have been missed in earlier reviews. In this event, an addendum can be issued correcting the discrepancy before bid date.

Mountaintown Warehouse

Special Floors	Palmer Floors, Inc.	Curt Flooring Co.	Rutherford Floor Co.	Fair cost estimate
1. Base slab	380,000	404,000	370,000	354,300
2. ¾ in. topping	160,000	150,000	150,000	144,000
3. Joints & other	32,000	40,000	50,000	52,000
Subtotal	572,000	594,000	570,000	550,300
Bond	3,600	4,000	4,000	incl.
Total bid price	575,600	598,000	574,000	550,300
Alternate Bids				
1. Expansive cement	(15,600)	—	6,000	
Addenda acknowledged	2	2	2	
Completion	165 days	170 days	per schedule	5 ½ mos.
Bid bond	yes	yes	yes	
Exclusions	none	none	based on availability of area by 11-1	
Evaluated bid	560,000	598,000	574,000	

Remarks: Award is recommended to Palmer Floors, Inc., based upon lowest price after architect approval of expansive cement alternate.

FIGURE 6-1
Bid tabulation (Mountaintown Warehouse).

In the event of widespread differences between estimate and bids, discussions with the low bidder can often pinpoint reasons for the difference, especially if a fair-cost estimate has been prepared as carefully as did the bidder.

Records of actual manpower employed on the contract are normally kept during the construction phase. Comparisons with estimated manpower can often provide quantitative productivity measurements.

Often, in preparing the estimate, the manager's estimator may spot alternate materials or methods that have escaped prior consideration. Exploration of these alternatives with the low bidder prior to award can often significantly add to savings developed in the value-engineering program.

One of the principal benefits from preparing a detailed fair-cost estimate is that of placing the professional construction manager or other owner's representative on a par with the successful bidder. Through this estimate, detailed job-site scheduling is encouraged, review of modifications or changes is facilitated, and in general the manager is able to deal with the successful contractor from a position of knowledge, strength, and mutual respect.

Analyses of Bids and Value Engineering

Once all proposals have been received, the professional construction manager first prepares a spread sheet tabulating all quotations and notes other pertinent factors such as qualifications, omissions, unit prices, and completion times. He then reviews alternates requested or proposed by the bidder.

If they are encouraged prior to bid and dealt with fairly in the bid evaluation, alternates proposed by the bidders themselves form an excellent value-engineering opportunity. Proposed alternates, either volunteered or requested, should always be reviewed jointly with the designer, and ideally, a joint recommendation should be made to the owner. In some cases, only the owner can evaluate the desirability of a saving resulting from a proposed modification.

One method of assuring that bidders are treated fairly in the evaluation of volunteered alternates provides the following guidelines:

1 When a bidder who is not low under the basic bid volunteers an alternate judged by the designer to be equal to the specified requirements, and thus becomes the new low bidder, the award will be made to that bidder; both the bidder and the owner gain from the bidder's ingenuity.

2 When, under similar conditions, a bidder submits an alternate that represents a sizable monetary saving but is not equal to the specified product, and if in this event the proposed substitution is acceptable to the owner because of the magnitude of the cost saving, all bidders in contention should be given an opportunity to quote on the acceptable but lower-quality alternate, and the award should be made to the subsequent low bidder.

3 When an alternate is volunteered by a high bidder who would not be low even after taking the alternate into account, and if the alternate is acceptable to the designer and the owner, the low bidder may be contacted and requested to quote on the proposed modification.

In all these proceedings, the professional construction manager must apply fairness and good judgement to avoid any practices bordering on unethical "bid shopping." Where alternates will be accepted, guidelines of the type given here must be clearly stated in the bidding documents.

Recommendation for Award

After evaluation of the bids, including alternates, a recommendation for award is made to the owner. After approval by the owner, actual award can be made either by the manager or directly by the owner.

In the event that the award is recommended to other than the low bidder, a full explanation of the reasons for such a recommendation should be made to the owner, and his approval should be obtained. Under an invited, prequalified bid list, the award should normally go to the low initial bidder unless schedule demands or other equally significant considerations indicate that the award should go to a bidder whose volunteered alternates or other qualifications make him a more cost-effective choice overall.

Some companies follow a policy of opening all bids in the presence of the bidders. Others will advise all bidders of the bid results for the base bid. Where the owner is in the continuing-workload classification, such policies can benefit the owner as well as the bidders.

Negotiating Contracts

Occasionally it is necessary to contact bidders to obtain clarification in order to evaluate the bids properly. At other times, drawing revisions approved shortly after the receipt of bids may require price changes. The manager must be thoroughly aware of the practices within the industry. When further discussions or negotiations are required, the manager must pay particular attention to conducting such discussions in an ethical manner. The manager is responsible to the industry and to the owner to avoid any taint of "bid shopping," as pointed out earlier. No rules can be laid down to cover all situations, but a qualified professional construction manager must know what is right.

APPLICATION OF CONTROLS

Controls during the bidding and award phase will include the following two main areas:

Procurement Schedule Control

The initial planning schedule specified the design, review, bidding, and award periods for each construction package. Actual progress can be monitored against the planning schedule so that current status is immediately apparent and planning can be revised where required to take care of delays or other revisions. If schedules are to be met, construction contracts must be awarded as required by the overall construction schedule. This simple, updated control schedule, showing actual accomplishments compared with programmed accomplishments, can be of great value in assuring timely procurement. See Figure 6-2 for an updated procurement schedule.

Construction Cost Control

As contracts are awarded for a phased construction program, the actual cost at completion becomes more certain. Comparing actual award prices with the preliminary and fair-cost estimates is important for both feedback and control. A similar cumulative comparison of all awards to date will indicate the current status of the project.

On individual contracts, any significant difference between the fair-cost estimate and the bids is cause for further investigation. Through review of the fair-cost estimate and discussions with the low bidder, this difference can be accounted for in many instances. In other cases, specifications can be modified. As a last resort, packages can be modified and rebid.

After the "buy out" is complete, the sum of the individual contract costs becomes a committed cost to the owner. Almost all jobs will require change orders due to

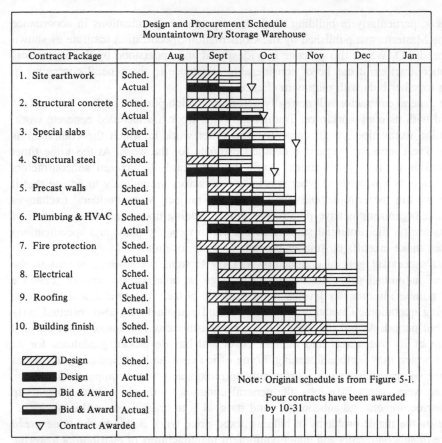

FIGURE 6-2
Design and procurement schedule.

changes, drawing errors, and other modifications. A contingency allowance for changes and minor omissions should thus be developed so that an overall estimated cost at completion can be determined. Allowances for changes will vary depending on the nature of the project, the requirements of the owner, and the accuracy of the designer. A contingency of 2 to 5 percent added to the sum of the individual contracts is normally sufficient for minor changes. Major modifications, if contemplated, should be separately evaluated.

BIDDING AND AWARD TO GENERAL CONTRACTORS

The major difference between bidding and award for a general contract and for construction management lies in the treatment of the specialty or trade contractors. The traditional general contractor has historically managed the subcontractors and has acted as a single entity in dealing with the owner or his representative. A number of architects, particularly in building work, have prepared specifications in accordance with the Masterformat published by the Construction Specification Institute as shown in Appendix D. Where this practice is widespread and understood by the contractors and subcontractors, it can be of considerable assistance to the general contractor in putting together bids with minimum chance of error.

The general contractor will normally determine which portions of the work it plans to build with its own workforce. Traditionally this work has included concrete work, foundation excavation, rough carpentry and other work in which the contractor is skilled. The contractor prepares a detailed estimate for the work. At the same time, it advertises in trade publications, sends out post cards to favored subcontractors and/or contacts a number of other specialty contractors individually to advise that it is interested in receiving subbids for a particular project. Many builders' exchanges or similar organizations have plan rooms where bidding documents for a number of projects are on file. Potential subcontractors can review the plans and specifications and even make takeoffs on smaller projects at these locations.

Most successful general contractors will try to agree in advance regarding the scope of individual subcontracts and will assign an approximate amount (called a "plug" number) in order to develop the magnitude of the project value and for use in reviewing quotations. Quotations for material and equipment are also solicited early in the bid period. As bid time nears, the general contractor has completed a detailed estimate of its own direct and indirect costs and has developed guidelines for its markup which may include financing, home office costs and anticipated profit. It has also received preliminary prices from subcontractors and major equipment suppliers. On many very competitive jobs, the general contractor will add little or no markup on major equipment and subcontracts and its goal is to break even on this work.

One of the major problems in bidding fixed-price work is due to the widespread practice of subcontractors and major equipment manufacturers or distributors lowering their price significantly in the last hour or less before bid time. Some general contractors will have arrangements with subcontractors which allow them to meet someone else's price (right of first refusal). Other generals may use the price of one subcontractor to negotiate with others. By delaying submitting their best bid until just before bid time, the subcontractors hope that the general will not have time to "peddle" or to "bid shop" their quotation. Problems arise when Brand X subcontractor who is unknown to the general comes in with a last minute verbal quotation over the telephone which is significantly lower than other quotes. The problems are compounded with bids received over fax machines that have long lists of modifications and exclusions. Does the contractor accept the unknown quotation without a detailed understanding of the full scope or of the financial and other qualifications of the bidder? On the one hand, it is afraid that its competitor will use the quotation and it may lose all chance of being low. On the other hand, if the subcontractor is not bondable or doesn't perform, it accepts considerable financial risk in the event of nonperformance.

The well organized general contractor formalizes its estimate in advance of bid time. It then prepares separate tabulations of overs and unders due to last minute quotations compared to its "plug number" or previous quotation. On unit-price bids, the contractor will finalize allmost all of its units except for one or two which it will leave open to take care of last minute adjustments. By leaving a lump-sum item open, it can easily add or subtract the approximate difference quickly and with minimum chance of error. If its bid is successful, he can prepare its working budget with the items placed in their respective places.

SUMMARY

The bidding and award phase establishes the framework for project construction; by encouraging contractor ingenuity, it can offer another major opportunity for improvements and cost savings on the project. Here, again, participation of the owner and designer with the owner's representative is important for review and decision making.

Success in this phase depends first upon clear and unambiguous development of construction work packages. The complete set of plans and specifications must be subdivided and categorized in such a way that no items are omitted, none is duplicated, and all contractors bidding on a given package are indeed bidding on the same scope of work. The professional construction manager's contractor-type knowledge of area and trade work practices is essential to this step. In general contractor bidding, the contractor sorts out the work to be performed by subcontractors and takes the risk of omissions or nonperformance.

Development of work scope is followed by preparation of bidding documents, which is again a joint effort of the designer, owner, and manager, or other owner representative. Items typically included are the invitation to bid; bid form; bid breakdown; construction contract; general conditions; special conditions; specifications, addendums, and drawings; supplemental provisions; owner-furnished items, and construction schedules. These basic steps are equally applicable to general-contractor contracts.

The selection of the best contractors will be aided by prequalification and by keeping the number of contractors small enough so that the best contractors will know they have a sufficiently reasonable chance of obtaining the work to make bidding worthwhile. The owner's representative should also prepare his own fair-cost estimates; these help him to evaluate bids more knowledgeably and to find previously undiscovered errors in the contract documents. By specifying fair and reasonable procedures for evaluation bidders can also be encouraged to submit worthwhile value-engineering proposals and volunteered alternates.

Application of controls in the bidding and award phase falls in two main areas: procurement schedule control, and construction cost control. Important documents for control at this stage include an updated procurement schedule and the preliminary and fair-cost estimates. Once the "buy out" is completed, the project is committed to construction.

CHAPTER 7

CONSTRUCTION

Once the planning, design, bidding, and award phases have been accomplished, the success of any construction project depends on completing the field construction phase on schedule and in compliance with plans and specifications. The initial planning forms the basis of the construction program. With this as a point of departure, the basic plan must be made current, and implementation of detailed planning and controls to accomplish overall planned objectives must begin. Detailed planning at the field office by all parties will continue throughout the project.

The initial planning was based upon broad assumptions, many of which will prove to be either incorrect or oversimplified as the project develops. Detailed field planning by all organizations must both anticipate and overcome possible problems in advance and help provide satisfactory solutions to current problems and delays if overall objectives and schedules are to be met.

OVERALL PLANNING AND CONTROL FOR CONSTRUCTION MANAGEMENT

The field construction manager and his staff are the keys to successful construction. Procedures, guidelines, rules, handbooks, and other aids can never replace the ingenuity of qualified construction professionals in anticipating and avoiding problems and in reacting to minimize the effects of unexpected developments.

To understand, manage, and coordinate the project properly, the field construction manager must plan and coordinate each of the major initial operations almost as would a general contractor.

For example, structural steel erection is usually on the critical path, and it is often the source of potential delays. The alert field construction manager will know the fabrication status in the vendor's plant long before the first deliveries are expected.

If potential delays become apparent, all resources at the manager's disposal can be brought to bear before significant delays are experienced. The professional will have planned well enough to be able to monitor early performance and react while a satisfactory solution can still be achieved. The nonprofessional will blame and threaten the fabricator long after the damage is done, and he will use such nonperformance as an excuse for not achieving project schedules.

Home Office Management Services

The basic requirement for home-office organization throughout construction is control. Field offices can be staffed to provide all, or a major part of, the necessary services during construction, and the home office must depend upon the field construction manager to manage the project in accordance with the predetermined plan. However, the overall responsibility and accountability to the owner for developing and monitoring a proper control system are at the top; they cannot be delegated entirely to the field construction manager.

Once the professional construction manager has chosen a sound, qualified on-site construction team and has developed a plan for achieving the objectives of the owner, his remaining duty is to know at all times whether the project is proceeding according to plan so that he can react quickly when necessary to modify, assist, and correct prior planning.

Apart from control responsibilities and accountability to the owner, most or all construction-phase management tasks can be delegated to the field construction manager and his assistants. On small- and medium-sized jobs, the minimum home-office staff will thus generally consist only of a part-time project manager or construction executive qualified to oversee project accomplishment, accept control responsibilities, and retain overall accountability to the owner. The home-office staff can, in addition, provide the field construction manager with part-time services that cannot economically be staffed at the job site. On the other hand, on certain large projects located in remote areas, many of the responsibilities for the preplanning, planning and design, and bidding and award phase outlined in preceding chapters can be best handled from the field construction site.

The extent of detailed work performed in the home office during construction will dictate organization requirements. Each project is unique, and for optimum results each will require different planning and assignment of responsibilities between the home office and the field construction office. The location of the owner, designer, and sources of supply will affect the balance for an individual project, as will the extent of delegation by the owner to the manager and the designer. As a general observation, for a project requiring full-time personnel, it is probably more economical and satisfactory to staff the project site than it is to perform the functions in the home office.

Certain services are commonly performed in whole or in part from the home office of the professional construction manager. They may include services that:

1 Provide management-level reporting to the owner through a straightforward quantitative description of project status.

2 Keep the owner informed of current and anticipated problems and their proposed or planned solutions.

3 Provide general supervision to assist, counsel, and direct field activity when necessary.

4 Monitor or administer project control systems; initiate remedial action when warranted.

5 Provide special assistance to the field construction manager where desirable. Such assistance may include preparation of schedules, estimates at completion, fair-cost estimates for extra work, claims negotiations, and expediting critical materials or equipment.

Field Management Services

Services normally performed in the field office will include the following:

1 Establishing field office, including provisions for general conditions items such as sanitary services, water supply, and temporary electrical and other items to be furnished to the contractors by the owner.

2 Hiring the testing laboratory and surveyor either jointly with, or with the approval of, the designer. Services needed include soils engineering, concrete inspection and testing, and other specialized requirements. On major projects, surveying and inspection often are provided directly by the professional construction manager. On medium-sized and smaller projects, it is usually more economical to contract for surveying services on an as-needed basis.

3 Obtaining necessary permits on behalf of the owner. Depending on project conditions, the designer or owner can share some or all of this responsibility.

4 Managing, coordinating, and inspecting the work of contractors to help achieve project cost, schedule, and quality objectives. Weekly contractor meetings with written minutes are often worthwhile.

5 Performing schedule, progress, and cost-control functions as needed for a particular project. Requesting home-office assistance for specialty items where indicated.

6 Maintaining job diaries, drawing registers, and other records to document the development of the project and promote a businesslike relationship with all contractors. These records will further assist in evaluating change-order requests and claims.

7 Initiating notice to proceed for individual contracts, preparing or approving progress-payment requests, and developing final contract closeout in accordance with owner and local requirements.

8 Maintaining progress-and-record photographs as part of progress and schedule controls, and to document potential claims, accidents, or similar occurrences.

9 Preparing input for the project control system by evaluating progress of each individual contract.

10 Maintaining job safety in accordance with contract and legal requirements. While safety is primarily the responsibility of each individual contractor, the manager has the duty to assist and to insist upon compliance with contract provisions. Recent legislative and judicial precedents have given the professional construction manager greater responsibilities in this area.

11 Maintaining liaison with the designer, requesting his assistance to interpret plans and specifications, and keeping him fully informed of the project status.

12 Obtaining or developing information for "as built" drawings, including maintenance of a current set of working drawings at the job site, available for all contractors and showing all current revisions and field changes.

13 Preparing field reports, including weekly progress reports, force reports, delay (or *force majeure*) reports, contract status reports, evaluation of claims, evaluation of requests for change orders, and reports covering other significant and periodic requirements.

TYPICAL ORGANIZATIONS FOR CONSTRUCTION MANAGEMENT

Each project is physically different, has different objectives, and is constructed under different local conditions. Depending upon project size, location, and other factors, certain functions can be performed in either the home office or the field project location. To illustrate a range of professional construction management approaches, additional organization charts are included here showing a phased construction program for a $3 million dry storage warehouse, a $13 million meat plant, a $60 million airport expansion program and a $100 million industrial project constructed under professional construction management.

$6 Million Dry Storage Warehouse

Figure 4-7 in Chapter 4 shows an organization chart for the example warehouse project. The owner and the professional construction manager's home offices are in one city, and the architect and the job location are in another.

$25 Million Meat Plant

Figure 7-1 shows an organization chart for a meat processing plant. The owner and the job site are located in one city, and the architect and professional construction manager in another.

$100 Million Airport Expansion

Figure 7-2 shows an organzation chart for an airport expansion program featuring a runway addition, a new terminal building, and appurtenant facilities.

$100 Million Hospital

Figure 7-3 shows an organization chart for a hospital. Construction is planned to utilize a separate foundation contract, a building construction contract, and separate mechanical and electrical contracts.

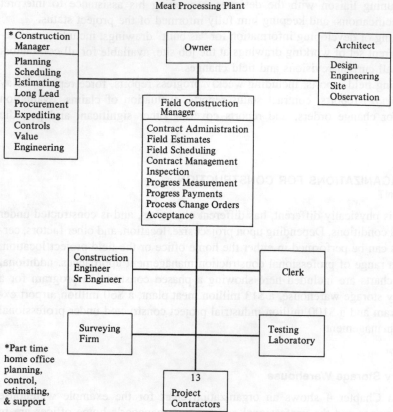

Organization Chart
Meat Processing Plant

*Part time
home office
planning,
control,
estimating,
& support

FIGURE 7-1
CM organization chart for $25 million meat processing plant.

$200 Million Industrial Project

Figure 7-4 shows a typical professional construction management organization for a major project where essentially all work except initial planning is performed in the field.

OVERALL PLANNING AND CONTROL FOR GENERAL CONTRACTING

With the general contractor approach, all of the requirements previously discussed remain although the performance may be handled by different organizations and methods.

The general contractor contracts to take full responsibility for dealing with subcontractors and in managing and coordinating overall job performance in order to achieve planned goals. The owner may assign administrative, inspection and testing responsi-

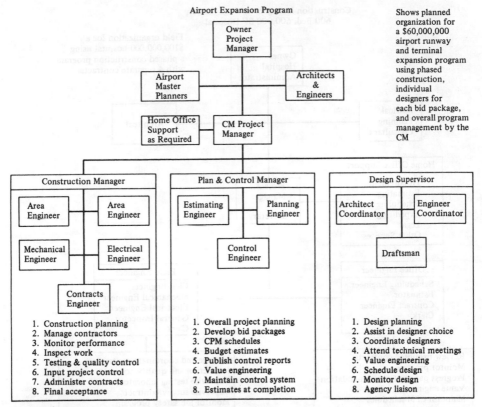

Airport Expansion Program

Owner
Project
Manager

Airport
Master
Planners

Architects
&
Engineers

Home Office
Support
as Required

CM Project
Manager

Shows planned
organization for
a $60,000,000
airport runway
and terminal
expansion program
using phased
construction,
individual
designers for
each bid package,
and overall program
management by the
CM

Construction Manager	Plan & Control Manager	Design Supervisor

Area
Engineer

Area
Engineer

Mechanical
Engineer

Electrical
Engineer

Contracts
Engineer

Estimating
Engineer

Planning
Engineer

Control
Engineer

Architect
Coordinator

Engineer
Coordinator

Draftsman

1. Construction planning	1. Overall project planning	1. Design planning
2. Manage contractors	2. Develop bid packages	2. Assist in designer choice
3. Monitor performance	3. CPM schedules	3. Coordinate designers
4. Inspect work	4. Budget estimates	4. Attend technical meetings
5. Testing & quality control	5. Publish control reports	5. Value engineering
6. Input project control	6. Value engineering	6. Schedule design
7. Administer contracts	7. Maintain control system	7. Monitor design
8. Final acceptance	8. Estimates at completion	8. Agency liaison

FIGURE 7-2
CM organization chart for $100 million airport expansion.

bility to a construction manager, to the designer, or it may perform such duties with its own employees.

Relationships between the owner's representative and the general contractor need not be adversarial and successful projects can fulfill the expectations of the owner while producing a fair profit for the contractor.

The next section reviews the construction project from the point of view of a single general contractor working on a fixed price contract.

Home Office Management Services

The major requirement for the contractor's home-office organization is to obtain new work and to manage and control current projects. The overall responsibility for profit and loss and for enhancing the contractor's position with the owner for possible repeat business cannot be delegated completely to the site superintendent or project manager.

Field projects fall into two distinct categories in the relationship between the job site and the home office. On smaller and regional projects the contractor may elect to

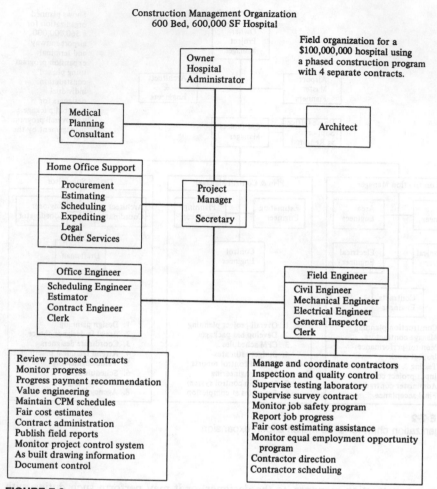

FIGURE 7-3
CM organization chart for $100 million hospital.

provide a substantial amount of detained support from the main or branch office. On larger and/or more remote job sites, the field office may be fully staffed to provide for all requirements except overall top management control.

For a project similar to the Mountaintown Warehouse, the following services might be performed in the head office:

1 Preparation of the craft payroll from field time cards, preparation of pay checks, federal and state tax returns and financial and cost accounting for the project.

2 Assignment of a project manager (often the estimator who bid the project) who will oversee all aspects of the business operations and monitor and manage the project control system.

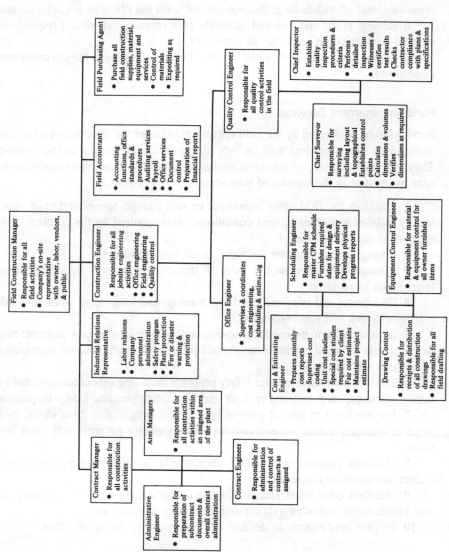

MAJOR INDUSTRIAL PROJECT
Construction Management Functional Organization

Field Construction Manager
- Responsible for all field activities
- Company's on-site representative with owner, labor, vendors, & public

Field Purchasing Agent
- Purchase all field construction supplies, material, equipment and services
- Control of materials
- Expediting as required

Contract Manager
- Responsible for all construction activities

Industrial Relations Representative
- Labor relations
- Company personnel administration
- Safety program
- Plant protection
- Fire or disaster warning & protection

Construction Engineer
- Responsible for all jobsite engineering activities
- Office engineering
- Field engineering
- Quality control

Field Accountant
- Accounting functions, office standards & procedures
- Auditing services
- Payroll
- Office services
- Document control
- Preparation of financial reports

Administrative Engineer
- Responsible for preparation of subcontract documents & overall contract administration

Area Managers
- Responsible for all construction activities within an assigned area of the plant

Office Engineer
- Supervises & coordinates cost engineering, scheduling & estimating

Quality Control Engineer
- Responsible for all quality control activities in the field

Contract Engineer
- Responsible for administration and control of contracts as assigned

Cost & Estimating Engineer
- Prepares monthly cost reports
- Supervises cost coding
- Unit cost studies
- Special cost studies required by client
- Fair cost estimates
- Maintains project estimate

Scheduling Engineer
- Responsible for master CPM schedule
- Furnishes required dates for design & equipment delivery
- Develops physical progress reports

Chief Surveyor
- Responsible for surveying including layout & topographical
- Establishes control points
- Calculates dimensions & volumes
- Verifies dimensions as required

Chief Inspector
- Establish quality inspection procedures & criteria
- Performs detailed inspection
- Witnesses & certifies test results
- Checks contractor compliance with plans & specifications

Drawing Control
- Responsible for receipts & distribution of all construction drawings
- Responsible for all field drafting

Equipment Control Engineer
- Responsible for material & equipment control for all owner furnished items

FIGURE 7-4

119

3 Keep contractor top management fully informed of financial and schedule performance and of current or expected problems and their proposed solutions.

4 Provide procurement services for major materials, subcontracts and equipment rentals. Smaller and incidental items may be purchased from the field office.

5 Review all new drawing revisions, prepare the estimated cost of change order proposals, and often handle price negotiations with the client or his representative.

6 Review profit and loss and schedule performance and initiate requests for additional compensation or claims as necessary.

7 Supply special assistance and oversight responsibilities to the job superintendent as required.

Field Management Services

Services to be performed by the field organization under the direction of a construction manager or superintendent will include all items not performed in the home-office. They will include many of the items performed by the professional construction manager as previously discussed and may include services that:

1 Establish the field office, construct or rent change houses and craft shops and supply utilities and other general conditions items required by the contract or for the support of the work force.

2 Manage and coordinate the work of all subcontractors.

3 Manage and supervise the work of the craft employees including hiring and termination as required.

4 Coordinate and integrate the daily, weekly, and long range work schedules for craft employees and subcontractors.

5 Initiate notice to proceed for subcontractors, prepare progress payment requests, perform tests required by the contract, and prepare contract closeout documents with owner and subcontractors.

6 Develop an overall job-site safety program including subcontractors, hold weekly safety meetings, keep required records and coordinate periodic safety inspections.

7 Prepare input to project control system including coding time cards, measurement of installed quantities, evaluation of subcontractor progress and other required tasks.

8 Maintain progress and record photographs to document progress, potential claims, accidents and other items.

9 Maintain good relationships with the owner's representative including submitting notifications and other submittals required by the contract.

10 Prepare field reports as defined and requested by the home-office.

TYPICAL ORGANIZATIONS FOR GENERAL CONTRACTING

As previously discussed for construction management, each project is somewhat unique. To illustrate the general contractor approach, additional organization charts are included here to show the differences in a manner comparable to Figures 4-7 and 7-4.

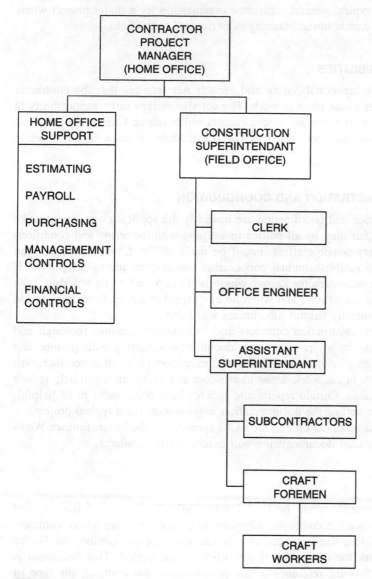

**MOUNTAINTOWN WAREHOUSE
CONTRACTOR ORGANIZATION**

CONTRACTOR
PROJECT
MANAGER
(HOME OFFICE)

HOME OFFICE
SUPPORT

ESTIMATING

PAYROLL

PURCHASING

MANAGEMEMNT
CONTROLS

FINANCIAL
CONTROLS

CONSTRUCTION
SUPERINTENDANT
(FIELD OFFICE)

CLERK

OFFICE ENGINEER

ASSISTANT
SUPERINTENDANT

SUBCONTRACTORS

CRAFT
FOREMEN

CRAFT
WORKERS

FIGURE 7-5
General contractor organization chart for Mountaintown warehouse.

$6 Million Dry Storage Warehouse

Figure 7-5 shows an organization chart for the example warehouse project as constructed by a general contractor. The contractor's main office is located in Mountaintown.

$200 Million Industrial Project

Figure 7-6 shows a typical general contractor organization for a major project where essentially all work except initial planning is performed in the field.

SAFETY RESPONSIBILITIES

The 1970 federal Occupational Safety and Health Act requires that the contractor furnish its employees a safe place to work. The act also assigns joint responsibility to the general contractor, to lower tier subcontractors and to others. Chapter 17 sets forth safety responsibilities and good practices for construction contractors, construction managers and others.

CONTRACT ADMINISTRATION AND COORDINATION

Contract administration and coordination are basically the application of businesslike common sense and fair play by all parties in keeping with the terms and conditions of the contract. Every possible effort should be made to assist each contractor and subcontractor and to establish mutual cooperation and respect among all parties. If each contractor and subcontractor knows where he is expected to be working, who will be working adjacent to him, and when he can expect others to finish their work, a cooperative and mutually helpful job climate will evolve.

Administration of construction contracts and subcontracts requires thorough and timely documentation by all parties. Such documentation, along with prompt and equitable adjudication of differences during the execution phase of a contract, will facilitate performance of the work, assist in closeout and acceptance, and help reduce the possibility of claims. Certain reports and records have been found to be helpful; those discussed here outline the documentation requirements for a typical project.

The following examples are chosen for a CM approach to the Mountaintown Warehouse. General contractor documentation will essentially be similar:

Notice to Proceed

The "Notice to Proceed" should be issued sufficiently in advance of the required starting date to provide the contractor adequate lead time. In cases where contracts are written to provide a fixed number of calendar days for completion, the Notice to Proceed will mark the beginning of the allotted time period. This document is generally a straightforward reference to the provisions of the contract, the type of work to be performed, and the date or dates that the work is to commence. Figure 7-7 shows a sample Notice to Proceed for the example warehouse project.

Change Orders

Change orders document changes from the original scope of the contract, confirm schedule revisions, and set forth other modifications; they are issued, whether or not the amount of compensation to be paid to the contractor will be affected. The change order

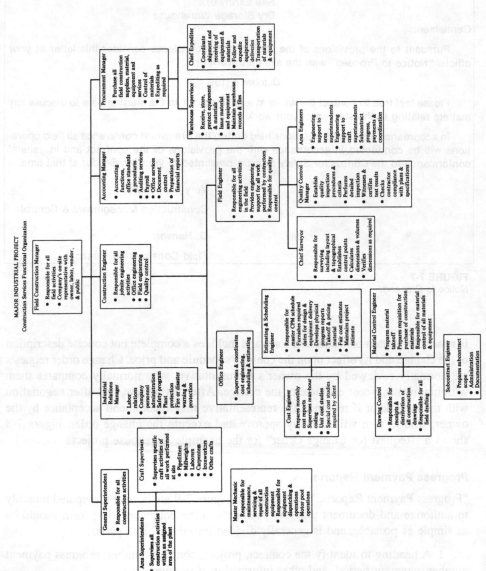

MAJOR INDUSTRIAL PROJECT
Construction Services Functional Organization

Field Construction Manager
- Responsible for all field activities
- Company's on-site representative with owner, labor, vendor, & public

Procurement Manager
- Purchase all field construction supplies, material, equipment and services
- Control of materials
- Expediting as required

Accounting Manager
- Accounting functions, office standards & procedures
- Auditing services
- Office services
- Document control
- Preparation of financial reports

Construction Engineer
- Responsible for all jobsite engineering activities
- Office engineering
- Field engineering
- Quality control

Warehouse Supervisor
- Receive, store, protect equipment & materials
- Issue material and equipment
- Maintain warehouse records & files

Chief Expediter
- Coordinate shipment and receiving of equipment & materials
- Follow up and expedite deliveries
- Transportation of materials & equipment

Field Engineer
- Responsible for all engineering activities in the field
- Provides engineering support to all work performed by contractors
- Responsible for quality control

Quality Control Manager
- Establish quality inspection procedures & criteria
- Performs detailed inspection
- Reviews & certifies test results
- Checks contractor compliance with plans & specifications

Area Engineers
- Engineering support to area superintendents
- Engineering support to craft superintendents
- Subcontract progress, payments & coordination

Chief Surveyor
- Responsible for surveying including layout & topographical
- Establishes control points
- Calculates dimensions & volumes
- Verifies dimensions as required

Estimating & Scheduling Engineer
- Responsible for master CPM schedule
- Furnishes required dates for design & equipment delivery
- Develop physical progress reports
- Takeoff & pricing material
- Fair cost estimates
- Maintains project estimate

Office Engineer
- Supervises & coordinates cost engineering, scheduling & estimating

Material Control Engineer
- Prepares material takeoff
- Prepares requisition for purchasing of construction materials
- Responsible for material control of all materials & equipment

Cost Engineer
- Prepares monthly cost reports
- Supervises man-hour coding
- Unit cost studies
- Special cost studies required by client

Drawing Control
- Responsible for receipts & distribution of all construction drawings
- Responsible for all field drafting

Subcontract Engineer
- Administration
- Documentation

Industrial Relations Manager
- Labor relations
- Company personnel administration
- Safety program
- Plant protection
- Fire or disaster warning & protection

General Superintendent
- Responsible for all construction activities

Area Superintendents
- Supervises all construction activities within an assigned area of the plant

Craft Supervisors
- Supervises specific craft activities of all work performed at site
  - Pipefitter
  - Millwrights
  - Laborers
  - Carpenters
  - Ironworkers
  - Other crafts

Master Mechanic
- Responsible for maintenance, servicing & repair of all construction equipment
- Responsible for equipment dispatching & operations
- Motor pool operations

FIGURE 7-6

Jensen Excavators   September 25, 199_
155 South First Street
Mountaintown, WestAmerica

Subject: Contract No. M-1
Site Earthwork
Dry Storage Warehouse

Gentlemen:

Pursuant to the provisions of the subject contract, please consider this letter as your official "Notice to Proceed" with the site earthwork on:

October 4, 199_

Please feel free to contact the writer at any time prior to the starting date to discuss any matters relating to the start-up of your work.

In accordance with our discussions held during the pre-award conference all field operations will be conducted in accordance with the provisions of the contract and in general conformance to the *Contractor Safety Bulletin* presented to your Mr. Snyder at that time.

Very truly yours,

Construction Management & Control

O. Hanson

Field Construction Manager

FIGURE 7-7
Notice to Proceed.

is generally written on a standard form and includes a complete but concise description of the change and its effect upon the contract schedule and price. Change order requests are normally reviewed by the owner's representative, who normally compares them with his own fair-cost estimate of the change. After his review, and after negotiation with the contractor if required, the representative may recommend acceptance by the owner, who in turn will formally approve and execute the change order. Figure 7-8 shows a "Request for Change Order" for the example warehouse project.

Progress Payment Reports

"Progress Payment Reports," such as that in Figure 7-9 are normally prepared monthly to authorize and document progress payments to the contractor. The form should be as simple as possible, and is generally divided into three basic parts:

1 A heading to identify the contract, project, contract number, progress payment number, payment period, and other information if required.

2 The body of the report organized to identify work completed during the period and to date. Columns are normally provided for recording quantities or percent complete, and dollar amounts earned for each line item. Provisions for deducting previous payments and the retainage are normally included so that the computation of the net amount earned during the period is shown.

CONTRACTOR:

ADDRESS:

REQUEST FOR CHANGE ORDER

TO:

No._____

Date_____

Project No.

Description:

Below listed are changes in our contract price. Until formal change order has been issued this request for change will be held in suspense status.

Application of increases (or decreases) in price of contract are as follows:

CODE REF.	DESCRIPTION OF WORK	INCREASE	DECREASE
	TOTAL THIS CHANGE REQUEST		
	NET CHANGE		
	CONTRACTOR'S FEE		
	TOTAL CHANGE		

REQUEST CONTRACT TIME EXTENSION (OR DECREASE) OF_____CALENDAR DAYS IF ABOVE ACCEPTED.

CONTRACTOR:

Approved_____ Date_____   By_____ Date_____

FIGURE 7-8
Request for change order.

Contractor: Henri Steel Company

Job No. _____ Location _____

Contract No. M-4
Progress Report No. 3-1 to 3-Final
Period 3-1 to 3-30

Cost account	Item no.	Description	Unit	Unit prices	Period Quantity or % complete	Period Amount	Cumulative Quantity or % complete	Cumulative Amount
M4.1	1	Fabricate & deliver structural steel	280T	1400.00			280T	392,000
M4.2	2	Fabricate & deliver joists	200T	808.00			200T	161,600
M4.3	3	Erect structural steel	480T	260.00			480T	124,800
M4.4	4	Furnish & erect normal roof deck	165,000SF	1.10	20,000	11,000	165,000SF	181,600
C.O.1	5	Substitute tube columns	L.S.	(5800)			100%	(5,800)

Total estimated contract price	$854,200	Totals	11,000	854,200
		Retained 10%	1,100	85,420
		Difference	9,900	768,780
		Less previous payments		758,880
		Net amount this payment		$9,900

Prepared by: _____ Date: _____
Checked by: _____ Date: _____
Approved by: _____ Date: _____
Approved by: _____ Date: _____

FIGURE 7-9
Contract progress payment order.

3 Information summarized at the bottom of the form may include total estimated contract price, signatures of the person who prepared the report and of the contractor, approvals, and other information.

Final Acceptance

Basic documentation for acceptance of contract work depends somewhat upon state law and individual preference, but will generally include the following:

Notice of Completion This form states that all work under the contract is complete, and that it is being processed for acceptance.

Completion and Acceptance Certificate This document certifies that all work is complete in accordance with plans and specifications. Normally signed by the trade contractor and the professional construction manager (or general contractor if applicable), it serves as documentation to the owner that the work is complete and ready for acceptance based upon the contract closeout procedures and requirements. The owner's signature will also be desirable for protection to the construction manager or general contractor. See Figure 7-10 for an example.

Release and Waiver of Lien An executed release and waiver of lien form, or other certification from the contractor, is normally required prior to releasing final payment. Current claims are normally exempted from the release.

Field Transmittal Memorandum

An FTM is useful in all field contract correspondence, transmittals, instructions, or other communications with the contractor. The FTMs should be consecutively numbered for ease in filing and documentation. Acknowledgment of receipt by the contractor can be required if desired. Figure 7-11 shows one form of FTM that has been successful.

Contract Status Reports

This report, normally issued monthly, can be helpful to show the current status of each awarded contract. Information contained in the report will include the original contract value, number and amounts of change orders in process, the revised estimated contract value at completion, and estimated completion dates. Figure 7-12 shows a sample "Contract Status Report." A general contractor's report might similarly show the status of all subcontractors.

Contract Logs

A separate contract log for each contract documents daily the work performed, conditions affecting progress of the work, delays or interferences, or other items of current interest or possible future significance. The contract log should be in a bound book

Contract Completion and Acceptance Certificate

This is to advise that the work covered by Contract No. _____ and all Change

Orders numbered _____, was completed as specified

below:

Description of Work	Account No.	Completion Date

Contractor's Certification

This is to certify that the work described by the contract has been fully completed in accordance with the Terms, Conditions, Plans and Specifications set forth in said Contract.

Construction Manager's Certification

This is to certify that the work as covered by this contract has been inspected under our supervision, and to the best of our knowledge and belief it has been completed according to the said contract and Plans and Specifications specified therein and is hereby recommended for acceptance subject to the Terms and Conditions of the Contract.

_____
(Contractor)

By _____

Title _____

Date _____

_____
(Construction Manager)

By _____

Title _____

Date _____

Owner's Acceptance

Based upon the above certification, the work is hereby accepted subject to the Terms and Conditions of the Contract.

_____
(Owner)

By _____

Title _____   Date _____

FIGURE 7-10
Contract completion and acceptance certificate.

with consecutively numbered pages, and entries should be made in ink. Particular emphasis should be placed on recording information that might prove helpful in evaluating or developing potential claims and in protecting the owner or contractor from unwarranted claims. Figure 7-13 shows a typical entry by a construction manager.

F.T.M. No. _____1_____

FIELD TRANSMITTAL MEMORANDUM

CONTRACTOR	CONTRACT NO.	Mo.	Day	Yr.
Jensen Excavators	M1	10	1	

Attached for your use as set forth in the Contract are four (4) sets of the Contract Specifications and four (4) sets of the Drawings listed therein. Please review these documents as stipulated under Article 16-A of the Terms and Conditions.

Also attached is one (1) reproducible, Contractor's Performance Schedule form which is to be completed by you in accordance with Article 15 of the Terms and Conditions and submitted for approval not later than October 8, 1976.

SAMPLE

Typical wording for transmitting the drawings and specifications on which the Contract is based. The number of copies of each should not be less than the quantity listed in the specification.

In the event it is desirable to have the Contractor use his own Performance Schedule Form then the last paragraph would be reworded accordingly.

(courtesy of R. C. Wilson who pioneered the adoption of this format)

RECEIVED FOR CONTRACTOR	Construction Manager
BY	BY
TITLE	TITLE

FIGURE 7-11
Field transmittal memorandum.

Dry Storage Warehouse Contract Status Report

Contract (No. and Description)	Contractor	Original Contract Price	Change Orders: Issued	Change Orders: Value	Change Orders: Est. Value of Pending	Revised Estimated Contract Value	% Com-plete	Completion Date: Contract Completion Date	Completion Date: Estimated Completion Date
1. Site earthwork	Jensen Excavators	206,000	1	3,000	—	209,000	100	12/31	12/22
2. Structural & slab concrete	Hulseman Construction	640,000	2	50,000	—	690,000	100	4/30	4/30
3. Interior special slabs	Palmer Floors	560,000	—		—	560,000	100	4/30	4/23
4. Structural steel & deck	Henri Steel	860,000	1	(5,800)		854,200	100	2/28	3/24
5. Precast double tees	Gurecki Precast	560,000	—		—	560,000	100	3/31	4/22
6. Plumbing, heating, mechanical	Orne Mechanical	440,000	3	38,000	—	478,000	100	5/31	5/29
7. Fire protection	Morschauser Inc.	350,000	—		—	350,000	100	5/31	5/21
8. Electrical	Jones Electrical	380,000	2	18,000		398,000	100	5/31	5/7
9. Roofing	Rocky Roofing	222,000	1	12,000		234,000	100	4/15	4/22
10. Building finish	Finsand Construction	604,000	4	56,000		660,000	100	5/31	5/31
Total		4,822,000	14	171,200	—	4,993,200			

Prepared by: Betty Willis
Construction Engineer

FIGURE 7-12
Contract status report

October 4, 199_

Jensen excavators moved on the site this date. The following equipment was delivered:

1 ea CAT No 14 Motor Patrol
1 ea CAT No D-8 Tractor
2 ea CAT 613B Scrapers
1 ea Water Wagon
1 ea CAT 815 Compactor

Superintendent Snyder advises that work would begin tomorrow with a total of 6 men.

Typical example of handwritten entry in log for site excavation contract. All important starts, completions, and other significant items which could be of importance in the event of future claims or disagreements should be recorded.

FIGURE 7-13
Typical log entry.

Contract Force Report

If schedules are to be accomplished, sufficient manpower must be provided by the individual contractors. See Figure 7-14 for a sample "Contract Force Report" useful for recording and appraising contractor manpower. It also can serve as an alternate evaluation of progress where a fair-cost estimate, including labor units, has been prepared. Figure 7-15 shows a weekly field activity report by a general contractor.

QUALITY CONTROL SERVICES

Some organizations believe that the owner can never fully delegate quality assurance and control responsibilities. However, some or all of these duties are conventionally delegated to or administered by the owner's representative.

Professional Construction Management

Each project will require a sound plan for inspection, testing-laboratory services, and basic survey control. On very large projects, all three services can be supplied by the professional construction manager. On smaller ones, the manager may perform some of the services to the extent that his staff will allow, and contract for the balance on a part-time or as-required basis. Normally outside professional service contracts for inspection, testing, and surveying should first be scoped jointly with the designer, and then a joint recommendation for award should be made to the owner.

The manager, owner, and designer should all be aware of the inspection, testing, and surveying responsibilities assumed by each member of the construction management team. The professional construction manager must recognize the professional responsibility of the designer. Conversely, the designer must recognize the responsibilities delegated to the manager by the owner.

On large professional construction management projects, inspection can be fully separated from contract management and administration. On smaller ones, the manager's limited staff can perform both functions.

INTEROFFICE MEMORANDUM

TO   Client Project Manager
AT   San Francisco

DATE   November 29, 199_

FROM

COPIES TO   Home Office

AT   Mountaintown, West America

JOB NO.

WEEKLY FORCE REPORT FOR PERIOD ENDING: November 28, 199_

SUBJECT

Company	Mon	Tue	Wed	Thu	Fri	Sat	Sun	Week Total	Cum Total	Estimate	%
Jensen excavators		3	3	3	3			12	117	184	64
Hulseman construction	1	2	2	2	4			11	207	966	21
Palmer floors	2	9	9	10	10			40	119	916	13
Orne mechanical	6	6	6	7	7			32	64	531	12
George sprinkler			3	3	3			9	9	403	2
Others										2,688	0
Total man-weeks	9	20	23	25	27			104	516	5,688	9
Total man-hours								832	4,128	45,500	9

Note: At this early stage, all of the contractors have not started work and some contracts have
not yet been awarded. For illustration, these have been grouped under others with an
estimated man-days. Note that if initial man-hour estimates are accurate, the job is
about 9% complete. This figure can be compared to the physical progress calculations
(Fig. 8.9 and 8.10). Discrepancies should be further explored.

FIGURE 7-14
CM weekly force report.

Testing laboratories can be hired by the owner, architect, or professional construction manager. In any event, a general scope should be developed and one member of the team must be selected to monitor and manage the contract. Reports should be distributed to all members of the team.

For survey control, the professional construction manager must at least provide basic line and grade for all contractors. In addition, and depending upon area practice, detailed construction layout may be provided for certain contracts, or check surveys may be performed for critical items. On large projects, a number of survey crews can be fully utilized by the professional construction manager, while on small and

GENERAL CONTRACTOR

WEEKLY FIELD ACTIVITY REPORT

JOB NO._____

REPORT NO. _____

WEEK ENDING_____

SUPERINTENDANT

JOB/LOCATION				
VISITORS	SUMMARY FORCE REPORT			
		CONTRACTOR	SUBS	TOTAL
	SUPERVISION			
	ENGINEERING			
	CLERICAL			
	CRAFT			
WEATHER	SUBCONTRACTOR			
	SUBCONTRACTOR			
	SUBCONTRACTOR			
	SUBCONTRACTOR			
	SUBCONTRACTOR			
	SUBCONTRACTOR			
	TOTAL			
PROGRESS & DELAY COMMENTS				
COSTS, DRAWING REVISIONS, PRODUCTIVITY COMMENTS				
OTHER REMARKS (ATTATCH ADDITIONAL PAGES AS REQUIRED)				

FIGURE 7-15
General contractor weekly force report.

medium-sized projects all surveying can be performed on a part-time basis through a professional services contract. The preferable solution will depend upon the job and the area practice as well as on overall economy to the owner.

General Contractor

Some government agencies are delegating quality control to the contractors on fixed-price contracts with only an audit or oversight role assigned to the owner or a representative. However, the conventional approach on fixed-price contracts is for the owner to retain inspection responsibilities either directly or through a designated representative.

On design-construct projects constructed on a reimbursable-cost basis, it is common practice for the owner to delegate most or all of the quality responsibilities to the design-constructor. Conversely, on a lump-sum or guaranteed-maximum-price design-construct contract, the owner usually retains a major involvement in this area.

START-UP AND FINAL CLOSEOUT

System Validation, Testing, and Start-Up

The professional construction manager or other owner's representative may be requested to assist the owner in system validation, testing, and start-up. The representative is often charged with the responsibility for performance testing of individual pieces of equipment or entire systems as required by applicable codes, drawings, and specifications. The startup manager often prepares a detail plan for unit and system start-up operation involving contractor, engineer, manufacturer's representative, and owner personnel. Development of a tagging procedure on equipment to ensure safety during initial operation is especially important.

Final Closeout

In addition to final acceptance documentation as previously discussed, the professional construction manager or general contractor may be responsible for substantial additional duties, either from the job site or from the main office after fieldwork has been completed. Some of these duties will include furnishing or obtaining guarantees, operating information, spare parts, instruction manuals, as-built drawings, bonds, maintenance agreements, inspection certifications, and other documents required under the contracts. Often an inspection is scheduled prior to the expiration of performance guarantees, to ensure that facilities are operating as specified. The submission of a final report showing documented project costs along with schedule performance can be of great value to the owner, designer, and CM firm or general contractor if sufficient information and explanations are included to form a planning base for other similar projects to be constructed at a future date.

LEGAL CONSIDERATIONS

For a more comprehensive discussion of the legal aspects of construction, see Chapter 20 (Claims, Liability and Dispute Resolution).

Claims and Backcharges (Owner's Consideration)

There are potential claims and backcharges in every construction project. In any contractual relationship, situations can develop whereby any or all parties to a contract may believe that they have legitimate claims.

Field personnel must be thoroughly familiar with the contract provisions, rights of the parties, and concepts of contractual relations. In addition, they must be alert to circumstances which may serve as the basis for claims, such as changes in the work, changes in conditions affecting the work, failure of a party to perform contractual obligations, or improper or inadequate performance.

Claims can be much more readily adjudicated when a contract has been thoroughly documented throughout its performance with daily logs indicating events and conditions affecting the work. Proper documentation will include, but not be limited to, the following:

An independent evaluation of the merits of each item outlined in the claim or backcharge, including applicable references to contract provisions; plus an estimate of applicable costs, including overhead and markup as provided by the contract

Minutes of any negotiation meetings, including additional facts or information that may develop

Prompt and timely notification to the other party or parties in accordance with contract provisions

Early notification to the owner and, if indicated, an early request for legal assistance and advice

Written approval of the claim settlement by the owner

Legal Aspects (Manager's Consideration)

Construction, like other businesses, is becoming increasingly legalistic. Architects have replaced the terms "inspection" or "supervision" by "observation" in an attempt to minimize responsibility under the law and thus to minimize insurance rates. Contractors and design firms are facing million-dollar suits for on-the-job accidents and deaths, as well as for design or construction deficiencies.

The position of the construction manager still remains somewhat unclear, and each contract will differ in its legal responsibilities. Basically, the usual position of the professional construction manager is that it has hired out his expertise as agent to the owner, and the owner should be responsible for business risks incurred by the project that are beyond the manager's control. The manager represents a reasonable degree of professional competence in acting on behalf of the owner, and it should be responsible to the owner for this representation. However, the prudent manager will protect itself against possible litigation and exposure by utilizing competent legal advice in developing the professional construction management services agreement and by maintaining a comprehensive liability insurance program.

Legal Aspects (Contractor's Viewpoint)

The construction contractor normally operates in a very competitive environment. Faced with tough competition, most contractors do their best to submit the lowest possible bid in order to win the award. Where specification items or drawings are subject to varying interpretation, the contractor often disagrees strongly with the owner's interpretation. In other cases the contractor may have underestimated or omitted work items either through his own action or through unclear plans and specifications. In either case the contractor will normally search for reasons why the owner should bear some or all of the responsibility. If the owner disagrees, too often initial polarization keeps the parties from reaching a reasonable compromise at an early date. In this situation, mediation is beginning to have some initial success in settling such disputes.

SUMMARY

Once the project's contracts have been awarded , work must be completed on schedule and in accordance with plans and specifications if success is to be achieved by all parties. The field construction manager and his staff are the ones to ensure that the objectives of the work plan are realized. To do this, they need sufficient authority to carry out these objectives; but these individuals must also be fully accountable to the home office for this delegation.

Should the project control system indicate that primary objectives are not being achieved, replanning is needed. Either initial goals must be modified or changes in contractor programs must be implemented.

Home-office personnel associated with planning generally stop gathering additional information when the "buy-out" phase is completed, and the field construction manager and his staff will have generally surpassed the home-office personnel's knowledge about the job by the time the job is well underway. Ideal management delegation for the construction phase will encourage and permit the field construction manager to take action when he is convinced of the right moves. When he is unsure, the home office should respond to his request for help so that a mutually responsible relationship may be provided.

After the contracts or subcontracts are awarded, the emphasis shifts to the field. Home-office personnel who insist upon retaining the authority to make detailed job decisions will find that problems which should be solved at the job level assume major proportions in the home office. On the other hand, home-office managers who fail to exercise overall control may one day discover that the objectives of the job can no longer be met.

CHAPTER 8

APPLICATION OF CONTROLS

Previous chapters discussed preparation of the overall plan for the project as well as its implementation. Throughout the project, the control system quantitatively measures actual performance against the plan and acts as an early warning system to diagnose major problems while management action can still be effective in achieving solutions. Development and application of a practical control system to measure progress and costs is one of the the the most important contributions of the professional construction manager and is critical to the success of general and specialty contractors.

Many of the management control tools utilized by a construction manager can also be used by the individual project contractors to manage their own work. Examples of a narrower scope and additional detail and other information used by the concrete contractor are presented.

The final portion of the chapter sets forth some of the different methods that a single general contractor might utilize on a fixed price contract with a number of subcontractors.

MANAGEMENT-LEVEL REPORTING BY CONSTRUCTION MANAGER

Management-level reporting must provide a straightforward statement of the work accomplished, predict future accomplishment in terms of the project cost and schedule, and measure actual accomplishments against goals set forth in the plan. It should also review current and potential problems and indicate management action underway to overcome the effects of the problems. These requirements are similar on projects ranging in size from $1 million to $100 million and more. They are also relatively independent of the sophistication of the techniques that measure accomplishments; these vary with project size and complexity.

A comprehensive "Monthly Progress Report" can convey this essential information. The contents of a sample report are as follows:

1 Summary of project status
2 Procurement status
3 Construction status
4 Schedule status
5 Cost report summary

Each of these items will be described briefly in the sections that follow.

Summary of Project Status

This item represents a short, overall summary of project status. It may contain a brief narrative description of the status of each major phase, provide quantitative information such as the physical percentage complete compared with scheduled completion, and forecast "at-completion" costs against budget.

Procurement Status

This item reviews contracts awarded during the period, contracts currently out for bid, and other significant information. A simple bar chart showing actual procurement status and contract awards compared with the original plan is often helpful.

Construction Status

This unit of the Progress Report should provide a description of work accomplished during the period, significant work to be accomplished in the next period, and a discussion of major problems, with solutions or proposed solutions. Quantitative information is more significant than general discussion.

Schedule Status

This item should contain the summary control schedules by contract and by facility, showing actual progress compared to early- and late-start schedules. Where contracts or facilities are behind schedule or are slipping, an explanation of the problems and the indicated solution or measures being adopted to solve the problems should be included.

Cost Report Summary

This summary should show actual recorded costs, committed costs, and estimated costs-to-complete. It should compare "at-completion" costs with project budgets and identify and explain changes from the previous report. An evaluated contingency should be included so that an overall estimate of actual costs at completion is provided. Professional construction management costs should appear in a similar manner. A summary of value-engineering savings to date and new items added during the period can be included.

OVERALL COST CONTROLS BY CONSTRUCTION MANAGER

Overall cost controls should be integrated with schedule controls. Computer-based systems with common data files facilitate this integration. Overall cost controls, designed to measure project status against budget, include the following:

Preliminary estimates
Fair-cost estimates
Definitive estimate
Cost report summary
Value-engineering studies
Value-engineering status
Other significant data

Each of these items is briefly introduced here with examples from the Mountaintown Warehouse Project, and will be discussed in greater detail in Part 3, Project Control Techniques for Managing Construction.

Preliminary Estimates

Preliminary estimates assist the overall cost-control program by serving as the first check against the budget, and by indicating cost overruns early enough for the project team to review the design for possible alternates. Since preliminary estimates are made prior to the completion of detail drawings, the margin for error is usually greater than for fair-cost estimates. Consequently, a larger contingency should be applied; this will vary with the amount of design information available and the extent of cost information obtainable from similar projects.

For a phased construction program, it is especially important to prepare preliminary estimates by contract package. By comparing actual contract awards with the preliminary estimate, a running total of the current status of the project is available. Indicated overruns can stimulate revision of the criteria for later work packages in order to preserve overall budgets.

Fair-Cost Estimates

Fair-cost estimates are best prepared from the actual bid documents provided to the bidders. Whenever possible, it is helpful to complete the fair-cost estimate well before receiving bids so that any discrepancies in plans and specifications, duplications in scope, and possible value-engineering alternates suggested by the estimator can be communicated to the bidders via addenda before bids are received. In the event of a major difference from the preliminary estimates, this lead time is always helpful so that an intensive review of possible alternates can be started.

Fair-cost estimates represent the professional construction manager's appraisal of the fair value of the bid package to the owner. Local conditions, such as materials prices, wage rates, labor productivity, and anticipated competition, are important in achieving a reasonable estimate for the area.

Fair-cost estimates in an integrated cost-progress control system will also develop significant additional information for upgrading the usefulness and accuracy of the schedule- and progress-control portion of the overall control system. Some of these items include the following:

An estimate of total man-hours of field effort required

Estimated quantities for major items

An estimate of reasonable unit costs for various components of the work

Information for allocation of contract costs for owner capitalization and tax considerations

In the event actual bids differ significantly from the fair-cost estimate, the manager can often meet with the low bidder and compare quantities and scope. Many times this comparison can pinpoint the reason for the discrepancy. Decisions regarding award, modification, or rejection of bids will be greatly assisted if the professional construction manager has prepared a careful fair-cost estimate, separately itemizing labor, materials, and equipment costs in a manner similar to that of the bidding contractor. This estimate can be of great value to the field construction manager in scheduling work, in reviewing change order requests, and in determining manpower requirements. Figure 8-1 shows a fair-cost estimate summary for Easyway's Mountaintown warehouse. Details of the estimate are included as Appendix A.

FIGURE 8-1
Fair cost estimate summary (Mountaintown Warehouse).

Cont. No.	Contracts	Labor Hours	Total Direct Cost	Overhead & Fee	Total
1.	Earthwork—site	1470	197,200	23,600	220,800
2.	Structural & yard concr.	7730	580,800	69,600	650,400
3.	Special slabs	7330	491,300	59,000	550,300
4.	Structural steel	3770	799,000	95,800	894,800
5.	Double tee walls	2640	497,400	59,600	557,000
6.	Mechanical—HVAC	4250	441,200	53,000	494,200
7.	Fire protection	3220	297,600	35,600	333,200
8.	Electrical	3470	354,000	42,000	396,000
9.	Roofing	3150	218,000	26,000	244,000
10.	Building finish	8470	566,300	68,000	634,300
	Total	45,500	4,442,800	532,200	4,975,000
	Contingency @ 5%				250,000
	Total estimate		151,600SF @ 34.47/SF = 5,225,000		

Estimate Criteria

1. Labor cost including fringe benefits, payroll taxes, workmen's compensation insurance, public liability & property damage insurance etc. is $30/HR.

2. Estimates are based upon drawings & specifications.

3. Construction management, survey, testing laboratory, & other owner's costs are not included.

4. Estimates are based upon an 8-month overall construction schedule.

5. Average manpower 34, estimated peak 68.

By comparing actual man-hours (or man-days) required through computation from the project force report, a measure of local productivity can be developed and compared with the manager's estimate; this has numerous project uses as well as long-range benefits to the manager and to the owner. Table 8-1 shows a manpower summary comparing actual manpower requirements with the estimated man-hours from the fair-cost estimates.

Definitive Estimates

Definitive estimates fix the anticipated cost of the project with little margin for error. As contracts are bid on a phased construction program, the overall estimated cost becomes more certain. When 90 percent of the contracts have been awarded, less contingency is required than at the 50 percent level. When 100 percent of the contracts have been awarded, contingency is generally limited to providing for plan changes due to interference or error, for omissions or conflicts, or for other business risks inherent in the project.

Several numerical and statistical methods have been proposed and applied for forecasting total cost underruns or overruns at various contract award percentages. However, each project is different. On some, a definitive estimate can be prepared with reasonable accuracy when 50 percent of the contracts have been awarded. On others, accurate definitive estimates must wait until almost all contracts have been awarded. The manager's knowledge of the area construction and bidding practices,

TABLE 8-1
PRODUCTIVITY SUMMARY—MOUNTAINTOWN WAREHOUSE

Cont. No.	Contract	Man-hrs from fair cost estimate	Man-hrs from change orders	Est. total man-hrs	Actual man-hrs	% Productivity
1	Site Earthwork	1,470	50	1,520	1,288	118
2	Structural & Yard Concrete	7,730	560	8,290	8,272	100
3	Special Slabs	7,330	—	7,330	7,232	101
4	Structural Steel	3,770	—	3,770	4,080	92
5	Double-Tee Walls	2,640	—	2,640	2,040	129
6	Mechanical HVAC	4,250	300	4,550	4,136	110
7	Fire Protection	3,220	—	3,220	2,736	118
8	Electrical	3,470	150	3,620	3,952	92
9	Roofing	3,150	100	3,250	2,808	116
10	Building Finish	8,470	600	9,070	8,688	104
	Total	45,500	1,760	47,260	45,232	104

Notes:
1. Productivity is defined as estimated man-hours from fair-cost estimates plus estimated man-hours for change orders divided by actual man-hours as measured from the force report.
2. Measured man-hours for special slabs has been increased by 1000 to allow for scheduled cement finisher overtime during topping phase.
3. See Figure 11-9 for manpower summary and actual manpower expended, as computed from the project force report, Figure 7-14.

area workload, design considerations, and estimating practices will indicate the point at which a reliable definitive estimate can be prepared.

Cost Report Summaries

Cost report summaries describe the actual and forecast status of the project; they generally commence with the preliminary estimate and end when the project is complete and all claims, if any, have been settled. In a normal program, cost reports showing estimated cost-at-completion can be prepared from the committed cost plus estimated costs-to-complete for the various contracts involved. Some owners prefer that the professional construction manager perform additional accounting for recorded costs when these are paid as progress payments. Others prefer to handle this phase themselves and are interested in the manager's report only to cover total commitments to date and estimated costs-to-complete. Similarly, some owners require continual cash-flow projections to accompany the cost reports, while others prefer to handle this themselves.

Figure 8-2 provides a sample summary cost report for the example project in its early stages, when only a few of the contracts have been awarded. The evaluated contingency reflects this early stage.

Figure 8-3 shows a similar report after nine contracts have been awarded and a fair-cost estimate has been completed for the balance of the work. The evaluated contingency reflects the definitive nature of this state of the project.

Figure 8-4 gives a comparison, at the time of preparation of Figure 8-2's summary cost report, between preliminary estimates, fair-cost estimates, and contract awards to date for the example warehouse project. Note that fair-cost estimates have been prepared for several packages for which bids have not yet been received.

Figure 8-5 shows the detailed evaluation of contingency for the early report. Figure 8-6 shows a similar evaluation after almost all contracts have been awarded.

Value Engineering Studies

Value-engineering studies help in determining the most economical approach prior to detailed design. If best results are to be obtained, value engineering must involve a partnership where the professional construction manager, designer, and owner all work together. Application of construction cost knowledge during design, and consideration of alternates proposed by the team or by the bidders themselves, can be of great benefit to the owner.

Figure 5-2 in Chapter 5 showed a simplified value-engineering study of alternate wall systems for the warehouse. The study was a joint effort by the professional construction manager and the architect, and resulted in the use of double-tees for the project.

Value Engineering Status

A report showing value-engineering savings approved to date by the owner can keep the results of the program clearly in focus, and can be of long-term benefit to all parties

Control account number	Description	Original commitment	Approved changes	Cumulative Total recorded & committed	Estimated Cost		Prelim. budget as of estimate	(under) or over budget
					To complete	At completion		
1	Site earthwork	206,000		206,000	3,000	209,000	208,000	1,000
2	Foundation & slab concrete	640,000	30,000	670,000	6,000	676,000	616,000	60,000
3	Special floors—interior	560,000		560,000		560,000	548,000	12,000
4	Structured steel	860,000	(5,800)	854,200		854,200	1,048,000	193,800
5	Precast walls	560,000		560,000		560,000	480,000	80,000
6	Plumbing & HVAC				594,200	594,200	380,000	214,200
7	Fire protection				333,200	333,200	300,000	33,200
8	Electrical				330,000	330,000	330,000	
9	Roofing				244,000	244,000	298,000	(54,000)
10	Building finish				590,000	590,000	590,000	
	Total direct cost	2,826,000	24,200	2,850,200	2,100,400	4,950,000	4,798,000	152,600
11	Field general conditions			28,400	143,600	172,000	172,000	
12	Home office fixed fee	200,000		200,000		200,000	200,000	
	Total indirect cost	200,000	28,400	228,400	143,600	372,000	372,000	
	Estimated total cost	3,026,000	52,600	3,078,600	2,244,000	5,322,600	5,170,000	152,600
	Contingency				257,400	257,400	482,000	(224,600)
	Total			3,078,600	2,501,400	5,480,000	5,652,000	72,000

Note: The control budget is based upon the preliminary estimate. At completion estimates are based upon contract awards (5), fair cost estimates (3), and preliminary estimates (2).

FIGURE 8-2
Summary cost report—early stages (Mountaintown Warehouse).

143

Control account number	Description	Original commitment	Approved changes	Cumulative Total recorded & committed	Estimated Cost		Prelim. budget as of estimate	(under) or over budget
					To complete	At completion		
1	Site earthwork	206,000	3,000	209,000		209,000	208,000	1,000
2	Foundation & slab concrete	640,000	50,000	690,000		690,000	616,000	74,000
3	Special floors—interior	560,000		560,000		560,000	548,000	12,000
4	Structural steel	860,000	(5,800)	854,200		854,200	1,048,000	193,800
5	Precast walls	560,000		560,000		560,000	480,000	80,000
6	Plumbing & HVAC	440,000	38,000	478,000		478,000	380,000	98,000
7	Fire protection	350,000		350,000		350,000	300,000	50,000
8	Electrical	380,000	18,000	398,000		398,000	330,000	68,000
9	Roofing	222,000	12,000	234,000		234,000	298,000	(64,000)
10	Building finish	604,000	56,000	660,000		660,000	590,000	70,000
	Total direct cost	4,822,000	171,200	4,993,200		4,993,200	4,798,000	195,200
11	Field general conditions			157,000	21,800	178,800	172,000	6,800
12	Home office fixed fee	200,000		200,000		200,000	200,000	
	Total indirect cost			357,000	21,800	378,800	372,000	6,800
	Estimated total cost			5,350,200	21,800	5,372,000	5,170,000	202,000
	Contingency				52,000	52,000	482,000	(430,000)
	Total			5,350,200	73,800	5,424,000	5,652,000	(228,000)

Note: See Figure 7-12 for contract status report and Figure 8-6 for contingency evaluation.

FIGURE 8-3
Summary cost report—late stages (Mountaintown Warehouse).

Contr. no.	Contract package	Contract price	Prelim. estimate Fig. 4.5	Fair cost estimate Fig. 8.1	Over (under) pre. est.	Over (under) F.C. est.
1.	Earthwork	206,000	208,000	220,800	(2,000)	(14,800)
2.	Concrete	640,000	616,000	650,400	24,000	(10,400)
3.	Special slabs	560,000	548,000	550,300	12,000	9,700
4.	Struct. steel	860,000	1,048,000	894,800	(188,000)	(34,800)
5.	TT walls	560,000	480,000	557,000	80,000	3,000
6.	Mechanical		380,000	494,200		
7.	Fire protect.		300,000	333,200		
8.	Electrical		330,000	—		
9.	Roofing		298,000	244,000		
10.	Building finish		590,000			
	Estimated		4,798,000			

	To Date	Cumul	Cumul	Cumul	Cumul	Cumul
1.	Earthwork	206,000	208,000	220,800	(2,000)	(14,800)
2.	Concrete	846,000	824,000	871,200	22,000	(25,200)
3.	Special slabs	1,406,000	1,372,000	1,421,500	34,000	(15,500)
4.	Struct. steel	2,266,000	2,420,000	2,316,300	(154,000)	(50,300)
5.	TT walls	2,826,000	2,900,000	2,873,300	(74,000)	(83,300)
6.	Mechanical		3,280,000			
7.	Fire protect.		3,580,000			
8.	Electrical		3,910,000			
9.	Roofing		4,208,000			
10.	Building finish		4,798,000			

Notes
1. As additional contracts are awarded, the missing blanks are filled in and the final estimated cost becomes more certain.

2. This exhibit compares actual contract price with preliminary and fair cost estimates. See Figure 8-2 summary cost report for estimated cost at completion.

FIGURE 8-4
Cost comparison before contingency (Mountaintown Warehouse).

on future projects. Figure 8-7 shows the approved value-engineering savings achieved on the warehouse project; it features an architect-manager-owner value-engineering program.

Field Cost Controls

Overall cost controls can be developed and administered either at the job site or in the home office, depending upon the particular project. However, evaluation of plan changes, claims, and other change-order requirements can often be done better at the job site.

Description	Amount	Factor	Evaluation
Open commitments	2,850,200	3.0%	85,600
Definitive estimates	1,071,400	7.5%	80,400
Preliminary estimates	920,000	10.0%	92,000
Estimated total			258,000
			Say 257,400 (Rounding Figure 8-2)

FIGURE 8-5
Contingency evaluation—early stages (Mountaintown Warehouse).

Description	Amount	Factor	Evaluation
General conditions	1 month	18,000	18,000
Building finish claim	44,000	50.0%	22,000
Electrical claim	24,000	50.0%	12,000
Estimated total			52,000

FIGURE 8-6
Contingency evaluation—late stages
(Mountaintown Warehouse).

The professional construction management firm's field construction manager is also the representative of the owner. When drawings are changed, an increase or a decrease in the contract price may be indicated. The adjustment should be fair to both the contractor and the owner.

In one helpful technique, the contract terms require an itemized breakdown by labor and materials to be supplied by the contractor for all changes, including applicable quantities. Preparation of an independent fair-cost estimate based upon an independent quantity takeoff by the field construction manager can often pinpoint differences, lead to reasonable agreements, and prevent later disputes.

Whenever possible, the price for a modification should be settled before the work is performed. However, in many cases the schedule demands immediate performance. Here, the contractor can be directed to proceed on a time-and-materials (force-account) basis, and a lump-sum change can be requested and negotiated soon thereafter. If agreement cannot be reached, the work can continue on force account in accord with contract compensation terms.

Probably the most troublesome changes to adjudicate involve work modifications that eliminate certain items and replace them with other more- or less-complicated items. Here, performance of the work on a time-and-materials basis is not possible unless a credit is negotiated for work not performed.

An important item of field cost control is scheduling contractors to avoid interference, delays, and other detrimental effects of one contractor's operations upon another's. In a professional construction management program, the owner through his manager is largely responsible for the coordination involved among site contractors.

1. Use of Double Tee wall panels in lieu of tilt-up as a result of predesign Value Engineering Study.

$ 33,000

2. Modification of specifications to include 6 in. stripping in lieu of 12 in. called for in standard earthwork specifications as a result of the site visit.

$ 25,000

3. Use of expansive cement in lieu of Portland cement based on alternate quotations for the base slab for interior special floor slabs. This savings is a result of eliminating expansion and control joints.

$ 14,000

4. Use of tube columns in lieu of pipe or WF shapes based upon alternate bids.

$ 5,800

5. Use of asbestos-cement pipe in straight runs for fire lines and domestic water mains based upon Value Engineering Study.

$ 10,400

6. Alternate Proposal by Electrical Contractor for use of XHHW wire submitted with the bid.

$ 2,400

7. Use of special patented soil pipe eliminating hub and spigot joints based upon alternate bid.

$ 1,800

Total Project Savings .... $ 94,000

$$\frac{\text{Total Project Savings}}{\text{Approximate Project Cost}} = \frac{94,000}{5,000,000} = 1.9\%$$

FIGURE 8-7
Value engineering savings (Mountain Warehouse Project).

SCHEDULE AND PROGRESS CONTROLS BY CONSTRUCTION MANAGER

Control schedules are developed and refined through preparation and revision of the overall plan. As project construction proceeds, it is evident that actual accomplishments must be compared with the overall plan if effective control is to be achieved.

Many items discussed in cost control are equally applicable to progress control. Cost-schedule-progress control systems integrated via a common computerized data base have been successful on very large projects. On smaller projects, the same integrated approach can be followed manually or with a microcomputer.

CPM Control Schedule

The **Critical Path Method** (CPM) is the foundation of the progress control system. A simplified CPM precedence diagram giving the construction schedule for the example project is shown in Appendix A. Figure 8-8 shows a bar-chart schedule, fully consistent with the CPM, upon which superimposed S curves show "cumulative per-cent completes" at the end of each month for both early- and late-start schedules. The double-S curves form an envelope. If actual performance is within the envelope, project goals have a good chance of being accomplished. If actual performance falls below the late-start schedule, the project will normally not be completed on schedule without a revised program.

Field effort in man-hours is the criterion for weighing the various components in the chart. This information is obtained from the fair-cost estimate. Basing progress upon estimated man-hours has numerous advantages in an integrated cost-schedule-progress control system. Manpower forecasts can be developed and compared with actual man-hours expended to accomplish key sections of the work. Contractor productivity can be calculated monthly.

Physical Progress Measurement

Figure 8-9 shows a worksheet for calculating the physical percentage complete for a single contract; it is based upon actual quantities completed during the period. The productivity of the construction forces compared with the manager's estimate can be calculated by comparing actual hours with calculated hours.

Figure 8-10 shows the summary of all contract data entered and plotted upon Figure 8-8. For a multifeature project, work can be grouped and tabulated by facility as well as by contract in order to determine the status of its various components.

The examples in Figures 8-8, 8-9, and 8-10 were chosen to illustrate a manual approach, but the system is the same whether or not it is computerized. The choice is one of project complexity; however, today it is usually more efficient to use the computer than to make the computations manually.

Field Schedules and Progress Controls

The preceding examples represent control schedules. If actual progress stays within the S-curve envelope, the project appears to be heading on a course that will achieve project goals. However, field scheduling must provide much more detail than can ever be shown on a precedence diagram prepared during the planning stages. The key milestones are set forth in the plan, but the field construction manager, his staff, and the contractors must develop more detailed schedules to accomplish short-term goals.

Weekly meetings with all contractor representatives are a must; they provide the opportunity to review the current status of the program and to develop key dates for "move in," "move out," and completion of critical items affecting several contractors. The minutes of the meetings should be promptly distributed to all participants.

Preparation and job-site posting of a detailed bar-chart schedule, listing key require-ments for each contractor, can be helpful. It may be revised frequently; but, if geared to detailed plans for accomplishing project objectives, it will assist all contractors and eliminate bottlenecks and interference.

Reporting by the field construction manager of actual quantities completed each month forms the heart of the overall progress control previously discussed. When an

Progress and Schedule Summary Sheet

Period Ending _____ November _____

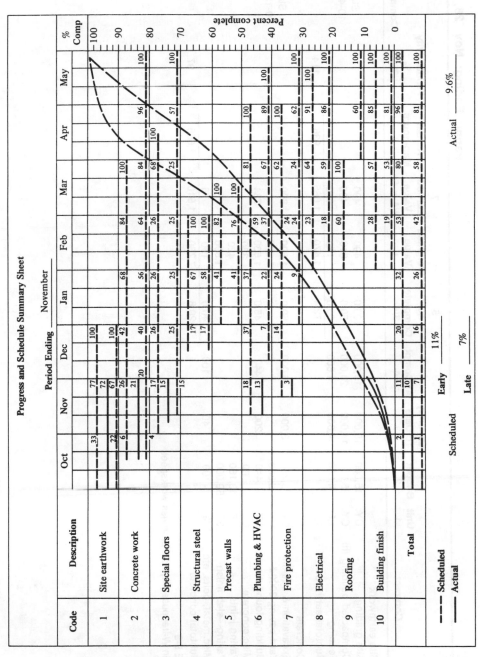

Code	Description	Oct	Nov	Dec	Jan	Feb	Mar	Apr	May	% Comp
1	Site earthwork	33 / 22	77 / 72	100						
2	Concrete work	6	67 / 26 / 21	100 / 42 / 40	68 / 56	84 / 64	100 / 84	96	100	
3	Special floors	4	17 / 15	26	26	26	68	100		
			15	25	25	25	25	57	100	
4	Structural steel			17 / 17	67 / 58	100 / 100				
5	Precast walls				41	82 / 76	100 / 100			
6	Plumbing & HVAC		18 / 13	37	41 / 37	59 / 37	81	100		
				7	22		67	89	100	
7	Fire protection		3	14	24 / 9	24 / 24	62	100		
						23	24	62	100	
8	Electrical					18	64	91	100	
							59	86	100	
9	Roofing					60	100			
10	Building finish	2	11 / 10	20	32	28	57	60	100	
						19	53	85	100	
						53	80	81	100	
							96	100		
	Total	1	7	16	26	42	58	81		

Percent complete: 100, 90, 80, 70, 60, 50, 40, 30, 20, 10, 0

---- Scheduled
—— Actual

Scheduled   Early _____ 11% _____
             Late _____ 7% _____

Actual _____ 9.6% _____

FIGURE 8-8

149

Period Ending: Nov. 28, 199_

	Contract	Unit 1	Budget 2	Quantity To Date 3	At Compl 4	5 % Complete 3/4	Weighted Value 6	7 Earned Value 3/4 × 6	Actual Man-hours 8	Productivity % 7/8	Comments
1.0	Site earthwork										
1.1	Site grading	CY	15,000	15,000	15,000	100.0	490	490			Start
1.2	Compacted building fill	CY	25,000	14,000	28,000	50.0	810	405			10-4-9_
1.3	Fencing	LF	4,680	4,800	4,800	100.0	170	170			
	Subtotal earthwork					72.4	1,470	1,065	936	114	
2.0	Concrete										
2.1	Excavation & backfill	CY	5,250	3,000	6,000	50.0	1,600	800			Start
2.2	Yard paving-3000#	CY	3,850				1,970				10-13-9_
2.3	Building foundations	CY	440	200	440	45.0	1,630	734			
2.4	Misc. concrete	CY	60				330				
2.5	Paving joints	LF	13,180				420				
2.6	Paving—slab finish	SF	159,000				1,590				
2.7	Misc. metal, bolts, etc	%	100	40	100	40.0	190	76			
	Subtotal concrete					20.8	7,730	1,610	1,656	97	

FIGURE 8-9
Mountain Warehouse physical progress worksheet.

Contract	Unit 1	Budget 2	Quantity To Date 3	At Compl 4	% Complete 3/4 5	Weighted Value 6	Earned Value 3/4 × 6 7	Actual Man hours 8	Productivity % 7/8	Comments
3.0 Interior special slabs										
3.1 Base slab—place 5¼"	CY	2,500	1,000	2,500	40.0	1,670	668			Start 11-8-9_
3.2 Reinforcing steel	#	140,000	60,000	140,000	43.0	930	397			
3.3 Screed & cure base slab	SF	144,500				770				
3.4 Topping incl finish ¾"	SF	144,500				2,890				
3.5 Curbs, misc concr., emb.	%	100				1,070				
Subtotal special slabs					14.5	7,330	1,065	952	112	
6.0 Mechanical—HVAC										
6.1 Storm sewers	LF	2,240	1,030	2,100	49.0	790	387			Start 11-15-9_
6.2 Sanitary sewers	LF	450	460	460	100.0	120	120			
6.3 Domestic water	LF	200	100	200	50.0	80	40			
6.4 Plumbing	%	100				1,130				
6.5 Sheet metal	%	100				1,000				
6.6 Heating	%	100				1,130				
Subtotal mechanical					12.9	4,250	547	512	107	
7.0 Fire protection										
7.1 Fire loop	LF	1,800	200	1,800	11.1	720	80			Start 11-24-9_
7.2 Interior sprinklers	SF	150,000				2,500				
Subtotal fire protection					2.5	3,220	80	72	111	
Total to date							4,367	4,128	106	

FIGURE 8-9
Mountain Warehouse physical progress worksheet (continued).

Acct no.		% Complete 1	Weighted value 2	Earned value 3	Start	Finish
1.0	Site earthwork	72.4	1,470	1,065	10-4-9_	
2.0	Concrete	20.8	7,730	1,610	10-13-9_	
3.0	Special slabs	14.5	7,330	1,065	11-8-9_	
4.0	Structural steel		3,770			
5.0	Double tee walls		2,640			
6.0	Mechanical	12.9	4,250	547	11-15-9_	
7.0	Fire protection	2.5	3,220	80	11-24-9_	
8.0	Electrical		3,470			
9.0	Roofing		3,150			
10.0	Building finish		8,470			
	Total	9.6	45,500	4,367		

$$\text{Productivity} = \frac{\text{Earned Value (Man-Hrs)}}{\text{Cumulative Actual Man-Hrs}} = \frac{4367}{(516)(8)} = 106\%$$

Notes

1. Shows computation of physical % complete for overall project based upon earned value compared to weighted value (man-hours) as determined from fair cost estimates.

2. Results are shown in Figure 8-8, construction schedule summary, expressed as overall % complete.

3. If original schedules are to be maintained the weighted value is not adjusted for actual overruns or underruns in quantities. Actual productivity computations must take these variations into account.

4. See Figure 8-9 for individual computations of % complete & earned value.

5. This report can be easily adapted to an integrated overall computer-assisted, management-control system with a common data base in man-hours. See Figure 7-14 for cumulative actual man-hours used in productivity calculation above.

FIGURE 8-10
Physical progress contract summary (Mountain Warehouse Project).

integrated system is being used, the computer can economically take much of the clerical and computational effort out of this task even on relatively small projects.

MANAGEMENT-LEVEL REPORTING BY CONCRETE CONTRACTOR

Individual concrete contractors on construction management projects can range from small to medium sized general or subcontractors who may often bid other portions of the work such as some or all of the interior finish or other qualified specialty. Smaller companies may minimize formal reporting methods by utilizing the owner or a traveling general superintendent who will oversee several projects using periodic job visits to manage and control the work. Other larger companies or those who operate in remote areas may develop reporting methods similar to those previously described for a construction manager but which will include increased details in productivity, unit cost and scheduling information.

At a minimum the on-site manager will normally fill out a daily report on a preprinted form showing the weather, number of craftsmen and other employees, significant starts and completions, problems or delays, and other information to either document the work or to request assistance. The home office may prepare a weekly unit cost report from payroll information and quantities supplied by the job superintendent or engineer. Monthly progress reports covering many of the items described in the previous section may also be issued. Up to a certain level of work hands on management through frequent visits can be very effective. However, as the organization grows larger more formal methods become necessary.

COST CONTROLS BY CONCRETE CONTRACTOR

Overall cost controls may or may not be integrated with schedule control. If the contractor utilizes man hours in his estimating, he can easily develop an integrated system tailored to his requirements.

Bid Estimate

The concrete contractors bid the job competitively to the construction manager. The contractor maintains an estimating capability in the home office sized to achieve his planned annual bidding volume. The estimator first develops a material take-off from the plans and specifications listing all of the required quantities for labor, material, construction equipment and subcontract pricing. Work to be performed by his own forces will be tabulated in detail for labor, material and construction equipment pricing. Some firms will make approximate estimates for subcontracts in order to compare with quotations or to utilize if competitive quotes are not obtained. Most firms will not make detailed take-offs for subcontract items. Quotations for materials, subcontracts, equipment rental and other items will be solicited. Figure 8-11 shows the concrete contractor's bid estimate summary for the Mountaintown Warehouse. See also Figure 11-3 for the structural steel contractor's bid estimate.

When a bid is successful the estimator often acts as the project manager to oversee the work from the home office. In this event his first task will be to "buy out" the job by awarding subcontracts and purchase orders for materials and other required items.

Cost Reports

The concrete contractor will utilize a weekly unit labor cost report to compare actual unit costs or man-hours with estimated units. Estimating in man-hours has considerable advantage to the contractor as discussed in Chapter 11. Figure 8-12 shows the contractor's weekly unit man-hour report for the period ending November 30 illustrating both progress and productivity.

The concrete contractor utilizes a monthly cost report to tabulate recorded and committed costs plus estimated costs to complete in order to forecast estimated costs

CONCRETE CONTRACTOR
BID ESTIMATE SUMMARY
MOUNTAINTOWN WAREHOUSE

DESCRIPTION	QUANT.	UNIT	MH	UNIT COST ($) M	L	EU	SC	TOTAL	LABOR HOURS	LABOR $	MAT'L $	EQ.USE $	SUB CON $	TOTAL $
131 STRUCTURAL EXCAVATION	4950	CY	0.10		3.00	1.21		4.21	495	14850		5990		20,840
132 HAND EXCAVATION	225	CY	0.10		30.00			30.00	225	6750		0		6,750
134 BACKFILL	4000	CY	0.15	6.00	4.50	0.60		11.10	600	18000	24000	2400		44,400
130 TOTAL EARTHWORK	5175	CY						13.91	1320	39600	24000	8390		71,990
211 FORMS-FOUNDATION	11000	SF	0.15	1.00	4.50	0.10		5.60	1650	49500	11000	1100		61,600
261 FORMS-MISC. CONCRETE	1650	SF	0.30	1.00	9.00	0.10		10.10	495	14850	1650	165		16,665
214 CONCRETE FOUNDATION	450	CY	0.80	56.00	24.00	1.00		81.00	360	10800	25200	450		36,450
264 CONCRETE-MISCELLANEOUS	55	CY	1.50	56.00	45.00	1.00		102.00	83	2475	3080	55		5,610
235 YARD PAVING (ALL COSTS)	3900	CY					82.00	82.00					319800	319,800
212 REINFORCING STEEL	30000	LBS					0.61	0.61					18300	18,300
217 GROUT WALL PANELS	500	CF	0.15	1.00	4.50			5.50	75	2250	500			2,750
218 GROUT BASE PLATES	108	EA	0.25	1.00	7.50			8.50	27	810	108			918
200 TOTAL CONCRETE	4405	CY						104.90	2690	80685	41538	1770	338100	462,093
311 PIPE GUARDS	10	EA	0.30		9.00	1.00		10.00	3	90	10			100
312 EMBEDDED RAIL	30	LF	0.20		6.00	0.15		6.15	6	180	5			185
313 ANCHOR BOLTS	432	EA	0.30	2.00	9.00			11.00	130	3888	864			4,752
300 TOTAL MISC. METAL	LOT								139	4158	879			5,037
TOTAL DIRECT COST									4149	124443	66417	10160	338100	539,119
911 SUPERINTENDANT	6	MO.			4000			4000		22000				22,000
912 ENGINEER	1	MO.			3000			3000		3000				3,000
915 CLERK	5	MO.			1500			1500		7500				7,500
SUB TOTAL LABOR										32500				32,500
919 BENEFITS,TAXES,INSURANCE	30	%		0.30				0.30		9750				9,750
910 TOTAL LABOR										42250				42,250
921 OFFICE EXPENSE	5	MO.		300				300			1500			1,500
923 REPRODUCTION	5	MO.		300				300			1500			1,500
924 TELEPHONE & FAX	5	MO.		300				300			1500			1,500
926 TRAVEL & EXPENSE	5	MO.		100				100			500			500
928 FIRST AID & SAFETY	5	MO.		250				250			1250			1,250
941 TOOLS & MINOR EQUIPMENT	5	MO.				1000		1000				5000		5,000
945 OFFICE TRAILER	5	MO.				300		300				1500		1,500
946 CHANGE TRAILER	5	MO.				125		125				625		625
948 PICKUP TRUCK	5	MO.				300		300				1500		1,500
900 TOTAL FIELD INDIRECTS				1250		1725		2975		42250	6250	8625		57,125
TOTAL ESTIMATED COST										166693	72667	18785	338100	596,244
MARKUP FOR HOME OFFICE OVERHEAD AND PROFIT														43,756
TOTAL BID PRICE														640,000

FIGURE 8-11
Bid estimate summary (Concrete Contractor).

UNIT MAN HOUR REPORT
CONCRETE CONTRACT
MOUNTAINTOWN WAREHOUSE

CODE	DESCRIPTION	BUDGET QUANTITY 1	UNIT	BUDGET MAN HOURS 2	BUDGET UNIT (2)/(1) 3	TO DATE QUANTITY 4	TO DATE MAN HOURS 5	TO DATE UNIT (5)/(4) 6	EARNED VALUE (4)(3) 7	JOB FACTOR (5)/(7)	PER CENT COMPLETE J10/E10
131	STRUCTURAL EXCAVATION	4950	CY	495	0.10	4000	475	0.12	400	1.19	80.8%
132	HAND EXCAVATION	225	CY	225	1.00						
133	BACKFILL	4000	CY	600	0.15	778	100	0.13	117	0.86	19.5%
	SUB TOTAL EARTHWORK	5175	CY	1320	0.26	4778	575	0.12	517	1.11	39.1%
211	FOUNDATION FORMS	11000	SF	1650	0.15	4950	800	0.16	743	1.08	45.0%
261	MISCELLANEOUS FORMS	1650	SF	495	0.30						
214	CONCRETE FOUNDATIONS	450	CY	420	0.93	200	180	0.90	186	0.97	44.3%
264	MISCELLANEOUS CONCRETE	55	CY	83	1.51						
217	GROUT WALL PANELS	500	CF	75	0.15						
218	GROUT BASE PLATES	108	EA	27	0.25						
	SUB TOTAL CONCRETE	505	CY	2750	5.45	200	980	4.90	929	1.06	33.8%
311	PIPE GUARDS	10	EA	3	0.30						
312	EMBEDDED RAIL	30	LF	6	0.20						
313	ANCHOR BOLTS	432	EA	130	0.30	172	48	0.28	52	0.93	39.8%
	SUB TOTAL MISC. METAL	LOT		139			48		52	0.93	37.2%
	TOTAL CONCRETE CONTRACT			4209			1603		1497	1.07	35.6%

JOB FACTOR = TO DATE MAN HOURS/EARNED VALUE

PER CENT COMPLETE = EARNED VALUE/BUDGET MAN HOURS

PRODUCTIVITY = 1/ JOB FACTOR = EARNED VALUE/TO DATE MAN HOURS

NOTES:

1. THE UNIT MAN HOUR REPORT ONLY INCLUDES WORK PERFORMED BY THE CONTRACTOR'S OWN FORCES AND DOES NOT INCLUDE THE PAVING SUBCONTRACTOR.

2. JOB FACTOR OF 1.06 INDICATES THAT WORK TO DATE HAS BEEN PERFORMED AT 7.0% OVER CONTRACTORS LABOR BUDGET.

3. THE WORK PLANNED TO BE COMPLETED BY THE CONTRACTOR'S FORCES IS 35.6% COMPLETE

FIGURE 8-12
Unit man-hour report (Concrete Contractor).

at completion. This figure is then compared to the bid price adjusted for change orders to determine anticipated profit and loss in a manner illustrated by Figure 8-2.

SCHEDULE AND PROGRESS CONTROLS BY CONCRETE CONTRACTOR

The concrete contractor prepared a simple bar-chart schedule at the time of bid as shown by Figure 8-13. Figure 8-12 shows the comparison of actual progress compared to scheduled progress as of November 30 from the contractor's viewpoint. Major quantities were agreed between the contractor and the construction manager monthly for use in measuring overall physical progress and for progress payments. Note that the contractor's figures are in considerably more detail than the professional construction manager's figures shown in Figure 8-9.

MANAGEMENT-LEVEL REPORTING BY GENERAL CONTRACTOR

If Easyway had chosen to bid the project to prequalified general contractors, internal management-level reporting for the project by the contractor might follow closely the format developed by the construction manager as described earlier in this chapter. If the contractor planned to perform the concrete and allied work with his own forces, unit cost and progress reports similar to Figure 8-12 would be in order. An overall summary monthly cost report might be similar to Figure 8-2. A Construction schedule summary could also be similar to the construction manager's report Figure 8-8.

COST CONTROLS BY GENERAL CONTRACTOR

Cost controls will be a combination of methods described under construction management for subcontract control supplemented by detailed comparison of costs and man hours to the bid estimate for own forces work. Productivity analysis will be enhanced if the contractor uses man-hours as the basis of his cost estimating. Man-hour units also form a valuable reference for bidding similar jobs in the future.

Bid Estimate

The general contractor's bid estimate will include a detailed estimate of work to be performed directly generally by elements including man hours, labor including payroll additives, materials and construction equipment. Industrial contractors will usually add an element for process equipment purchase which forms a major portion of industrial work. Heavy construction contractors will add elements for construction equipment operation which represent a large portion of such work. A detailed estimate for field and home office indirect costs similar to Figure 4-6 will be prepared including additional items required due to performing work with own forces. The general contractor's indirect costs for the Mountaintown Warehouse will include items for small tools, consumable supplies, and other items required to support his craft employees similar to

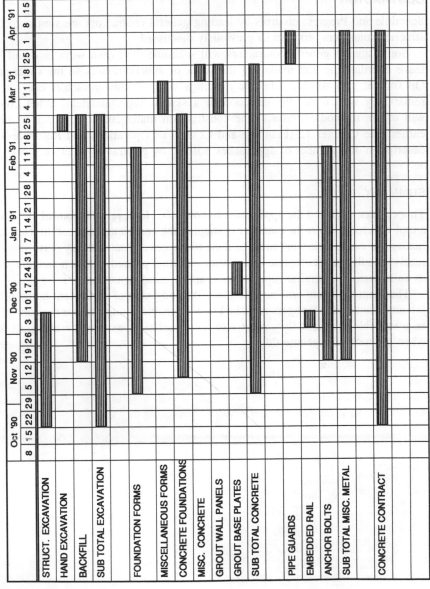

FIGURE 8-13
Bar chart schedule (Concrete Contractor).

the concrete contractor's estimate as shown in Figure 8-11. See Figures and Tables in Chapter 11 for examples of summary bid sheets prepared by general contractors.

SCHEDULE AND COST CONTROLS BY GENERAL CONTRACTOR

Schedule and Progress controls by leading general contractors are similar to the construction managers program as previously discussed.

Some smaller general contractors will schedule their own work in detail but will leave coordinating and managing the work of subcontractors to the superintendent. On smaller jobs weekly schedule meetings with all subcontractors can often work remarkably well under the direction of a skilled superintendent. Other contractors will utilize methods similar to the construction management approach as discussed earlier. See Chapter 12 (Planning and Control of Operations and Resources) for the description of a number of approaches utilized by general contractors.

SUMMARY

This chapter reviewed the application of controls. However, the control system is only a tool which can assist the manager in shaping the completed product. In itself it cannot manage; it cannot tell what must be done to improve unsatisfactory performances; if it receives faulty or incomplete information, its final printed output nonetheless appears as authoritative as similar reports based upon sound input data.

The control system will never replace the judgment of competent home-office and job-site managers. Managers must be knowledgeable enough to use the control system as a valuable tool; but they must also have sufficient experience and skill to know when the tool has become dull.

9

SELECTING A PROFESSIONAL CONSTRUCTION MANAGER

Selecting a professional construction manager contractor or design-constructor under some form of reimbursable contract is in many ways similar to selecting an architect, consulting engineer, contractor or design-constructor. Criteria may include overall experience, understanding of the project, preliminary plans submitted for completing the project, and price. These factors may be weighed differently by different owners, but all are important in selecting the right professional construction manager or other management entity for a new project. By outlining a proven approach to selection, this chapter will tie together many of the concepts that have been introduced earlier in this book.

BASIC QUALIFICATIONS

Certain basic qualifications for selection of construction managers, general or subcontractors under a negotiated type of contract should almost always be investigated. Among the most important are the following:

Overall experience
    General contracting
    Professional construction management
    Project planning and control
    Inspection (If desired)
    Value engineering
    Scheduling methods
    Estimating methods
    Design knowledge
    Other applicable experience

159

Financing status
   Adequacy for project under consideration
   Overall financial strength

Depth of organization
   Present workload
   Available personnel
   Key personnel
   Recruiting requirements

Specialized experience
   Area familiarity
   Special knowledge
   Industry experience

References
   Current clients
   Prior clients

Understanding project requirements
   Anticipation of problems
   Understanding of special features
   Understanding owner objectives
   Understanding project features
   Understanding area practices
   Understanding site conditions

Preliminary plans for implementation
   Overall approach
   Preliminary schedules
   Preliminary costs
   Proposed organization
   Services to be provided
   Services not to be provided

Price and compensation
   Reimbursable costs
   Nonreimbursable costs
   Fee basis
   Other basis

TYPICAL SELECTION METHODS

Various owners have developed different methods for selecting a professional construction manager. Some are very formal, such as those based upon a numerical grading system; others are very informal, leaving the presentation to be largely determined by the proposer. The following sections will illustrate some common approaches.

General Services Administration (GSA)

A booklet, *The GSA System for Construction Management*,[1] contains a suggested "Construction Management Project Notice" which generally outlines this agency's initial method of selecting a construction manager. This section is quoted below:

CONSTRUCTION MANAGEMENT PROJECT NOTICE
FOR ISSUANCE IN THE COMMERCE BUSINESS DAILY

Y—Construction Management Services

The General Services Administration seeks construction management services for the proposed (insert building name and location) to provide approximately (insert) gross square feet within an estimated cost range between $(insert) and $(insert) million. Design and construction will be concurrently phased with separate construction contracts awarded as segments of the design are completed by the architect-engineer.

Consideration will be given to firms or joint ventures generally meeting the following requirements:
(1) Experience as a Construction Manager or potential competence to perform construction management services; (2) Financial ability to provide the services required by the Government; (3) Competence in civil, mechanical, electrical and structural engineering; construction estimating, cost accounting and control; tenant coordination; project management; contract negotiation and administration; construction superintendence and inspection; and other related fields; (4) Experience in constructing buildings in the general geographic area of this project, or good recent knowledge of local conditions in the project area, or ability to retain others with such knowledge; (5) Proven competence in the implementation and maintenance of network-based construction management systems and in the application of systematic cost control throughout the design and construction process; (6) Good professional and business reputation, and an on-time and within-budget performance record; and (7) Ability to provide professionally qualified key personnel with a minimum of 12 years' satisfactory experience in the design and construction industry. Satisfactory experience should include:
a Eight years in work related specifically to the duties to be performed in the designated position for this project; and
b Four years in positions with requirements equal to those for the designated position of this project.

Prospective construction management firms or joint ventures who are interested in the project are invited to ask for Request for Qualifications Submission which will be issued by the office below on or by (insert date). Qualifications will be received until (insert date) at the office below, and then evaluated on the basis of the requirements and criteria contained in the Request for Qualifications Submission. Request for Priced Proposal will be subsequently issued to only those firms or joint ventures whose Qualifications have

---

[1]General Services Administration, Public Buildings Service, Washington, D.C., April 1975, rev. ed.

been determined by GSA as being within a competitive range. Only Priced Proposals specifically requested by the Government will be considered.

Associated General Contractors of America (AGC)

An AGC report titled *Construction Management Guidelines for Use by AGC Members*[2] includes the following section illustrating the AGC selection recommendations:

HOW IS THE CONSTRUCTION MANAGER MEMBER OF THE CONSTRUCTION TEAM SELECTED?

The Construction Manager will be selected on the basis of an objective analysis of his professional and general contracting qualifications. In this selection, major considerations will be given to:

a Its success in performing the normal general contractor's function on projects of comparable type, scope, and complexity.

b Its financial strength, bonding capacity, insurability, and ability to assume a financial risk if the owner requires it.

c Its in-house staff capability and the qualifications of the person who will manage the project.

d Its record for completing projects on time and within the budget.

e Its demonstrated ability to work cooperatively with the Owner and the Architect-Engineer throughout the project, and to display leadership and initiative in performing his tasks as a member of the Construction Team.

Airport Expansion Program

A western city requested construction management proposals for a $40 million terminal addition as a part of an overall master plan. Here the city initially screened firms and selected four to submit proposals. The general scope of the facilities to be constructed was spelled out, and relationships and criteria for operation under the city engineer and airports director were set forth. The notice was general, and included the statement, "What we will want to learn from you is what service you can provide to the City, how would you organize to perform the work, and on what basis would we expect to pay for such Construction Management Services." A copy of the master plan and development program was attached. Arrangements were also made to interview firms submitting proposals, and a $1\frac{1}{2}$-hour period was tentatively selected. Within this framework, selected firms were free to develop their own programs for consideration.

Negotiated Construction Contracts

A novel approach, called "Management Contracting," was developed some years ago at the University of California. Other universities in Alaska and Colorado and a number of private owners have used a similar approach for selecting general contractors when

---

[2]The Associated General Contractors of America, 1957 E Street, N.W., Washington, D.C., Feb. 15, 1972, p. 3.

work must begin prior to design completion. In summary, the method provides the opportunity for the owner to achieve shortest overall design-construction time under a negotiated contract while complying with competitive bid requirements imposed by a state or by company policy. Bidders are first rated on experience through a points evaluation program to ensure prequalification. Selected firms then bid on individual items of cost such as workman's compensation premiums, small tools, equipment rental, home office overhead and profit on a lump-sum or unit-price basis. All other costs are reimbursable up to a Guaranteed Maximum Price which is negotiated near the close of design work. Any savings below the GMP are split between the contractor and the owner based upon the percentages set forth in the contractor's bid. Other owners will choose the selected contractor or design-constructor and will negotiate a fixed fee for profit and general overhead with all other direct costs being reimbursable.

Other Examples

Private construction owners commonly adopt portions of all the above approaches on negotiated work. Requests for proposals for constuction management, design-construction and general contractor services can range from an individually designed prequalification requirement, coupled with specific quotation information including pricing, to a simple verbal or written request leaving the content of the proposal largely up to the ingenuity of the bidder.

RECOMMENDED METHODS

What is the best method for the owner to use when selecting a professional construction manager or other management entity on negotiated and competitively bid projects? The ideal selection process would first develop a short list of qualified firms based upon prior accomplishment and financial reliability.

For negotiated work on incomplete design information, each of these firms would be given an opportunity to show its ingenuity, planning ability, and estimating and scheduling abilities by submitting a preliminary program or work plan for the new project. Finally, questions of costs to the owner, such as fees and services to be included and not included, would be taken into consideration.

On competitively bid fixed-price work from prequalified contractors on completed plans and specifications, the low bidder is normally awarded the contract.

Were the authors to be requested by a construction owner to develop criteria for the selection of a professional construction manager, or other cost reimbursable management method, the program would evolve along the following lines:

Prequalification

Initially, interested firms would be requested to submit *overall experience* qualifications as previously set forth. A standard form can readily be developed so the required information can be inserted or attached by the prospective professional construction manager or contractor. Upon receiving all forms, the owner should check references, especially present and past clients, and make an overall judgment of the manager's qualifications. Firms designated as qualified would proceed to the next step in the selection process, and those designated as not qualified would be so notified. No statistical or numerical order would be developed on the basis of past experience. Once the owner determines which firms are qualified, the selection process continues in order to identify the one that appears to be best suited for the particular project under consideration. Government agencies are increasingly utilizing negotiation from a short list of prequalified contractors on critical defense work or other important projects where construction work must begin prior to design completion.

Owners should be primarily interested in what a construction manager or contractor can do for them. One hundred successful past jobs for others will be of no benefit to the owner if the new job is not completed on schedule and within budget. The next stage of the selection process is thus the most important; it is designed to obtain a project work plan from each of qualified candidates so that the best and most practical plan can be selected by the owner for implementation.

Request for Proposals

The request for proposals should be carefully designed to give all candidates a basic amount of information so that all proposed plans are founded on common criteria. The more information about the project that can be given the candidates, the better the plan that can be expected.

A request for a proposal for such services might be outlined as follows:

1 Owner's Basic Criteria   The owner assembles and includes all pertinent, current information about the proposed project as developed from his feasibility study or other preplanning information. While some of the information will not be available or desirable to include, such criteria may include the following:

Feasibility studies
General layout or preliminary drawings
Preliminary specifications or design criteria
Owner's operating requirements
Owner's contracting requirements
Design schedules
Completion requirements
Location of job site
Designer, if selected
Appropriation estimate
Other owner requirements

2 Proposed Work Plan   Within these parameters, considerable leeway should be given to each candidate in developing its implementation plan. The purpose of this request is to determine the candidate's understanding of the project requirements and his ingenuity and skill in developing a preliminary work plan for implementing those requirements. Some of the items that the owner can request candidates to include in their work plans are:

Description of overall approach
Services to be provided in field office
Services to be provided in home office
Proposed contract package, scope, and general contracting program
Preliminary procurement schedule
Proposed value-engineering program
Preliminary construction schedule
Cost-and-progress control system
Additional construction cost estimates (if sufficient information available in criteria)
Estimates of professional construction management costs or contractor overhead and fixed fee (or other compensation requirements)
Definitions of reimbursable and nonreimbursable costs if applicable
Proposed project organization, and résumés of key personnel

3 Technical Evaluation of Proposals   Proposals can now be evaluated on the basis of the candidates' demonstrated knowledge and ingenuity in developing a program to implement the basic criteria available. While evaluation criteria will be different for all projects, some key questions important in determining the interest, understanding, ingenuity, practicality, and skill of the candidate include:

Are the schedules and programs outlined in the proposal based upon the particular project criteria, and do they reflect an effort to avoid problems and to present a practical plan tailored to fit the proposed project conditions?

Has the plan been individually developed by practical construction professionals, or is it a modification of standard documents adapted by sales personnel and featuring "buzz words" and other impressive but undefined techniques?

Is the method for control of costs clearly identified and explained?

Is the method for control of schedules and progress clearly identified and explained?

On the basis of the owner's interviews with the proposed home-office project manager, the field construction manager, and other key personnel, what does he believe was their part in preparing the proposals, what does he think of their background, and does he believe them to be individually suitable for this project?

What part will people who developed the proposal play in the project implementation?

Is the value-engineering program described specifically enough for it to be evaluated under anticipated project conditions?

Is the overall tone of the proposal aimed at presenting a well-thought plan to achieve a successful program for the project under consideration, or is it generally aimed at listing past accomplishments, past solutions of problems, and broad overall skills?

4 Final Selection   After the technical evaluation has been completed, an overall evaluation of costs and fees is needed. Fees should be reasonably related to the services to be provided. In the real world, price is certainly a factor, but in view of the construction manager's level of influence on costs, the potential savings associated with the choice of the best-qualified professional would appear to be greater than a small difference in fee.

COMPENSATION AND FEES ON COST-REIMBURSABLE PROJECTS

The subject of fees for professional construction management and other cost reimbursable projects has been one of considerable misunderstanding. Since each job is different, comparison is difficult. In some cases, fees are defined to include all costs of construction or construction management including home-office costs, field costs, and profit. In other cases, fees are defined to be profit only and do not include home-office or field-office costs.

Current Practices

Field and home-office responsibilities can vary widely for different projects. Field costs are largely determined by the extent of the field organization coupled with the extent of general conditions services that the manager or contractor is expected to provide.

Surveying, inspection, testing laboratory, utility bills, sanitary facilities, security services, and other items broadly set forth as general conditions can be included or omitted depending upon the particular project. Therefore, it probably is not feasible to develop a meaningful comparison of overall charges for different projects.

Many negotiated general construction, as well as professional construction management, contracts provide for a fee to cover home-office cost and profit. The fee can be expressed as a lump sum, as a percentage of total estimated construction cost, or as a percentage of actual cost. Other contracts provide for full reimbursement of field and home-office costs plus a fixed fee for profit.

Representative Fees

Figure 9-1 gives two curves which show the low and high ranges of home-office costs plus profit for approximately 50 actual or proposed jobs performed in the United States and Canada, many of which were professional construction management projects. In order to obtain sufficient data to draw the curves, a number of negotiated contracts for general contractor services, where the preponderance of work has been subcontracted, have also been included. The project data were obtained from publicly reported projects and from reported data obtained from sources considered to be reliable; they cover projects constructed for different owners by different managers and contractors. The curves are not intended to be fee curves for any proposed projects. Rather, they are intended simply to indicate the range of fees for profit and home-office services that have actually been charged for a representative number of projects. Reasonable fees expressed in this manner may range from the high curve to the low curve depending upon project duration, the nature of the project, the extent of the services,

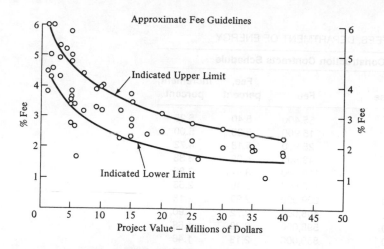

FIGURE 9-1
Approximate fee guidelines.

Notes
1. Fee includes all home office costs including profit & overhead.
2. All field expenses are reimbursable.
3. Based upon 52 construction management and negotiated general contract projects throughout the U.S. & Canada.
4. Includes 4 projects for USGS & DOE; balance are for public & private owners.
5. Fees can vary widely dependent upon profit & amount of work done in the home office.

and other factors. While the fees shown in Figure 9-1 were developed in the late 1970s, comparison with projects in the late 1980s indicates that on a percentage basis fees have generally remained within the ranges shown. Table 9-1 shows maximum fees covering home office costs and profit for negotiated cost reimbursable construction management and general contractor contracts for the Department of Energy as of April 1989.

Fees for negotiated design-construct projects would include a design fee in addition to construction related fees shown in Figure 9-1 and Table 9-1. Design fees can also be be treated as a reimbursable cost with a profit added as a part of the overall fixed fee. Other design-construct projects may include design fees as a part of the lump sum for home office services and profit.

CONTRACTOR SELECTION BY COMPETITIVE BIDDING

Selection of contractors on competitively bid projects varies between the public and private sectors.

Public Sector Projects

Public sector projects normally require bid, payment, and performance bonds to guarantee payment and performance upon the part of the contractor. Traditionally, the

TABLE 9-1
MAXIMUM FEES, DEPARTMENT OF ENERGY

Construction Contracts Schedule

Fee base	Fee	Fee, percent	Increment, percent
$100,000	$5,400	5.40	5.25
$300,000	15,900	5.30	5.00
$500,000	25,900	5.18	4.72
$1,000,000	49,500	4.95	3.33
$3,000,000	116,100	3.87	2.82
$5,000,000	172,500	3.45	2.53
$10,000,000	299,000	2.99	2.18
$15,000,000	408,000	2.72	1.90
$25,000,000	598,000	2.39	1.68
$40,000,000	850,000	2.13	1.46
$60,000,000	1,142,000	1.90	1.28
$80,000,000	1,398,000	1.75	1.24
$100,000,000	1,646,000	1.65	1.03
$150,000,000	2,161,000	1.44	.80
$200,000,000	2,561,000	1.28	.65
$300,000,000	3,211,000	1.07	.50
$400,000,000	3,711,000	.93	.35
$500,000,000	4,061,000	.81	
$Over 500,000,000	4,061,000		[1] .35

[1] 0.35 percent excess over $500 million.

Construction Management Contracts Schedule

Fee base	Fee	Fee, percent	Increment, percent
$100,000	$5,400	5.40	5.25
$300,000	15,900	5.30	5.00
$500,000	25,900	5.18	4.72
$1,000,000	49,500	4.95	3.33
$3,000,000	116,100	3.87	2.82
$5,000,000	172,500	3.45	2.53
$10,000,000	299,000	2.99	2.18
$15,000,000	408,000	2.72	1.90
$25,000,000	598,000	2.39	1.68
$40,000,000	850,000	2.13	1.46
$60,000,000	1,142,000	1.90	1.28
$80,000,000	1,398,000	1.75	1.24
$100,000,000	1,646,000	1.65	1.03
$Over 100,000,000	1,646,000		[1] 1.03

[1] 1.03 percent over $100 million.

Source: Department of Energy Acquisition Regulation, Amendment 4, April, 1989. See source document for applicable conditions affecting fee computations.

public sector is open to all bidders and the awarding agency often relies upon bonding in lieu of prequalification. Some repeat public owners will prequalify contractors in advance based upon experience, special expertise or financial qualifications. In this case, only prequalified contractors are issued bidding documents.

Some agencies utilize a two-step prequalification process on major projects which will require particular skills and experience and which often present unusual risks to the contractor. An example could be a major transportation project with complex geological conditions, critical schedule dependencies, and other unusual conditions. Here the agency would first prequalify all firms meeting a designated criteria which may include applicable experience and other special requirements. The selected firms would be issued bidding documents. Bids would normally be opened in the presence of the bidders and the low bidder would receive the award in the absence of irregularity.

Private Sector Projects

Private sector owners often prequalify contractors either directly or through an architect/engineer or construction manager. Many awarding entities will prequalify only a limited number of bidders in order to make the proposed contract attractive to the most qualified contractors. Many private owners require potential contractors to submit a substantial amount of financial information along with experience records and references from past clients. Based upon the information received the owner can decide whether or not to require performance and payment bonds. Many private owners do not hold formal bid openings that are common in the public sector.

The selection of general contractors is very similar to the selection of individual trade contractors by competitive bidding under a construction management program as previously discussed.

MARKUPS FOR FIXED PRICE AND GUARANTEED MAXIMUM PRICE PROJECTS

The term *markup* is often used to include performance risk, home office costs, general overhead, financing costs and planned profit. A previous discussion is applicable to cost-reimbursable contracts where the contractor or construction manager assumes little or no price risk. GMP and lump-sum contracts can be negotiated for design-construct and general contractor projects A GMP can also be negotiated for construction management contracts as defined by the AGC. (See Appendix C.) These methods should receive consideration where earliest completion is important and a risk sharing program between owner and contractor is both desirable and equitable to both parties.

Fixing the GMP or negotiated lump-sum price too early often leads to later arguments regarding scope. An alternate method involves selecting the contractor, design-constructor or construction manager based upon qualifications. A fixed fee for profit may be negotiated based upon early estimates. The project can begin in a manner similar to a cost-plus-fee program utilizing a phased schedule. Subcontracts and material procurements can be let competitively. When drawings near completion, the GMP or lump-sum can be negotiated based upon a new estimate prepared when the scope is almost fully defined. This type of estimate is often called a definitive estimate. Savings below a GMP will be shared with the owner. On the lump-sum alternative, the owner has transferred all of the price risk for remaining work to the contractor who will also receive full benefit from any underrun. Changing the scope during construction, however, will result in negotiations for extra work on both types of contract.

GMP project estimates will normally include a contingency allowance for the price risk to both owner and contractor based upon furthur detail development of the conceptual scope. The magnitude of the contingency will depend upon the the status of design completion at the time of negotiations. A fixed fee will be treated as a cost to the owner. Price risk to the contractor will normally be less than for a lump-sum contract since the contractor has the opportunity to negotiate costs without competition and the contingency allowance can minimize price risk and provide for an additional profit under favorable circumstances. On the other hand, the work is not fully defined when fixing the GMP. Here design growth or minor scope changes may cause adversarial relationships when the owner considers such work to be within the original scope and the contractor disagrees. Some contractors prefer a negotiated lump sum since they believe that a GMP provides for taking all of the upside risk but shares underruns with the owner.

Markups on competitively bid work are the most challenging to the contractor. If it sets the price too high, it seldom receives any work. If it bids too low he may destroy many years of successful work with one bad bid. Heavy construction work usually carries the most risk to the contractor and markups tend to be somewhat higher than for building and commercial work where a large number of subcontract bids limit the general contractor's risk. In fact, during very competitive periods, many contractors will put little or no markup on subbids and materials in order to obtain work. Contractors vary in their approach to determining markups. Some heavy construction contractors will look upon markup as a function of the labor cost for work which they plan to perform with their own forces. Other contractors will develop certain percentages for components such as labor, materials, subcontracts, indirects, etc. In any event, to be successful over the years a general contractor must recover all costs, including home office overhead and estimating expense, the cost of financing the project, depreciation on its equipment and all other direct and indirect costs plus a reasonable profit.

Table 9-2 includes selected financial data based upon a survey of 517 contractors including specialty trades, heavy and highway, single family and residential, industrial and other non-residential firms. After-tax income averaged only 1.2 percent for industrial and non-residential companies based upon a gross profit at the job level of 5.9 percent. All contractors as a group averaged 1.0 percent after-tax income from a gross profit of 8.2 percent.

Considering the risks involved, the after-tax return for construction companies as a percent of sales is extremely low compared to other businesses. Return on equity and invested capital is often highly leveraged and can be relatively favorable for the best managed construction firms.

OWNER RESPONSIBILITIES

If professional construction management is to prosper, not only does the manager have responsibilities to both the designer and the owner, but the owner has definite

TABLE 9-2
SELECTED FINANCIAL DATA (FROM BUILDING DESIGN &
CONSTRUCTION, JAN. 1990, PG. 17)

**Selected financial data for all contractors
vs. Industrial and Nonresidential Contractors
(thousands of dollars)**

	All companies	Industrial and Nonres. companies
Number of respondents	517	208
Assets	$23,872	$30,349
Liabilities	$17,717	$23,101
Net worth	$6,155	$7,248
% net worth to assets	25.8	23.9
Revenues	$61,295	$83,528
Gross profit	$4,996	$4,955
% gross profit	8.2	5.9
Gen'l/admin. expense	$4,113	$4,568
Net income	$588	$969
% net income	1.0	1.2
Return on assets (%)	2.5	3.2
Return on equity (%)	9.6	13.4

Source: CFMA Financial Survey.

responsibilities to the designer and the manager. On negotiated work using a general contractor or a design-constructor, the owner has similar responsibilities. Some of these responsibilities are the following:

Define Responsibilities

The owner must clearly develop or agree to a delineation of the responsibilities borne by all three members of the professional construction management team. Each must understand his own responsibilities as well as the extent of responsibilities delegated to the others. The owner must clearly spell out to the professional construction manager the extent of his procurement and cost authorities. The manager must be fully accountable to the owner to the extent of such delegations.

Maintain Professional Relationships

The designer and the professional construction manager must respect the professional relationships of each other, and the owner must also respect the professional integrity of each. The owner should place particular emphasis upon respecting the manager's position in dealing with all project contractors. Although the owner may normally retain the final approval, he should refrain from entering into negotiations separately except in unusual circumstances. Nothing can destroy the professional relationship faster than an owner who will deal directly with project contractors and conclude agreements or modifications directly in the absence of the manager.

Make Timely Decisions

Where owner decisions are required, the owner has a right to expect alternative evaluations and recommendations from the manager. In turn, if the manager is to be held accountable for the project schedules and costs, the owner must make decisions he has reserved for himself in a manner that is timely and that will not prejudice the program.

For professional construction management to work, the three-party team of owner, designer, and professional construction manager must truly function as a team with all parties working for the overall benefit of both the owner's project and the industry. This fact must be kept clearly in mind when selecting a designer as well as a professional construction manager.

MARKETING CM AND NEGOTIATED CONTRACTS

Marketing of CM services or negotiated work as a general contractor, developer or design-constructor varies according to both the potential client and the individual firm.

Public Clients

Public potential clients generally have a list of qualifications and requirements similar to the GSA requirement previously discussed. Firms are usually asked to submit qualification proposals, and through a system of ranking and scoring individual firms, the top three or so firms are "short-listed" to proceed with the final selection process, which may involve further qualification requests, development of a proposed organization, and nomination of the proposed project personnel, as well as a proposed plant for implementing the program.

Fees are usually quoted in the final selection process, and the final choice is sometimes dependent upon the fee as well as qualification.

Other agencies have relatively standard methods of negotiating fees, and the selection process is restricted to the determination of the best qualified firm. A contract is then negotiated with the chosen firm.

Marketing success for public clients depends highly upon the proposal response, the quality and believability of the proposed project personnel, and the impression given during a personal interview with proposed team members and top management as a part of the selection process.

Private Clients

Considerable ingenuity can be developed in marketing services to private clients. Personal contact with potential clients well in advance of a particular project inception can often turn up opportunities for negotiating a project with minimum or no competition. Offering to provide preliminary work plans, schedules, and even conceptual cost estimates is often very helpful to the owner preparing funding requests. Many times such "free services" can result in a negotiated project where the construction manager has the opportunity to carry out his proposed program. One of the best marketing tools is to perform an outstanding job on the first project, thus obtaining repeat business with minimal or no competition.

In the private sector, price is always a consideration. However, ingenuity by the construction manager in preparing programs that may save substantial sums in construction costs is a very effective basis of marketing and can lead to substantial future work if such savings are in fact achieved.

Marketing CM or other type negotiated work is similar to marketing other professional services. It helps to understand the owner objectives, and to be able to effectively propose programs to acheive these objectives as well as to present the firms' qualifications in the best possible light.

SUMMARY

In outlining an approach to the selection of a professional construction manager and other contractual services for negotiated work, this chapter in effect has reviewed several of the major advantages and disadvantages of each of the management approaches. Important qualifications for the manager or contractor include overall experience, financial status, depth of organization, experience in the project locality and type of work, an understanding of the owner's requirements, references, the preliminary plan for implementation, and the reimbursement and fee structure. The selection process should involve, in the order given, prequalification, request for proposals, evaluation of proposals, and final selection including fee considerations. Fees themselves will vary depending on the type and scope of the project, the services to be provided by the professional construction manager, and the allocation of costs to the reimbursable and nonreimbursable categories.

This chapter has concluded with guidelines regarding the owner's responsibilities. To make professional construction management or other negotiated type contract work successfully, it is important that owner, designer, and professional construction manager or contractor, are indeed united in a nonadversary team structured to serve the needs of the project.

On fixed-price projects it is important that risks assigned to the various parties bear a reasonable resemblance to the ability to control such risks. Contract documents that are fair to all, along with equitable contract administration during construction in a climate of mutual respect, can also help to minimize contractor claims and disputes.

THREE

METHODS IN PROJECT MANAGEMENT

10

CONCEPTS OF PROJECT PLANNING AND CONTROL

This chapter first describes interrelationships between engineering, design, construction, and operation costs for a facility and shows how the level of control over these costs decreases as a project evolves. An analogy and a model then introduce basic concepts of the information feedback-control process used on well-managed projects. The objective here is to provide a broad perspective against which to reference specific subsystems discussed in the subsequent seven chapters of Part 3; these will include estimating, project planning and scheduling, cost engineering, materials procurement and tracking, value engineering, quality assurance, and safety. Key components of the feedback control process to be examined in greater detail include: means for measuring and controlling progress; methods for information processing; requirements for effective reporting; and guidelines for taking corrective action to keep a project on target. Finally, given the important and increasingly integral nature of computer applications in project planning and control, the last section will briefly give an overview of this topic.

DESIGNING TO REDUCE CONSTRUCTION COSTS[1]

The concept to be explained in this section lies at the heart of professional construction management. The basic idea is not new. Variously described as "level of influence," "percent of effective control," "possible cost savings," "ability to control" and "degree of effectiveness," it has been well understood in some sectors of industry for many years, particularly in manufacturing, in heavy-industrial design-construct work, and more recently by general contractors interested in professional construction manage-

---

[1]This section is based on Boyd C. Paulson, Jr., "Designing to Reduce Construction Costs," *Journal of the Construction Division*, ASCE, vol. 102, no. CO4, December 1976, pp. 587–592.

ment. This chapter will adopt the term "level of influence" and will explore some of its implications in more detail.

Level of Influence on Project Costs

Figure 10-1 illustrates essential features of the level-of-influence concept. The lower portion simplifies the life of a project to a three-activity bar chart consisting of (1) engineering and design, (2) procurement and construction, and (3) utilization or operation. The upper portion plots two main curves. The curve ascending to the right-hand ordinate tracks cumulative project expenditures. The curve descending from the left-hand ordinate shows the decreasing level of influence. The bar chart and both curves are plotted against the same horizontal abscissa: project time.

FIGURE 10-1
Level of influence on project costs. (From Boyd C. Paulson, Jr., "Designing to Reduce Construction Costs," *Journal of the Construction Division*, ASCE, vol. 102, no. CO4, December 1976, p. 588.)

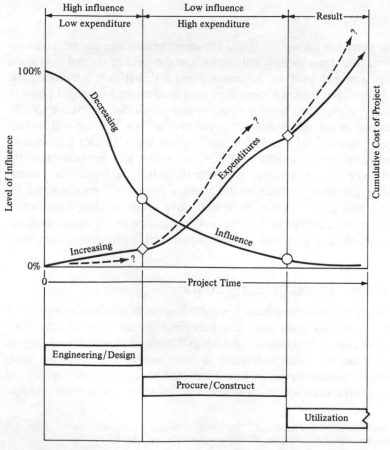

The parts of the figure interrelate as follows: In the early phases of a project, that is, during feasibility studies, preliminary design, and even detail design, the relative expenditures are small compared with those of the project as a whole. Typically, engineering and design fees amount to well under 10 percent of total construction costs. Similarly, capital costs invested by the time that construction is completed often are but a small fraction of the operation and maintenance costs associated with a project's complete life cycle. However, although actual expenditures during the early phases of a project are comparatively small, decisions and commitments made during that period have far greater influence on what later expenditures will in fact be.

On the first day, management has a 100 percent level of influence in determining future expenditures. Paraphrased simply, the question is: To build or not to build? A decision of not to build requires no expenditure at all for the project. A decision to build requires more decision making, but initially at a very broad level. For example, shall a new power plant use conventional or co-generation (co-gen) technology? If co-gen, how many units and how big? As engineering and design continue, decisions become more detailed, but the implications are no less significant. Shall we use a thicket of number 11 (3.5 cm) bars or more widely spaced number 18's (5.7 cm) in reinforcing this part of the turbine pedestal? And so on until the final drawing change is stamped, signed, and approved.

As these decisions evolve and commitments are made, the remaining level of influence on what the project costs will ultimately become drops precipitously. For example, a rough but educated guess would put the remaining level of influence at about 25 percent of the original by the time field construction commences on a new petroleum refinery. This 25 percent represents the control that construction contractors have through productive use of labor, innovative uses of equipment and methods, and wise materials procurement practices. But it is the designer who has more radically influenced construction costs. It is the designer who may or may not have packed the reinforcing steel so densely that concrete cannot be placed by economical procedures. It is the designer who may or may not have specified nonstandard sizes, impossible formwork configurations, techniques requiring incompatible mixtures of labor crafts, bronze fittings where galvanized were more than adequate, etc. All too often such decisions are made without the slightest notion as to their impact on construction costs, but they influence the costs nonetheless. "Value-engineering" clauses in construction contracts are at best after-the-fact remedies for such fundamental oversights.

In like manner, decisions made during construction, even within the 25 percent or so remaining level of influence, can greatly impact the costs of operating and maintaining a facility. Skimping on quality in workmanship or substituting inferior materials may save a few dollars in construction costs, and in contracts with a "profit-sharing" clause even the owner may be pleased in the short run, but costs resulting from excessive maintenance, downtime, and inefficient operations can consume those "savings" many times over. By the time construction is completed, however, the die is cast. What little influence remains after project start-up often takes the form of shutting the facility down for expensive rework, modifications, or retrofitting.

Costs to Whom?

Many problems associated with the level-of-influence concept result from cost suboptimization inherent in the contractual structure for a project. As a hypothetical example, assume that an owner desiring a new ore-processing facility has obtained a fixed- fee contract with an engineering firm and plans to let a competitively bid construction contract once designs are complete. Assume also that the engineer has designed several similar plants in the past, and has thus negotiated his fee, say $800,000 including 20 percent ($160,000) profit and general overhead, on the assumption that much of the design can be adapted from earlier drawings.

Two of the required drawings describe the electrical instrumentation and controls for a crushing and grinding circuit. Since it is similar to one done before, the engineering budget provided for it as follows:

$$80 \text{ design-hours} \times \$60/\text{design-hour} = \$4800$$

However, assume that these drawings represent $640,000 worth of equipment, materials, and field construction labor.

When the electrical engineer was about 60 percent complete on these drawings, he got into a discussion with a vendor who described a new system for this application that, through the use of solid-state technology, could save approximately 20 percent of the costs of purchasing and installing a conventional facility, or $128,000 in this case. The vendor had the facts to back up his numbers.

After this discussion, the electrical engineer went to his supervisor to suggest that this change be incorporated. His supervisor denied the request with the following reasoning:

Costs to date for adapting conventional design:	
60 percent of $4800 =	$2,880
Estimate for revised design using new technology:	
200 design-hours × $60/design-hour	$12,000
Subtotal	$14,880
Initial estimate for drawings E-247 and I-186	($4,800)
Design cost overrun:	$10,080

The fact that this potential 210 percent overrun on these two drawings represented over 6 percent of the profit and overhead for the whole job was but one problem. It might have been negotiated as a change with the owner. However, the design as a whole was already two months behind schedule, and the client was thought to be in no mood to even hear of, let alone approve, such a change at this time, or so rationalized the supervisor to the electrical engineer. Also, he argued, "This new technology is not well proven yet. Stick with conventional designs that we know will work."

Similar decisions are made every day in practice. The one just described suboptimized costs at the level of the engineering and design firm at the expense of potentially much greater savings ($128,000 versus $10,080) at the capital-cost level. Analogously, capital costs are too often suboptimized at the expense of life-cycle costs. For example:

Unit A costs $250,000 and operates at $1.20/ton.
Unit B costs $300,000 and operates at $1.15/ton.
Annual production is 800,000 tons.

The $0.05 per ton saved by unit B times 800,000 tons per year represents annual savings in operation costs of $40,000. This should quickly recoup the extra $50,000 capital cost, but again, if a design-construct contract puts too strong an incentive on reduced capital costs, a bad economic decision for the owner may result.

Even at the owner's level, costs can be suboptimized at the expense of industry as a whole or of society. An owner sometimes moves into an area to construct a large project where "time is of the essence" and construction cost is secondary. Perhaps the objective is a factory to produce a new small car to stem the tide of foreign competition, or a pipeline of high national priority to deal with the energy crisis. Too often, however, the indirect costs resulting from the distortion of the labor market, the economic and social impact on nearby communities, and the disruption to other firms competing for the same scarce labor and material resources far exceed the savings to the individual owner.

The point here is that contractual structures can be adjusted to minimize the consequences of suboptimization of the type described above. The first prerequisite, however, is an understanding of some of the economic forces involved. The level-of-influence concept as shown in Figure 10-1 may help toward this end.

Contractual Implications

Two important conclusions can be drawn from the level-of-influence concept. First, owing to the tremendous impact that design decisions have on construction and operation costs, contractual arrangements should be drawn to assure that construction and, where appropriate, operations thinking is strongly injected in the conceptual, preliminary, and detail design processes. Second, efforts to suboptimize design costs alone, for example by requiring competitive bidding for professional engineering and architectural services, can have disastrous consequences for the owner's budget when construction and utilization costs are considered.

Injecting Construction Knowledge In the first case, several current contractual arrangements, if properly applied, can at least inject construction thinking into the design phase. Major examples include professional construction management and also design-construct or turnkey contracts. The name alone, however, does not guarantee results. For example, an architect may offer "professional construction management" services. If by this he means that, acting as agent for the owner, he will let and administer separate construction contracts, possibly on a phased construction basis, one has to assume that he has a wealth of contractor-type construction knowledge if there are to be any savings at all. Knowing how to package separate construction contracts along recognized trade and jurisdictional boundaries, as well as accurate knowledge for estimating time and costs for the different operations, are essential. Few design consultants really have these capabilities.

Design-construct has its pitfalls also. First, third-party objectivity and interaction at the design-construct interface are lost. Even in a fully professional and highly ethical firm, organizational inertia can perpetuate obsolete practices to the exclusion of innovative thinking. Where separate design and professional construction management firms are teamed in different combinations, innovations are more likely to be transferred. Another problem is that even in design-construct firms, one too often finds people on the drawing boards who rarely if ever get to the field to see the physical results of their decisions, for better or worse.

It is also important to reemphasize that no one contractual arrangement is best for all situations. For example, one large private university increasingly uses professional construction management only in the design phase, and it sees considerable benefits from injecting construction thinking there. However, the university discerns much less benefit for professional construction management during construction, where it lets conventional competitively bid general contracts instead. Some professional construction management enthusiasts would object that the latter forgoes the time savings resulting from phased construction. Where there are revenue-generating time pressures for beneficial occupancy, phased construction's uncertainty and the possible increase in direct costs are often a risk worth taking. In the university's case, however, risk of overrun when the later contracts are let is unacceptable because fund limitations are often absolute. Also, time pressures for beneficial occupancy are much lower, and potential short-run economic returns are more difficult to quantify, since the size of the student body, faculty, and staff is held constant anyway. New facilities serve mainly for replacement or enhancement rather than for generating new revenue.

Competitive Bidding The second case, that of competitive bidding and related procurement techniques for professional services, is more insidious. Volumes of material have been written on this subject in recent years, and much of it has been read into the *Congressional Record*. It is nevertheless important to view the subject again in light of the level-of-influence concept.

The assumption that one can save money by choosing an architect/engineer solely on the basis of lowest design fee is false economy of the worst sort. On the other hand, this need not imply that there is a direct linear correlation between the amount spent on design and the quality or utility of the structure. But knowledgeable owners and agencies need the authority to evaluate alternative professional firms through selection procedures that will engage the firm that can most competently produce a structure of maximum utility for the lowest overall costs, including social and environmental costs as well as design, construction, and operation costs. Selection solely on the basis of lowest fee is likely to perpetuate obsolete designs based upon drawings long since filed away and to force the use of other short-cuts for cost shaving.

PROJECT PLANNING AND CONTROL[2]

Once the professional construction manager has been chosen, and even when fair and equitable contractual agreements have united all members of the project team and

---

[2]This section is based on Boyd C. Paulson, Jr., "Concepts of Project Planning and Control," *Journal of the Construction Division*, ASCE, vol. 102, no. CO1, March 1976, pp. 67–80.

oriented them toward the owner's goals, the team will still face the most challenging part of project management: planning and control to bring the project to completion on schedule, within budget, and in accordance with the owner's functional objectives. For this they will need the fullest understanding of the planning and control process and all the practical tools that can be put at their disposal.

How will they cope? To begin, consider an analogy.

Planning and Control: An Analogy

In many ways, planning and control principles applied on a complex engineering and construction project resemble those needed in planning and taking a trip in a car. Assume, for example, that a family intended to travel from Chicago to Los Angeles for a two-week Disneyland vacation. At one extreme, they could start with no plans; they would simply set out without maps, without a planned budget, and with only the general notion that Disneyland is near Los Angeles, which, in turn, is somewhere west or southwest of Chicago. After ending up lost, broke, out of gas, tired, and distraught on a lonely highway somewhere between Riddle, Idaho, and Wild Horse, Nevada, the trip—their "project"—would most likely be declared an unmitigated disaster, or perhaps, more charitably, a "unique experience."

Now consider the other extreme. Assume that several months before starting, they engaged a licensed highway engineer to study the most feasible alternative routes, select the best one, then literally survey that route and prepare a log documenting, at 20-foot intervals, all stop signs, intersections, potholes, curves in the road, speed limits, gas stations, motels, and other details. The engineer would then prepare a detailed list of instructions telling the speeds to maintain, when to apply the brakes, when to turn, where to stop for the night, and so forth. Once underway at 65 miles per hour, this detailed information would come so fast and in such great volumes that the whole family would constantly have to be reading the plan to figure out what was happening. They would be so engrossed in the documentation of what had gone past that they might miss unexpected opportunities en route, or, worse yet, the driver would not be prepared to deal with unexpected emergencies, such as a child running into the street, a detour ahead, or a truck overturned. Clearly, this approach to planning and control is another formula for disaster. Not only would the "information system" be inordinately expensive, but it would also prove so distracting that it not only would fail in its basic purpose but would inhibit human judgment and reasoning as well.

Both approaches have been and continue to be used in construction, and both cause analogous results. What is needed is something between these two extremes. One should have good basic planning before commencing a project as well as a journey, but the approach taken must also allow management the flexibility to respond to, and even turn to advantage, the unexpected changes and events that will inevitably occur. A project should have a budget; its designs should be on paper; it should have a schedule which in turn forecasts the requirements for resources of labor, equipment, and materials; but it also needs a dynamic and responsive feedback-control system to cope with the operations underway.

Consider again the analogy from a moment ago. A car with its driver is, in effect, such a dynamic feedback control system. The car in motion is the project. The driver looks down the road, that is, into the future, and receives information: the road curves to the right; there is a stop sign; the car ahead brakes suddenly to miss a deer. The driver takes this information, analyzes it, and through her body and the mechanism of the car, takes responsive action to keep the vehicle safely on the road; that is, she controls her project.

Now take this analogy of the car and do a rather odd thing. Paint out the windshield so that the driver can only look to the sides to see where she is, and to the rear to see where she has been. This may not be as absurd as it sounds. If the driver has driven the road hundreds of times, say from the family farm to town, with caution she might well be able to start the engine and proceed slowly to her destination. The information from the sides is better than none at all, and it at least keeps her on the road. This is analogous to the type of information system in construction that does a good job of documenting how the project has progressed to date, but has little or no provision for forecasting where it is going or what its needs will be in the future. If a foreman arrives at work and finds that he is out of bricks, he will order more at that time. If delivery takes 3 weeks, the project is delayed but will nonetheless be completed, slowly and inefficiently, but eventually. Many contractors have managed to operate this way for years.

Now successively paint out the side windows and the rear window of the car, until there is no outside vision at all. Perhaps by the feel of an old familiar road, the woman from the farm may still find her way to town, but the hazards will be great. Clearly, if she finds herself in this situation at 65 miles per hour on a crowded urban freeway, that is, on a larger, unfamiliar project, she is really in trouble. This, again, is analogous to construction planning and control systems with successively slower and slower feedback of less and less information, until finally either the bank or the Internal Revenue Service informs the contractor (the driver's counterpart) that she was bankrupt some time ago.

Although the extremes in the analogy of the automobile may seem a little absurd to any reasonable construction contractor, it is nonetheless astonishing how many of these same contractors continue to run their projects at the analogous extremes. If this book brings slightly more order to this chaos, it will have served its purpose.

Objectives

General objectives for an information system designed to aid management in the planning and control of engineering and construction projects may be stated as follows:

1 To provide an organized and efficient means of measuring, collecting, verifying, and quantifying data reflecting the progress and status of operations on the project with respect to schedule, cost, resources, procurement, and quality.

2 To provide standards against which to measure or compare progress and status. Examples of standards include CPM schedules, control budgets, procurement schedules, quality control specifications, and construction working drawings.

3 To provide an organized, accurate, and efficient means of converting the data from the operations into information. The information system should be realistic and should recognize (*a*) the means of processing the information (e.g., manual versus computer), (*b*) the skills available, and (*c*) the value of the information compared with the cost of obtaining it.

4 To report the correct and necessary information in a form which can best be interpreted by management, and at a level of detail most appropriate for the individual managers or supervisors who will be using it.

In keeping with the principles of management by exception, the following two objectives should be added:

5 To identify and isolate the most important and critical information for a given situation, and to get it to the correct managers and supervisors, that is, those in a position to make best use of it.

6 To deliver the information to them in time for consideration and decision making so that, if necessary, corrective action may be taken on those operations that generated the data in the first place.

Project Planning and Control: A Model

The flowchart in Figure 10-2 models the operations, flow of information, and decision-making processes characteristic of a feedback control system appropriate for a medium to large sized engineering construction project. It has been designed to reflect the objectives stated in the preceding section.

Note that although this flowchart applies equally well in the conventional design and construction process where the two phases are largely separated, a control system of this type can have its greatest impact in the professional construction management or design- construct approaches where there is a strong interplay between all aspects of the system: concept; design; procurement; and construction. These approaches are especially prevalent in the large heavy-industrial projects, such as refineries, mining developments and nuclear power plants, and in the commercial building field, with projects such as the New York World Trade Center and the Chicago Sears Tower. In structures of this type, engineers involved in design and those in construction interact continuously to optimize the facilities from both points of view.

Components In the flowchart, the project is initiated according to a predefined plan (box 1) and operations get underway (box 2). The plans also become reference standards for control purposes (box 5). As operations continue, external factors (box 3) such as recently imposed standards or newly available materials in design, or bad weather, strikes, procurement delays, foundation excavation problems, or even unexpectedly good conditions on the site, may cause the course of operations to differ from the plan, or may provide opportunities for improving on the plan. The operations underway generate indicators or progress (quantities in place, elapsed time, money expended, or resources consumed) which may be measured (box 4) and fed as data into a system (box 6) to produce information for decision makers. This in-

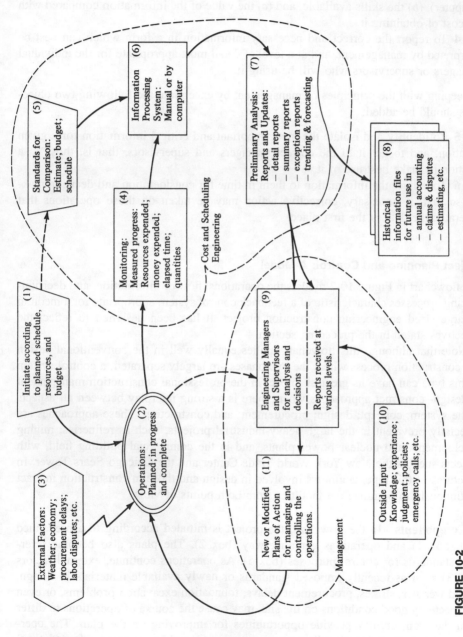

FIGURE 10-2
Flow chart of project control system. (From Boyd C. Paulson, Jr., "Concepts of Project Planning and Control," *Journal of the Construction Division*, ASCE, vol. 102, no. CO1, March 1976, p. 71.)

The following labels appear within the flow chart:

(1) Initiate according to planned schedule, resources, and budget

(2) Operations: Planned; in progress and complete

(3) External Factors: Weather; economy; procurement delays; labor disputes; etc.

(4) Monitoring: Measured progress; resources expended; money expended; elapsed time; quantities

(5) Standards for Comparison: Estimate; budget; schedule

(6) Information Processing System: Manual or by computer

(7) Preliminary Analysis: Reports and Updates:
  – detail reports
  – summary reports
  – exception reports
  – trending & forecasting

(8) Historical information files for future use in
  – annual accounting
  – claims & disputes
  – estimating, etc.

(9) Engineering Managers and Supervisors for analysis and decisions. Reports received at various levels.

(10) Outside Input: Knowledge; experience; judgment; policies; emergency calls; etc.

(11) New or Modified Plans of Action for managing and controlling the operations.

Cost and Scheduling Engineering

Management

formation processing system refers to planned standards (box 5), such as schedules and budgets, to show deviations, variances, and trends. The information is analyzed and made available through reports (box 7), which may be stored for future reference (box 8), or given to engineering managers and supervisors for their further analysis and decision making (box 9), or both. They combine and compare this information with their own knowledge, experience, policies, and other qualitative and quantitative information and judgement (box 10) in order to produce new or modified plans for continuing and controlling the project operations (box 11).

Feedback   This is a feedback control system, and it operates continuously throughout the life of a project. Associated with it is a feedback time. Ideally, the time through parts 4, 6, and 7 should be as short as possible so that engineering managers and supervisors can receive accurate and up-to-date information in time to make decisions and formulate plans of action so as to have maximum impact in controlling those operations which are generating the information in the first place.

On a small project, it is possible to short-circuit the path from box 2 to box 9 and provide direct feedback. A master builder who works with his own tools in constructing custom suburban homes is a common example of this. If something is wrong, he knows about it immediately.

On a large project, such as the design and construction of a rapid transit system or a nuclear power plant, direct feedback to the decision makers of all information on all activities is no longer possible. One needs a staff and an organized system to measure, process, analyze, and report the most important information to the decision makers. In engineering design and construction, this staff consists largely of scheduling engineers and cost engineers. Nevertheless, the goal remains to provide feedback to decision makers in minimum time for maximum impact in controlling operations. Needless to say, however, it is here that the industry experiences some of its greatest difficulties.

A major need in project planning and control is significantly to improve and expedite the operations represented by boxes 4, 5, 6, and 7 on the flowchart in order to help resolve these difficulties and improve the quality of information available to decision makers. On larger projects, some improvements are being made through computer applications. Indeed, like a neighborhood power outage on a cold winter's night, it almost takes a computer failure these days to dramatize the extent to which such projects are coming to take their computers for granted.

The following sections will focus on some of the key components of the system that has been described and will amplify some of the concepts presented in the model.

Status and Progress

Numerous measures can be taken to determine the progress or status of operations on a project. Quantities of work units in place can be physically surveyed and compared with those shown on the drawings. Elapsed *time* can be compared with the estimated activity or project durations. *Money* committed or expended can be compared with the estimated budget. *Resource usage* can be plotted versus expected requirements for labor, materials, and equipment. Finally, an experienced professional construction manager can simply apply *judgment* to estimate the percentage completed on individual activities or on the project as a whole.

Each of these measures has its advantages and disadvantages. For example, field measurements may be more accurate than judgment estimates of percentage complete, but it is expensive to use a surveying crew to obtain these data. Judgment, in turn, can reflect qualitative factors, not evident in the quantities themselves. Just as pulse, temperature, blood tests, and x-rays give several different readings on the condition of one's body, each of the aforementioned measures tells something different about the project. All of them are necessary to gain a full understanding of the status and progress of the operations.

Nonlinear Relationships   In applying such measures, it is important to recognize nonlinear relationships among them. For example, there may be a nonlinear relationship between quantities in place and elapsed time. To illustrate, if the bulk of the work is scheduled to be completed earlier in the activity's scheduled duration, then when the time is 50 percent elapsed, the work might actually and correctly be 60 percent complete. Similar nonlinear relationships apply among the other measures. The time at which money is expended on materials, for example, might be only loosely related to the actual time those materials are used.

When comparing the expenditure of labor resources over time, one can also often recognize nonlinear "learning-curve" effects. Learning curves relate time, resources consumed, and quantities produced. Their basic principle is that skill and productivity in performing tasks improve with experience and practice. The nonlinear implications of learning curves are different for planning, or estimating, than they are for control. Chapter 11 will discuss these concepts in greater detail.

Source of Data   Data reflecting status and progress come from numerous sources. In the formal information system, sources include labor and equipment time cards, purchase orders, invoices, field quantity reports, quality control reports, and so forth. In all cases, accuracy, timeliness, and completeness are important. Human considerations are particularly essential at this point if good management information is to be produced.

In addition to the formal sources, there are numerous other inputs to management, some of which short-circuit most of the regular steps. If there is a serious accident on the job, the superintendent or project manager is told about it almost immediately. Similarly, a manager simply cannot wait for a computer printout to tell him the cofferdam is about to be overtopped. However, if the more routine aspects of planning and control are organized into an accurate and effective information system, management is in an even better position to cope with the unexpected events that inevitably occur.

Information Processing

Later chapters will go into the methods and techniques that are applied in analyzing data and converting them into useful information. Broad categories include activity and resource scheduling tools such as bar charts and critical path networks, cost engineering budgets and cash flows, materials procurement and tracking systems, and statistical quality control.

In concept, information processing systems take progress and status data, compare them against reference standards such as budgets or schedules, and convert the results to information needed by the managers and supervisors on the project. As stated in the objectives, the level of detail, the variety, and the frequency of reports to be produced should be appropriate to the people who will use them, should be feasible for the means of processing the information (manual or computer), should recognize the skills available, and should realistically assess the value of the information compared with the cost of obtaining it. Finally, the system should be fast, efficient, and accurate.

In practice, several related subsystems are needed fully to plan and control projects. Examples include activity and resource scheduling and control, cost engineering, materials procurement and tracking, and quality control. Each of these systems is important, but if fully integrated into one system, the sheer volume of data would dominate and obscure the vital information that is needed from any one of them. An interrelated modular system is thus essential. That is, each subsystem should be largely self-sufficient, but it should be logically coordinated and compatible with the others.

Consider the whole process from the point of view of a network-based subsystem for activity and resource scheduling and control. Costs, materials, and quality functions can also be identified with activities, so it is possible to use the activities as a means to tie into other systems. In summary, an information processing system for project planning and control should recognize that there are many subsystems involved in the process, and it should further recognize the interrelationships among those subsystems.

Reporting

Reporting can take many forms, ranging from conversations and telephone calls through tabular presentations of cost information and graphical presentations on bar charts, cumulative progress ("S") curves and CPM diagrams, to up-to-the-minute reports from computers transmitted via microwave or satellite telecommunication links to sophisticated microcomputers in the field. Certain basic principles should guide each of these, however, if the reporting is to be effective for control purposes.

Content   Regardless of the form, in order to be effective for control purposes, a complete report should have five main components:

1 *Estimates* either total, to-date, or this period, that provide a reference standard against which to compare actual or forecast results

2 *Actuals* what has already happened, either this period or to-date

3 *Forecasts* based on the best knowledge at hand, what is expected to happen to the project and its elements in the future

4 *Variances* how far actual and forecast results differ from those which were planned or estimated

5 *Reasons* anticipated or unexpected circumstances that account for the actual and forecast behavior of the project and its operations, and especially that explain significant variances from the plan

The following paragraphs discuss some related reporting principles in more detail.

Selectivity and Subreporting One of the objectives stated previously was to report the correct and necessary information in a form which can best be interpreted by management and at a level of detail most appropriate for the individual managers who will be using the information. Selectivity and subreporting are important here. Since time is among their scarcest resources, construction managers and supervisors simply cannot afford to wade through piles of extraneous data to obtain the information they need. The concrete superintendent should have reports focusing on concrete operations. The project manager should have summary reports as well as logically coordinated detail reports to back them up.

Variances Reports for control purposes should calculate variances to show which operations are relatively more in need of attention than others. "Variance" is used here to mean a deviation from a planned or budgeted item. The variances, in turn, should be expressible in both relative (percentage) and absolute (quantities, dollars, etc.) terms. For example, is it more important for a manager to focus attention on a $100,000 operation with an absolute variance of plus $2,000 (overrun) and a relative variance of plus 2 percent, or on a $10,000 operation with a relative variance of plus 15 percent and an absolute variance of $1,500? With both types of variance information, the manager can apply her judgment as she thinks best.

Management by Exception By showing only those operations with variances or other parameters exceeding certain predefined limits, exception reports focus management attention directly upon those operations most in need of control. The principle here is to identify and isolate the most important and critical information for a given situation, and to give it to the right person as quickly as possible for consideration, decisions, and action. To be truly effective, however, it is particularly important that exceptions be related to standards that are indeed accurate.

Forecasting and Trending If management is to have clear vision ahead and be able to anticipate problems before they arise, reports must look to the future as well as document the past. Forecasting and trending are two means by which this is done. In network-based schedule and resource control, the network logic itself provides a vehicle for determining what effect a change in one operation will have on the project as a whole. Procurement and tracking systems should be similarly designed so that a superintendent does not arrive at work one morning to find that a critical 6-month lead-time item of equipment to be installed that day has not even been ordered yet. Related principles apply to other systems.

Feedback Time In all the aforementioned cases, the information reported must be received in time so that, if necessary, corrective action may be taken on those operations that generated the information in the first place.

Example Figure 10-3 shows an example report that illustrates many of the ideas mentioned here. First of all, note that the report is selective in showing only those

EXCEPTION REPORT BASED ON COST VARIANCES

ACTIVITIES SHOWN HAVE VARIANCES GREATER THAN 10%

ACTIVITIES ARE SORTED AS FOLLOWS:
ES PRIMARY SORT
DUR SECONDARY SORT

PROJECT NUMBER: 198301
PROJECT TITLE: EXAMPLE CONCRETE BRIDGE
PROJECT LOCATION: STANFORD, CALIFORNIA

START DATE: 04/21/83
DATA DATE: 07/11/83
REPORT DATE: 07/14/83

LAB. CODE / ACTIVITY DESCRIPTION		UNIT	ESTI-MATED	ACTUAL TO DATE	EST. TO COMPL.	EST. TTL @ COMPL.	FORECAST VARIANCE	PERCENT COMPL.	PERCENT VARIANCE	TREND
CFF. 0310 CONCRETE FORMWORK FOR FOUNDATION		SF								
EST START: 06/02/83	DURATION:		34	29	9	38	4	76%	11.8%	HIGH
ACT START: 06/02/83	COSTS:		$49800	$47600	$14900	$62500	$12700	76%	25.5%	HIGH
EST FINISH: 07/18/83	QUANTITIES:		15860	13600	4150	17750	1890	77%	11.9%	HIGH
ACT FINISH: / /	UNIT COSTS:		$3.14	$3.50	$3.59	$3.52	$0.38		12.1%	HIGH
CRF. 0320 CONCRETE RE-STEEL FOR FOUNDATION		LB								
EST START: 06/09/83	DURATION:		29	21	9	30	1	70%	3.5%	HIGH
ACT START: 06/12/83	COSTS:		$39400	$31700	$13610	$45310	$5910	70%	15.0%	HIGH
EST FINISH: 07/18/83	QUANTITIES:		82400	74410	31890	106300	23900	70%	29.0%	HIGH
ACT FINISH: / /	UNIT COSTS:		$0.48	$0.43	$0.43	$0.43	−$0.05		−10.4%	LOW
CPF. 0330 CONCRETE PLACING FOR FOUNDATION		CT								
EST START: 06/16/83	DURATION		34	16	21	37	3	43%	8.8%	HIGH
ACT START: 06/19/83	COSTS:		$73400	$35540	$46660	$82200	$8800	43%	12.0%	HIGH
EST FINISH: 08/01/83	QUANTITIES		1200	690	850	1540	340	45%	28.0%	HIGH
ACT FINISH: / /	UNIT COSTS:		$61.17	$51.51	$54.89	$53.38	−$7.79		−12.7%	LOW
CRW. 0320 CONCRETE RE-STEEL FOR WALLS		LB								
EST START: 06/30/83	DURATION:		34	9	26	35	1	26%	2.9%	LOW
ACT START: 06/30/83	COSTS:		$37800	$8320	$24940	$33260	−$4540	25%	−12.0%	LOW
EST FINISH: 08/15/83	QUANTITIES:		78750	19730	59170	78900	150	25%	0.2%	LOW
ACT FINISH: / /	UNIT COSTS:		$0.48	$0.42	$0.42	$0.42	−$0.06		−12.2%	LOW

- - - END OF REPORT - - -

FIGURE 10-3

Example control report. (Adapted from Boyd C. Paulson, Jr., "Concepts of Project Planning and Control," *Journal of the Construction Division*, ASCE, vol. 102, no. CO1, March 1976, p. 77.)

operations that are of interest to the concrete superintendent. Second, it is an exception report showing only those items with variances exceeding 10 percent. The actual cutoff used here is variable. A great deal of information is shown for each operation, including estimated, actual, and projected schedule times, costs, quantities, and unit costs, as well as variances and trends. To some users, this may prove confusing and would therefore need to be simplified. However, if the format were standardized and applied not only for exception reports, but also for detail reports, subreports, summary reports, and so forth, users would quickly become accustomed to it and would learn to benefit from all the information shown. The complete status for an operation is all conveniently in one place.

The report uses multiple lines for each operation rather than having one long line with some 30 or 40 columns, as is typical of many manually prepared job progress reports. An obvious reason for this use of multiple lines is that the printers used by computers are generally limited to 120- or 132-character widths. However, certain advantages are inherent in this approach that may not be readily apparent. First of all, if additional information were to be added, such as work-hours by craft or equipment usage hours, or a breakdown of cost into labor, materials, equipment, and subcontracts, it could easily be inserted as additional lines where desired, yet still preserve the same basic and familiar appearance. Second, the close grouping of information in this way assists in a pattern-recognition process that will analyze reasons for apparent inconsistencies. For example, on a given item, the cost may show a 12 percent overrun and the quantity may show a 28 percent overrun, both high and possibly bad, yet the unit cost will be under the estimate, which ought to be good. If this is a unit-price item, all is well. If it is a lump-sum item, there apparently are problems. The format helps a manager to apply this type of reasoning quickly and take the appropriate action.

This report is presented only as an example and not as a recommended standard. It illustrates the type of information that might be provided to a manager for control purposes. The next section considers what a manager might do upon receiving this information.

Corrective Action

When a control report indicates that something is "wrong" with an operation, that is, that its measures are deviating significantly from the plan, management should first investigate to find and understand the reasons behind the symptoms reported. Assuming that the source of the problem can be identified, one alternative course of action, and often the best one, is basically to do nothing except update the reporting system to reflect the reality on the job. Because of the dynamics of the operations on the project, the situation as it exists may actually be better in some ways than that which was planned. From day one onward, project management has more information than the planner had originally. The point here is that one should not take corrective action merely for the sake of making the job conform to the original plans. That is, budgets, schedules, and related standards are tools to be used by management. Returning to the earlier analogy, the driver, not the steering wheel, should control a car. Managers should use their control system as a guide, but they need flexibility and should be prepared to take advantage of and adapt to new conditions as they arise. Chapter 12 will have more to say on this subject.

COMPUTER APPLICATIONS IN PROJECT PLANNING AND CONTROL

In electronic technologies, the last few years have been like a time warp. Technical terms that once separated computer professionals from mere users have become household words. Whimsically named companies founded by child prodigies with capital from hocked momentos of the counterculture have grown by megabucks to become producers of mass consumer goods. Little machines more powerful than prestigious blue-chip business processors of two decades ago are mistaken for—indeed, used as—toys.

The "toy" that plays spacewar and adventure games in thousands of homes is also keeping business records for national organizations. Hard-headed contractors who would not have touched a keyboard five years ago are now overheard comparing notes with others on PRINT USING statements and DOS files, though still somewhat clandestinely lest the unconverted think they have substituted electronics for judgment and experience. Their colleagues do indeed whisper about them, but the whisperers are becoming more wistful and will continue to do so until they too figure out how the little machines can be used to cut through office drudgery.

The motor car was a toy of the elite, a rich man's luxury, until the Model T made it accessible to every person and also made a driver's license a socio-economic necessity. The microprocessor has become the Model T of computers, and proficiency in its application in individual professional activities will be essential to the productivity and success of future construction managers. Those equipped with skills in word processing for job-site correspondence, electronic communications, and tools like spreadsheets for estimates and payment requests will leave the unprepared drowning in paper.

Low-cost computers have arrived none too soon if construction professionals are to cope with the growing administrative burdens imposed by internal and external demands. Already, many major construction companies have software far more sophisticated than that available a decade ago, especially in cost control, procurement, estimating, and scheduling. Until recently, however, computer hardware acquisitions in smaller and medium-sized firms did not keep pace with the level of technology available in industry. It is catch-up ball, but microcomputers at least enable smaller firms to stay in the game. Central computer centers become increasingly obsolete, and $100,000 minicomputers are out of the question for most small companies, but many do seem to be able to break loose $2000 to $5000 from tight-fisted controllers, so the micros are cropping up all over.

What does the money buy? Mostly hardware. Typical acquisitions involve one or more Apple Macintoshes, IBM PCs or PC "clones," with a million or more characters of memory, a hard disk drive, one or two floppy disk drives, a small printer, a high- resolution CRT monitor, the disk operating system (DOS), and one or more applications (wordprocessor, spreadsheet, database, etc.). It is surprising how many people will willingly spend several thousand dollars for hardware and software, but not invest a few hundred more and invest the time for training on how to best use these powerful tools. It took only money to leave the Model T parked in the driveway for the neighbors to admire, but learning how to drive the car required an investment of time and effort. It also takes time and effort to learn how to use software—and develop it—but that is the key to productive use of computing machines. Whether acquired or developed, software must be the focus of computer efforts in construction management and engineering.

This section will next review present and potential construction application areas. Computers can be of assistance in all aspects of project planning and control. Topics corresponding to major sections of this book include estimating, cost control, scheduling, quality assurance, procurement, and the general subjects of administration and productivity analysis. Each topic is addressed briefly in the following sections.

Estimating

Computers can help estimators in most phases of their work. For example, a good application of smaller computers is to help with the partial automation of a quantity takeoff. In this case, an estimator can use a stylus or cursor on a drawing laid over a digitizer board to directly input measurements and item counts. An electronic image of the project's geometry can be built up and even displayed on graphical CRTs. Simultaneously, the computer can be building a file containing the bill of materials.

In developing crews and evaluating the productivity of labor and equipment, computers can first provide data from files of past projects and second assist with specialized engineering programs for calculations such as cableway cycles, earth-moving fleet simulation, formwork calculations, and so on. In the costing phase, a computer can maintain files of current cost information, and with remote access it is now becoming possible to tie into various data bases maintained by traditional estimating firms. As the estimate nears completion, a computer can be particularly valuable for developing spread sheets and in bid-sheet preparation. This is especially important in the final hectic hours of competitive bidding, when newly arrived subcontractor bids and materials quotations can make frequent recalculations of the estimate necessary. With the calculations accurately and reliably under control, management can also indulge in sensitivity analysis to determine which aspects of the bid contain the greatest risk and thus more intelligently determine the best markup. Computer programs providing a rough cash flow can also enable the markup to be based upon desired rate of return rather than simply a percentage of gross profit.

Cost Control

In larger construction companies, computers were often first installed under the control of the accounting department. Therefore, it was not surprising that cost control systems were generally extensions of the payroll and accounting and finance people. Over the years, however, job costing and cost-engineering systems have evolved that, although perhaps connected to the company's accounting system through a data base, are deliberately oriented toward the needs of company and project management. Some of the more sophisticated systems apply the principles of engineering economy to include the time value of money in project decision making. Other advanced systems

integrate with the schedule, again possibly through a data base, for cash-flow forecasting. An important aspect of any cost control system is to make the data available
to management in a timely manner, so there is an increasing trend to interactive applications using terminals or microcomputers directly in the hands of managers and
supervisors.

Scheduling

Critical path scheduling was one of the earliest applications of computers in construction, but it has taken time for it to gain wide acceptance. Like cost control, this
is an application that should be directly in the hands of management, so the slow
acceptance may be a result of the fact that earlier computers were not easily used by
non- dataprocessing people. Powerful scheduling tools are now becoming available
even on personal microcomputers, however, so the trend toward wider acceptance of
quantitative scheduling methods can only be expected to accelerate in the future.

Scheduling programs on computers typically at least have basic CPM computations,
and most will at least produce a bar chart on a line printer. As the level of sophistication
increases, the programs include resource loading and aggregation (where resources can
be assigned to activities and their total usage by period can be accumulated); resource
allocation and leveling, where the computer program attempts to do some rescheduling
to balance the utilization of resources; network-based cost control, which can either
be integrated with the job-costing system or used for time-cost trade-off applications;
and there is an increasing trend toward utilization of graphic output on interactive
CRTs and on plotters.

Quality Assurance

Quality assurance applications can begin with the online retrieval of specifications,
codes, and standards. Quality assurance systems also assist in the documentation of
procedures and testing requirements and in the reporting of test results and completion
of administrative steps to various interested agencies and parties. Some of the most
advanced applications involve not only administrative procedures but direct production
control. For example, modern automated concrete-batch plants enable the operator to
call up any of several predefined mixes; the computer then operates the plant until the
correct mix is discharged into a waiting concrete truck; batch information is printed out
and copies are given to the truck driver to take to the point of delivery for an inspector's
confirmation and approval before the concrete goes into the pour. Copies are also
attached to samples made at the pour site, and the loop is closed following testing,
when sample results are logged and sent back to the quality assurance department.
Similar applications are gaining acceptance in welding, asphalt paving, and so forth.

Procurement

Simpler procurement systems are just extensions of the accounting department's accounts payable program. However, in larger organizations, there are separate programs

for procurement scheduling and expediting to be sure that steps such as requisitions and shop drawings are not overlooked or to bring problems and delays to management's attention before they become too acute. Materials and procurement systems can also include simple or sophisticated inventory control systems for job materials, tools, and supplies. When linked to a data base, the procurement applications can interface to the quality assurance application for testing and documentation, to the accounts payable system for multiproject vendor correlations, and to the scheduling system to assess the impact of procurement delays on overall project status.

Administration

Even the simpler personal microcomputers can almost immediately pay for themselves in various kinds of administrative applications. For example, numerous lists must be maintained on projects for drawing logs, tool inventories, safety equipment, and so on. Any number of microcomputer file systems can handle such applications. Word processing software on small or large computers can assist with letters and the documentation of transmittals, claims, and so forth. Tools such as spreadsheets can be of immense benefit for financial planning and budgeting as the project evolves. In effect, most of the administrative drudgery that is typically assigned to junior engineers and office clerks can be greatly mitigated through the intelligent application of microcomputers on the job site.

Productivity

Deliberate and systematic efforts to improve productivity on construction projects are becoming increasingly common. Computers can assist with the statistical analysis of questionnaires distributed to workers and supervisors, with the simulation of operations before they are implemented, and so on. This is a new area, but with the improved availability of computers on job sites, it is expected to become increasingly common in the future.

SUMMARY

This chapter began by showing how the level of influence in determining and controlling costs drops rapidly as a project evolves from preliminary and detail design, through procurement and construction, to beneficial operation or utilization. The level of influence is by far the greatest during engineering and design, while actual expenditures at that stage are relatively small.

Understanding the level-of-influence concept can be helpful in forming contractual arrangements that minimize the suboptimization of costs for one party at the expense of overall project costs and benefits. Contractual arrangements should be drawn so as to be sure that current construction and even operations knowledge will be injected in the design process. Professional construction management and design-construct are two forms that, if appropriately tailored to the needs of the particular situation, can be useful for this purpose.

A second important conclusion is that efforts to suboptimize design costs by requiring competitive bidding for professional services are likely to produce much higher project costs in the long run. Owners need much more flexibility than this approach allows in selecting those professionals who can design structures producing maximum benefits for the lowest overall costs.

Both improved methods and the better application of existing principles are needed in the planning and control of engineering and construction projects. Using an analogy and a model, this chapter went on to explain basic concepts of the information feedback-control process used on well-managed projects. The objective was to provide a broad perspective against which to reference specific subsystems such as project planning and scheduling, cost engineering, materials procurement and tracking, and quality control. The chapter then amplified several key components of the feedback-control model, including means for measuring and controlling progress, information processing, requirements for effective reporting, and guidelines for taking corrective action.

Computer applications are now an integral part of project planning and control. The related computer technology is moving so rapidly that the number and variety of applications in construction will have increased markedly even by the time this book is published. A promising trend at present is that through microcomputers and user-oriented software, the computer tools are finally getting directly into the hands of the project managers and supervisors who need them. With this occurrence, however, is the need for such people to become more knowledgeable about the capabilities and limitations of computers and their related software for construction planning and control.

The range of possible applications includes estimating, cost control, scheduling, quality assurance, procurement, administration, and productivity. No one contractor, large or small, has the best in all these areas, but sophisticated applications can be found in both large and small firms for many of them. If intelligently used, all applications can improve the efficiency and effectiveness of project management and reduce the drudgery that has become an increasing burden in administrative applications.

All that can be certain in computer applications at present is that there will be rapid and continuing change. Today's managers and supervisors must be sensitive to these changes and alert for applications that will improve their effectiveness and efficiency.

The remaining seven chapters of Part 3 will explain in more detail each of the major methods and procedures available to the professional construction manager for planning and controlling projects. These include estimating project costs (Chapter 11), schedule and resource planning and control (Chapter 12), cost engineering (Chapter 13), procurement (Chapter 14), value engineering (Chapter 15), quality assurance (Chapter 16), and safety and health (Chapter 17). Although these subjects are developed in separate chapters, the reader should recognize by now that they are each highly interrelated and interdependent parts, each focusing on different aspects of the same overall project planning and control system.

11

ESTIMATING
PROJECT COSTS

Chapter 10 showed how the costs of projects can be evaluated at several levels. These range from the construction contractor's costs for the time and resources consumed in building the structure itself, through the owner's overall costs not only for design and construction but for long-term operation and maintenance, and more broadly to the socioeconomic costs to society as a whole. This chapter will focus on costs associated with the capital facility; these primarily consist of design, procurement, and construction costs.

INTRODUCTION AND OVERVIEW

There are numerous methods and levels of accuracy for preparing capital cost estimates for a construction project. Each method has its appropriate applications and limitations, but it is important to recognize and emphasize that all estimates are approximations based upon judgment and experience. Even the final reported cost figures on completed projects will differ in detail from what the true costs really were, because considerable judgment is required in recording and allocating cost figures while operations are in progress.

Estimates range in scope and detail from "educated guesses" to contractor bid estimates. The latter are based on a relatively complete set of plans and specifications, and they involve much more than simply applying historical unit costs to computed quantities. Indeed, a thorough, accurate and detailed estimate, especially in heavy and industrial construction, is much broader in concept than merely determining costs. To get the costs, the estimator must practically build the project on paper. He must assess quantities not only of the contract materials reflected in the drawings, but also of the temporary materials, such as formwork for concrete and temporary plant. The latter estimates, in turn, require that the estimator hypothesize alternative methods that could be used to build the different components of the project, determine the resources of labor, equipment, and materials that would be required by each method, evaluate the productivity and costs, and select those methods which, taken together, will complete the project on schedule and at the lowest overall cost.

It is also important to note that several different types of estimates are required as a project evolves. Clearly, a detailed estimate based on computed quantities cannot be made at the concept, feasibility study, or preliminary design stage, because the project itself is not yet defined in terms of the plans and specifications upon which computations of quantities are based. Furthermore, the estimating process itself becomes increasingly expensive as more detailed and accurate techniques are applied. Estimates for large projects sometimes cost hundreds of thousands of dollars. When the detail and accuracy are not required, simpler forms of estimating can suffice. Figure 11-1 shows a tabulation of estimate types and the level of information required for each.

This chapter will discuss the types of estimates that can be used during the evolution of a project by the designer, construction manager, design-constructor, general contractor or subcontractor. We will examine conceptual estimates prepared early in the project prior to engineering design completion, detailed estimates prepared from completed plans and specifications and definitive estimates which forecast the project cost within allowable limits from a combination of conceptual and detailed information often including partial contract or other procurement awards.

1 Conceptual and preliminary estimates
2 Detailed estimates
3 Definitive estimates

The chapter will next present a more detailed discussion about estimating and controlling labor costs. It will review some of the differences and common practices in estimating for building, industrial and heavy construction projects. We shall then introduce a new European approach to estimating that has promise for increasing the accuracy and reducing the time required for detailed estimates. A practical program for evaluating uncertainty through range estimating will also be discussed.

Estimates, in turn, become the reference standard for cost control when the project is executed, but for many good reasons they are not always used directly. The last portion of this chapter will therefore address considerations for converting an estimate into a control budget for a project.

CONCEPTUAL AND PRELIMINARY ESTIMATES[1]

Conceptual and preliminary estimates, as the name implies, are generally made in the early phases of a project. Initially they tell an owner whether a contemplated project scope is anywhere near to being economically feasible. Once under way, say in the preliminary and detailed engineering phases, successively refined estimating

---

[1]The background for parts of this section, particularly the cost-capacity factor and component ratios, came from O. T. Zimmerman, "Capital Investment Cost Estimation," in chap. 15 in F. C. Jelen (ed.), *Cost and Optimization Engineering*, McGraw-Hill Book Company, New York, 1970.

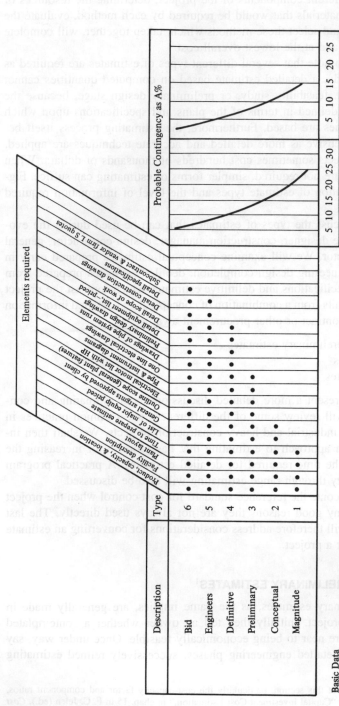

FIGURE 11-1
Estimate types. (From *Building Construction Handbook*, 2d ed., 1975, McGraw-Hill Book Company.)

techniques incorporate new information and thus keep a continuously updated estimate or budget available for control purposes. This in turn provides feedback to design to keep the overall project within budget.

Conceptual and preliminary estimating methods vary considerably from one type of construction to another. Generally the most sophisticated and accurate procedures have been developed for large projects in the industrial construction sector by the design-construct firms. However, high levels of sophistication are also found in individual design firms, construction contractors, professional construction managers, and repeat owners in other sectors of construction as well.

Most of the existing conceptual and preliminary estimating methods fall into one or more of the following categories:

1 Time-referenced cost indices
2 Cost-capacity factors
3 Component ratios
4 Parameter costs

They are listed generally in ascending order of accuracy, and correspondingly in ascending order of cost and complexity to produce. The sections that follow will introduce each of these categories in turn.

Cost Indices

Cost indices show changes in cost over time. Some types also reflect changes in technology, methods, and productivity as well as inflationary trends. Generally they are applied to the construction phase of projects, though they can account for the total design-construction package as well. Many of these indices are published periodically in the technical press. The "Quarterly Cost Roundup," published by the *Engineering News-Record*, lists over a dozen of these, including its own well-known "Construction Cost Index" and "Building Cost Index."

Types of Cost Indices The cost indices themselves are derived from two quite different approaches. The first periodically reprices and totals a constant package of resources that serve as *input* to a typical construction project. The index is computed by dividing this cost by the cost of the resources in a base reference period.

For example, the *Engineering News-Record*'s "Building Cost Index" is computed as follows:

Components: 1088 board feet of lumber (2 × 4, S4S, 20-city average)
2500 pounds of structural-steel shapes, base mill price
2256 pounds of portland cement (bulk, 20-city average)
68.3 hours of skilled labor (20-city average)

We can convert from one base period to another as demonstrated below:

$$\text{"Current cost"} = 2730 \text{ (assumed)}$$

$$\text{Base cost (1967)} = 676$$

$$\text{Index on 1967 base} = \frac{\text{index to be converted}}{\substack{\text{index of 1967 base} \\ \text{relative to original} \\ (1913 = 100) \text{ base}}} \tag{11-1}$$

$$= \frac{2730}{676} \times 100\% = 404\%$$

The second approach is based on the cost of a completed construction project, such as a particular building, or else on a survey of completed unit costs of selected components of projects, such as concrete pavement for highways. In both cases, the index is based on the *output* of the construction process. From there on, however, the process of computing the index is similar to that of the index based on input components. That is, the cost of the project or unit under current conditions is divided by the cost in a base reference period. Both types of indices can be computed as national averages, or they can be computed for particular geographic areas. Examples of published output-type cost indices include those of the American Appraisal Company, the Austin Company, the George A. Fuller Company, and the Port Authority of New York and New Jersey. These and others are reported by the *Engineering News-Record*.

Limitations   Before applying cost indices, it is important to understand how they are derived, their limitations, and the differences in the basic methods. For example, there are obvious problems if the proportions of the input components in an input-type cost index do not reflect the resources used on the project in question. To illustrate, about 40 percent of the costs in a petrochemical project is in piping, yet neither pipe nor pipe fitters are included in the most widely used general cost indices. Similarly, where the index is based on the final (output) costs of a particular building, the project under consideration may have very little in common with that structure. One must also know whether factors such as land, interest on financing, and contractor's profit are included.

Other limitations become evident when one compares advantages and disadvantages of the two types of indices. For example, indices based on input components do not consider factors such as productivity, changes in technology, and competitiveness of contractors. These factors are reflected to some extent in indices based on project outputs, or completed structures. On the other hand, the output type of indices are usually much more narrow in scope, and it is difficult to interpret one based on, say, commercial office buildings, to apply it to another type of work, such as a concrete dam. The input type of indices is much more general and can thus be applied to a broader range of construction projects. In both indices, it is important to recognize their geographic and demographic bases. Both prices and productivity can vary radically around the country and around the world; competitive market conditions for suppliers and contractors can be strong in one type of construction, such as industrial, at the same time that they are weak in another, say, building.

For these and many other reasons, owners, designers, contractors and professional construction managers must use careful judgment and draw upon well-documented personal experience before applying any type of index for purposes of conceptual estimating. Wise firms maintain their own records and do their own studies as well as use published sources. Properly applied, however, such indices can yield accuracies within 20 to 30 percent of actual costs and can provide this information with almost negligible time and effort. Such information can be valuable for policy and planning decisions early in the life of a project.

Example To illustrate the use of a cost index, let us apply *Engineering News-Record*'s "Building Cost Index" to a warehouse. Assume that we have an estimate on file for a similar structure that we completed in 1978 including design and owners expense for a cost of $4,200,000. We were planning to build the new warehouse in 1991.

The ENR index for 1978, relative to a base date of 1967, was 1674/676 = 2.48 or 248 percent. The projected index for 1991 was 2730/676 = 4.04, or 404 percent. Therefore, the estimated cost for a project of similar size and quality will be

$$\frac{404\%}{248\%} \times \$4,200,000 = \$6,840,000$$

Recognizing the level of accuracy, we might round this to $6,900,000.

Cost-Capacity Factor

Whereas cost indices focus on cost changes over time, cost-capacity factors apply to changes in size, scope, or capacity of projects of similar types. They reflect the nonlinear increase in cost with size, as a result of economies of scale. In simple analytical terms, the cost-capacity factor is expressed by the following exponential equation:

$$C_2 = C_1 \left(\frac{Q_2}{Q_1} \right)^x \tag{11-2}$$

where C_2 = estimated cost of new facility of capacity Q_2
$\quad\quad C_1$ = known cost of facility of capacity Q_1
and the exponent X = the cost-capacity factor for this type of work.

The exponents represented by X are empirically derived factors based on well-documented historical records for different kinds of projects. The capacities, represented by Q, are some parameter that reasonably reflects the size of the facility, such as maximum barrels per day produced by a refinery or tons of steel per day produced by a steel mill operating at capacity. In a structure such as a warehouse, gross floor area or enclosed volume might be a reasonable measure of capacity.

Cost-capacity factors have been most widely used in the petrochemical sector of the industrial construction industry. The "six-tenths factor rule" ($X = 0.6$) is typical, and applies fairly accurately to some types of plants. Table 11-1 lists some other typical factors.

TABLE 11-1
EXAMPLE COST-CAPACITY FACTORS*

Process	Capacity	Unit	Cost-Capacity factor	Capacity range
Acetylene	10	Tons/day	0.73	3.5 – 250
Aluminum (from alumina)	100M	Metric Tons/yr	0.76	20M – 200M
Ammonia (by steam-methane reforming)	100	Tons/day	0.72	100 – 3M
Butadiene	10M	Tons/yr	0.65	5M – 300M
Butyl alcohol	100MM	Lb/yr	0.55	8.5MM – 700MM
Carbon black	1	Tons/day	0.53	1 – 150
Chlorine	100	Tons/day	0.62	10 – 800
Ethanol, synthetic	10MM	Gal/yr	0.60	3MM – 200MM
Ethylene	100M	Tons/yr	0.72	20M – 800M
Hydrogen (from refinery gases)	10MM	Cu ft/day	0.64	500M – 10MM
Methanol	10MM	Gal/yr	0.83	5MM – 100MM
Nitric acid (50-60%)	100	Tons/day	0.66	100 – 1M
Oxygen	100	Tons/day	0.72	1 – 1.5M
Power plants, coal	100	Mw(elec)	0.88	100 – 1M
Nuclear	100	Mw(elec)	0.68	100 – 4M
Styrene	10M	Tons/yr	0.68	4M – 200M
Sulfuric acid (100%)	100	Tons/day	0.67	100 – 1M
Urea	250	Tons/day	0.67	100 – 250
Urea	250	Tons/day	0.20	250 – 500

*Adapted from O. T. Zimmerman and I. Lavine, *Cost Eng.*, vol. 6, July 1961, pp. 16–18; O. T. Zimmerman, *Cost Eng.*, vol. 12, October 1967, pp. 12–19; and F. C. Jelen (ed), *Cost and Optimization Engineering*, McGraw-Hill Book Company, New York, 1970, p. 312.

For our warehouse, cost varies fairly closely with floor area, so, for example purposes, assume first that a cost-capacity factor of $X = 0.8$ is representative of this type of work. Second, assume that we have a current estimate for a similar warehouse, located nearby, with a usable area of 120,000 square feet. Let us further assume that the prospective owner for the new warehouse wants a structure with a usable area of 150,000 square feet. Our estimate of the cost-capacity factor for the new warehouse can then be computed as follows:

$$C_2 = \$4,200,000 \left(\frac{150,000}{120,000} \right)^{0.8}$$

$$= \$5,020,000$$

$$\cong \$5,000,000$$

This is a very approximate and very preliminary estimate, but it would at least give the client an idea of the order of magnitude of what the cost of the new warehouse might be. Properly applied, and assuming well-documented empirical records, such estimates of the cost-capacity factor can be accurate to within 15 to 20 percent of actual costs.

It should be obvious at this stage that both cost indices and cost-capacity factors can be combined to take into account changes in both time and capacity. Our analytical formula can then be modified as follows:

$$C_2 = C_1 \times \left(\frac{I_2}{I_1}\right) \times \left(\frac{Q_2}{Q_1}\right)^x \tag{11-3}$$

where I_1 and I_2 are the relative cost indices for the times associated with the known and proposed facilities, respectively.

Assuming that $I_1 = 2.48$ and $I_2 = 4.04$ for the warehouse in question, thus putting off construction of the warehouse to the 1991 time frame, the estimated cost will become

$$C_2 = \$4,200,000 \times \left(\frac{4.04}{2.48}\right) \times \left(\frac{150,000}{120,000}\right)^{0.8}$$

$$= \$8,176,000$$

$$\cong \$8,200,000$$

As with either approach taken alone, both extreme caution and judgement should be exercised in applying and interpreting estimates based on a combination of the two.

Component Ratios

As engineering and design progress, more information can be obtained about a project and its elements. Once the size and type of major items of installed equipment are identified, the designer or professional construction manager is in a position to solicit price quotations from the manufacturers of these components. Examples of equipment or plant components include compressors, pumps, furnaces, refrigeration units, belt conveyors, and turbine generators. Given good price quotations, designers and constructors in many sectors of industry, again especially in industrial construction, have good historical documentation and analytical techniques that enable them to improve the accuracy of their earlier conceptual estimates. To do this, they use techniques such as "equipment-installation-cost-ratios" or "plant-cost-ratios." We shall refer to both of these as "component ratios."

The first approach, equipment cost ratios, multiplies the purchase cost of the equipment by an empirically documented factor to estimate the installation cost of that equipment, including shipping, erection labor, and ancillary fittings and supplies. Table 11-2 lists some example equipment installation factors. These factors average around 50 percent of f.o.b. cost, but they range widely. With good records to base them on, this type of estimate can be accurate to within 10 to 20 percent of final costs.

TABLE 11-2
TYPICAL EQUIPMENT INSTALLATION
FACTORS*

Item	Installation cost, %
Belt conveyors	20 – 25
Bucket elevators	25 – 40
Centrifugals, disk or bowl	5 – 6
Top suspended	30 – 40
Continuous	10 – 25
Crystallizers	30 – 50
Dryers, continuous drum	100[†]
Vacuum rotary	150 – 200[†]
Rotary	50 – 100[†]
Dust collectors, wet	220 – 450[†]
Dry	10 – 200[†]
Electrostatic precipitators	33 – 100[†]
Electric motors plus controls	60
Filters	25 – 45
Gas producers	45 – 250
Instruments	6 – 300
Ion exchangers	30 – 275[†]
Towers	25 – 50
Turbine generators	10 – 30

*Adapted from F. C. Jelen (ed.), *Cost and Optimization Engineering*, McGraw-Hill Book Company, New York, 1970, p. 316.
[†] Includes accessories.

Plant cost ratios, examples of which are given in Table 11-3, use equipment-vendor-price-quotations as a basis for determining the cost of the whole constructed facility. Two approaches can be used. In the first, the estimator adds the costs of all major items of equipment, and then multiplies this sum by a single ratio found to be appropriate for the type of project being constructed. For example, if the f.o.b. cost for all the equipment for a fluid process plant adds to $4,000,000, and a good historically valid plant cost ratio for this type of project is 4.5, then the estimated total cost will be

$$\$4,000,000 \times 4.5 = \$18,000,000$$

Clearly, the degree to which shop fabrication is used on the equipment items, the cost and productivity of field labor, and numerous other factors can seriously affect the accuracy of such estimates if the estimator lacks the judgement and experience to take them into account.

A variation on the plant cost ratio takes the cost of each major item of equipment separately, multiplies each by its own ratio, then takes the sum of the factored

TABLE 11-3
PROCESS-PLANT COST RATIO FROM INDIVIDUAL
EQUIPMENT*

Equipment	Factor[†]
Blender	2.0
Blowers and fans (including motor)	2.5
Centrifuges (process)	2.0
Compressors:	
Centrifugals, motor-driven (less motor)	2.0
Steam turbine (including turbine)	2.0
Reciprocating, steam and gas	2.3
Motor-driven (less motor)	2.3
Ejectors (vacuum units)	2.5
Furnaces (package units)	2.0
Heat exchangers	4.8
Instruments	4.1
Motors, electric	8.5
Pumps:	
Centrifugal, motor-driven (less motor)	7.0
Steam turbine (including turbine)	6.5
Positive displacement (less motor)	5.0
Reactors—factor as approximate equivalent type of equipment	
Refrigeration (package unit)	2.5
Tanks:	
Process	4.1
Storage	3.5
Fabricated and field-erected (50,000 + gal)	2.0
Towers (columns)	4.0

*From W. F. Wroth, "Factors in Cost Estimation," *Chem. Eng.*, vol. 67, October, 1960, p. 204; and F. C. Jelen (ed.), *Cost and Optimization Engineering*, McGraw-Hill Book Company, New York, 1970, p. 317.

[†] Multiply purchase cost by factor to obtain installed cost, including cost of site development, buildings, electrical installations, carpentry, painting, contractor's fee and rentals, foundations, structures, piping, installation, engineering, overhead, and supervision.

components. An example follows:[2]

Item	Cost	Factor	Plant cost
Blowers and fans	$ 10,000	× 2.5	$ 25,000
Compressors	50,000	× 2.3	115,000
Furnaces	100,000	× 2.0	200,000
Heat exchangers	80,000	× 4.8	384,000
Instruments	50,000	× 4.1	205,000
Motors, electric	60,000	× 8.5	510,000
Pumps	20,000	× 7.0	140,000
Tanks	125,000	× 2.4	260,000
Towers	200,000	× 4.0	800,000
Total	$685,000		$2,639,000

[2] Adapted from Zimmerman, op. cit., pp. 317–318.

This approach allows the estimator to apply at a more detailed level judgment regarding such things as the degree of shop fabrication, and it can thus produce greater accuracy. This assumes, of course, that good historical data are available for developing the individual factors for each item of equipment.

Parameter Costs

Conceptual estimates based on parameter costs are most commonly used in building construction. *Engineering News-Record* occasionally publishes examples in some of its "Quarterly Cost Roundup" issues. R. S. Means publishes "Means Square Foot Costs" annually which includes unit costs for a number of building types as well as for individual construction tasks. The parameter cost approach relates all costs of a project to just a few physical measures, or "parameters," that reflect the size or scope of that project. For example, the "gross enclosed floor area" would be a typical overall parameter for a structure such as a warehouse as illustrated in Figure 11-2a reported by R. S. Means.

For the warehouse, some of the unit costs are expressed in terms related to the gross enclosed floor area such as mechanical and electrical work; others are related to parameters such as the square foot of wall area for interior construction work items such as concrete block walls and wall finishes. Note that some costs are expressed as square feet of the component itself (such as masonry) and others in relation to the building floor area as a whole (such as electrical). Nevertheless, with good historical records on comparable structures, parameter costing can give reasonable levels of accuracy for preliminary estimates.

Figure 11-2b illustrates parameter cost data for medical office buildings. Note that this type of estimate requires at least schematic drawings sufficient for computing these few parameters. However, a parameter cost estimate can be prepared long before detailed drawings are complete. With this approach, an experienced estimator with access to well-documented records can quickly prepare an estimate and budget that will help influence the design and control costs in the early phases of a project.

DETAILED ESTIMATES

After conceptual design has been approved and after most or all of the detail design work is complete, approximate estimates are generally supplemented by detailed estimates. These normally require a careful tabulation of all the quantities for a project or portion of a project; this is called a "quantity takeoff." These quantities are then multiplied by selected or developed unit costs, and the resulting sum represents the estimated direct cost of the facility. The addition of indirect costs, plant and equipment, home-office overhead, profit, escalation, and contingency will develop the total estimated project cost. Since a careful takeoff can minimize or eliminate the unknowns regarding the amount of work to be performed, the margin for error is considerably reduced. Contingency requirements decrease since the cost of the work is the major variable left to the estimator's judgement.

Two types of detailed estimates will be explored in this section: the fair-cost estimate and the contractor's bid estimate. Referring again to Figure 11-1, the fair-cost

COMMERCIAL/INDUSTRIAL INSTITUTIONAL	2.690	Warehouse

Model costs calculated for a 1 story building with 24 foot story height and 30,000 square feet of floor area

NO.	SYSTEM/COMPONENT	SPECIFICATIONS		UNIT	UNIT COST	COST PER S.F.
1.0 FOUNDATIONS						
.1	Footings & Foundations	Poured concrete; strip and spread footings and 4' foundation wall		S.F. Ground	2.11	2.11
.4	Piles & Caissons	N/A		—	—	—
.9	Excavation & Backfill	Site preparation for slab and trench for foundation wall and footing		S.F. Ground	.73	.73
2.0 SUBSTRUCTURE						
.1	Slab on Grade	5" reinforced concrete with vapor barrier and granular base		S.F. Slab	5.33	5.33
.2	Special Substructures	N/A		—	—	—
3.0 SUPERSTRUCTURE						
.1	Columns & Beams	Steel columns included in 3.5 and 3.7		—	—	—
.4	Structural Walls	N/A		—	—	—
.5	Elevated Floors	Open web steel joists, slab form, concrete beams, columns, (above office)	10% of area	S.F. Elev. Floor	7.26	.94
.7	Roof	Metal deck, open web steel joists, beams, columns		S.F. Roof	2.98	2.98
.9	Stairs	Steel gate with rails		Flight	3960	.26
4.0 EXTERIOR CLOSURE						
.1	Walls	Concrete block	95% of wall	S.F. Wall	6.19	3.29
.5	Exterior Wall Finishes	N/A		—	—	—
.6	Doors	Steel overhead, hollow metal	5% of wall	Each	1527	.56
.7	Windows & Glazed Walls	N/A		—	—	—
5.0 ROOFING						
.1	Roof Coverings	Built-up tar and gravel with flashing		S.F. Roof	1.48	1.48
.7	Insulation	Perlite/urethane composite		S.F. Roof	1.24	1.24
.8	Openings & Specialties	Gravel stop, hatches and skylight		S.F. Roof	.25	.25
6.0 INTERIOR CONSTRUCTION						
.1	Partitions	Concrete block (office and washrooms)	100 S.F. Floor/L.F. Partition	S.F. Partition	4.59	.37
.4	Interior Doors	Single leaf hollow metal	5000 S.F. Floor/Door	Each	485	.10
.5	Wall Finishes	Paint		S.F. Surface	.75	.12
.6	Floor Finishes	90% hardener, 10% vinyl composition tile		S.F. Floor	.84	.84
.7	Ceiling Finishes	Suspended mineral tile on zee channels in office area	10% of area	S.F. Ceiling	3.87	.39
.9	Interior Surface/Exterior Wall	Paint	95% of wall	S.F. Wall	1.05	.56
7.0 CONVEYING						
.1	Elevators	N/A		—	—	—
.2	Special Conveyors	N/A		—	—	—
8.0 MECHANICAL						
.1	Plumbing	Toilet and service fixtures, supply and drainage	1 Fixture/2500 S.F. Floor	Each	2825	1.13
.2	Fire Protection	Sprinklers, ordinary hazard		S.F. Floor	1.46	1.46
.3	Heating	Oil fired hot water, unit heaters	90% of area	S.F. Floor	2.45	2.45
.4	Cooling	Single zone unit gas heating, electric cooling	10% of area	S.F. Floor	.57	.57
.5	Special Systems	N/A		—	—	—
9.0 ELECTRICAL						
.1	Service & Distribution	200 ampere service, panel board and feeders		S.F. Floor	.23	.23
.2	Lighting & Power	Fluorescent fixtures, receptacles, switches and misc. power		S.F. Floor	2.27	2.27
.4	Special Electrical	Alarm system		S.F. Floor	.21	.21
11.0 SPECIAL CONSTRUCTION						
.1	Specialties	Dock boards, dock levelers		S.F. Floor	1.32	1.32
12.0 SITEWORK						
.1	Earthwork	N/A		—	—	—
.3	Utilities	N/A		—	—	—
.5	Roads & Parking	N/A		—	—	—
.7	Site Improvements	N/A		—	—	—
			SUB-TOTAL			31.19
	GENERAL CONDITIONS (Overhead and Profit)				15%	4.69
	ARCHITECT FEES				7%	2.52
			TOTAL BUILDING COST			38.40

BUILDING TYPES

FIGURE 11-2a
Warehouse parameter costs, from "Means Square Foot Costs," R. S. Means Company, Inc., Kingston, Maine.

estimate is equivalent to the Type 5 engineer's estimate, and the contractor bid estimate is equivalent to the Type 6 bid estimate.

Proper evaluation of labor productivity, effects of local practices, market competitiveness, weather conditions, and completeness of plans and specifications are extremely important in the preparation of detailed estimates. Significantly different

COMMERCIAL/INDUSTRIAL INSTITUTIONAL	2.410	Medical Office, 2 Story

Model costs calculated for a 2 story building with 10 foot story height and 7,000 square feet of floor area

NO.	SYSTEM/COMPONENT	SPECIFICATIONS		UNIT	UNIT COST	COST PER S.F.
1.0	**FOUNDATIONS**					
.1	Footings & Foundations	Poured concrete; strip and spread footings and 4' foundation wall		S.F. Ground	5.14	2.57
.4	Piles & Caissons	N/A		—	—	—
.9	Excavation & Backfill	Site preparation for slab and trench for foundation wall and footing		S.F. Ground	.74	.37
2.0	**SUBSTRUCTURE**					
.1	Slab on Grade	4" reinforced concrete with vapor barrier and granular base		S.F. Slab	2.62	1.31
.2	Special Substructures	N/A		—	—	—
3.0	**SUPERSTRUCTURE**					
.1	Columns & Beams	Included in 3.5 and 3.7		—	—	—
.4	Structural Walls	N/A		—	—	—
.5	Elevated Floors	Open web steel joists, slab form, concrete, columns		S.F. Floor	5.92	2.96
.7	Roof	Metal deck, open web steel joists, beams, columns		S.F. Roof	2.54	1.27
.9	Stairs	Concrete filled metal pan		Flight	4120	1.18
4.0	**EXTERIOR CLOSURE**					
.1	Walls	Concrete block, insulated	70% of wall	S.F. Floor	3.81	3.81
.5	Exterior Wall Finishes	Stucco on concrete block	70% of wall	S.F. Wall	7.99	3.84
.6	Doors	Aluminum and glass doors with transoms		Each	3315	.95
.7	Windows & Glazed Walls	Outward projecting metal	30% of wall	Each	183	2.51
5.0	**ROOFING**					
.1	Roof Coverings	Built-up tar and gravel with flashing		S.F. Roof	1.40	.70
.7	Insulation	Perlite/urethane composite		S.F. Roof	1.24	.62
.8	Openings & Specialties	Gravel stop and hatches		S.F. Roof	.44	.22
6.0	**INTERIOR CONSTRUCTION**					
.1	Partitions	Gypsum bd. & sound deadening bd. on wood studs w/insul.	6 S.F. Floor/L.F. Partition	S.F. Partition	4.02	5.36
.4	Interior Doors	Single leaf wood	60 S.F. Floor/Door	Each	305	5.08
.5	Wall Finishes	50% paint, 50% vinyl wall coating		S.F. Surface	.84	2.22
.6	Floor Finishes	50% carpet, 50% vinyl asbestos tile		S.F. Floor	3.44	3.44
.7	Ceiling Finishes	Mineral fiber tile on concealed zee bars		S.F. Ceiling	2.73	2.73
.9	Interior Surface/Exterior Wall	Painted gypsum board on furring	70% of wall	S.F. Wall	2.38	1.14
7.0	**CONVEYING**					
.1	Elevators	N/A		—	—	—
.2	Special Conveyors	N/A		—	—	—
8.0	**MECHANICAL**					
.1	Plumbing	Toilet and service fixtures, supply and drainage	1 Fixture/160 S.F. Floor	Each	998	6.13
.2	Fire Protection	N/A		—	—	—
.3	Heating	Included in 8.4		—	—	—
.4	Cooling	Multizone unit, gas heating, electric cooling		S.F. Floor	7.92	7.92
.5	Special Systems	N/A		—	—	—
9.0	**ELECTRICAL**					
.1	Service & Distribution	100 ampere service, panel board and feeders		S.F. Floor	.46	.46
.2	Lighting & Power	Fluorescent fixtures, receptacles, switches and misc. power		S.F. Floor	3.37	3.37
.4	Special Electrical	Alarm systems and emergency lighting		S.F. Floor	.74	.74
11.0	**SPECIAL CONSTRUCTION**					
.1	Specialties	N/A		—	—	—
12.0	**SITEWORK**					
.1	Earthwork	N/A		—	—	—
.3	Utilities	N/A		—	—	—
.5	Roads & Parking	N/A		—	—	—
.7	Site Improvements	N/A		—	—	—
			SUB-TOTAL			60.90
	GENERAL CONDITIONS (Overhead and Profit)				15%	9.14
	ARCHITECT FEES				9%	6.31
			TOTAL BUILDING COST			76.35

153

FIGURE 11-2b
Medical office parameter costs, from "Means Square Foot Costs," R. S. Means Company, Inc., Kingston, Maine.

appraisals of these factors can result in sizable differences in finished estimates based upon exactly the same quantity takeoff.

Fair-Cost Estimates

As mentioned in Chapter 8, fair-cost estimates for construction projects are best prepared from the actual bid documents provided to the bidders (before award) and are used by the owner's representative to evaluate changes (after award). It is helpful to complete the fair-cost estimate well before receiving bids or before performance of the changed work. For an example of the fair-cost estimates for the Mountaintown Warehouse Project, refer back to Appendix A.

The major differences between estimates prepared by construction managers, engineers or other owner's representative and a contractor's bid estimate are (1) the absence of lump-sum subcontract quotations, and (2) a somewhat simplified number of line items. For example, in Appendix A, the markup included indirect costs, corporate overhead, profit, and contingency. A contractor estimate for the work would normally estimate the field indirect costs, might estimate or allocate office costs, and could establish separate profit and contingency (or risk) figures. The contractor estimate might also make separate provisions for material and subcontractor quotations, and for outside and company-owned equipment rentals as elements of cost.

However, a knowledgeable professional construction manager or other owner's representative will prepare an equally accurate quantity takeoff and will choose the number of line items to be estimated on the basis of the objectives of the particular project and the level of detail required to achieve these objectives. On a professional construction management or design construct project, the fair-cost estimate is one of the primary tools in establishing a basis for measuring job progress (earned value) and for schedule and cost control discussed in Chapters 12 and 13. When properly applied, it can also result in productivity evaluation, permit comparison with contractor estimates, and assist in continually updating the professional construction manager's estimating knowledge and skills.

Contractor's Bid Estimate

The contractor's bid estimate is his foundation for a successful project. He must bid low enough to obtain the work, yet high enough to make a profit.

Many people in the construction industry think of estimating as a more or less structured undertaking like engineering design. But a look at bids received for a typical project in a competitive area will sometimes show more than a 50 percent difference between the low and high bidders. When large amounts of subbids are involved, the spread tends to be smaller because most bidders will use the most competitive subcontractors.

For many years, numerous successful smaller contractors were able to compete effectively using the unit-cost system in which overall unit costs, including costs of labor, material, equipment, and overhead and profit, were applied directly to actual quantity takeoffs. However, with the sharp price changes for all components in recent years, this method is becoming increasingly rare in successful companies. Almost all successful contractors now estimate new projects with separate categories and evaluations for work hours, labor, materials, equipment usage, and subcontractors.

In one way, bid estimates are sometimes less detailed than fair-cost estimates. Subcontractors often account for some 30 to 80 percent of the project. The general or prime contractor does not usually prepare a detailed cost estimate for this work, but rather, merely incorporates the low bidders' quotations in his own proposal. Design-constructors and general contractors acting as professional construction managers or

Title **Mountaintown Warehouse**
Client **Easyway Grocery Co.**   Location **M'taintown, WA**
Subject **Structural Steel Estimate Summary**

Code	Description	Quantity	Man-hours	Labor	Material	Equipment
				Units		
	Fabricate in Shop			@ 24.00/Hr		
	Beams & columns	280 Ton	10 Hrs/Ton	240	620	100
	Roof joists-purch.	200 Ton	—	—	780	
	Detailing, etc.	300 Hr		20.00		
	Subtotal directs	480 Ton				
	Shop overhead	80% Labor		1.60		
	Estimated cost					
	Profit & overhead	12.5%				
	Estimated shop price					
	Erection & Field					
	Steel Erection Crew					
1	Foreman	1 @ 35.00		35.00		
4	Ironworkers	4 @ 31.50		126.00		
1	Operator	1 @ 35.00		35.00		
1	Oiler	1 @ 29.50		29.50		
2	Ironworkers	2 @ 31.50		63.00		
9	Total Crew		9 Hr @ 32.06 = 288.50			
	Furnish Fab Steel	480 T		L.S.		
	Erect struct. Steel	280 T	7 Hrs/Ton	224.00		
	Erect joists Steel	200 T	5 Hrs/Ton	160.00		
	Equipment Usage	2.0 Mos.				12000
	Misc. supp. & tools	10% Labor				
	Deck—subcontract	L.S.				
	Estimated cost					
	Profit & overhead			20%	3%	—
	Estimated price					
	Final bid price					

FIGURE 11-3
Structural steel-bid estimate summary.

who perform cost-reimbursable work, however, must develop in-house estimating capability for electrical, plumbing, piping, roofing, and other specialty work which is not normally estimated by the traditional general contractor.

Figure 11-3 shows the low bidder's competitive contractor estimate for the structural steel work on the Mountaintown Warehouse, and can be compared with the construction manager's fair-cost estimate shown in Appendix A.

DEFINITIVE ESTIMATES

As the overall project evolves from the owner's standpoint, initial approximate estimates become more refined and more accurate as additional information is developed.

Job No. _____

Date _____ By _____

Sheet _____ Of _____

Man-hours	Labor	Material & subs	Equipment	Total
2,800	67,200	173,600	28,000	268,800
		156,000		156,000
300	6,000			6,000
3,100	73,200	329,600	28,000	430,800
	58,600			58,600
	131,800	329,600	28,000	489,400
				30,600
				550,600
			Say	550,000

Schedule Computations

$$\frac{2960 \text{ Hrs}}{72 \text{ Hrs/crew day}} = 41 \text{ days}$$

$$= 8.2 \text{ weeks}$$

$$\text{duration} = 2.0 \text{ months}$$

		550,000		550,000
1,960	62,800			62,800
1,000	32,000			32,000
			24,000	24,000
		9,400		9,400
		142,000		142,000
2,960	94,800	701,400	24,000	820,200
	19,000	21,000		40,000
				860,200
			Use	860,000

FIGURE 11-3
(continued).

Finally, there comes a time when a definitive estimate can be prepared that will forecast the final project cost with little margin for error. This error can be minimized through the proper addition of an evaluated contingency. Fair cost estimates prepared from detailed plans and specifications can complete the process.

Projects can be separated into four broad categories for purposes of reviewing definitive estimates:

1 Unit-price projects
2 Traditional
3 Design-construct
4 Professional construction management

Each of these classifications will be discussed in the following subsections.

Unit-Price Projects

These projects usually encompass heavy construction jobs such as dams, tunnels, highways, and airports. Here the prices have been set constant, while quantities vary within limits inherent in the nature of the work. Quantities may overrun or underrun owing to a number of potential causes, such as additional foundation excavation to solid rock, poor ground conditions, excessive water in tunnels, or other factors usually associated with the anticipated accuracy of geological and geophysical interpretation.

In the event of favorable readings on market conditions, project location, history of similar projects, project duration, duration of collective bargaining agreements, and numerous other factors, a true definitive estimate can be developed before going out for bids and prior to completion of the detailed design. On the other hand, in the absence of firm collective bargaining agreements, with sketchy geological data and an unsettled price structure, a truly definitive estimate may not be possible until the project is well under way.

For heavy construction work featuring reasonably accurate geological exploration accompanied by accurate interpretations and quantity information, the estimate may be considered to be definitive at the time bids are received. Even here, however, the estimate should be accompanied by an evaluated contingency to allow for potential quantity increases. On the other hand, if geological work was either inadequate or was subject to faulty interpretations, and if this factor can cause significant differences in the cost and time involved in performing the work, a reasonably accurate definitive estimate from the owner's standpoint may not be obtainable until after the major quantities are known and agreement to a price change for changed work is achieved.

In the extreme, where there is lack of agreement on whether or not the character of the work has changed, the final costs may not be known accurately until the results of lengthy and costly litigation are known.

Traditional Projects

Projects in this category include lump-sum, guaranteed maximum price, and cost-plus-a-fee negotiated contracts.

On lump-sum projects, the definitive estimate can be developed, in the absence of changed conditions, using the low bidder's quotation plus an evaluated contingency to cover anticipated changes.

On negotiated fast track projects featuring a guaranteed maximum price, this price is often fixed at a point somewhere between the requirements of a Type 4 or Type 5 estimate, in the terminology of Figure 11-1. Many, but not all, of the detailed drawings are usually complete; firm subcontracts and material prices have been obtained for a sizable portion of the work; prices for major equipment are known, and some of the work will have been completed; an evaluated contingency is added to allow for the remaining unknowns.

On negotiated cost-plus-a-fixed-fee fast track construction projects, a definitive estimate can generally be prepared at approximately the same period as was possible for the guaranteed maximum price.

Design-Construct Projects

Design-construct projects can be generally divided into lump-sum, guaranteed maximum-price, and cost-plus-a-fixed-fee categories similar to the traditional approach.

Lump-sum contracts on design-construct projects can be extremely misleading to an unknowledgeable owner. Basing his decisions on performance criteria, a designated amount of floor space, or other parameters, the design-constructor (or engineer-contractor) agrees to provide a facility for a fixed price. Unless the details of the components of the facility are fully described and specified in the contract, the owner may find that he has purchased a facility which, while meeting the contract conditions, is much less than he thought he was getting. On the other hand, if the lump sum is fixed after substantial design work is complete, much of this objection disappears and the situation becomes quite similar to the guaranteed maximum price condition.

On negotiated projects featuring a guaranteed maximum price, the definitive estimate can generally be prepared at about the same time as on a negotiated cost-plus-a-fixed-fee project. Because one entity is performing both design and construction, it may be able to develop such an estimate with less detail design than in the traditional approach.

On heavy industrial projects where the process equipment forms a major share of the cost, a definitive estimate can often be prepared with reasonable accuracy after a detailed scope of work has been developed and after all major equipment has been purchased. This situation is illustrated in Figure 11-1 as a Type 4 estimate.

On negotiated projects featuring a cost-plus-a-fixed-fee contract, the definitive estimate can generally be prepared at about the same time as in the guaranteed maximum-price situation.

Professional Construction Management Projects

Definitive estimates for professional construction management projects can be accurately prepared about the same time as the guaranteed-maximum or cost-plus-a-fixed-fee option under the traditional or the design-construct approach. Because of the interrelationships among progress measurement, schedule, and cost control, it is very helpful to have fair-cost estimates for a majority of the contracts prior to completing the definitive estimate so that the base for the control system, as well as the overall cost estimate, will be as accurate as possible.

The definitive estimate for the Mountaintown warehouse project whose Summary Cost Report was shown in Figure 8-3 was issued when physical progress on the project was 9.6 percent complete. At this time eight fair-cost estimates had been prepared and eight contracts had been awarded. Detailed design was about 95 percent complete.

The application of a proper contingency is an important part of any definitive estimate. Figure 11-4 shows the development of an evaluated contingency for a fast track industrial project. Figure 11-5 shows a project cost summary comparing forecast "at completion" costs to the budget definitive estimate.

Description	Amount	Factor	Evaluation
Open commitments	$7,259,128	3	$ 216,000
Estimate to complete:			
Direct accounts			
Earthwork	706,000	20	141,200
Concrete	878,000	20	175,600
Architectural	604,000	20	120,800
Mechanical	1,160,000	10	116,000
Piping	732,000	10	73,200
Electrical	715,600	10	71,600
Equipment purchase	366,000	10	36,600
Contractors field overhead	701,200	10	70,000
Construction plant	21,000	10	2,000
Engineering, Supervision, and procurement		Allow	30,000
Escalation	170,000	10	16,000
Start		Allow	400,000
Special exposures			
Contractor ABC claim		Allow	95,000
Total			$1,564,000

Contingency Evaluation
Format for evaluating the cost exposures (unknown) not covered or anticipated in the current estimate to complete. Evaluation is based on applying "experience factors" to the remaining work and estimated cost.

FIGURE 11-4
Typical contingency evaluation (as of October).

ESTIMATING AND CONTROLLING CONSTRUCTION LABOR COSTS[3]

One of the most difficult aspects of preparing a fair-cost estimate, a detailed definitive estimate, or a control budget based on the estimate is the labor component. This section therefore outlines basic principles and concepts for estimating and controlling field labor costs on construction projects. The basic approach is to divide labor costs into

[3]From B. C. Paulson, Jr., "Estimating and Controlling Construction Labor Costs," *Journal of the Construction Division*, ASCE, vol. 101, no. CO3, September 1975, pp. 623–633.

Control account number	Description	Recorded costs		Cumulative total recorded & committed	Estimated Cost			Under or over budget
		Current period	Cumulative to date	Open commitments	To complete	At completion	Budget	
	Direct Cost								
1000.0000	Site development and improvements	$ 64,430	$ 2,558,396	$ 264,332	$ 2,822,728	$ 202,994	$ 3,025,722	$ 3,075,600	$(49,878)
2000.0000	Buildings and structures	1,204,598	12,551,752	495,366	13,047,118	1,254,954	14,302,072	14,470,000	(167,928)
3000.0000	Process equipment and systems	457,672	3,462,266	3,401,458	6,863,724	1,676,758	8,540,482	8,720,000	(179,518)
4000.0000	Utilities distribution	221,120	3,650,100	2,505,204	6,155,304	1,506,696	7,662,000	7,631,800	30,200
6000.0000	Distributable directs	82,456	613,378	592,768	1,206,146	690,244	1,896,390	2,000,600	(104,210)
	Total direct cost	$2,030,276	$22,835,892	$7,259,128	$30,095,020	$5,331,646	$35,426,666	$35,898,000	$(471,334)
	Indirect Cost								
7100.0000	Contractors field services	$ 51,578	$ 677,790	$ 21,000	$ 698,790	$ 701,210	$ 1,400,000	$ 1,446,000	$(46,000)
7200.0000	Construction plant	22,914	128,546	6,334	134,880	21,120	156,000	156,000	0
7300.0000	Construction equipment	19,228	95,498	2,502	98,000	0	98,000	100,000	(2,000)
	Total indirect cost	$ 93,720	$ 901,834	$ 29,836	$ 931,670	$ 722,330	$ 1,654,000	$ 1,702,000	$(48,000)
	Total construction cost	$2,123,996	$23,737,726	$7,288,964	$31,026,690	$6,053,976	$37,080,666	$37,600,000	$(519,334)
8100.0000	Engineering, supervision and procurement	$ 34,674	$ 1,898,790	0	$ 1,898,790	$ 201,210	$ 2,100,000	$ 2,200,000	$(100,000)
8900.0000	Contingency	0	0	0	0	1,564,000	1,564,000	3,600,000	$(2,036,000)
9000.0000	Clearings	(2,608)	7,142	4,060	11,202	(11,202)	0	0	0
	Total project	$2,156,062	$25,643,658	$7,293,024	$32,936,682	$7,807,984	$40,744,666	$43,400,000	$(2,655,334)

This report is a monthly summary (highest level) of cost, estimates, and comparisons to budget prepared for distribution

FIGURE 11-5
Project cost summary.

two main components and develop them separately. These components are (1) prices in money terms and (2) productivity. Both are essential to determining labor costs. For purposes of comparison, the concepts of "estimating" and "control" will be developed in parallel. Chapters 12 and 13 will deal more extensively with the control phase itself.

Components of Labor Costs

Two major factors determine labor costs in construction work. The first is the money or prices associated with hourly wages, fringe benefits, payroll insurance and taxes, and wage premiums. Though calculations of the money components can be complex, most of the parameters can at least be readily and accurately quantified. The second factor is productivity, the amount of work that a worker or crew can accomplish in a defined period of time. Of the two main factors, productivity is by far the harder to determine. While wages and other money components may stay essentially constant over the duration of an operation, productivity can fluctuate wildly. To estimate and control productivity, one not only needs accurate, consistent, and up-to-date records, but a great deal of experience and judgment as well.

The basic mathematics for labor costs is quite simple. For example, assume the following:

Price of all money elements = P ($/hour)
Productivity of labor = q (units/hour)
Combining these and dividing gives:

$$\text{Unit labor cost} = \frac{P \ \$/\text{hour}}{q \ \text{units/hour}}$$

$$= P/q \ (\$/\text{unit}) \tag{11-4}$$

Productivity can also be expressed in terms of worker-hours per unit of output (W hours/unit, where $W = 1/q$), in which case one would multiply the two elements to obtain unit cost:

$$\text{Unit labor cost} = P\frac{\$}{\text{hour}} \times W\frac{\text{hours}}{\text{unit}}$$

$$= P \times W \ (\$/\text{unit}) \tag{11-5}$$

By multiplying the unit cost times the total quantity (Q) of work associated with the operation, one obtains the total labor cost for the operation:

$$\text{Total labor cost} = Q \times P/q \ (\$) \tag{11-6}$$

or

$$= Q \times P \times W \ (\$) \tag{11-7}$$

The mathematics at this stage is trivial and thus belies the complexity of estimating and controlling labor costs. The difficulty, of course, is in finding and calculating the correct prices of the money elements (P), and in determining the productivity (q or W) of labor. The following sections will discuss these subjects in much greater detail.

Estimating and Controlling the Money Component

Estimating the money component of labor costs is more difficult in construction than in any other United States industry. Reasons for this situation include the scope and variety of the work involved, the craft structure of labor unions, and the regional and local autonomy of labor and employer collective bargaining units. There are literally thousands of different wage rates, fringe benefits, insurance rates, and work rules, and there are exceptions to almost all of them. Superimposed upon this are federal, state, and local laws, taxes, and special programs such as wage-and-price control.

Even if given identical money elements for labor, contractors vary widely in the way they analyze, combine, and distribute them for estimating and control purposes. Most now prefer to put all elements into a direct hourly rate. Others still prefer to split off various elements into the "indirect" or "overhead" category. Some estimate by craft and some by crew. Some combine scheduled overtime with straight time to produce an "average" hourly rate, and some keep them separate. There is not one, but several, correct methods, and there are generally good reasons why a particular contractor uses a specific method in his own type of work. Therefore, rather than present a single method, this section will outline the major money elements which should be recognized in any approach. They include basic wages, fringe benefits, payroll insurance and taxes, and wage premiums. We will also point out some of the more subtle factors which are often overlooked in calculating these elements.

This material will be presented mostly in terms of conventional union-labor construction work. Many of the principles will apply equally well to "open-shop" construction.

Basic Wages Basic wages vary by location, by craft, and, in many cases, by type of work within each craft. They also vary with time, both from increases scheduled within existing labor agreements and through increases to be negotiated in future contracts. For purposes of estimating, the contractor must determine the applicable location and labor agreement(s), the type of craft(s) to perform the desired work (this is often difficult in borderline jurisdictional cases), the appropriate classifications within the craft(s), and the wage rate(s) applicable at the time the work is to be done. This last item may itself have to be estimated if an agreement is not yet available for the time planned.

For purposes of control, the union contractor cannot have much direct effect on the basic wage, except to be sure that he has the proper craftsmen and classifications for the work being done. That is, if all other things are equal, he should not use a higher-paid worker where a lower-paid worker is permitted. The contractor can have some indirect effect by working through and supporting his collective bargaining association when new labor agreements are being negotiated. Nonunion or open-shop contractors have similar constraints depending upon the availability and skill of the nonunion labor force in the area.

Fringe Benefits Fringe benefits paid to workers variously include contributions to funds for health insurance, vacations, pensions, dental plans, apprenticeship training, industry advancement, and numerous others. They vary widely by region and by craft. As with basic wages, for estimating, the contractor must determine the applicable location and agreement(s), crafts and classifications, and timing of the work. Similarly, for control, the contractor cannot have much direct effect on the fringes once established, and must be sure that he complies with, but does not unwittingly exceed, the agreement.

Insurance Based on Payroll Several kinds of insurance and related programs are based directly on payroll. These include workmen's compensation, public liability, property damage (WC, PL, and PD), social security benefits (FICA, employer's contribution), and state disability and unemployment insurance (SDI, employer's contributions).

WC, PL, and PD premiums are quoted on the basis of rate per $100 of direct (straight-time equivalent) payroll. Standard manual rates for WC vary from about $2 to over $40 per $100. Union contractors would like to see premiums quoted upon an hours-worked basis. Since union wages are generally higher than nonunion, overall WC insurance costs can be lower to nonunion contractors under the present system.

For estimating and controlling WC, PL and PD premiums, two important facts should be recognized. First, the premiums charged to a particular contractor vary significantly with the frequency, severity, and size of risk characterized in his accident record. His actual rates can vary from less than half to more than double the manual rate. Therefore, efforts made to minimize accidents can yield direct financial returns in lower insurance premiums and accident costs. Indirect benefits have been estimated to be 4 to 7 times this amount.[4,5]

The second important factor regarding WC, PL and PD rates is that they vary widely with the classification (and hazardousness) of work. Records which document the type of work being done by individual employees can be used to significantly reduce costs. For example, the rate for the classification "high-rise concrete building construction" might be $18 per $100 base payroll. However, a carpenter building prefabricated forms in an on-site, grade-level shop on the same job might qualify for a much lower rate—say $4 per $100. If the contractor's cost system could document this fact, he could save $14 per $100 on this individual's WC insurance premium. Since both WC and PL and PD insurance is payroll based, it may be convenient to place both with the same company.

---

[4]H. Knox, "Construction Safety as It Relates to Insurance Costs," *AACE Bulletin*, vol. 16, no. 3, June 1974, pp. 71–73.

[5]Michael R. Robinson, *Accident Cost Accounting as a Means of Improving Construction Safety*, Technical Report No. 242, Stanford University, Dept. of Civil Engineering, The Construction Institute, Stanford, Calif., August 1979.

FICA and most SDI contributions are calculated as a percentage (7.65 percent for FICA in early 1990) of gross wages, including premium time. This amount is both subtracted from the employee's pay and added to the employer's payroll burden, so the net contribution is double the percentage shown. The important thing to recognize here about the FICA and SDI contributions is that there is an upper cutoff value above which payments no longer must be made. For example, the FICA ceiling in early 1990 was $51,300. If a construction craftsman moves from one employer to another, he may wind up paying more than required. However, like any other taxpayer, he can reclaim any excess on his income tax return. His employer, unfortunately, lacks the ability to reclaim his excess payment. However, had the employer been able to offer continuous employment, he can probably reduce or avoid excess payments.

Taxes Based on Payroll   Taxes based on payroll include federal income tax, state income tax, and sometimes local taxes. For the most part, these are deductions withheld from the employee's earnings and, unlike FICA, there is no matching employer's contribution. Therefore, such taxes are more a matter for payroll accounting than they are for estimating and control.

Wage Premiums   Wage premiums include extra money paid for overtime work, shift-work differentials, and premiums for hazardous or unusually arduous work. Overtime work in construction is paid at a minimum of time and a half (150 percent of base), is often paid at double time (200 percent), and in some cases, especially on holidays, at triple time (300 percent). Ordinarily, overtime is paid for work in excess of a 40-hour, 5-day week, but the base period may be less. For example, in some areas unionized electricians have in past years been paid overtime for work in excess of a 32-hour week.

There are several ways of paying a premium for multishift work. One way is to pay an extra percentage of the base wage for afternoon shift (swing) and night shift (graveyard) work. A common method encountered in the western states, as an example, is to give 8-hours' pay for all shifts. The day shift works eight hours; the swing shift works 7.5 hours; and the graveyard shift works 7 hours.

Premiums for hazardous or unusually arduous work are commonly paid as a fixed increase over the base wage rate. For example, cement masons working on a swinging scaffold more than 25 feet above grade might receive an extra $0.70 per hour. Electricians might get an extra $1.20 per hour when working underground. A crane operator might get an extra $0.80 per hour if the boom is over 185 feet long. These rates would in turn be multiplied along with the base wage for overtime and multishift work.

In estimating premium costs, the most important thing is to thoroughly understand all the implications of the relevant labor agreements and government legislation. When it comes to control, there is a great deal that can be done by the contractor. Discussion of most of this area will be deferred to the section on productivity. However, to illustrate briefly the money side for overtime and shift differentials, the alternatives are shown on Figure 11-6. Curve *A* shows the average hourly wages for working 8 hours straight time at $10 per hour. Curve *B* shows the increasing average hourly

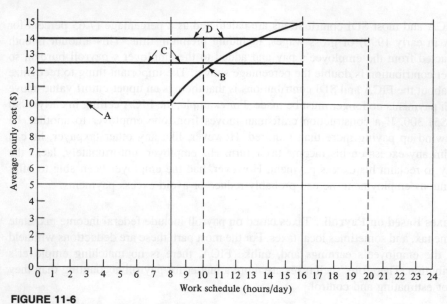

FIGURE 11-6
Impact of wage premiums on average labor costs. (From Boyd C. Paulson, Jr., "Estimation and Control of Construction Labor Costs," *Journal of the Construction Division*, ASCE, vol. 101, no. CO3, September 1975, p. 627.)

wages for working 9, 10, up to 16 hours, with double time paid for overtime in excess of 8 hours. Curve *C* shows the average hourly cost for working two 8-hour shifts under the premium clause: "pay 8 hours for the first 7 hours' work, and double time thereafter." Curve *D* shows the average hourly cost for working two 10-hour shifts under the conditions defined for curve *C*. These simplified examples use base wages and premiums only. They ignore fringes, insurance, etc. Most important, they ignore the reduced productivity which is encountered with premium work. Nevertheless, they do illustrate several important concepts. Readers may be interested to calculate the percentage increases incurred and the break-even points encountered under various conditions when using the terms of their own local labor contracts.

Estimating and Controlling Labor Productivity

In contrast with the money component of labor costs, productivity is much more difficult to estimate. Many of the factors influencing labor productivity are highly qualitative in nature, and a great deal of experience and judgement is needed to develop the type of quantitative information that is required. However, the productivity component also offers the contractor by far the greatest opportunity to control his labor costs, assuming that he has some basic understanding for the factors that influence this variable in the equation.

This section will discuss some of these factors and the principles and concepts related to them. It includes the effect of location and regional variations, the learning curve, work schedule (overtime and multishift), work rules, weather and other environmental effects, experience of the craftsmen employed, and management factors such as job morale, safety, and motivation.

Regional Variations   Apart from environmental and other effects discussed below, two factors related more directly to labor cause considerable regional variation in productivity. These include (1) the training, experience, and skill of the local labor force in the various crafts and (2) the work rules which are negotiated between employers and unions. These factors can cause productivity in some parts of the country to be more than double that in others. When working overseas they can be even more significant. Other things being equal, labor costs will vary accordingly.

Clearly, for both estimating and control a contractor working in more than one area must take regional variations into account. This is generally done by establishing base productivity levels for various crafts or activities in one region, and then applying index multipliers (or job factors) to ratio the base to other areas. If properly documented, a contractor's records will be his best source for developing such multipliers. Otherwise, he will be obliged to rely upon the experience of others as reflected in published sources. Precautions for this situation are mentioned briefly later in this chapter.

The amount of training and experience certainly varies from worker to worker even within a single craft in a local area. Formal distinctions are drawn between "apprentice" and "journeyman," but of course there are wide ranges within these. Although the estimator usually cannot forecast which craftsmen will work on which tasks, this is a factor that is very much subject to control by project management. Within the prescribed work rules, good superintendents can apply selective hiring, firing, and assignment of craftsmen in order to improve productivity. On a grander scale, contractors can actually set up their own training programs to enhance the skills of their employees. This has happened on many large overseas projects and is now being done in several "open-shop" areas in the United States.

Environmental Effects   The environment affects productivity on many levels. The weather, terrain, topography, and similar natural phenomena have obvious implications which need not be belabored here. The physical locations and working conditions of individual craftsmen can be equally significant. These include height above grade, heat, noise, light, constrictions, stability of work station, dust, and several others. Clearly, the productivity of an ironworker laboring outdoors during the Illinois winter would be different from one working in the Southern California sunshine. A carpenter erecting small form panels for an outside wall on the tenth story of a building would most likely produce less than if he were working in an on-site shop assembling the same panels into large ganged forms which would then be hoisted into place by crane.

Estimators generally take environmental conditions into account in preparing their estimates. Thorough preplanning can minimize the impact of many of these. Careful scheduling of outdoor operations with respect to seasons can help reduce weather effects, as can the appropriate use of enclosures. Thoughtful layout of construction facilities can offset the effects of difficult topography. A good comparative analysis of alternative methods of accomplishing specific operations can provide on-site working conditions conducive to higher productivity.

For controlling labor costs, management can also have considerable influence on the environmental effects on labor productivity. A monograph by Russo[6] shows how contractors can use readily available weather information services to improve their operations markedly. Oglesby, Parker, and Howell[7] offer numerous hints for analyzing and improving work methods in order to significantly increase labor productivity. The construction trade and professional literature provide many other sources of information in this area.

Learning Curves   The basic principle of the "learning curve" is that skill and productivity in performing tasks improve with experience and practice. For example, the tenth of ten identical concrete footing pours should take less time and be done more skillfully than the first.

It is not the purpose here to set forth the theory and mathematics of learning curves. These are adequately explained in construction textbooks such as that by Oglesby, Parker, and Howell.[8] Here we will show how these concepts apply to estimating and controlling labor costs.

Consider the example learning curve shown in Figure 11-7. For estimating purposes, the curve should be integrated through the number of units to be constructed, here defined as n, to obtain the total number of worker-hours required. Graphically, this may be expressed as the shaded area under the curve. Dividing the total number of worker-hours by the number of units gives the *average* worker-hours required per unit as shown by w on Figure 11-7b. This is the number that should be used for estimating. Note that if the number of units is greater, as shown by n', the average worker-hours per unit should be less, as shown by w'. It is therefore not sufficient simply to take average figures from one project and apply them directly to similar operations on a project being estimated. This fact also should be recognized by the estimator.

For control purposes, management should recognize that the worker-hours required for the first few of a number of repetitive operations should be expected to be higher than the average given by the estimator. However, as the operations continue, the worker-hours per unit required could drop below the estimated average so that the actual completed average might be less than, or equal to, the estimated average. This concept is illustrated on Figure 11-7c. However, many projects find a decrease in the productivity for the last 5 to 10 percent of the operation as confined spaces and fear of layoffs contribute to craft output. Another important feature of the learning curve that should be recognized for control is that if repetitive operations are interrupted or otherwise interfered with, an "unlearning curve" effect takes place which can cause

---

[6]J. A. Russo, Jr., *The Complete Money-Saving Guide to Weather for Contractors*, Environmental Information Services Associates, Newington, Conn., September 1971.

[7]Clarkson H. Oglesby, Henry W. Parker, and Gregory Howell, *Productivity Improvement in Construction*, McGraw-Hill Book Company, New York, 1989.

[8]Ibid.

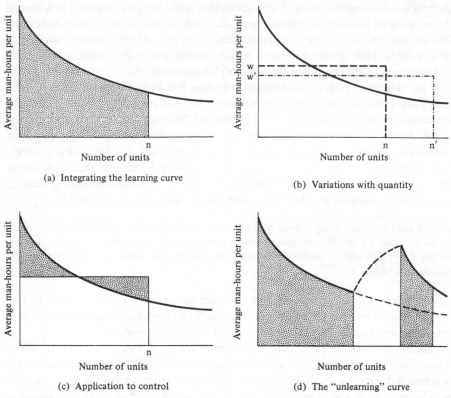

(a) Integrating the learning curve

(b) Variations with quantity

(c) Application to control

(d) The "unlearning" curve

FIGURE 11-7
Learning curve relationships in estimating and control. (From Boyd C. Paulson, Jr., "Estimation and Control of Construction Labor Costs," *Journal of the Construciton Division,* ASCE, vol. 101, no. CO3, September 1975, p. 627.)

the estimate to be exceeded. This is shown in Figure 11-7d. Therefore, for controlling labor costs it is generally good practice to provide continuity of work on repetitive operations.

At least one large engineering and construction firm has built a reference to learning curves into the cost codes for project control reports. Several learning curves have been defined and the one most appropriate to the operation concerned is chosen. Therefore, rather than make linear projections of expected productivity based on work completed to date (which may be at the higher end of the learning curve), the learning curve is included in the calculations (done by computer) in an attempt to make a more accurate forecast for control purposes.

Work Schedule   "Work schedule" here refers to using variations on straight time only, scheduled overtime, or multishift work for accomplishing project objectives. Note that scheduled overtime refers to the situation where operations are regularly scheduled to exceed the normal 8-hour day, 40-hour week. It does not refer to the occasional use of overtime to finish operations that could not be completed in a normal workday, such as finishing a concrete pour that took longer to set up than expected.

The section on the money component dealt with the strictly financial aspects of work schedules. This section will focus mainly on the productivity side, but will also show how the two are combined for the total labor-cost impact.

The self-defeating effect of scheduled overtime has been documented in a study by the Task Force of the Construction Users Anti-Inflation Roundtable under the chairmanship of Weldon McGlaun.[9,10] Findings of this study are summarized in Figure 11-8. Note that as the cumulative effects of scheduled overtime begin to set in, the actual total output for a 50- or 60-hour week drops below that for a 40-hour week. Specific consequences include reduced effectiveness due to fatigue, increased absenteeism, attraction of less-qualified workers, disruption of daily operations, reduced work pace and increased accident rates. The study further concluded that:

> Placing field construction operations of a project on a *scheduled* overtime basis is disruptive to the economy of the affected area, magnifies any apparent labor shortages, reduces labor productivity, and creates excessive inflation of construction labor cost with no material benefit in schedule.[11]

This, combined with the 50 to 100 percent increase in labor costs reflected in Figure 11-8b, should provide sobering second thoughts to owners and contractors hoping to save time and money by putting projects on scheduled overtime.

Scheduling projects on a multishift basis can avoid some but not all of the ill effects of scheduled overtime. This assumes, of course, that the shifts themselves do not run on a scheduled overtime basis. Nevertheless, as seen earlier, multishift work can also incur a significant financial premium, and it can introduce productivity problems as well.

Oglesby, Parker, and Howell point out that when shifts are regularly rotated (say on a weekly or biweekly basis), the natural bodily rhythms of the workers are continuously disrupted and the workers are therefore kept well below their peak efficiency.[12] The effect is not unlike that caused by the frequent changes of time zones encountered by international travelers. Body functions affected include temperature, kidney activity, hormone level, and corticosteroid production. Some adjustments can effectively take place within 1 or 2 days, but others may take several months.

Another factor is the research finding that some people are indeed "day" people and others are "night" people. They actually perform much better when their work fits their physiological schedule.

From these two findings it would appear that multishift work could be made more productive if (1) shifts were not regularly rotated, and (2) an effort were made to match employees to the shift on which they would be likely to perform the best.

[9]Construction Users Anti-Inflation Roundtable, "Effect of Scheduled Overtime on Construction Projects," *AACE Bulletin*, vol. 15, no. 5, October 1973, pp. 155–160.

[10]Weldon McGlaun, "Overtime in Construction," *AACE Bulletin*, vol. 15, no. 5, October 1973, pp. 141–143.

[11]Construction Users Anti-Inflation Roundtable, op. cit.

[12]Oglesby, Parker, and Howell, op. cit.

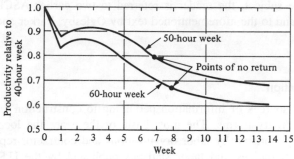

(a) Cumulative effect of overtime on productivity

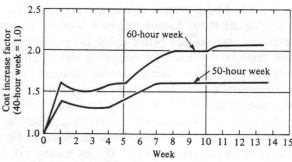

(b) Cumulative cost of overtime (including productivity & premium losses)

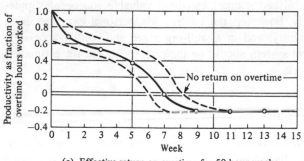

(c) Effective return on overtime for 50-hour weeks

FIGURE 11-8
Scheduled overtime versus productivity. (From "Effect of scheduled Overtime on Construction Projects," *AACE Bulletin*, vol. 15, no. 5, October 1973, pp. 155–160.)

Management

Possibly the most difficult of all the components to analyze is the interrelationship of labor and management. This alone can produce orders of magnitude variations in labor productivity. Factors involved include such things as management philosophy; motivation and morale; safety policies; employee participation in planning, incentives, and rewards; relationships with union locals; and many others. Entire books have been written on the subject and have barely scratched the surface. Little can be added here. For a quick introduction to the subject, the reader is referred to two earlier ASCE articles by Charles Schrader[13] and to the aforementioned text by Oglesby, Parker and Howell.[14]

Using Published Cost Information

There are numerous published sources of information pertaining to construction labor costs and man-hour productivity norms. The most basic of these are the local and national labor agreements negotiated by labor organizations and contractor representatives. Others include cost indices and labor statistics published by the U.S. Department of Labor, state and local governments, and numerous construction magazines and professional organizations. Annual reference works with unit-price and productivity information are also published by a number of business firms.

No attempt is made here to catalog all these various reference sources. Rather, the important thing to recognize at this stage is that before using any published information, be it a cost index, unit price or whatever, for estimating and controlling construction work, one must thoroughly understand exactly what the information does and does not include. For example, consider a unit-labor price for laying brick. Which of the various money components does it include? Is it just the base wage, or does it comprise fringes, insurance, and premiums? What does it assume for productivity? Does it consider regional differences, learning-curve variations with quantity, environmental effects, work schedule, or management approach? The increasing use of worker hours for control purposes helps eliminate much of this confusion.

When tempered with a contractor's own experience and judgment on his own company's operations, such published sources can be a valuable source of supplementary information. Ultimately, however, the contractor's personal experiences as documented accurately, consistently, and in a well-organized and readily-accessible manner by his labor-cost control system should be the primary source for estimating and controlling construction work.

The Importance of Worker-Hours

Hourly wages and fringe benefits continue to increase in the construction industry. Different sections of the country have vastly different basic hourly rates and varying fringe benefits. An electrician or other skilled craftsman may make 50 percent more in high-cost areas than he does in low-cost areas, and overtime premiums distort unit-labor costs. It has become almost a hopeless task to compare unit-labor costs from one section of the country to another, and it is equally difficult to compare today's costs with those of a few years ago.

---

[13]C. R. Schrader, "Motivation of Construction Craftsmen," *Journal of the Construction Division*, ASCE, vol. 98, no. CO2, September 1972, pp. 257–273; and C. R. Schrader, "Boosting Construction-Worker Productivity," *Civil Engineering*, vol. 42, no. 10, October 1972, pp. 61–63.

[14]Oglesby, Parker, and Howell, op. cit.

By keeping productivity records in worker-hours, however, one neutralizes the money component. For example, a contractor may determine that unit productivity for foundation forms is anticipated to be 0.10 unit hours per square foot (s.f.). If labor costs are $30.00/hr including fringes, the unit-labor cost is $3.00/s.f. If labor costs are $20.00/hr, the unit cost becomes $2.00/s.f. If the contractor has kept his cost-accounting records in man-hours as well as in unit costs, he can review performance in all projects and can compare actual productivity with estimated productivity in a straightforward manner.

When a truly integrated management control system is utilized, actual man-hours can be compared to estimated man-hours, and actual productivity for components and for the entire project can be easily measured. Forecasts of individual craft requirementst as well as labor costs to complete will automatically be available from an integrated system based upon estimated work hours. These forecasts can be updated periodically based upon estimated productivity at completion.

Even today some contractors keep unit-labor cost or unit-cost records in dollars rather than man-hours. These contractors are therefore finding it increasingly difficult to prepare accurate labor-cost estimates in an ever-changing construction-cost climate. To illustrate, Figure 11-9 shows the actual manpower compared with the professional construction manager's fair-cost estimate for each contract in the actual Mountaintown warehouse project. Table 8-1 shows productivity calculated from this information.

ESTIMATING DIFFERENT TYPES OF CONSTRUCTION

Estimators typically specialize in building construction, heavy engineering construction and industrial construction. Labor productivity and estimating methods can be substantially different for each category. Productivity can also vary widely within industry categories based upon management, job conditions, labor skills and other individual project factors.

Building Construction

A standard reference for building construction estimators is "Means Building Construction Cost Data," published annually.[15] Richardson's "General Construction Estimating Standards" presents a more detailed approach to building construction and with the use of good judgement can be expanded to be helpful to the industrial estimator as well.[16] Figure 11-10 illustrates the use of the unit costs and prices from Means. Data from both systems is coded in accordance with the CSI Masterformat illustrated in Appendix D. Figure 11-11 shows a typical page from Means showing structural formwork. Means divides cost elements into unit hours, material, labor and equipment. Many building estimators will add a column for subcontracts for ease in both bidding and later cost control on successful projects.

Contractor estimators will use Means as a reference for bare costs and unit hours. Unit labor costs for the contractors own locale can be easily developed by multiply-

---

[15]R. S. Means Company, Inc., PO Box 800, Kingston, MA 02364.
[16]Richardson Engineering Services, Inc., PO Box 9103, Mesa, AZ 85214.

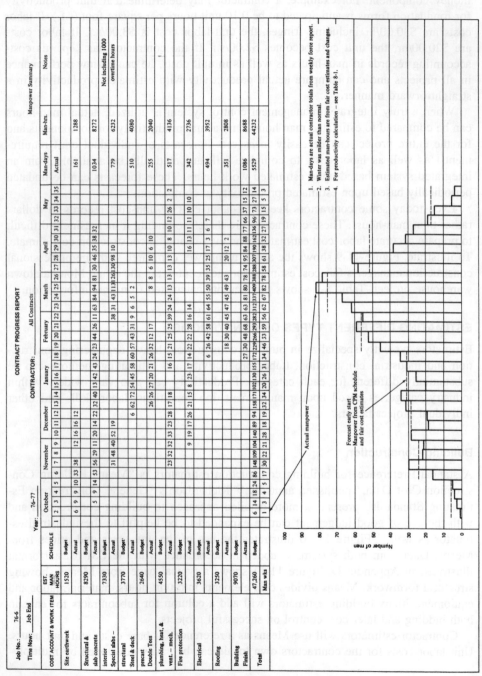

FIGURE 11-9

031 | Concrete Formwork

	031 100	Struct C.I.P. Formwork	CREW	DAILY OUTPUT	MAN-HOURS	UNIT	BARE COSTS MAT.	LABOR	EQUIP.	TOTAL	TOTAL INCL O&P	
170	1050	3 piece (see also edge forms)	C-1	400	.080	L.F.	.53	1.66	.06	2.25	3.22	170
	1100	4 piece		350	.091	"	.70	1.90	.07	2.67	3.79	
	2000	Curb forms, wood, 6" to 12" high, on grade, 1 use		215	.149	SFCA	1.30	3.09	.11	4.50	6.35	
	2050	2 use		250	.128		.70	2.66	.09	3.45	4.99	
	2100	3 use		265	.121		.54	2.51	.09	3.14	4.57	
	2150	4 use		275	.116		.49	2.42	.09	3	4.38	
	3000	Edge forms, to 6" high, 4 use, on grade		600	.053	L.F.	.22	1.11	.04	1.37	2	
	3050	7" to 12" high, 4 use, on grade		435	.074	SFCA	.68	1.53	.05	2.26	3.18	
	3500	For depressed slabs, 4 use, to 12" high		300	.107	L.F.	.50	2.21	.08	2.79	4.07	
	3550	To 24" high		175	.183		.63	3.80	.13	4.56	6.70	
	4000	For slab blockouts, 1 use to 12" high		200	.160		.50	3.32	.12	3.94	5.85	
	4050	To 24" high		120	.267		.63	5.55	.20	6.38	9.50	
	5000	Screed, 24 ga. metal key joint										
	5020	Wood, incl. wood stakes, 1" x 3"	C-1	900	.036	L.F.	.30	.74	.03	1.07	1.50	
	5050	2" x 4"		900	.036	"	.85	.74	.03	1.62	2.11	
	6000	Trench forms in floor, 1 use		160	.200	SFCA	1.63	4.15	.15	5.93	8.40	
	6050	2 use		175	.183		.90	3.80	.13	4.83	7	
	6100	3 use		180	.178		.66	3.69	.13	4.48	6.60	
	6150	4 use		185	.173		.53	3.59	.13	4.25	6.30	
174	0010	FORMS IN PLACE, STAIRS (Slant length x width), 1 use (33)	C-2	165	.291	S.F.	2.46	6.20	.19	8.85	12.55	174
	0050	2 use		170	.282		1.43	6.05	.18	7.66	11.10	
	0100	3 use		180	.267		1.09	5.70	.17	6.96	10.20	
	0150	4 use		190	.253		.91	5.40	.16	6.47	9.55	
	1000	Alternate pricing method (0.7 L.F./S.F.), 1 use		100	.480	LF Rsr	3.62	10.25	.31	14.18	20	
	1050	2 use		105	.457		2.18	9.75	.30	12.23	17.85	
	1100	3 use		110	.436		1.69	9.30	.28	11.27	16.60	
	1150	4 use		115	.417		1.47	8.90	.27	10.64	15.75	
	2000	Stairs, cast on sloping ground (length x width), 1 use		220	.218	S.F.	1.74	4.66	.14	6.54	9.30	
	2100	4 use		240	.200	"	1.01	4.27	.13	5.41	7.85	
182	0010	FORMS IN PLACE, WALLS										182
	0020											
	0100	Box out for wall openings, to 16" thick, to 10 S.F.	C-2	24	2	Ea.	14.20	43	1.30	58.50	83	
	0150	Over 10 S.F. (use perimeter)	"	280	.171	L.F.	1.35	3.66	.11	5.12	7.30	
	0250	Brick shelf, 4" wide, add to wall forms, use wall area										
	0260	above shelf, 1 use	C-2	240	.200	SFCA	1.40	4.27	.13	5.80	8.30	
	0300	2 use		275	.175		.81	3.73	.11	4.65	6.80	
	0350	4 use		300	.160		.56	3.42	.10	4.08	6.05	
	0500	Bulkhead forms for walls, with keyway, 1 use, 2 piece		265	.181	L.F.	1.70	3.87	.12	5.69	8	
	0550	3 piece		175	.274	"	2.15	5.85	.18	8.18	11.65	
	0700	Buttress forms, to 8' high, 1 use		350	.137	SFCA	1.60	2.93	.09	4.62	6.40	
	0750	2 use		430	.112		.95	2.38	.07	3.40	4.82	
	0800	3 use		460	.104		.71	2.23	.07	3.01	4.31	
	0850	4 use		480	.100		.60	2.14	.07	2.81	4.04	
	1000	Corbel (haunch) forms, up to 12" wide, add to wall forms, 1 use		150	.320	L.F.	1.67	6.85	.21	8.73	12.65	
	1050	2 use		170	.282		.94	6.05	.18	7.17	10.60	
	1100	3 use		175	.274		.71	5.85	.18	6.74	10.05	
	1150	4 use		180	.267		.60	5.70	.17	6.47	9.70	
	2000	Job built plyform wall forms, to 8' high, 1 use (33)		370	.130	SFCA	1.46	2.77	.08	4.31	6	
	2050	2 use		435	.110		.90	2.36	.07	3.33	4.72	
	2100	3 use		495	.097		.71	2.07	.06	2.84	4.06	
	2150	4 use		505	.095		.62	2.03	.06	2.71	3.89	
	2400	Over 8' to 16' high, 1 use		280	.171		1.65	3.66	.11	5.42	7.60	
	2450	2 use		345	.139		.91	2.97	.09	3.97	5.70	
	2500	3 use		375	.128		.71	2.73	.08	3.52	5.10	
	2550	4 use		395	.122		.64	2.60	.08	3.32	4.81	
	2700	Over 16' high, 1 use		235	.204		1.82	4.36	.13	6.31	8.90	
	2750	2 use		290	.166		1.06	3.54	.11	4.71	6.75	

78

For expanded coverage of these items see *Means Concrete Cost Data 1990*

FIGURE 11-10
Concrete Formwork from "Means Building Construction Cost Data," 1990.

ing local wage rates including burden and fringes by the unit hours. The contractor may utilize the Total column including overhead and profit to develop a preliminary price for subcontracts (plug prices) pending recipt of actual bids. Many architects and engineers on the other hand will prepare estimates using the Total column and will not separate the individual components for their preliminary or fair cost estimates.

032 | Concrete Reinforcement

032 100 | Reinforcing Steel

		Description	CREW	DAILY OUTPUT	MAN-HOURS	UNIT	MAT.	LABOR	EQUIP.	TOTAL	TOTAL INCL O&P	
102	3900	¾" diameter, 3-½" high				C	345			345	380	102
	3950	12" high					775			775	855	
	4200	Subgrade stakes (SCSGS) ½" diameter, 16" long					210			210	230	
	4250	24" long					300			300	330	
	4300	¾" diameter, 16" long					230			230	255	
	4350	28" long					345			345	380	
	4500	Tie wire, 16 ga. annealed steel, under 500 lbs.				Cwt.	73			73	80	
	4520	2,000 to 4,000 lbs.				*	68			68	75	
	4550	Tie wire holder, plastic case				Ea.	32			32	35	
	4600	Aluminum case				*	34			34	37	
104	0010	COATED REINFORCING Add to material										104
	0100	Epoxy coated				Cwt.	15			15	16.50	
	0150	Galvanized, #3					23.30			23.30	26	
	0200	#4					22.25			22.25	24	
	0250	#5					19.50			19.50	21	
	0300	#6 or over					18.20			18.20	20	
	1000	For over 20 tons, #6 or larger, minimum					16.90			16.90	18.60	
	1500	Maximum					18.90			18.90	21	
107	0010	REINFORCING IN PLACE A615 Grade 60										107
	0100	Beams & Girders, #3 to #7	4 Rodm	1.60	20	Ton	595	490		1,085	1,450	
	0150	#8 to #18		2.70	11.850		585	290		875	1,125	
	0200	Columns, #3 to #7 (38)		1.50	21.330		595	525		1,120	1,500	
	0250	#8 to #18		2.30	13.910		585	340		925	1,200	
	0300	Spirals, hot rolled, 8" to 15" diameter		2.20	14.550		1,240	360		1,600	1,950	
	0320	15" to 24" diameter		2.20	14.550		1,175	360		1,535	1,875	
	0330	24" to 36" diameter		2.30	13.910		1,150	340		1,490	1,825	
	0340	36" to 48" diameter		2.40	13.330		1,175	330		1,505	1,825	
	0360	48" to 64" diameter		2.50	12.800		1,240	315		1,555	1,875	
	0380	64" to 84" diameter		2.60	12.310		1,330	305		1,635	1,950	
	0390	84" to 96" diameter		2.70	11.850		1,330	290		1,620	1,950	
	0400	Elevated slabs, #4 to #7		2.90	11.030		590	270		860	1,100	
	0500	Footings, #4 to #7		2.10	15.240		585	375		960	1,250	
	0550	#8 to #18		3.60	8.890		575	220		795	990	
	0600	Slab on grade, #3 to #7		2.30	13.910		595	340		935	1,225	
	0700	Walls, #3 to #7		3	10.670		595	260		855	1,075	
	0750	#8 to #18		4	8		580	195		775	960	
	1000	Typical in place, 10 ton lots, average (39)		1.70	18.820		600	465		1,065	1,425	
	1100	Over 50 ton lots, average		2.30	13.910		575	340		915	1,200	
	1200	High strength, Grade 75, #14 bars only, add					52			52	57	
	2000	Unloading & sorting, add to above	C-5	100	.560			13.75	4.76	18.51	27	
	2200	Crane cost for handling, add to above, minimum		135	.415			10.20	3.53	13.73	20	
	2210	Average		92	.609			14.95	5.15	20.10	30	
	2220	Maximum		35	1.600			39	13.60	52.60	78	
	2400	Dowels, 2 feet long, deformed, #3 bar	2 Rodm	140	.114	Ea.	.85	2.81		3.66	5.55	
	2410	#4 bar		125	.128		1.01	3.15	*	4.16	6.25	
	2420	#5 bar		110	.145		1.23	3.58		4.81	7.20	
	2430	#6 bar		105	.152		1.66	3.75		5.41	7.95	
	2450	Longer and heavier dowels		450	.036	Lb.	.40	.87		1.27	1.87	
	2500	Smooth dowels, 12" long, ¼" or ⅜" diameter		140	.114	Ea.	.64	2.81		3.45	5.30	
	2520	⅝" diameter		125	.128		1.13	3.15		4.28	6.40	
	2530	¾" diameter		110	.145		1.39	3.58		4.97	7.35	
	2550											
	2700	Dowel caps, 5" long, ½" to ¾" diameter	2 Rodm	800	.020	Ea.	.06	.49		.55	.87	
	2720	1-¼" diameter	*	750	.021	*	.08	.52		.60	.94	
109	0010	SPLICING REINFORCING BARS Incl. holding bars in (44)										109
	0020	place while splicing										
	0100	Butt weld columns #4 bars	C-5	190	.295	Ea.	.89	7.25	2.51	10.65	15.30	
	0110	#6 bars	*	150	.373	*	1.31	9.15	3.17	13.63	19.60	

84

For expanded coverage of these items see *Means Concrete Cost Data 1990*

FIGURE 11-11
Concrete Reinforcement from "Means Building Construction Cost Data," 1990.

Table 11-4 lists relative productivity factors for building work based upon different conditions. Based upon 1.0 as optimum output for a skilled fully equipped small crew with a working foreman and no organized coffee breaks, typical overall productivity on building work could typically range beween 1.0 to 1.75 optimum on commercial building projects.

TABLE 11-4

PRODUCTIVITY JOB FACTORS FOR
COMPARABLE INSTALLATIONS

Building construction	
Nonworking supervision	1.0 to 1.15
Craft skill	1.0 to 1.20
Job conditions	1.0 to 1.20
Work conditions	1.0 to 1.20
Shift work	1.0 to 1.20
Total building range	1.0 to 2.00

Industrial construction	
Light industrial	1.25 to 2.25
Heavy industrial	1.50 to 3.00
Total industrial range	1.25 to 3.00

Nuclear power plants	
Pre-Three Mile Island	2.25 to 4.00
Post-Three Mile Island	3.00 to 5.00
Total nuclear power plant range	2.25 to 5.00

Notes:
   1. Ranges are based upon evaluation from records kept by D. S. Barrie from 1975 to 1985. Totals represent approximate arithmetic averages of individual factors illustrated under Building construction.
   2. Comparable installations are based upon physical installations of like work such as one square foot of wall form, one ton of reinforcing steel, one diameter-inch of like-sized large bore pipe spool field weld, one lineal foot of like-sized rigid conduit installation, etc.
   3. Specification requirements and productivity factors are reflected in overall figures for industrial and nuclear work.
   4. Job Factor of 1.0 for building work represents a skilled small crew with a working foreman or leadman based upon 60 minutes per hour effective work, normal building specifications, no shift work and under ideal conditions.

Specialty subcontractors will perform most of the work on most building projects. The general contractor will not normally estimate in detail the work of the subs and will depend upon the lowest qualified bidder for his price. Thus general contractor markups are typically lower than subcontractor markups since a major share of the risk is passed on to the sub. Typical work performed by a general contractor in the building sector might include building foundations and miscellaneous concrete, structural concrete, rough and finish carpentry along with other specialties for which the contractor is qualified.

Industrial Construction

No standard references for industrial estimating comparable to Means and Richardson exist due to the widespread size and complexity variations peculiar to industrial construction work. Labor productivity for basic units of work tends to be lower than for building work due to many factors including nonworking foremen, larger crews, and other factors.

Most industrial contractors and design-constructors will have developed their own data base for unit-labor costs which is almost universally expressed in unit hours. Standard crew makeups are developed and unit labor and other cost elements for extension purposes are normally calculated by the computer based upon crew makeup, hourly rates for wages and fringes, and production rates as established by the estimator, who can selectively override data base information when desirable. See Figure 11-12 for a computerized report showing unit costs prepared upon a microcomputer for a major project. Figure 11-13 from the same project shows the extensions for direct cost accounts by element of work in CSI format. Figure 11-14 illustrates a direct cost summary by process area. Figure 11-15 illustrates a typical form crew showing crew makeup, equipment requirements and overall crew and equipment hourly rates which form the basis for the extension units. This estimate was prepared at approximately 20 percent design completion and might be classified as about Type 3.5 on the scale shown on Figure 11-1. Indirect costs were estimated rationally as a line item and can be optionally spread to the direct accounts through the overhead and profit element if warranted. Markup can be similarly spread. On a major project involving several contractors the use of a contract code in the work breakdown structure can permit tabulation of the estimate by proposed contract including indirect costs and markup.

Richardson's "General Construction Estimating Standards" and "Process Plant Construction Estimate Standards" can be particularly helpful to industrial estimators in developing their own structural, piping and electrical units. Some of the leading design-constructors initially adopted the Richardson system for basic mechanical and electrical units adding their own productivity modifier more applicable to their particular class of work. As the years developed, the individual systems became tailored through feedback to the particular company experience and may now have little resemblance to the current Richardson.

Table 11-4 compares overall ranges for industrial unit hourly rates compared to the building units tabulated previously. Industrial units may range from about 1.25 to 3.0 times optimum units described under building construction. Mean unit rates on nuclear power plants based upon a study of 17 plants have ranged from 3.0 to 5.5 dependent upon the year measured, upon individual site conditions, and other factors. A controlled study by the Foothill Electric Corporation in the early 1980s indicated that overall weighted electrical units on nuclear work as shown for the 17 projects was 3.4 times similarly weighted units for raceways, cable installation and terminations on a $30 million nonnuclear experimental power plant. These comparisons are indicated to illustrate the wide variations possible in industrial work. Such work requires substantial experience and skill upon the part of the estimator if estimates are to be realistic for the actual conditions to be encountered.

Heavy Engineering Construction

Heavy construction estimating for dams, tunnels, mass grading, etc. makes much less use of standard unit hours for functional operations than is typical of both building

KAL KRISHNAN CONSULTING SERVICES, INC.
ESTIMATE RECONCILIATION - DRAFT

WORK ELEMENT SUMMARY REPORT

WORK ELEM	DESCRIPTION	UN	UNIT MH	AVG. WAGE RATE	UNIT LABOR	UNIT MATERIAL	UNIT EQUIP	UNIT SUB-CONT	UNIT OTHER	TOTAL UNIT COST	CREW CODE	CREW SIZE	UTS/ CREW HOUR
02	SITEWORK												
0201000	SUBSURFACE INVESTIGATION	LT	40.000	25.92	1036.80	0.00	3800.00	0.00	0.00	4836.80		0.00	0.000
0201501	MOVE-IN & SITE PREPARATION	LT	500	24.00	12000.00	9500.00	4500.00	0.00	0.00	26000.00		0.00	0.000
0211001	SITE CLEARING & GRUBBING	SF	0.003	24.00	0.07	0.00	0.05	0.00	0.00	0.12		0.00	0.000
0220003	MASS EXCAVATION	CY	0.068	26.07	1.77	0.00	4.33	0.00	0.00	6.10	STMSEW	39.00	573.529
0220005	EXCAVATION STOCKPILE	CY	0.036	26.08	0.94	0.00	2.29	0.00	0.00	3.23	STMSEW	39.00	1083.333
0220007	DISPOSAL	CY	0.000	0.00	0.00	2.00	0.00	0.00	0.00	2.00		0.00	0.000
0220009	BACKFILL & COMPACT-STRUCTURAL	CY	0.060	23.63	1.42	0.00	0.98	0.00	0.00	2.40		0.00	0.000
0220010	GRADING, SITE	SF	0.002	24.13	0.05	0.00	0.17	0.00	0.00	0.22		0.00	0.000
0220011	DRAINAGE & TRENCHING, SITE	LF	0.150	25.36	3.80	0.00	1.70	0.00	0.00	5.50		0.00	0.000
0220013	MASS EXCAVATION -wall	CY	0.071	26.10	1.85	0.00	5.18	0.00	0.00	7.03	RWSEXB	25.00	352.113
0220014	HAUL IN STRUCTURAL BACKFILL	CY	0.080	26.08	2.09	20.00	5.09	0.00	0.00	27.18	STMSEW	39.00	487.500
0220017	DISPOSAL- wall	CY	0.000	0.00	0.00	2.00	0.00	0.00	0.00	2.00		0.00	0.000
0221004	EXCAVATION, SPREAD FOOTING	CY	0.178	26.90	4.79	0.00	4.78	0.00	0.00	9.57		0.00	0.000
0221009	MASS BACKFILL & COMPACTION	CY	0.060	24.27	1.46	9.46	2.79	0.00	0.00	13.70	STRBF	12.00	200.000
0222101	FILL, IMPORTED	CY	0.250	23.63	5.91	9.25	2.00	0.00	0.00	17.16		0.00	0.000
0222105	FILL, IMPORTED DRAIN GRAVEL	CY	0.250	33.75	8.44	16.65	1.08	0.00	0.00	26.17		0.00	0.000
0222106	TOP SOIL, IMPORTED	SF	0.002	23.63	0.05	0.17	0.02	0.00	0.00	0.24		0.00	0.000
0236601	DRILL AND DRIVE 18" DIA. PRECAST PILE	LF	0.000	0.00	0.00	0.00	0.00	26.50	0.00	26.50		0.00	0.000
0236603	DRILL AND DRIVE 24" DIA. PRECAST PILE	LF	0.000	0.00	0.00	0.00	0.00	35.00	0.00	35.00		0.00	0.000
0252700	RESET RIP RAP	SF	0.040	23.47	0.94	0.00	3.56	0.00	0.00	4.50		0.00	0.000
0255003	GRADE & COMPACT, PRIMARY TRAIL	LF	0.014	24.13	0.34	0.00	0.31	0.00	0.00	0.65		0.00	0.000

FIGURE 11-12
Unit cost data. (Courtesy of Kal Krishnan Consulting Services, Oakland, Calif.)

KAL KRISHNAN CONSULTING SERVICES, INC.
WESTPOINT SECONDARY TREATMENT FACILITY
ESTIMATE RECONCILIATION - DRAFT

-- ACCOUNT NUMBER -- AREA.WRKELEM.PACKG	ITEM	MANHOURS	LABOR	PERMANENT MATERIAL	CONST EQUIP	SUB-CONTRACT	PERMANENT EQUIPMENT	OVERHEAD & PROFIT	TOTAL COST
02	SITEWORK	378,827	8,041,693	13,002,206	4,242,432	54,646,817	0	0	70,933,148
03	CONCRETE	1,573,729	40,415,202	36,134,202	13,369,327	2,574,801	0	0	92,493,532
04	MASONRY	4,461	104,769	64,555	21,406	3,680	0	0	194,410
05	METALS	120,119	3,271,138	16,727,544	1,283,457	180,539	0	0	21,462,678
06	WOOD & PLASTICS	1,023	25,521	46,508	2,000	97,218	0	0	171,247
07	MOISTURE PROTECTION	151,510	3,838,614	1,960,707	216,442	595,377	0	0	6,611,140
08	DOORS, WINDOWS & GLASS	3,484	88,348	193,268	43,844	86,200	0	0	411,660
09	FINISHES	31,908	507,990	637,218	28,635	2,298,728	0	0	3,472,571
10	SPECIALTIES	1,232	30,707	48,908	30	1,500	0	0	81,145
11	PROCESS EQUIPMENT	137,363	3,440,365	117,773	2,440,859	1,003	54,000,000	0	60,000,000
12	FURNISHINGS	0	0	0	0	1,280	0	0	1,280
13	SPECIAL CONSTRUCTION	32	920	800	612	2,072,000	0	0	2,074,332
14	CONVEYING SYSTEMS	8,732	218,377	119,602	139,157	254,532	1,045,416	0	1,777,084
15	MECHANICAL	318,248	9,925,677	17,295,393	6,111,013	9,397,361	2,470,435	0	45,199,879
16	ELECTRICAL	300,787	8,699,453	7,740,579	5,364,066	3,637,277	10,656,760	0	36,100,135
17	INSTRUMENTATION & CONTROLS	38,995	1,192,980	3,779,661	640,242	0	14,387,235	0	20,000,118
*** TOTAL ***		3,070,450	79,801,754	97,868,924	33,905,522	75,848,313	82,559,846	0	369,984,359

FIGURE 11-13
Direct cost summary by element of work. (Courtesy of Kal Krishnan Consulting Services, Oakland, Calif.)

OCTOBER 04, 1990 04:51
BASE MP01 LAYOUT

KAL KRISHNAN CONSULTING SERVICES, INC.
WESTPOINT SECONDARY TREATMENT FACILITY
ESTIMATE RECONCILIATION - DRAFT

PAGE 1
Job No: WESTPNI3
BY DON BARRIE

ACCOUNT NUMBER -- AREA.MKELEM.PACKG	ITEM	MANHOURS	LABOR	PERMANENT MATERIAL	CONST EQUIP	SUB-CONTRACT	PERMANENT EQUIPMENT	OVERHEAD & PROFIT	TOTAL COST
01	ADMINISTRATION BUILDING	86,825	1,938,828	2,399,320	453,873	2,230,799	42,500	0	7,065,320
02	MAINTENANCE BUILDING	61,835	1,634,905	1,538,111	666,964	2,280,712	48,000	0	6,168,692
03	INFLUENT CONTROL STRUCTURE	8,072	219,807	474,157	94,494	158,973	1,235,152	0	2,182,583
04	RAW SEWAGE PUMPING	31,416	892,102	782,128	569,090	47,076	4,584,640	0	6,875,116
05	GRIT/SCREENS HANDLING	21,627	746,178	593,031	359,274	215,268	1,969,334	0	3,883,085
06	PRIMARY CLARIFIERS	175,132	4,676,138	6,291,967	1,744,154	207,129	3,386,976	0	16,306,364
07	EFFLUENT PUMPING	130,568	3,522,616	4,886,164	1,536,187	49,086	4,416,450	0	14,410,503
08	NEW/EXISTING DIGESTERS	169,791	4,652,987	5,730,233	2,258,760	1,095,483	8,138,444	0	21,875,207
09	EXISTING/NEW GALLERIES	93,755	2,833,671	4,323,694	1,339,612	2,444,325	0	0	10,941,302
10	AERATION TANKS	391,999	10,253,864	10,242,659	3,057,748	599,587	6,701,830	0	31,655,690
11	OXYGEN PRODUCTION	84,804	2,232,622	1,683,043	871,154	193,308	9,841,964	0	14,822,091
12	SECONDARY CLARIFIERS	541,876	14,706,432	16,046,468	6,143,369	326,975	7,887,928	0	45,111,372
13	CHLORINE GENERATION	56,059	1,523,142	1,408,454	817,354	577,368	3,456,515	0	7,762,833
14	CHLORINE CONTACT	173,971	4,533,818	3,604,324	1,319,929	82,113	1,369,092	0	10,909,276
15	SOLIDS HANDLING	206,470	5,510,253	5,119,335	2,476,839	764,584	11,991,902	0	25,862,913
16	ODOR CONTROL	60,924	1,659,919	2,070,977	716,619	1,425,776	1,070,779	0	6,943,870
17	COGENERATION	55,406	1,487,891	1,192,536	720,513	100,642	4,470,390	0	7,971,992
18	FACILITY SERVICES	5,094	146,639	411,075	44,917	295,666	42,500	0	942,797
19	OPERATIONS CENTER	0	0	1,000,000	0	0	0	0	1,000,000
20	ANNEX	3,410	91,879	83,531	38,526	188,684	22,950	0	425,570
21	FLOW DIVERSION STRUCTURE	33,759	884,748	1,541,694	244,446	45	9,300	0	2,680,233
22	ELECTRICAL SUBSTATION	7,501	214,769	23,659	140,356	171,180	1,125,000	0	1,676,964
23	RETAINING WALL	4,260	111,180	120,000	310,620	25,621,600	0	0	26,163,400
51	ACCESS ROADWAY	0	0	0	0	2,540,851	0	0	2,540,851
52	CANAL FEEDER	0	0	0	0	3,100,000	0	0	3,100,000
53	SITEWORK	24,319	655,874	4,796,006	714,016	25,992,490	680,000	0	32,838,386
54	EMERGENCY BYPASS	0	0	0	0	68,000	0	0	68,000
55	PLANT SITE UTILITIES	45,721	1,374,988	3,977,522	1,148,806	234,127	0	0	6,737,443
56	DISTRIBUTED CONTROL SYSTEM	45,392	1,309,450	207,640	842,770	10,068,000	0	0	12,427,860
57	MITIGATION	499,332	11,155,372	17,320,396	4,020,532	848,246	0	0	33,344,546
58	TEMPORARY OPERATIONS FACILITY	0	0	0	0	1,000,000	0	0	1,000,000
59	TEMPORARY MAINTENANCE FACILITY	0	0	0	0	1,000,000	0	0	1,000,000
60	ASBESTOS REMOVAL	0	0	0	0	1,300,000	0	0	1,300,000
61	DEMOLITION	33,132	828,300	0	452,800	688,200	0	0	1,969,300
*** TOTAL ***		3,070,450	79,801,754	97,848,924	33,905,522	75,846,313	82,559,846	0	369,084,359

FIGURE 11-14

Direct cost summary by process area. (Courtesy of Kal Krishnan Consulting Services, Oakland, Calif.)

OCTOBER 04, 1990 04:51
BASE HP01 LAYOUT

KAL KRISHNAN CONSULTING SERVICES, INC.
WESTPOINT SECONDARY TREATMENT FACILITY
ESTIMATE RECONCILIATION - DRAFT
CREW DETAIL REPORT

PAGE 1
Job No: WESTPNT3
BY DON BARRIE

CREW CODE: C2106, FORMS WITH CRANE USAGE

CODE	QUANTITY	DESCRIPTION	RATE	$/CH
LABOR				
CARFORE	1.00	CARPENTER-FOREMAN	26.64	26.64
LAB4	2.00	LABORER - JOURNEYMAN	21.72	43.44
CARPNTR	4.00	CARPENTER-JOURNEYMAN	25.51	102.04
OPERENG	1.00	OPERATING ENGINEER 1	25.92	25.92
TOTAL	8.00		24.76	198.04
EQUIPMENT				
E11103	1.00	PORTABLE GEN,ONAN 10 KW,14 HP,GAS	2.31	2.31
E13101	0.10	HYDRAULIC CRANE, 65 TON CAPACITY	15.80	1.58
TOTAL	1.10		3.54	3.89
CREW TOTAL PER HOUR				201.93

FIGURE 11-15
Typical form crew. (Courtesy of Kal Krishnan Consulting Services, Oakland, Calif.)

and industrial construction. Heavy construction applications such as site grading, mass excavation, paving and other similar items are also utilized on major building and industrial projects either by specialty subcontractors or by a qualified general contractor. Heavy construction estimating methods treat each bid item as a separate entity and the individual unit bid prices are based upon the use of specific construction equipment, plant facilities, labor rates, labor and equipment productivity assumptions and other working conditions applicable to each individual project and operation. Most project bid forms feature a combination of lump sum and unit price bid items. The bid form sets forth the owner's estimate of quantities and the contractor supplies unit prices or lump sum bids for each individual item. The overall bid price which determines contract award is totaled from the sum of the individual extensions. During construction, individual quantities are measured for the unit price items which when extended by the bid price determine actual revenue to be received by the contractor.

Unit direct costs are computed by the estimator who develops a crew for each major operation, chooses the equipment to be utilized, and develops an hourly or shift cost for both direct labor and equipment operating cost. The estimator then estimates the unit productivity for the individual operation based upon the output of the equipment under the assumed job conditions and develops hourly or shift unit costs, total direct costs and production. A standard industry reference for construction equipment production is the *Caterpillar Performance Handbook*, latest edition, published by the Caterpillar Tractor Co., Peoria, Illinois. Other manufacturers will issue similar information for their own products. Units can be developed from theoretical output modified by a productivity factor or from the contractor's actual records on past projects corrected for any perceived differences. Plant and equipment ownership costs along with other indirect costs are usually estimated as line items and along with an allowance for profit are spread back to the respective bid items to develop the unit bid prices.

Figure 11-16 illustrates estimated maximum hourly production of power shovels based upon 100 percent efficiency and other factors. Caterpillar recommends use of the following formula for calculating estimated production for shovels and other similar types of equipment:

Production = Max. Production × Job Efficiency Factor

$$\times \text{ Swing \& Depth Factor} \times \text{Bucket Load Factor}$$

A summary of the Sierra Tunnel direct cost Estimate is shown in Figure 11-17 to illustrate the cost elements commonly utilized for heavy construction work. Since plant and equipment has multiple usage throughout the project and represents a major share of contractor cost, such costs are estimated as individual line items of capital equipment and plant facilities and are later spread to the direct costs less an assumed salvage value. Operating labor and fuel, oil and gas are estimated at the direct cost level based upon estimated hourly usage in each account.

On major projects contractors may form a joint venture to spread the risk and to lower individual company bonding requirements. A joint venture meeting under the direction of the sponsoring company, reviews the individual estimates and develops an agreed estimate of total cost as shown in Figure 11-18, including plant and equip-

Excavators — Front Shovels | Rock Loading Production — E450 and E650 Front Shovels
- Tons Per 60 Min. Hour*
- Shot Rock
- Estimated Density —
  1600 kg/Lm³ or 2700 lb/LCY (1.35 ton/LCY)

METRIC TONS PER 60 MIN. HR.

ESTIMATED CYCLE TIME		ESTIMATED BUCKET PAYLOAD — LOOSE CUBIC METER**					ESTIMATED CYCLES	
Cycle Time (Sec)	Cycle Time (Min)	2.6 m³	3.8 m³	7.5 m³	8.8 m³	12 m³	Cycles/ Minute	Cycles/ Hour
15	.25	998	1459				4.0	240
18	.30	832	1216				3.0	200
21	.35	711	1040	2052	2406	3283	2.9	171
25	.42	599	876	1728	2028	2765	2.5	144
32	.53	470	687	1356	1591	2170	1.9	113
40	.67	374	547	1080	1267	1728	1.5	90
45	.75			980	1128	1536	1.3	80
50	.83			864	1014	1382	1.2	72

U.S. TONS PER 60 MIN. HR.

ESTIMATED CYCLE TIME		ESTIMATED BUCKET PAYLOAD — LOOSE CUBIC YARD**					ESTIMATED CYCLES	
Cycle Time (Sec)	Cycle Time (Min)	3.4 Yd³	5 Yd³	9.75 Yd³	11.5 Yd³	15.75 Yd³	Cycles/ Minute	Cycles/ Hour
15	.25	1102	1620				4.0	240
18	.30	918	1350				3.0	200
21	.35	785	1154	2251	2655	3636	2.9	171
25	.42	661	972	1895	2236	3062	2.5	144
32	.53	519	783	1487	1754	2403	1.9	113
40	.67	413	608	1185	1397	1914	1.5	90
45	.75			1053	1242	1701	1.3	80
50	.83			948	1118	1531	1.2	72

*Actual Hourly Production = (60 Min. Hr. Production) x (Job Efficiency Factor)
**Estimated Bucket Payload = (Heaped Bucket Capacity) x (Bucket Fill Factor)
These tables are calculated using a 100% bucket fill factor.
See bucket fill factors prior to the rock loading production charts.

FIGURE 11-16
Estimated hourly production dipper type power shovels from *Caterpillar Performance Handbook*.

ment, indirect and escalation costs. Figure 11-19 shows the final bid tabulation after spreading markup and making adjustments due to final quotations from subcontractors and material suppliers.

SUCCESSIVE ESTIMATING[17]

In this chapter it is worth introducing one additional approach to estimating that is being increasingly used in Northern Europe. Known there as "successive estimating" or "successive planning," it was developed by Prof. Steen Lichtenberg of the Technical University of Denmark for network-based planning and scheduling as well as for estimating.

Successive estimating uses statistical principles both to improve the accuracy and greatly to reduce the amount of effort required for estimating the cost of a project.

[17]For additional information, see Steen Lichtenberg, "Project Management Systems—Monsters or Assistants to the Manager," in Proceedings of the 8th Annual Seminar/Symposium of the Project Management Institute, Montreal, October 1976, pp. 152–158.

SIERRA TUNNEL ESTIMATE: *SUMMARY OF DIRECT COST*

Bid Item	Description	Quantity	Labor Unit	Labor Amount	Supplies Unit	Supplies Amount	Permanent Materials Unit	Permanent Materials Amount	Subcontracts Unit	Subcontracts Amount	Total Direct Cost Unit	Total Direct Cost Amount
Tunnel excavation:												
1	Tunnel excavation	431,000 yd³	$ 17.42	$7,508,020	$ 9.10	$3,922,100					$ 26.52	$11,430,120
2	Steel sets	2,290,000 lb	0.13	297,700			0.48	1,099,200			0.61	1,396,900
3	Timber	400 MBM	125.00	50,000	320.00	128,000	320.00	128,000			765.00	306,000
4	Roof bolts	12,000 lin ft	2.41	28,920	0.59	7,080	1.60	19,200			4.60	55,200
5	Drill exploratory holes	300 lin ft	10.82	3,246	5.30	1,590					16.12	4,836
6	Drill grout holes as aid to tunnel driving	2,400 lin ft	1.63	3,912	5.30	12,720					6.93	16,632
7	Grout as aid to tunnel driving	400 ft³	12.98	5,192	6.62	2,648					19.60	7,840
	Total tunnel excavation			7,896,990		4,074,138		1,246,400				13,217,528
Shaft excavation:												
8	Shaft excavation	5,300 yd³	49.69	263,357	18.51	98,103					68.20	361,460
Concrete:												
9	Concrete tunnel lining	111,000 yd³	42.16	4,679,760	10.99	1,219,890			14.06	1,560,660	67.21	7,460,310
10	Shaft concrete	1,900 yd³	67.18	127,642	38.16	72,504			12.70	24,130	118.04	224,276
11	Cement for concrete and low-pressure grout	163,000 bbl					16.00	2,608,000			16.00	2,608,000
12	Steel reinforcement in shaft	285,000 lb							0.47	133,950	0.47	133,950
13	Embedded anchor bolts	3,200 lb	2.50	8,000	0.20	640	1.60	5,120			4.30	13,760
14	Embedded vent pipe	600 lb	2.00	1,200	0.15	90	1.10	660			3.25	1,950
	Total concrete			4,816,602		1,293,124		2,613,780		1,718,740		10,442,246
Consolidation grouting:												
15	Drill grout holes	4,000 lin ft							10.00	40,000	10.00	40,000
16	Consolidation grouting	10,000 ft³							10.00	100,000	10.00	100,000
17	Cement for grouting	5,000 sacks							5.00	25,000	5.00	25,000
	Total grouting									165,000		165,000
	TOTAL DIRECT COST			12,976,949		5,465,365		3,860,180		1,883,740		24,186,234

FIGURE 11-17
Sierra tunnel estimate: Summary of direct cost. (From Parker, Barrie, and Snyder, *Planning and Estimating Heavy Construction*, McGraw-Hill Book Company, 1984.

SIERRA TUNNEL BID PREPARATION: SPREADING PLANT AND EQUIPMENT, INDIRECT, AND ESCALATION COSTS TO ARRIVE AT TOTAL COST

Bid Item	Description	Pay Quantity	Agreed Direct Cost Unit	Agreed Direct Cost Amount	Agreed Plant and Equipment Cost Unit	Agreed Plant and Equipment Cost Amount	Agreed Indirect Cost Unit	Agreed Indirect Cost Amount	Agreed Escalation Cost Unit	Agreed Escalation Cost Amount	Agreed Total Cost Unit	Agreed Total Cost Amount
Tunnel excavation:												
1	Tunnel excavation	431,000 yd³	$ 27.66	$11,921,460	$10.21	$4,400,510	$ 3.61	$1,555,910	$0.82	$ 353,420	$ 42.30	$18,231,300
2	Steel sets	2,290,00 lb	0.61	1,396,900			0.08	183,200			.69	1,580,100
3	Timber	400 MBM	765.00	306,000			99.98	39,992			864.98	345,992
4	Roof bolts	12,000 lin ft	4.60	55,200			0.60	7,200			5.20	62,400
5	Drill exploratory holes	300 lin ft	16.12	4,836			2.11	633			18.23	5,469
6	Drill grout holes as aid to tunnel driving	2,400 lin ft	6.93	16,632			0.91	2,184			7.84	18,816
7	Grout as aid to tunnel driving	400 ft³	19.60	7,840			2.56	1,024			22.16	8,864
	Total			$13,708,868		$4,400,510		$1,790,143		$ 353,420		$20,252,941
Shaft excavation:												
8	Shaft excavation	5,300 yd³	68.20	361,460	19.13	101,389	8.91	47,223			96.24	510,072
Concrete:												
9	Tunnel lining	111,000 yd³	66.71	7,404,810	18.50	2,053,500	8.73	969,030	7.06	783,660	101.00	11,211,000
10	Shaft concrete	1,900 yd³	118.04	224,276	36.96	70,224	15.43	29,317			170.43	323,817
11	Cement for concrete and low-pressure grout	163,000 bbl	16.00	2,608,000							16.00	2,608,000
12	Steel reinforcing in shaft	285,000 lb	0.47	133,950							0.47	133,950
13	Embedded anchor bolts	3,200 lb	4.30	13,760			0.56	1,792			4.86	15,552
14	Embedded vent pipe	600 lb	3.25	1,950			0.42	252			3.67	2,202
	Total			$10,386,746		$2,123,724		$1,000,391		$ 783,660		$14,294,521
Consolidation grouting:												
15	Drilling grout holes	4,000 lin ft	10.00	40,000							10.00	40,000
16	Consolidation grouting	10,000 ft³	10.00	100,000							10.00	100,000
17	Cement for grouting	5,000 sacks	5.00	25,000							5.00	25,000
	Total			$ 165,000							$ 165,000	
	Total from extensions*			$24,622,074		$6,625,623		$2,837,757		$1,137,080		$35,222,534
	Total from agreed estimate and instructions from principals†			$24,622,976		$6,624,471		$2,838,000		$1,138,172		$35,223,619

*Totals are adjusted from agreed direct cost to give even unit prices.
†It is not necessary to bid exactly the agreed amount as bid extensions can seldom be make to total the agreed bid.

FIGURE 11-18

Sierra tunnel bid estimate: Summary of total cost. (From Parker, Barrie, and Snyder, *Planning and Esti-*

SIERRA TUNNEL BID PREPARATION: SPREADING MARKUP AND MAKING QUOTATION ADJUSTMENTS TO BID ITEMS

Bid Item	Description	Pay Quantity	Total Cost Unit	Total Cost Amount	Markup* Unit	Markup* Amount	Bid Price Unit	Bid Price Amount	Quote Adjustments Unit	Quote Adjustments Amount	Adjusted Bid Price Unit	Adjusted Bid Price Amount
Tunnel excavation:												
1	Tunnel excavation	431,000 yd³	$ 42.30	$18,231,300	$ 11.42	$ 4,922,020	$ 60.27	$25,976,370	$+0.54	$+232,740	$ 60.81	$26,209,110
2	Steel sets	2,290,000 lb	0.69	1,580,100	0.19	435,100	1.00	2,290,000			1.00	2,290,000
3	Timber	400 MBM	864.98	345,992	233.66	93,464	1,230.00	492,000			1,230.00	492,000
4	Roof bolts	12,000 lin ft	5.20	62,400	1.40	16,800	7.50	90,000			7.50	90,000
5	Drill exploratory holes	300 lin ft	18.23	5,469	4.92	1,476	26.00	7,800			26.00	7,800
6	Drill grout holes as aid to tunnel driving	2,400 lin ft	7.84	18,816	2.12	5,088	11.00	26,400			11.00	26,400
7	Grout as aid to tunnel driving	400 ft³	22.16	8,864	5.99	2,396	32.00	12,800			32.00	12,800
	Total			$20,252,941		$5,476,344		$28,895,370		$+232,740		$29,128,110
Shaft excavation:												
8	Shaft excavation	5,300 yd³	96.24	510,072	26.00	137,800	136.65	724,245			136.65	724,245
Concrete:												
9	Tunnel lining	111,000 yd³	101.00	11,211,000	27.25	3,024,750	101.00	11,211,000	-1.29	-143,190	99.71	11,067,810
10	Shaft concrete	1,900 yd³	170.43	323,817	46.04	87,476	171.00	324,900			171.00	324,900
11	Cement for concrete and low-pressure grout	163,000 bbl	16.00	2,608,000	0.80	130,400	16.00	2,608,000			16.00	2,608,000
12	Steel reinforcing in shaft	285,000 lb	0.47	133,950	0.02	5,700	0.50	142,500			0.50	142,500
13	Embedded anchor bolts	3,200 lb	4.86	15,552	1.31	4,192	5.00	16,000			5.00	16,000
14	Embedded vent pipes	600 lb	3.67	2,202	0.99	594	4.00	2,400			4.00	2,400
	Total			$14,294,521		$3,253,112		$14,304,800		$-143,190		$14,161,610
Consolidation grouting:												
15	Drilling grout holes	4,000 lin ft	10.00	40,000	0.50	2,000	10.50	42,000			10.50	42,000
16	Consolidation grouting	10,000 ft³	10.00	100,000	0.50	5,000	10.50	105,000			10.50	105,000
17	Cement for grouting	5,000 sacks	5.00	25,000	0.25	1,250	5.25	26,250			5.25	26,250
	Total		$	165,000	$	8,250	$	173,250			$	173,250
	Total from extension			35,222,534		8,875,506		44,097,665				44,187,215
	Total from agreement			$35,223,619		$8,874,049		$44,097,668			$	44,186,620

*Includes 20 percent markup and financing costs.

FIGURE 11-19

Sierra tunnel estimate: Summary of bid price. (From Parker, Barrie, and Snyder, *Planning and Estimating Heavy Construction*, McGraw-Hill Book Company, 1984.

The reduction in time and effort can also permit estimators to explore and refine many more alternative approaches to project execution, particularly in the conceptual phases.

In essence, the successive approach requires that an estimator not only estimate the cost for each of the elements of a project, but also assess the uncertainty associated with the estimate for each element. By expressing the elemental uncertainties as standard deviations, squaring these to get variances, then taking the square root of the sum of the variances, one can quantify the uncertainty for the estimate as a whole. The estimator then focuses on the element that most adversely affected the uncertainty for the project as a whole: he subdivides and analyzes that element in greater detail to reduce the uncertainty associated with it, then recomputes the variances, or uncertainty, for the project as a whole. This process is repeated until either the estimator is satisfied with the level of accuracy of the whole estimate, or the estimate reaches an irreducible level of uncertainty about which the estimator can do nothing. For example, there is no point in computing the exact cost of bolting connections if one does not even yet know for sure whether the designer will specify welding instead of high-strength bolting.

To illustrate, consider a hypothetical project subdivided initially into the six elements shown in Figure 11-20. Each box lists the estimated cost of the element, such as $100 for element A, a range of accuracy of the cost, in this case $\pm\$20$, and the variance (400) which is the square of the range. The "inevitable uncertainty" might be due to factors beyond the control of the project, such as contingencies associated with inflation or changes in environmental regulations. In this case, they are quantified as $\$100 \pm \$70$, with a variance of 4900. The total cost for the project is

$$\$100 + \$500 + \$80 + \$150 + \$100 + \$50 + \$100 = \$1,080$$

The variance is

$$(20)^2 + (80)^2 + (15)^2 + (30)^2 + (30)^2 + (10)^2 + (70)^2 = 13,825$$

The standard deviation for the project as a whole is

$$\sqrt{13,825} = \pm 118$$

The cost estimate at this stage is thus:

$$\$1,080 \pm 118$$

Clearly, the estimate at this point is most adversely affected by the $\pm\$80$ uncertainty in element B. It will thus be further subdivided for analysis in greater detail. Assume it can be broken down as shown in Figure 11-21. In this way, element B's estimate is refined from $500 to $440, and its uncertainty is reduced to $\pm\$44$, which alters the uncertainty for the whole project as follows:

$$(20)^2 + (44)^2 + (15)^2 + (30)^2 + (30)^2 + (10)^2 + (70)^2 = 9361$$
$$\sqrt{9361} = \pm\$97$$

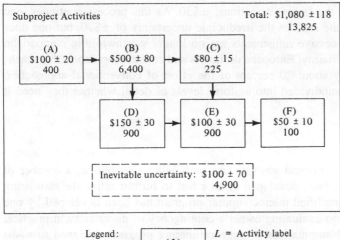

FIGURE 11-20
Subproject activities. (From Steen Lichtenberg, *Successive Planning*, Technical University of Denmark, June 1971.)

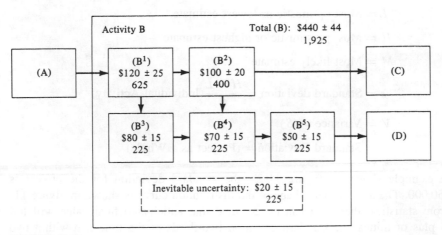

FIGURE 11-21
Activity "B" subdivision.

giving a new estimate of $1020 = \pm97$. This moves much closer than ±118 to the irreducible uncertainty of ±70.

Note that in this case element B as a subunit was analyzed as if it were a project in its own right, so that its summary statistics (440 ± 44) can still be treated at the higher level. Further to refine the estimate, attention would now focus on elements D and E, which now have the largest deviations, ±$30. As this process continues, one approaches but never quite reaches the irreducible uncertainty of ±$70, but one does reach a point where successive refinements are no longer worthwhile in comparison with the irreducible uncertainty. European designers and constructors report that reaching this point takes only about 20 percent of the effort of conventional approaches, where all elements are subdivided into uniform levels of detail whether they need it or not.

RANGE ESTIMATING

Using the principles of statistical analysis and probability distribution, a number of computer programs have been developed in an effort to furthur refine the estimating process. A somewhat simplified microcomputer program has been developed by one of the authors as an aid to evaluating owner's contingency or uncertainty in practical situations. Figure 11-22 summarizes a range estimating program designed to assist in determining contingency from the owner's standpoint for a major project. In this case, the estimators were requested to supply a lowest estimate and a highest estimate of expected bid price in addition to the most likely value. The expected value and standard deviation can be approximated from statistical theory as follows:

$$E = \text{Expected Value} = \frac{L + 4M + H}{6}$$

$L = $ Most optimistic or lowest estimate

$H = $ Most pessimistic or highest estimate

$M = $ Most likely estimate

$$s = \text{Standard deviation} = \frac{H - L}{6}\text{(Individual Activity)}$$

$V = \text{Variance} = (S)$

$S = \text{Standard Deviation} = \text{(Project as a Whole)}$

In the example shown in Figure 11-22, the standard deviation for the project is $14,150,000. The approximate shape of the distribution curve is shown in Figure 11-23. From statistical theory, it can be assumed that anticipated final value will fall within plus or minus one standard deviation two-thirds of the time and within two standard deviations within 95 percent of the time. The example shows that by adding one standard deviation the work can be expected to be completed within the chosen amount 67 percent of the time and correspondingly within 95 percent of the time if two standard deviations are added. While standard deviations are considered to be plus or minus, in developing proposed owner contingencies only the plus amounts are utilized. A factual basis for this assumption is that estimators can often omit items of cost but are much less likely to add unneccessary items.

    Utilization of other confidence factors can be developed in accordance with statistical theory and as illustrated by Figure 11-23. Further simplicity can be developed

LEVEL 1 SUMMARY
TABLE B: RANGE ESTIMATING INPUT SUMMARY
WEST POINT SECONDARY TREATMENT PLANT
ALTERNATIVE HPO-1

COMBINED RANGE ESTIMATE
SUB TOTAL CONSTRUCTION HPO-1 ($1988)
(NOT INCLUDING ESCALATION OR ALLIED COST)

Date: 15-Jun-90
By: D.S. BARRIE

FACIL. NO.	FACILITY DESCRIPTION	LOWEST ($000's)	MOST LIKELY ($000's)	HIGHEST ($000's)	EXPECTED VALUE ($000's)	SIGMA STD. DEV. ($000's)	SIGMA SQUARED ($000,000's)	68% CONFIDENCE ($000's)	95% CONFIDENCE ($000's)
01	General Conditions	0	4,000	8,100	4,017	1,350	1,822,500		
02	Administration	9,782	8,909	12,800	9,703	503	253,009		
03	Maintenance	2,300	9,005	10,402	8,120	1,350	1,822,500		
04	Influent Control Structure	1,800	3,106	3,654	2,980	309	95,481		
05	Raw Sewage Pumping	9,485	9,790	13,200	10,308	619	383,161		
	Grit/Screenings Handlings	2,600	5,589	6,450	5,234	642	412,164		
06	West/East Primary Sediment.	11,200	24,009	27,761	22,500	2,760	7,617,600		
07	Effluent Pumping	13,100	21,061	25,746	20,515	2,108	4,443,664		
08	Existing/New Digesters	22,400	31,727	51,931	33,540	4,922	24,226,084		
09	Existing/New Galleries	1,700	16,069	20,411	14,398	3,119	9,728,161		
10	Aeration Tanks	28,100	46,738	54,201	44,875	4,350	18,922,500		
11	Oxygen Production	17,600	21,143	25,721	21,316	1,354	1,833,316		
12	Secondary Sedimentation	39,200	66,754	76,517	63,789	6,220	38,688,400		
13	Chlorine Generation	10,068	11,224	20,400	12,561	1,722	2,965,284		
14	Chlorine Contact/Seawater Pmp	11,000	16,300	19,421	15,937	1,404	1,971,216		
15	Solids Handling	29,000	37,396	45,136	37,287	2,689	7,230,721		
16	Odor Control	5,100	10,122	12,417	9,668	1,220	1,488,400		
17	Cogeneration	9,100	11,460	12,860	11,300	627	393,129		
18	Facility Service	700	1,364	1,498	1,276	133	17,689		
19	Operations Center	1,366	1,450	2,400	1,594	172	29,584		
20	Annex	588	615	1,200	708	102	10,404		
21	Flow Diversion Structure	1,400	3,997	4,876	3,711	579	335,241		
22	Electrical Substation	2,261	2,377	8,000	3,295	956	913,936		
23	Retaining Wall	31,969	36,339	48,300	37,604	2,722	7,409,284		
51	Access Roadway	2,600	3,524	4,018	3,453	236	55,696		
52	Canal Feeder	4,233	4,300	6,700	4,689	411	168,921		
53	Sitework	32,000	46,021	64,768	46,809	5,461	29,822,521		
54	Emergency Bypass	98	95	2,800	546	450	202,500		
55	Plant Site Utilities	2,800	9,929	12,590	9,184	1,632	2,663,424		
56	Distributed Control System	16,900	17,553	20,933	18,008	672	451,584		
57	Mitigation	27,500	49,709	62,314	48,108	5,802	33,663,204		
58	Temp. Oper. Facility(On-Site)	300	1,387	1,581	1,238	214	45,796		
59	Temp. Maint. Facility(Off-Site)	300	1,387	1,581	1,238	214	45,796		
60	Asbestos Removal	1,300	1,803	2,429	1,824	188	35,344		
61	Demolition	2,690	2,917	3,900	3,043	202	40,804		
	Total Construction Cost HPO-1 ($ 1988)	352,539	539,171	697,016	534,376	14,150	200,209,018	548,526	562,676

FIGURE 11-22
Range estimate summary courtesy C M Consultants Incorporated, Danville, Calif.

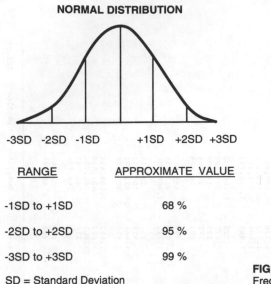

NORMAL DISTRIBUTION

-3SD -2SD -1SD   +1SD +2SD +3SD

RANGE	APPROXIMATE VALUE
-1SD to +1SD	68 %
-2SD to +2SD	95 %
-3SD to +3SD	99 %

SD = Standard Deviation

FIGURE 11-23
Frequency distribution curve.

by fixing minor costs or costs with minimum variation at the expected value. Pareto's principle suggests that 20 percent of the items are responsible for 80 percent of the cost. In the chosen example, 7 percent of the elements were responsible for 86 percent of the difference between the two estimates. Successive Estimating techniques might also have been utilized to decrease the uncertainty as previously discussed.

CONVERTING ESTIMATE TO CONTROL BUDGET

Chapter 13, which deals with cost engineering, will discuss control budgets in greater detail after it has introduced the subject of cost codes. At this stage, however, it is worth pointing out some differences in philosophy and content that should be recognized in converting an estimate to a control budget.

Estimates often organize costs in different categories and in different levels of detail from those most appropriate for cost-control purposes. For example, for a unit-price bid, the estimate would most probably be prepared against the schedule of quantities in the owner's bidding documents. This organization might be convenient for the owner, but quite likely would not be the way the contractor's historical costs and job costs are kept. Thus, once the job is awarded, the contractor might reorganize the costs into a form suitable for his cost control system while preserving his ability to compare costs and earned value with the bid format.

Estimators are encouraged to make detailed take-offs to promote estimating accuracy. However, there is a practical limitation to the ability and cost effectiveness of measuring to this degree of detail in the field. Therefore, in budget preparation, many companies will consolidate a number of minor costs into summary or parameter accounts which can be easily measured. For example, many piping contractors recognizing the futility of keeping track of small bore piping labor by individual fitting size or lineal feet, will use a parameter such as diameter inches of weld or joint which can be easily measured in the field to measure performance against the budget. Many contractors separate piping into designated size groups by specification such as small bore piping (2.5 in. and under diam.) and large bore piping (over 2.5 in. diam.). Small bore piping can be reported by the craft foreman with time cards as 5 each 2-inch welds completed or 10 diameter inches. Large bore piping can be reported as number of diameter inches of field welds plus number or lineal footage of prefabricated spool pieces erected. In either case all estimated hours are recognized in the parameter budget and included in the measured units.

Structural steel and reinforcing steel are taken off from drawings in full detail by individual piece in order to price material costs accurately and to enable the estimator to apply an estimated labor factor. However, for control purposes the work is normally separated into several groupings by size or weight and within each category an average unit hour per pound or ton is utilized for budgetary control during fabrication and installation operations.

As also discussed in the section on control budgets in Chapter 13, estimates are based upon averages. Some firms believe that for control purposes somewhat tighter standards should be used to control field operations. Figure 11-23 illustrates a frequency distribution curve for a construction activity. Construction estimates are often based upon the average or 50th percentile where there is a 50 percent chance of overrunning or underrunning the estimate. If budgets are based upon the 33rd percentile, there is one chance in 3 of equalling or "beating" the estimate. Needless to say, this method requires a positive approach if it is to achieve motivation rather than frustration.

Other firms and most field superintendents believe that cost and schedule control is most effective if actual estimating and scheduling targets from the bid or proposal are utilized and that all objectives must clearly be perceived to be achievable. On some projects, monetary production bonuses have proven effective while on others nonmonetary incentives have also been successful. The award fee type of contract has often proved beneficial to both contractor, owner and supervisor when properly implemented.

A number of other progressive firms are achieving productivity enhancement through nonquantified and nonmonetary incentives such as recognition at the craft level through consideration of human factors. A comprehensive treatment of productivity improvement in construction is contained in Oglesby, Parker, and Howell.[18]

In any event owing to inflation and other factors, it is best if budgets for job-site labor control and for historical records are kept in terms of worker-hours per unit of output, rather than in terms of monetary unit costs.

SUMMARY

This chapter has examined several approaches to estimating the costs of a construction project. They include conceptual estimating techniques that require little data and effort

---

[18]Oglesby, Parker, and Howell, op. cit.

to give approximate estimates that can provide budget and planning guidance in the early phases of a project, and detail estimates and definitive estimates to provide more accuracy and control as the project evolves further. Owing to its complexity and importance, labor-cost estimating was examined in even greater detail. Similarities and differences between prevailing estimating practice in building, industrial and heavy construction applications were discussed. "Successive estimating," a technique now used in Europe, was introduced as an approach that can improve accuracy and reduce the corresponding level of effort in preparing estimates. Range estimating is being increasingly utilized by a number of construction owners in an effort to improve estimating accuracy. Finally, considerations for converting estimates to control budgets were briefly described.

Techniques examined for conceptual and preliminary estimates were (1) cost indices, (2) cost-capacity factors, (3) component ratios, and (4) parameter-cost estimates. Cost indices mainly reflect time-dependent changes in costs resulting from inflation, changes in technology, market competition, the state of the economy, etc. Cost indices can also reflect regional variations in costs. Cost-capacity factors account for nonlinear "economies of scale" when comparing projects of similar types but of differing magnitudes. Component ratios proceed from vendor price quotations on major items of equipment, and factor these quotations to reflect either installed equipment costs or completed project costs. Parameter costs relate all cost elements of a project to just a few basic parameters that characterize its type and scope. All these techniques depend on good historical documentation and on the estimator's experience-based judgment.

Fair-cost estimates are best prepared from completed plans and specifications. They are based upon actual quantity takeoffs which are multiplied by unit prices developed by the estimator.

Contractor's-bid estimates are based upon similar information, but may be developed in considerably more detail depending upon the contractor's own procedures. Bid estimates typically include lump-sum or unit-price material and subcontract quotations.

Professional construction managers, design-constructors, and owner representatives must develop sufficient in-house estimating ability in mechanical, electrical, and other specialty items so that they can effectively forecast the cost of the work, prepare budgets and determine the fair cost of changes and modifications.

The concept of a definitive estimate as developed in this chapter represents the earliest stage at which the final project cost can be forecast with little margin for error. This error can be minimized through the addition of an evaluated contingency. The determination of the proper time to classify an estimate as "definitive" will vary according to the nature of the project, the anticipated accuracy of the underlying information, and the degree of risk the owner is prepared to accept.

The labor-cost section outlined basic principles for estimating and controlling field labor costs on construction projects. It divided labor costs into two main categories. The first dealt with the strictly financial aspects. These included basic wages, fringe benefits, insurance, taxes, and wage premiums. Most of these are readily quantifiable. The second category introduced factors related to productivity. These included regional variations in skill and work rules; environmental effects ranging from weather and topography to the immediate working conditions of individual craftsmen; learning-curve relationships; project work schedule; and interrelationships of labor and management. Productivity factors are difficult to quantify and require much experience and judgment to do so. However, they offer the contractor the greatest opportunity for control. In both the financial and productivity categories, several of the more subtle cost implications of the various factors were explained. Brief precautionary reference was also made to the many published sources of labor-cost data.

Estimators typically specialize in either building, industrial or heavy construction although many individuals can operate successfully in more than one area. Some of the favored methods in each branch are investigated and similarities and differences are explored. Several examples illustrating typical practices in each area are presented.

The "successive estimating" approach developed by Steen Lichtenberg of Denmark provides a potentially useful alternative to traditional estimating methods. By incorporating assessments of uncertainty with the cost estimates for the various elements of a project, it uses statistical concepts to allow selective detailing to refine the accuracy of the estimate. Rather than prepare the whole estimate at an arbitrarily uniform level of detail, this approach can produce equal or better levels of accuracy with less time and effort. This time saving can enable estimators to explore more alternatives within a project, particularly during the conceptual phases.

Range estimating in which the estimator develops a most optimistic, most pessimistic and most likely figure has been utilized by a number of owners in an attempt to refine the estimating process and to achieve desired confidence levels through application of statistical methods.

In converting an estimate to a control budget, two important differences should be considered. First, the organization and categorization of costs suitable for preparing an estimate are often not compatible with realistic field cost control. Second, estimates necessarily must deal in averages, whereas tighter standards are sometimes desirable for control purposes. For these and other reasons, careful thought should go into converting estimates to control budgets. Chapter 13 will explore this subject further.

This chapter has set forth some basic concepts for estimating and controlling costs of projects. There are many legitimate approaches for estimating project costs, and each has its appropriate applications and limitations. Of importance in any method, however, is that (1) it correctly accounts for all the various cost and productivity factors in designing and constructing a project, and (2) it is applied uniformly and consistently from one project to another.

This chapter is not intended as a substitute for a textbook or work manual devoted entirely to the subject. Some of the references at the end of this book provide access to further reading and study. Most important for developing good estimators, however, is experience in construction, and thorough, well-organized documentation of work that has been done.

CHAPTER **12**

PLANNING AND CONTROL OF OPERATIONS AND RESOURCES

Planning, scheduling, and control of the functions, operations, and resources of a project are among the most challenging tasks faced by a professional construction manager. Normally, this responsibility involves coordinating design with construction to produce the necessary plans and specifications, to package them along recognized trade and subcontractor boundaries, and to contract with the construction organizations best qualified to carry out their work efficiently and economically in conjunction with other contractors on the site. In the construction phase, the professional construction manager or general contractor normally provides the overall planning, scheduling, and control needed to sequence operations properly and to allocate efficiently the resources involved.

This chapter deals with the methods and procedures available to the construction manager for accomplishing these objectives and offers practical guidelines for their effective application. The critical path method will be presented in this chapter as but one of many possible control tools that may best suit the needs of a particular project or a particular management situation. However, because network-based critical path methods (**Program Evaluation and Review Technique** (PERT) and CPM) are among the most powerful tools available, this chapter will explain these techniques and their methods of computation in more detail. Additional emphasis will be placed on their application in the control phase.

In addition to discussing methods and procedures for planning, scheduling, and control, this chapter will introduce some concepts and guidelines for the contractual implications, including documentation and analysis of change orders, changed conditions, delays, claims, and disputes. Although it may not be obvious at this stage, the many impact costs associated with these contractual problems will enable their discussion to build quite logically upon the earlier parts of this chapter.

ALTERNATIVE PLANNING AND CONTROL TOOLS

Besides the critical path method, there are many different analytical tools and graphical techniques for the planning, scheduling, and control of operations and resources. Before proceeding to CPM, a few alternatives that will be introduced here include bar charts, progress curves, matrix schedules, and linear balance charts. It is important to emphasize at the outset, however, that none of these is in and of itself the *plan* for the project. The complete plan, if it exists at all, exists only in the minds of the planners. All the tools that have been mentioned are merely abstract means to aid the planners in *organizing* and *documenting* their thinking and assumptions and in *communicating* that thinking to those persons responsible for putting the plan into action. These tools succeed only to the extent that they at least accurately document the major parameters of the plan and effectively communicate the planners' intentions to others.

The situation here is somewhat analogous to the process of composing, documenting, and playing music.[1] The composer, be he Beethoven, John Lennon, or whoever, is the planner, and initially the plan exists only in his head. Our standard form of musical notation enables him to document his thoughts as he goes so that (1) he does not have to keep everything in mind at once, and (2) he can refer to what he has documented for modifications and revisions while the composition as a whole takes shape. Once the composer is satisfied with his plan, possibly having tested it as he went, the final plan, or sheet music, can be neatly rendered and possibly published. Eventually a musician gets hold of this plan, reads and comprehends it, and, analogous to the execution of a project, carries out the planner-composer's intentions in a concert or for personal enjoyment.

In contrast with all existing "notations" for documenting the planners' intentions for the execution of a construction project, musical notation is remarkably clear and comprehensive. It transcends language barriers to communicate effectively with musicians in Europe, Asia, the Americas, Africa, Australia—wherever people have acquired knowledge of its basic principles. It also communicates effectively through time. A Bach concerto can be rendered today precisely as the maestro intended, and deviations from the plan are intentional.

No existing tool for documenting and communicating plans for engineering and construction even approaches the capabilities of musical notation. Even network-based methods capture only the barest outlines of the planner's thinking, and all the tools discussed in this chapter should be viewed in the context of these limitations. Expecting too much from existing techniques accounts for many of their so-called failures in practical applications.

Each tool should be evaluated with respect to its suitability for documenting the characteristics of the planned project, the knowledge and level of sophistication of those who are expected to use it, the desired level of detail, and the means available for updating and revision. One should not immediately discard simpler tools when a more powerful one seems appealing. It is often the very simplicity of such tools as bar charts and progress curves that makes them more effective as a means of

---

[1]This analogy is from Prof. D. W. Halpin of Purdue University.

communication. Similarly, even the more powerful techniques can be mismatched to the physical and managerial characteristics of certain projects. For example, CPM is often a poor choice for linear or repetitive operations, such as a pipeline or tunnel, and sometimes even for major earthmoving operations, where artificially strict logic constraints can critically impair the better judgment of a good superintendent. As another example, for both documentation and communication, matrix schedules and linear balance charts can be more effective than other techniques for coordinating the subcontractors on a high-rise building. Several of these applications and limitations will be explored in the context of the techniques introduced here.

Bar Charts

Bar charts date back at least to the Gantt Charts developed by Henry L. Gantt in the early part of this century. Technically speaking, there are a number of differences between the two, but in this chapter all forms of these diagrams will be called "bar charts."

A bar chart graphically describes a project consisting of a well-defined collection of tasks or activities, the completion of which marks its end. An *activity* is a task or closely related group of tasks whose performance contributes to completion of the overall project. A typical activity noted in a bar chart for a building project could be "Excavate foundation."

A bar chart is generally organized so that all activities are listed in a column at the left side of the diagram. A horizontal time scale extends to the right of the list, with a line corresponding to each activity in the list. A bar representing the progress of each activity is drawn between its corresponding scheduled start and finish times along its horizontal line. A simple bar chart for a small concrete gravity-arch dam is shown in Figure 12-1.

FIGURE 12-1
Bar chart for concrete gravity-arch dam.

Item No.	Description	First Year												Second Year											
		J	F	M	A	M	J	J	A	S	O	N	D	J	F	M	A	M	J	J	A	S	O	N	D
M-10	Mobilization																								
E-10	Foundation excavation																								
D-10	Diversion stage − 1																								
D-20	Diversion stage − 2																								
G-40	Foundation grouting																								
C-10	Dam concrete																								
I-20	Install outlet gates																								
I-30	Install trash racks																								
P-10	Prestress																								
R-80	Radial gates																								
S-50	Spillway bridge																								
G-60	Curtain grout																								
L-90	Dismantle plant, clean up																								

Bar charts differ in the way they show *planned* progress on the horizontal scale, in the way they *report* progress, and in numerous details of diagrammatic style. Although these differences may at first seem trivial, they have important but subtle implications that can create serious misunderstandings for people who are unfamiliar with them. Three of the more common types of bar charts are therefore discussed in some detail below. Since they have no known standard names, they are arbitrarily called Type I, Type II, and Type III bar charts in this chapter.

Type I: Linear Time-scaled for Planning; Linear Progress-scaled for Reporting
Type I, a common form of bar chart, assumes that progress on an activity is a direct linear function of elapsed time. Therefore, in *planning*, no attempt is made to show the physical percentage completion at any point on the bar representing an activity. The basic form is the open bar shown on Figure 12-2*a*.

In order to report progress, a parallel bar is sometimes placed immediately below the plan bar, and is initially open also. Then, as the job progresses, it is shaded in direct proportion to *physical work* (not necessarily elapsed time) completed on the activity. This is shown on Figure 12-2*b*. Alternatively, a narrow, shaded reporting bar could be superimposed on an open plan bar, as shown on Figure 12-2*c*. Other

FIGURE 12-2
Type I bar charts.

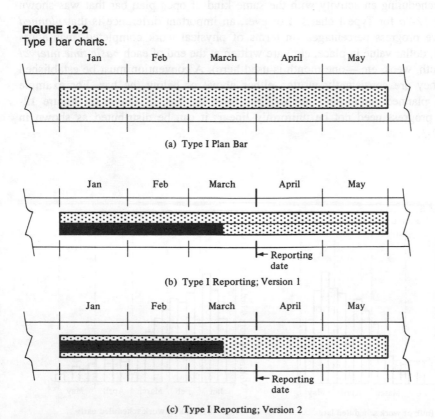

(a) Type I Plan Bar

(b) Type I Reporting; Version 1

(c) Type I Reporting; Version 2

variations are also used. Note that the current physical progress, a work function, does not necessarily coincide with the current reporting date, a time function. By comparing the shaded reporting bar with the open plan bar and with the current date, one obtains only a rough indication of whether the activity is behind or ahead of schedule.

The example in Figure 12-2 shows that 5 months were originally scheduled for the activity (shown by the open plan bar) and that 60 percent of the time (3 months out of 5) has elapsed by the reporting period. However, the shaded bar reports that only 50 percent of the physical work in the activity has been completed. On first appearance, it may seem that the activity is about one-half month, or 10 percent, behind schedule. This may or may not be true. It is quite possible that the bulk of the resource hours and dollars was scheduled to be expended during the latter half of the activity's duration, as shown in Figure 12-3a. In this case, the activity may actually be on or ahead of schedule. On the other hand, if the bulk of the effort has been planned as in Figure 12-3b, the activity may be much more than 10 percent behind schedule. The point is, the Type I bar chart accurately reflects activity status only when cumulative progress is indeed a direct linear function of time. This knowledge is important in avoiding serious misinterpretations of the diagram.

Type II: Time-scaled for Planning; Time-scaled for Reporting   Type II bar charts start by scheduling an activity with the same kind of open plan bar that was shown in Figure 12-2a for Type I charts. However, an important difference is that *planned* cumulative progress percentages (in terms of physical work completed, man-hours expended, dollar value in place, etc.) are written at the end of each *basic time interval* (day, month, week, etc.—one month is used here). A convention must be established so that they are consistently written either above or below the bar. For example purposes, planned percentages are written above the bar, as shown on Figure 12-4a. This progress need not be uniformly linear; it can be distributed as shown in Figure 12-3.

FIGURE 12-3
Activities scheduled with nonuniform workloads.

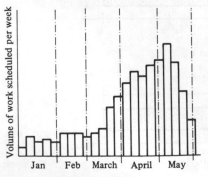

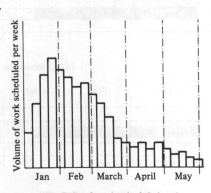

(a) Bulk of work scheduled late   (b) Bulk of work scheduled early

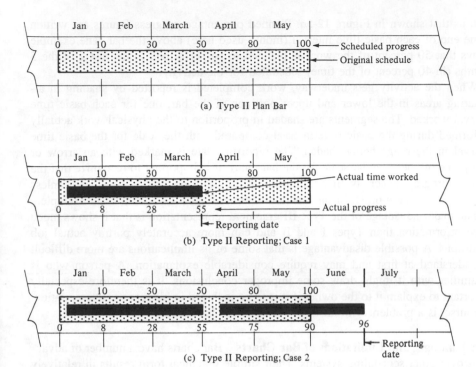

FIGURE 12-4
Type II bar charts.

Progress may be reported on Type II bar charts by using either of the graphical conventions explained for Type I, using Figure 12-2b and c. Figures 12-4b and c use the superimposed shaded-bar convention. Note, however, that there are important differences in what the reporting bar says in Type II. It is shaded to show the actual *time* worked on the activity up to the current date or to completion, whichever is earlier. Figures giving the *actual* percentage cumulative progress are written on the opposite side of the bar (lower side, here) from the planned progress.

As with the Type I bar chart, it is important that subtleties and limitations of the Type II chart be clearly understood. Because the reporting bar is shaded to the current date or to completion, the reporting bar in itself gives no indication of whether a current activity is ahead of or behind schedule. This progress is shown only by writing the *actual* cumulative percentage by the bar on the opposite side of the planned percentages. Comparisons of actual and planned percentages is the only way to evaluate the status of a current activity, so it is essential that these figures be written on the diagram. This convention has some graphical disadvantages, but it does show more vividly what happened on activities which are already complete. This, of course, has little relevance for project control purposes.

Type III: Time-scaled for Planning; Variable Progress-scaled for Reporting Type III bar charts start by representing an activity with a horizontally divided open bar such as that shown in Figure 12-5a. Planned percentage progress figures are written at the end of each basic time interval (month used here) above the bar. The example shows that 50 percent of the work is planned to be performed in the last 2 of the 5 months (in 40 percent of the time) scheduled for the activity.

When the activity gets underway, work completed is reported by shading in alternating areas in the lower and upper portions of the bar, one for each basic time interval worked. The segments are shaded in proportion to the physical work actually performed during the basic time interval compared with the scale for the basic time interval in the range being shaded. The reporting date is marked with an arrow or heavy line on the calendar scale for the bar chart. It is important to recognize that the scale of progress generally changes during each basic time interval considered unless progress is indeed a direct linear function of time. Figure 12-5b shows an example.

The major advantage of the Type III graphical representation is that it shows much more information than Types I and II and can more accurately portray actual job conditions. A possible disadvantage is that some of its implications are more difficult to understand at first and may require considerable explanation. A person who is unfamiliar with it might draw some improper conclusions. It is therefore especially important to explain it to the owner if it is to be included in his reports. Interpretation, of course, is a problem with all three forms of the bar chart.

Advantages and Limitations of Bar Charts   Bar charts have a number of advantages over other scheduling systems. Their simple graphical form results in relatively easy general comprehension. This, in turn, has led to their common acceptance and widespread use as a good form of communication in industry, with a basic understanding usually found at all levels of management. Also, they are fairly broad planning and scheduling tools, so they require less revision and updating than more sophisti-

FIGURE 12-5
Type III bar charts.

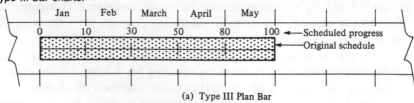

(a) Type III Plan Bar

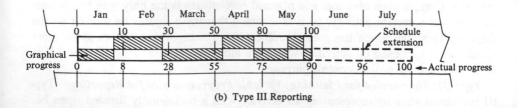

(b) Type III Reporting

cated systems. This feature is especially helpful in the turbulent early stages of an engineering and construction project when frequent changes and revisions are a fact of life.

In addition to the specific problems of interpretation and misunderstanding mentioned with each of the three specific formats discussed in this chapter, the use of bar charts has a number of general limitations. First, because of their broad planning nature, they become very cumbersome as the number of line activities, or bars, increases. If several sheets are required, logical interconnections are difficult to comprehend. Second, although the planner who prepared the bar chart undoubtedly considers the logical interconnections and constraints of the various activities in the project, this logic is not expressed in the diagram. It therefore becomes very difficult for another individual to reconstruct the logic and to recognize sequence constraints unless a substantial amount of documentation is included with the chart. Third, although the bar chart is a good planning and reporting tool, it is difficult to use it for forecasting the effects that changes in a particular activity will have on the overall schedule, or even to project the progress of an individual activity. It is therefore limited as a control tool. With due regard to these and other limitations, bar charts will nevertheless continue to be valuable assets in project management. An understanding of their limitations, however, is important to their effective and appropriate application.

Progress Curves

General Principles   Progress curves, also called S curves, graphically plot some measure of cumulative progress on the vertical axis against time on the horizontal axis. Progress can be measured in terms of money expended, quantity surveys of work in place, man-hours expended, or any other measure which makes sense. Any of these can be expressed either in terms of actual units (dollars, cubic meters, etc.) or as a percentage of the estimated total quantity to be measured.

The shape of a typical S curve results from integrating progress per unit of time (day, week, month, etc.) in order to obtain cumulative progress. On most projects, expenditures of resources per unit time tend to start slowly, build up to a peak, then taper off near the end. This causes the slope of the cumulative curve to start low, increase during the middle, then flatten near the top.

These principles will be illustrated by using cubic meters excavated to measure progress on a 10,000 cubic meter, 10-day earth-moving activity. Assume that daily excavation quantities are as shown on Figure 12-6a. Summing all the daily excavation quantities through any particular day gives the cumulative quantity by that day. For example, by the end of day 4, the cumulative quantity is the sum of excavation on days 1, 2, 3, and 4. That is,

$$3200 = 200 + 600 + 1000 + 1400$$

The shape of the S curve can be seen by connecting the points at the end of each day's cumulative production, as shown on Figure 12-6b.

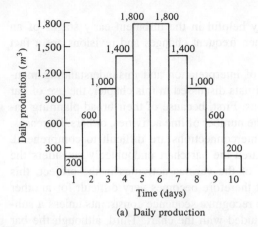

(a) Daily production

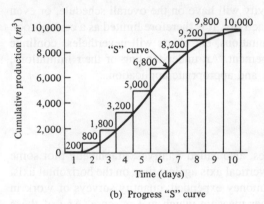

(b) Progress "S" curve

FIGURE 12-6
Development of a progress curve.

Planning and Reporting Progress Like bar charts, progress curves can express some aspects of project plans. Once the project is underway, actual progress can be plotted and compared with that which was planned. It is then possible to make projections based on the slope of the actual progress curve. Such projections, however, should neither be made nor interpreted without a good understanding of the reasons for deviations, if any, from planned progress, and of the current and future plans of project management. Basic concepts of planning, reporting, comparing, and projecting progress are shown on Figure 12-7.

Early, Late, and Actual Progress This section requires some understanding of the critical path network concepts discussed later in this chapter.

Free float and total float on noncritical activities give managers considerable flexibility in rescheduling activities without delaying the overall project. The scheduling of an activity has a close correlation to the timing of its resource expenditures (money, labor, materials, etc.) and, of course, to its accomplishment. It follows that if all activities in a project are scheduled as early as possible (an early-start schedule), the

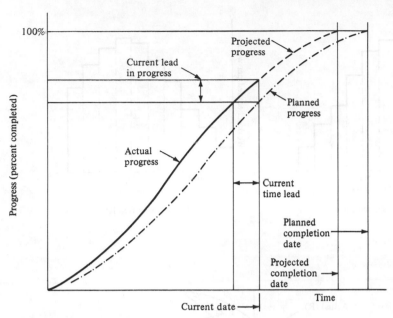

FIGURE 12-7
Planning and reporting progress.

progress will take place and be reported earlier. Conversely, if all activities are scheduled at their late starts, progress takes place and is reported later. These ideas are illustrated in Figure 12-8.

Figure 12-8a shows total progress measured during each basic time interval of the project with activities scheduled at their early starts. Curve *ES* on Figure 12-8c is the corresponding cumulative progress curve. Figure 12-8b shows total progress in each period for the late-start schedule. Curve *LS* on Figure 12-8c is the cumulative late-start progress curve. Note that the two curves start and end at the same point, but that at any other point the *LS* curve falls below or to the right of the *ES* curve. Actual planned and reported progress should most likely fall between these two extremes, as shown by curves *P* and *R* respectively on Figure 12-8c.

Cash Flows Cash flows may be shown graphically by plotting one progress curve for expenditures on the same graph with a second curve for income. A third curve representing the financing required or cash surplus at any time may then be plotted by subtracting the expenditures ordinate from the income ordinate at each point in time. This idea is shown on Figure 12-9.

Superimposing Progress Curves on Bar Charts Planned and actual progress curves can be superimposed on a bar chart to make a useful hybrid report. An example report of this kind was shown on Figure 8-8. Note that although the bar chart and the progress curves share a common horizontal time scale, there is generally no correlation

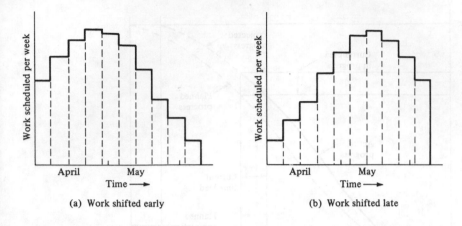

(a) Work shifted early   (b) Work shifted late

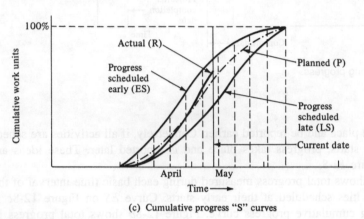

(c) Cumulative progress "S" curves

FIGURE 12-8
Early, late, and actual progress. (Adapted from Boyd C. Paulson, Jr., "Concepts of Project Planning and Control," *Journal of the Construction Division*, ASCE, vol. 102, no. CO1, March 1976, p. 74.)

between the vertical scale of the progress curves, usually shown on the right, and the vertical list of activities on the left. In this kind of report, any type of bar chart can be combined with any type of progress chart.

Matrix Schedules

Matrix schedules are a tool that has evolved and become fairly common on high-rise buildings with successive floors repeating essentially the same plan. The technique is fairly narrow in its application, but it does serve to illustrate how, in the right context, a simple idea can be quite effective for documenting and communicating a plan.

A typical schedule, partially rendered, is shown in Figure 12-10. It is no coincidence that at first glance the schedule looks like a cross section or elevation of the building

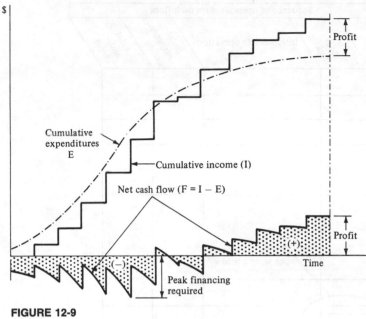

FIGURE 12-9
Cash flows.

itself. The horizontal rows on the schedule do indeed correspond to floors within the building, starting with one or more basement levels at the bottom and working up to the top floor. The vertical columns, however, are not structural features, but rather, they correspond to the operations to be performed on each floor. They read from left to right, and list the operations on each floor roughly in chronological order. The descriptions are given at the top of the columns. In a sense, the building's schedule then proceeds from the lower left-hand corner of the matrix to the upper right-hand corner. A typical sequence of operations on a floor might include such items as:

Erect columns and girders for structural frame
Place decking
Place floor inserts
Place lightweight concrete
Intermediate operations
Paint and carpet

Each operation is scheduled by a box like the typical example shown in the insert in Figure 12-10. The box is subdivided to show the scheduled start and finish dates and the expected duration, and it provides space for the actual dates to be entered in the field. Additional graphics can be provided by coloring the boxes as they are completed, so that just a quick look shows the status of the building.

This type of schedule has numerous advantages when it comes to communication on a project of this type. First, the vertical correlation of floors to rows is immediately

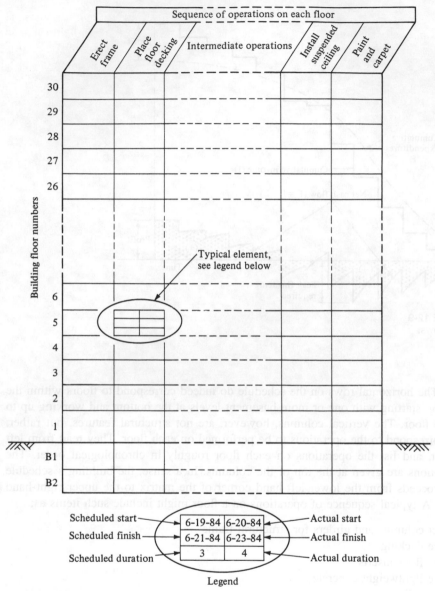

FIGURE 12-10
Matrix schedule for high-rise building.

obvious to anyone and requires no explanation. Contrast this with the debates between "arrows" and "circle" notation in CPM. Second, the chronological, left-to-right flow of each floor's operations is easy to see. The logical interrelationships among operations are also more obvious than in a bar chart. And, very important, with some forethought the vertical columns can be made to correspond to the specialty subcontractors; thus, there is built-in selectivity in reporting. When a subcontractor walks into the office to discuss his work, all his operations are shown in one or a few adjacent columns, so he does not have to sort his activities from a maze of others. Also, the relationships to the other subcontractors that most affect him or are affected by his work are readily apparent. All this information is right there in one compact, easily understood, one-page schedule.

The "Horse Blanket"

A variation on the matrix schedule used in high-rise buildings has been used on some major rail-rapid transit systems. An example is shown on Figure 12-11. Essentially,

FIGURE 12-11
"Horse blanket" schedule for part of a rail rapid transit system. (Adapted from schedules used on Washington, D.C., and Atlanta, Ga., rapid transit projects.)

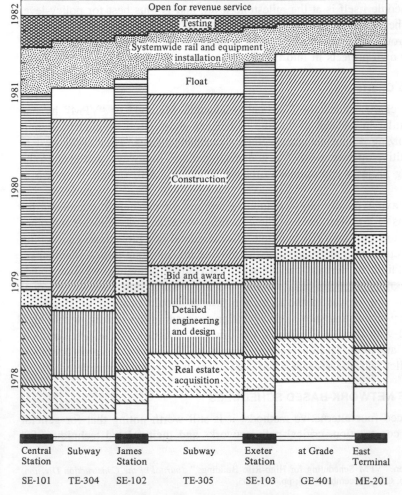

what it shows is a section of the system intended to go "on-line" as a unit, at the same date, for revenue service. The section may consist of several stations and connecting sections of grade, subway, or aerial rail line, and will normally be contracted for design and construction in several different segments. The horizontal axis corresponds to the line itself and its several contractual subdivisions. The vertical axis, working chronologically from the bottom upward, shows the major phases for each contractual section, including planning, real estate acquisition, preliminary design, detail engineering, key reviews and approvals, bidding, construction of the main structures, and construction of system-wide components (track, train controls, etc.). The amount of leeway left at the top of various sections gives management a good idea of which parts are most critical to the project's scheduled start-up.

The name "horse blanket" was given to this schedule because the schedules that have been produced use brilliant colors for the various phases of each contractual section, and the overall matrix therefore has a patchwork pattern reminiscent of the brightly colored horse blankets used by some Indian tribes in the Western United States. The schedule itself is at the milestone level, and is thus best for policy-level planning at higher levels of management. Nonetheless, when copies are displayed at many offices around the system, designers, contractors, and others can also readily see where their own projects fit into the "grand scheme of things."

Linear Balance Charts

Linear balance charts, called the "Vertical Production Method" or "VPM" by one author,[2] are similar in concept to the line of balance charts used by industrial engineers for optimizing output on manufacturing production lines. They apply best to linear and repetitive operations, such as tunnels, pipelines, highways, and even to the type of building projects appropriate to matrix schedules. An example is shown in Figure 12-12.

The vertical axis typically plots cumulative progress or percentage completed for different systems of a project, such as the structural, electrical, mechanical, and other trade subcontractors on a high-rise building. The horizontal axis plots time. The sloping lines can each represent trade subcontractors moving up from one floor to another on a high-rise building, or perhaps the clearing, excavation, stringing, welding, pipe-laying, and backfill operations on a pipeline. As long as the slopes are either equal or decreasing as one moves to the right, the project should proceed satisfactorily. However, if early scheduling shows one operation proceeding too rapidly, with a high slope compared with those preceding it, the time and location of the first conflicts become rapidly apparent. To illustrate this, Figure 12-12 shows the eighth operation starting to conflict with the seventh when each is about 70 percent complete.

CONCEPTS OF NETWORK-BASED SCHEDULES

Having introduced four alternative methods, it is still worth noting that for general applications in construction, critical path networks and their related techniques for

---

[2]James J. O'Brien, "VPM Scheduling for High-Rise Buildings," *Journal of the Construction Division*, ASCE, vol. 101, no. CO4, December 1975, pp. 895–905.

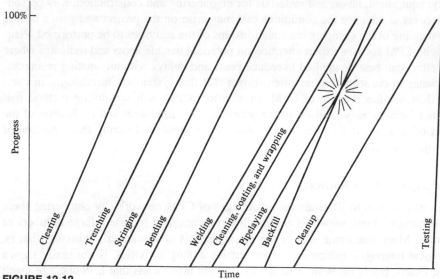

FIGURE 12-12
Example linear balance chart for pipeline.

schedule, resource, and cost analysis are still by far the most powerful analytical tools that we have for project planning and control. This section will briefly review their basic concepts and compare them with other methods. The remaining sections will focus on networks for demonstrating more general ideas in project control, and also for the analysis and documentation of claims and disputes.

Background

The critical path method is a graphical network-based scheduling technique that evolved from a research effort initiated in late 1956 by the Engineering Services Department of the E. I. Du Pont de Nemours Company. They were assisted in this effort by a computer group from Remington Rand UNIVAC. Their objective was to explore the use of computer-aided systems in planning, scheduling, monitoring, and controlling Du Pont's engineering projects. The research was coordinated by Morgan R. Walker of Du Pont and James E. Kelley, Jr., of Remington Rand.

Although the original technique was developed around the computer, the computer is by no means necessary for the successful implementation of CPM on many projects. Simplified graphical representations such as that developed by John Fondahl at Stanford University have made it possible to handle some fairly complex schedules manually. CPM has therefore been effectively employed by small as well as large firms in the construction industry. When applied intelligently, it has achieved considerable success.

CPM enables planners and managers to thoroughly analyze the timing and sequential logic of all operations required to complete a project before committing time, money, equipment, labor, and materials for engineering and construction. Key personnel discover in advance the conditions that may arise on the project and gain a deeper understanding of the complex interrelationships of the activities to be performed. Planning with CPM focuses expert attention on potential trouble spots and indicates where extra effort can best be applied to reduce costs and delays without wasting resources. Experience in construction has often shown that this systematic forethought in itself more than justifies the use of CPM. In addition, CPM is a scheduling method that permits relatively easy revision of the schedule and simulation and evaluation of the impact of changes. It thus becomes an excellent control tool during the execution of the project as well.

Advantages of CPM Networks

It is most common to illustrate the advantages of CPM networks by comparing them with bar charts. First, networks can much more concisely represent large numbers of activities. More important, one of the major aspects of a planner's thinking, that is, the logical interrelationships and dependencies among activities, is not really shown on bar charts, but is inherent in networks. This in turn means that networks are much more useful for forecasting and control. For example, the impact on the whole project of a delay in an activity, a change in an activity's scope, or the addition or deletion of activities is readily transmitted, by network logic and computations, through the whole schedule.

Networks are also said to encourage a higher level of logical discipline in the planning, scheduling, and control functions, and to stimulate more attention to both long-range and detailed planning. Assuming that a standard notation and diagramming convention can be established at least within an organization and that there are willing and knowledgeable users, networks provide a more powerful means for documenting and communicating project plans, schedules, and performance. Finally, it is worth reemphasizing that, in contrast to other techniques, they identify the most critical elements in the project schedule and thus allow management to set priorities and focus attention on them.

Most of the advantages of networks over bar charts also hold in comparing them with the other techniques we have described. Even in matrix schedules, the logic is only imprecisely implied, and there really is no assessment of criticality. Nevertheless, returning to the importance of documentation and communication, all the various techniques do have their appropriate and best applications, and none should be discarded—at least not until something with the power analogous to musical notation comes along.

Network Fundamentals

CPM is a project-oriented scheduling technique. The *project* has an intentionally limited life span and should consist of a well-defined collection of tasks or activities that, when completed, mark the end of the project. Examples of projects include the design and/or construction of dams, tunnels, ports, refineries, and buildings. Each may actually be a component of a larger project.

Essential elements in almost all project networks are *activities*, their *durations*, and the *logical interrelationships* among them. Given these, one can compute each activity's early start, late start, early finish, late finish, total float (slack), and free float. These computations also yield the total expected duration for the project, and, of great importance, they focus attention upon the most *critical activities* and hence the *critical path* for the project. This is a powerful concept that greatly aids management in setting its priorities for allocating resources to operations.

An *activity* is a task or closely related group of tasks whose performance contributes to the completion of the overall project. An activity should be so sufficiently well-integrated that it can be rescheduled as a unit. One example of an activity could be "construct column footing." Most activities, including this one, can be further subdivided into component activities. The degree of breakdown depends upon the size and type of project, its requirements, and the purpose for which the schedule is intended. In this example, component activities could include: "excavate," "fabricate forms," "assemble rebar cage," "set forms," "fine grade," "set rebar cage," "place and finish concrete," "cure concrete," and "strip forms." These activities consume both time and resources. In scheduling, activities might also be used to represent administrative procedures, such as "client approval of plant layout"; to show delays that require time but no resources, such as "winter shutdown" and "spring floods"; and to allow for shipment and delivery of equipment and materials, such as "order ready-mix."

In the arrow diagram form of the CPM, directed lines, called *arrows*, represent activities. The direction of the arrow indicates the direction of progress. Nodes are placed at the beginning and end of each arrow, and an alphabetic and/or numeric label is assigned to each node to symbolically identify the activity. Figure 12-13 is an example activity using such notation.

Activities may be combined in a logical manner determined by the construction sequence of the operation being performed, the methods used, the time allowed, and the resources available. The combination of activities may be represented in a graphical form called a *network*. Figure 12-14 is a possible arrow network for the concrete footing example. This network itself may be but a small *subnetwork* of the overall project network.

Precedence diagramming (sometimes called the "activity-on-node" or "circle and connecting line" method) is an alternative way of representing a critical path network and has a number of advantages. It is the opposite of arrow diagramming in that the nodes represent activities and the arrows or connecting lines show the logical relationships between them. Figure 12-15 shows the activity "excavate footing" in precedence notation. A major advantage of this system is that it eliminates the need for dummies to correctly represent the logic. Precedence diagrams are not nearly

F_1 ——Excavate Footing——▸ F_2

FIGURE 12-13
Example arrow activity.

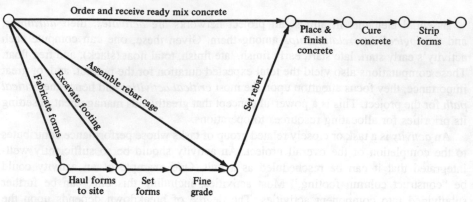

FIGURE 12-14
Arrow diagram for concrete footing construction.

as strict in regard to the positioning of activities and are therefore much easier to construct and modify: One simply puts the nodes representing activities on paper and then draws the lines to connect them. Figure 12-16 is a precedence diagram equivalent to the arrow network in Figure 12-14.

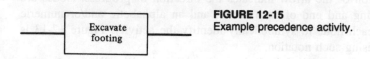

FIGURE 12-15
Example precedence activity.

FIGURE 12-16
Precedence diagram for concrete footing construction.

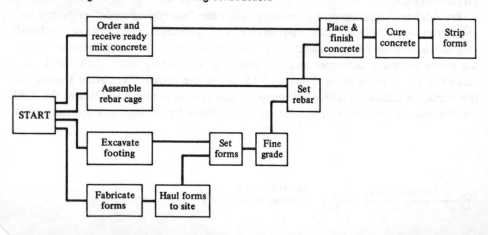

Logical properties of networks include *precedence, succession,* and *concurrence.* The start of a particular activity is permitted by the completion of all preceding activities or by the start of the project. In the example, both "order and receive ready-mix concrete" and "set rebar" precede or are *predecessors* of "place and finish concrete." Both predecessors must be completed before "place and finish concrete" can start. Similarly, the completion of an activity either permits the start of a following activity or marks the end of the project. In the example, completion of "place and finish concrete" permits its direct *successor,* or *follower,* "cure concrete," to commence. Other activities may logically proceed simultaneously, in which case they are logically *concurrent.* In the example, "set forms" and "assemble rebar cage" are logically concurrent activities. Planners must logically think through these relationships to construct a network. The process of constructing a network, in turn, is a strong aid in such logical thought processes.

Constructing Arrow Networks

Here and in the following section, a symbolic example illustrates basic procedures for constructing arrow networks and making CPM calculations. Table 12-1 defines the sequential logic of the example project.

On an actual project, development of this logic would generally parallel initial rough drafting of the network. The full logic is defined here at the beginning to serve as a point of departure for explaining network construction techniques.

To begin, note that activities A, B, and C have no predecessors. They may therefore share a common starting node. The starting node for an activity is commonly called its *i node.* Here the initial i node is labeled 1. A *j node* is placed at the end of each activity and labeled. Each j node, in turn, becomes the i node for activities immediately following. Each set of *i-j numbers* should uniquely define one and only one activity. The logic described thus far may be shown as in Figure 12-17.

A node such as 1, which serves as an i node for more than one activity, is sometimes called a *burst node.*

TABLE 12-1
EXAMPLE NETWORK LOGIC
DEFINITION

Activity	Predecessors	Followers
A	—	D, E
B	—	G, H, K
C	—	F
D	A	L
E	A	G, H
F	C	K
G	B, E	L, M
H	B, E	L, M
K	B, F	—
L	D, G, H	—
M	G, H	—

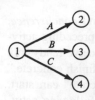

FIGURE 12-17
Beginning of arrow network.

Referring to the Followers column in Table 12-1, activity A is followed by D and E. Activity A, in turn, is the sole predecessor of D and E. Similarly, F follows C, and C is the sole predecessor of F. The partial network now appears as in Figure 12-18.

Node numbers are shown a bit out of sequence here to be consistent with the final form of the network. Ordinarily, the only restriction is that each one be unique. So far, the i-j numbers for activities A to F are 1-2, 1-3, 1-4, 2-9, 2-5, and 4-6, respectively.

Activity G follows both B and E. For the moment, this logic may be shown by terminating, or *merging*, activities B and E at the same j node. This could be done by deleting node 3 and connecting B to node 5. A node such as 5, which is the end node to two or more activities, is sometimes called a *merge node*. Node 5, in turn, becomes the i node for activity G.

Note that G and H both have the same predecessors and the same followers. One might be inclined to show this parallel relationship as in Figure 12-19.

It can be seen, however, that although the logic is correct, each activity is identified by the same i-j numbers. To uniquely identify each activity, an additional element, called a "dummy activity" or a "dummy arrow," is introduced. A *dummy* requires neither time nor resources, but is required to properly show the logic of an arrow network or to provide unique labeling. With a dummy, the logic of this section of the network may correctly be shown in one of the forms in Figure 12-20.

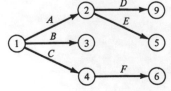

FIGURE 12-18
Second stage of network construction.

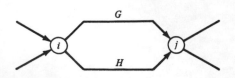

FIGURE 12-19
Incorrect notation for concurrent activities.

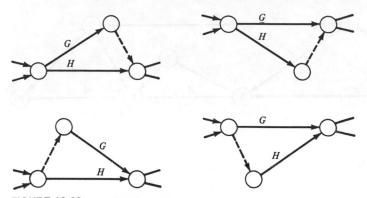

FIGURE 12-20
Four possibilities of G-H with dummy arrow.

In this case, where identification is the problem, the dummy requires an additional node. The partial network may now be sketched as in Figure 12-21.

Referring back to the logic table, activity M follows both G and H, which in turn are the sole predecessors of M. Node 8 may therefore correctly be used as the i node for M. Referring again to the table, activity L follows G, H, and D. At first it might appear that this could be shown by also terminating activity D at node 8 and using this as the i node for L. This would indeed show the correct predecessors for activity L. However, it would also show that activity D precedes activity M, which is incorrect. This is a situation in which it is necessary to introduce a dummy to show the correct network logic. By putting a dummy from node 8 to node 9, the logic of this portion of the network may be correctly shown as in Figure 12-22. Since neither L nor M has any followers, they are terminated at a common end node, labeled 10.

Activity K follows both B and F. One might try to show this logic by ending F at node 5 and starting K at node 5. This course would incorrectly show E as a

FIGURE 12-21
Partial network including dummy activities.

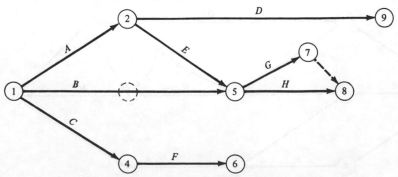

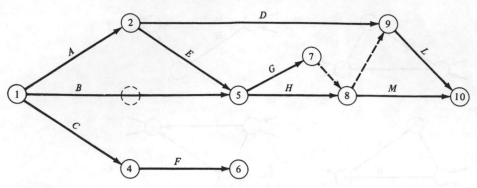

FIGURE 12-22
Use of dummy 8-9 to preserve correct logic.

predecessor of K and F as a predecessor of G and H, also incorrect. Apparently a dummy is needed, but where? Putting a dummy from node 5 to node 6 improves the situation by making F no longer preceding G and H, but it still leaves E incorrectly preceding K. To correctly show the logic in this case, two dummies are needed. Node 3 is reinstated as the j node for activity B. A dummy is then inserted from node 3 to node 5 to show B preceding G and H, and another dummy from node 3 to node 6 shows B preceding K. Node 6 is the i node for K. Since K has no followers, it is merged with L and M at j node 10, the last node in the network. The completed network in shown in Figure 12-23.

This example illustrates most of the basic concepts in arrow network logic. One additional pitfall that should be pointed out is the unintentional creation of *loops* in the network. For example, if the direction of the arrow representing activity D were reversed, there would be one loop through activities 2-5, 5-8, 8-9, and 9-2 and another loop through 2-5, 5-7, 7-8, 8-9, and 9-2. Loops such as these have no logical beginning or end; their presence indicates an error in a CPM network. In this case, the error is

FIGURE 12-23
Complete network.

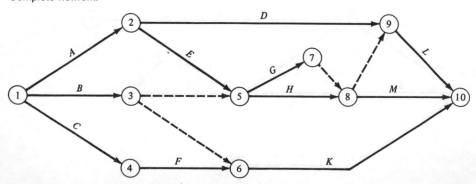

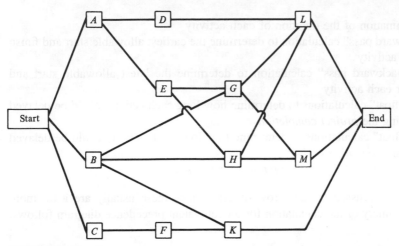

FIGURE 12-24
Equivalent precedence network.

easy to spot, but some fairly lengthy loops can creep into more complex networks and be fairly difficult to detect. Most CPM computer programs have routines to search out and trace any loops that may be present. Loops must be corrected before processing can continue.

Constructing Precedence Networks

Figure 12-24 shows the logic of precedence network equivalent to Figure 12-23. Although not mandatory except in some computer scheduling systems, common "start" and "end" activities have been shown to tie the logic together.

Since no special consideration need be given to dummy activities, all that really need be done to construct a draft of a precedence network is to put the activities down on paper and draw lines to show the precedence logic. However, for clarity, it is preferable that the logic flow left to right, so a second draft may be necessary. This of course is true also in arrow networks, although the arrow tips help clarify the logic also.

Another problem that occurs in both arrow and precedence diagrams is that sometimes the activity or logic lines unavoidably cross over each other. In this precedence diagram, B-G crosses E-H and H-L crosses G-M. Redrafting can minimize the number of crossing lines, but this criterion should not normally take priority over factors such as left-to-right logic flow, grouping of related activities, and so on. Rather, a small jump symbol ($\curvearrowright$), such as the two shown on B-G and G-M, is normally sufficient to preserve clear graphic communication.

CPM Time Calculations

So far, only the logic element in CPM networks has been discussed. The second main element is time. In general, these schedule calculations involve the following steps:

1 Determination of the duration of each activity

2 A "forward pass" calculation to determine the earliest allowable start and finish time for each activity

3 The "backward pass" calculation to determine the latest allowable start and finish time for each activity

4 "Total-float" calculations to determine how long each activity could be delayed without delaying the *project completion*

5 "Free-float" calculations to determine how long each activity could be delayed without delaying *any other activity*

6 Determination of the *critical path(s) for the network*

Each step is discussed for an arrow diagram since these usually are a bit more difficult. A summary of the calculation for an equivalent precedence diagram follows.

Estimating Durations   Once all activities in a project have been defined and organized into a logical CPM network, their durations must be estimated. An activity's *duration* is the expected amount of time, expressed in consistent time units, that will be required to complete the activity from start to finish. The time units may be days, weeks, or even hours or minutes, just so that all activities use the same units.

The importance of accurate estimates of durations cannot be overstated. On actual projects, these numbers are not just pulled out of the air; each duration estimate is the product of careful thinking involving the methods by which the activity will be accomplished, the resources (labor, equipment, material, financing) that are available, productivity, external constraints, and so forth. This process closely parallels that for making cost estimates.

For purposes of illustration, assume that duration estimates in weeks for the example from Figure 12-23 have been carefully made. These durations are shown beneath the label for each activity on the network in Figure 12-25.

FIGURE 12-25
Network with time units.

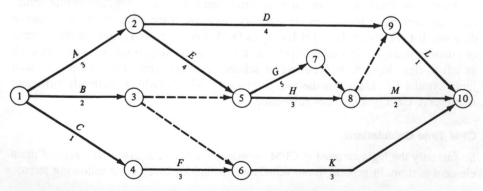

Forward Pass[3] The calculation procedure called the *forward pass* establishes the earliest expected start and finish times for each activity in the network. The following nomenclature is used in the discussion of the forward pass:

$$D(x) = \text{Estimate of } duration \text{ for activity x}$$

$$ES(x) = Earliest \text{ (expected) } start \text{ time for activity x}$$

$$EF(x) = Earliest \text{ (expected) } finish \text{ time for activity x}$$

$$S = \text{Project } start \text{ time}$$

Initially, a convention must be established for S, the project start time. In this discussion, S will be 1; that is, the project commences at the start of time period 1. An alternative convention often used is to set S equal to 0; that is, the project is ready to go by the end of time period 0. Actually, any number may be used for S; it is merely a reference point for the schedule.

The set of rules that defines the procedure for the forward pass calculations is called an *algorithm*. Only three rules are required:

1 The early start (ES) of all activities with no predecessors is equal to the project start time (S).

2 No activity may commence until all its preceding activities have been completed. Therefore, the early start time of any activity other than starting activities is equal to the maximum of the early finish (EF) times of its predecessors.

3 The early finish (EF) of an activity is equal to its early start (ES) plus its duration (D).

In mathematical notation, these rules can be expressed as follows:

$$ES \text{ (initial activities)} = S$$

$$ES(x) = \text{Maximum (EF(all predecessors of x))}$$

$$EF(x) = ES(x) + D(x)$$

To illustrate these rules with the example, Figure 12-26 shows the diagram notation convention used.

---

[3]The form of the notation used in this chapter is based upon *A Management Guide to PERT/CPM*, Jerome D. Wiest and Ferdinand K. Levy, Prentice-Hall, Englewood Cliffs, N.J., 1969, p. 31.

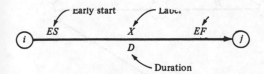

FIGURE 12-26
Notation for arrow diagram forward pass calculations.

To begin, the early start of activities A, B, and C is 1. The early finish for each activity is determined as follows:

$$EF(A) = ES(A) + D(A) = 1 + 3 = 4$$

$$EF(B) = ES(B) + D(B) = 1 + 2 = 3$$

$$EF(C) = ES(C) + D(C) = 1 + 1 = 2$$

The early finish for A of 4 may be interpreted to mean that the activity is completed by the beginning of time period 4. On the network, the calculations appear as in Figure 12-27.

The completion of A permits D and E to commence since they have no other predecessors. Their early starts are therefore set equal to 4. Similarly, F can commence once C is complete and thus has an early start of 2. The calculations for these activities in mathematical form and on the network (Figure 12-28) are:

$$ES(D) = ES(E) = EF(A) = 4$$

$$EF(D) = ES(D) + D(D) = 4 + 4 = 8$$

$$EF(E) = ES(E) + D(E) = 4 + 4 = 8$$

$$ES(F) = EF(C) = 2$$

$$EF(F) = ES(F) + D(F) = 2 + 3 = 5$$

The commencement of activity G requires that both B and E be complete. Therefore, the early start of G is the maximum of the early finishes of B and E:

$$ES(G) = Max(EF(B), EF(E)) = Max(3,8) = 8$$

FIGURE 12-27
Network with first step of forward pass calculations.

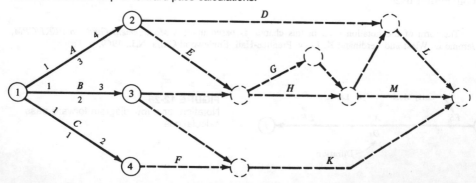

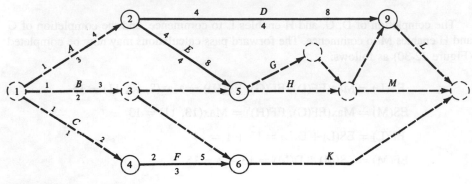

FIGURE 12-28
Second step of forward pass.

The same applies to activity H:

$$ES(H) = Max(EF(B), EF(E)) = Max(3,8) = 8$$

Similarly, the commencement of K requires the completion of both B and F.

$$ES(K) = Max(EF(B), EF(F)) = Max(3,5) = 5$$

Now early finish values for G, H, and K can be computed.

$$EF(G) = ES(G) + D(G) = 8 + 5 = 13$$
$$EF(H) = ES(H) + D(H) = 8 + 3 = 11$$
$$EF(K) = ES(K) + D(K) = 5 + 3 = 8$$

FIGURE 12-29
Third step of forward pass.

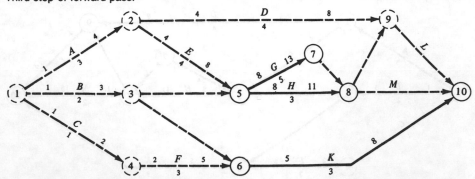

The completion of D, G, and H enables L to commence, and the completion of G and H enables M to commence. The forward pass calculations may now be completed (Figure 12-30) as follows:

$$ES(L) = Max(EF(D), EF(G), EF(H)) = Max(8, 13, 11) = 13$$

$$ES(M) = Max(EF(G), EF(H)) = Max(13, 11) = 13$$

$$EF(L) = ES(L) + D(L) = 13 + 1 = 14$$

$$EF(M) = ES(M) + D(M) = 13 + 2 = 15$$

The early finish of M is the maximum early finish in the project and thus becomes the earliest completion date of the project. The total project duration is therefore calculated to be 14 time units. That is, project duration $= Max(EF) - S = 15 - 1 = 14$. This completes the forward pass calculations.

Backward Pass   The calculation procedure called the *backward pass* establishes the latest allowable start and finish times for each activity that will still permit the overall project to be completed without delaying beyond the scheduled completion date. The following nomenclature is used in the discussion of the backward pass:

$$D(x) = \text{Estimate of } Duration \text{ for activity x}$$

$$LS(x) = Latest \text{ allowable } Start \text{ time for activity x}$$

$$LF(x) = Latest \text{ allowable } Finish \text{ time for activity x}$$

$$T = Target \text{ project completion time}$$

The project completion time T is generally taken as the early project completion time from the forward pass; this will be done here. However, any date may be taken as this reference point. For example, a contractual completion date could be used.

FIGURE 12-30
Forward pass complete.

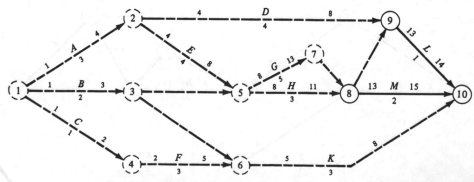

The following rules define the algorithm for the backward pass:

1 The latest allowable finish (LF) of all activities with no followers is equal to the target project completion time (T).

2 The latest allowable finish time (LF) for any other activity is equal to the earliest of the latest allowable start times of its successors.

3 The latest allowable start time (LS) for any activity is equal to its latest allowable finish (LF) minus its duration (D).

In mathematical notation these rules can be expressed as follows:

$$LF(\text{end activities}) = T$$

$$LF(x) = Min(LS(\text{all followers of } x))$$

$$LS(x) = LF(x) - D(x)$$

To illustrate these rules with the example, the notation in Figure 12-31 is added to the forward pass diagram convention.

To begin the calculations, the late finish of activities K, L, and M, which have no followers, may be taken as 15, the project completion time. The late start for each activity is determined as follows:

$$LS(K) = LF(K) - D(K) = 15 - 3 = 12$$

$$LS(L) = LF(L) - D(L) = 15 - 1 = 14$$

$$LS(M) = LF(M) - D(M) = 15 - 2 = 13$$

On the network, the calculations appear as in Figure 12-32.

The late start for L, the only follower of D, becomes the late finish of D. The minimum of the late starts of L and M becomes the late finish for G and H. These calculations are:

$$LF(D) = LS(L) = 14$$

$$LF(G) = LF(H) = Min(LS(L), LS(M)) = Min(14, 13) = 13$$

$$LS(D) = LF(D) - D(D) = 14 - 4 = 10$$

$$LS(G) = LF(G) - D(G) = 13 - 5 = 8$$

$$LS(H) = LF(H) - D(H) = 13 - 3 = 10$$

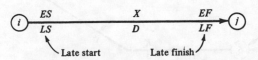

FIGURE 12-31
Supplementary notation for backward pass.

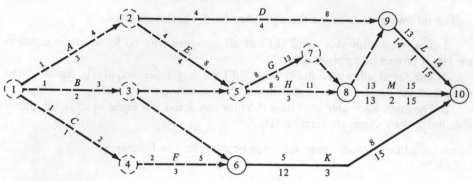

FIGURE 12-32
First step of backward pass.

The late finish of activity B is the minimum of the late starts of activities G, H, and K. Similarly, the late finish for activity E is the minimum of the late starts of activities G and H. The late start of activity K, the only follower of activity F, becomes the late finish for F. This stage of the backward pass is calculated as follows:

$$LF(B) = Min(LS(G), LS(H), LS(K)) = Min(8, 10, 12) = 8$$

$$LF(E) = Min(LS(G), LS(H)) = Min(8, 10) = 8$$

$$LF(F) = LS(K) = 12$$

$$LS(B) = LF(B) - D(B) = 8 - 2 = 6$$

$$LS(E) = LF(E) - D(E) = 8 - 4 = 4$$

$$LS(F) = LF(F) - D(F) = 12 - 3 = 9$$

The minimum of the late starts of D and E becomes the late finish of A. The late start of F is the late finish of C. The backward pass calculations may now be completed:

FIGURE 12-33
Second step of backward pass.

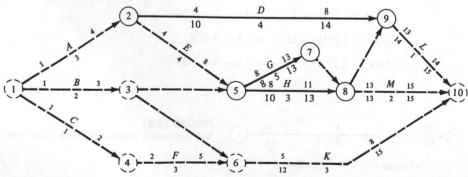

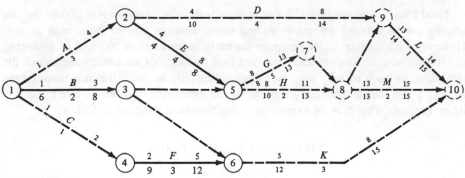

FIGURE 12-34
Third step of backward pass.

$$LF(A) = Min(LS(D), LS(E)) = Min(10, 4) = 4$$

$$LF(C) = LS(F) = 9$$

$$LS(A) = LF(A) - D(A) = 4 - 3 = 1$$

$$LS(C) = LF(C) - D(C) = 9 - 1 = 8$$

Since this backward pass started with the completion time from the forward pass, the minimum late start of all the activities **that is, LS(A) = 1** should equal the start time S of the project. This serves as a check on the calculations and completes the backward pass. If another value of T were used, the minimum late start of all the activities would equal the difference between this and the completion time from the forward pass. For example, if 19 days were available for the project, the last day could be set at 20, and on the backward pass all late starts, late finishes, and total floats would be 5 days higher. The critical activities would have the minimum total float—5 days instead of 0.

FIGURE 12-35
Backward pass complete.

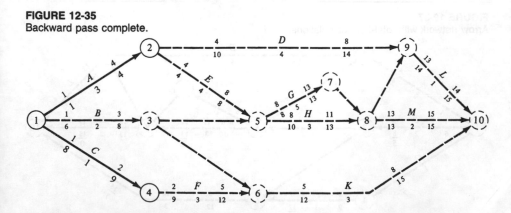

Total Float The *total float*[4] for an activity is the maximum amount of time that the activity can be delayed without extending the completion time of the overall project. However, such a delay might postpone the early start of one or more of its following activities. Once the forward and backward pass calculations have been completed, the total float for each activity may be calculated directly as the difference between the activity's late start and its early start, or as the difference between its late finish and its early finish. This may be expressed in mathematical notation as follows:

$$TF(x) = \text{total float for activity x}$$
$$= LS(x) - ES(x)$$
$$= LF(x) - EF(x)$$

It may be added to the diagram notation in either of the forms shown in Figure 12-36. On the example, total floats are calculated as in Figure 12-37.

[4]Float is sometimes called slack.

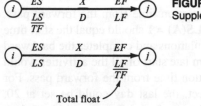

FIGURE 12-36
Supplementary notation for total float.

Total float

FIGURE 12-37
Arrow network with total-float calculations.

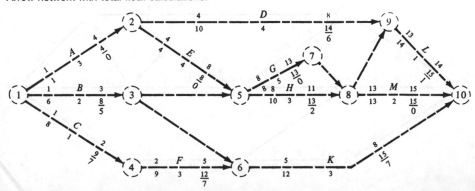

Free Float *Free float* is the maximum amount of time an activity can be delayed without delaying the early start of any of its followers. It also follows directly from the forward and backward pass calculations, and it may be determined as the minimum of the early starts of all the activity's immediate followers minus the activity's early finish. This may be expressed mathematically as follows:

$$FF(x) = \text{Free float for activity } x$$

$$= \text{Min (ES(all immediate followers of } x)) - EF(x)$$

On the example, activities B and D have positive free float, which differs from their total float. The free floats for these activities are calculated:

$$FF(B) = \text{Min}(ES(G), ES(H), ES(K)) - EF(B)$$

$$= \text{Min}(8, 8, 5) - 3 = 5 - 3 = 2$$

$$FF(D) = ES(L) - EF(D) = 13 - 8 = 5$$

The positive free float of activities H, K, and L equals their total float and is determined in a manner similar to that for B and D. Note that activity C has 7 units of total float but no free float. All the other activities are on the critical path and thus have zero float—total or free. Also note that the free float for any activity is always less than or equal to the total float.

Critical Path A *critical path* is a continuous chain of activities from the beginning to the end of a network with the minimum float value. In the case where the target project completion time is set equal to the early project completion time, a critical path is a chain of activities with zero float. By summing activity durations, the critical path is the longest path through the network. The critical path for the example network is shown in Figure 12-38 with a heavy line.

FIGURE 12-38
Critical path.

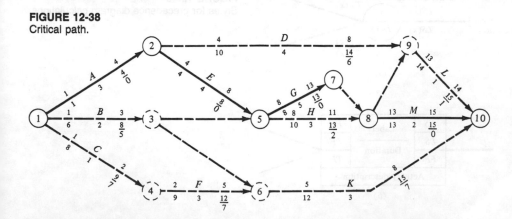

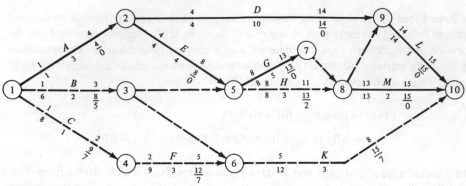

FIGURE 12-39
Multiple critical paths.

There may be more than one critical path in various parts of the network. For example, if the duration of activity D were increased to 10, there would be an additional path through activities D and L, as shown in Figure 12-39.

Equivalent Calculations on a Precedence Diagram   The calculation procedure for a precedence network is identical to that for an arrow network, except that one need not consider dummies. One of the main issues in precedence networks, however, is the style in which the calculations and related activity information are shown. Two main alternatives are to put the calculations outside the activity symbol (Figure 12-40a) and within the activity symbol (Figure 12-40b). Using the style in Figure 12-40a, the calculations for the network diagram would be as shown in Figure 12-41.

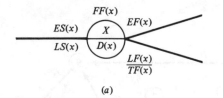

(a)

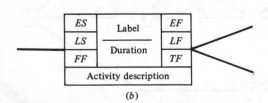

(b)

FIGURE 12-40
Styles for precedence diagram calculations.

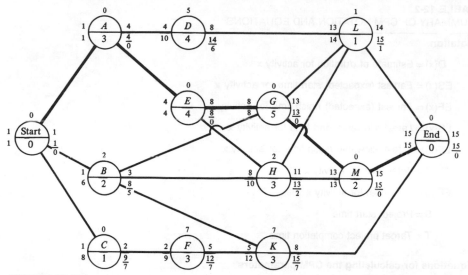

FIGURE 12-41
Precedence network CPM calculations.

Summary of Notation and Equations Regardless of whether the logic is shown by arrow or precedence notation, the algorithms for network computations are the same. The notation and formulas in Table 12-2 provide all that is required for standard CPM network calculations.

Calendar-Scaled Diagrams

This section describes a method of constructing calendar-scaled CPM diagrams that are similar to the plotter-drawn networks produced by many computer systems. The basic unit of time for this purpose is 1 day.

To convert the *working-day schedule* to a *calendar-dated schedule*, the start day is designated on the calendar. Succeeding parts of the working-day schedule are then compared to the calendar on a day-by-day basis. Extra days are inserted for non-workdays, such as weekends and holidays. These non-workdays must be specified in advance. The overall procedure is quite straightforward and yields corresponding calendar dates for each workday in the schedule.

In the arrow type of calendar-scaled diagram, extra dummy arrows and nodes are inserted so that all time and resource-consuming activities can be shown horizontally and each has a unique i node. In this way the interpretation of the horizontal arrows is analogous to bars on a bar chart. The diagram has therefore been found to be easily understood in the field. Activities are generally shown scheduled at their early start times by a solid line, and a contrasting dashed line shows the remaining free float, if any.

TABLE 12-2
SUMMARY OF CPM NOTATION AND EQUATIONS

Notation

$D(x) =$ Estimate of duration for activity x

$ES(x) =$ Earliest (expected) start time for activity x

$EF(x) =$ Earliest (expected) finish time for activity x

$LS(x) =$ Latest allowable start time for activity x

$LF(x) =$ Latest allowable finish time for activity x

$TF(x) =$ Total float for activity x

$FF(x) =$ Free float for activity x

$S =$ Project start time

$T =$ Target project completion time

Equations for calculating the CPM parameters

Forward pass

$ES(x) = S$ for beginning activities, or

$ES(x) = $ Max (EF (all predecessors of activity x))

$EF(x) = ES(x) + D(x)$

Backward pass

$LF(x) = T$ for ending activities, or

$LF(x) = $ Min (LS (all followers of activity x))

$LS(x) = LF(x) - D(x)$

Floats

$TF(x) = LS(x) - ES(x)$

$= LF(x) - EF(x)$

$FF(x) = $ Min (ES (all immediate followers of activity x)) $- EF(x)$

Critical Path

A critical path is a continuous chain of activities with the minimum total float value. By summing activity durations, it is the longest duration path through the network. There may be more than one critical path in various parts of the network.

To illustrate the construction of a calendar-scaled arrow diagram, assume that the example project from the previous sections is to be started on Monday, July 3, 1989. Assume also that a standard 40-hour, 5-day week will be worked. The calendar for July 1989 is as follows:

Sun.	Mon.	Tue.	Wed.	Thur.	Fri.	Sat.
						1
2	3	<4>	5	6	7	8
9	10	11	12	13	14	15
16	17	18	19	20	21	22
23	24	25	26	27	28	29
30	31					

A holiday is taken on the 4th. The 1st, 2d, 8th, 9th, 15th, 16th, 22d, 23d, 29th, and 30th fall on weekends. The remaining days are available for work

For arrow network construction, the line symbols in Figure 12-42 are used. With this notation, the calendar-scaled network may be constructed. The general procedure is:

1 Lay out a calendar grid with a duration greater than that of the project.
2 Block out non-workdays.
3 Plot activities to the same scale.
   a Show continuation through non-workdays (e.g., activities A and B).
   b Where an i node is common to two or more activities, such as node 2 for D and E, insert a dummy, such as 2a → 2b to separate them vertically In the case of start nodes, use new separate nodes for each starting activity: 1a, 1b, and 1c.
   c If another activity does not immediately follow, the interval until the start of the earliest follower is filled with a dashed line representing free float, as shown on activities D, H, and K in the following example.
4 Label the activities.

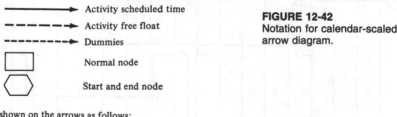

→ Activity scheduled time
--→ Activity free float
----→ Dummies
☐ Normal node
⬡ Start and end node

FIGURE 12-42
Notation for calendar-scaled arrow diagram.

Data is shown on the arrows as follows:

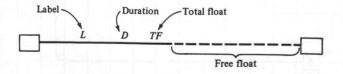

Label — Duration — Total float
L D TF
Free float

The calendar-scaled CPM diagram may be drawn directly from the tabular logic, or it may be converted from the conventional CPM diagram.

The CPM solution for the example network is repeated on Figure 12-43a; Figure 12-43b shows its corresponding calendar-scaled arrow diagram. Readers familiar with precedence diagramming may be interested to note a number of similarities between the calendar-scaled diagram and a precedence network.

Extensions

There have been numerous useful extensions to the basic elements of activities, time, and logic in networks. The strict logic requirements of basic CPM have been re-

FIGURE 12-43
Conversion to calendar-scaled arrow network.

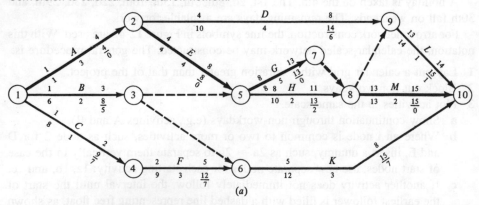

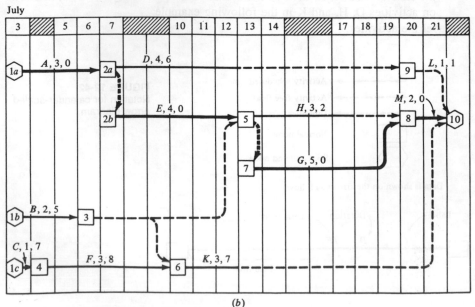

laxed somewhat with techniques permitting activities to be overlapped or to have delays inserted between them. The methods of probability and statistics have also been applied to both time and logic. Other enhancements include addition of costs and resources. Cost enhancements include analytical procedures for time-cost trade-off analysis, network-based cash flows, and network-based cost control. By identifying resources of labor, materials, and equipment with activities, planners and managers can take advantage of powerful analytical techniques for resource allocation to assure that the project can be completed within finite resource limits, and of resource leveling to ensure the efficient and continuous utilization of resources.

The full development of these concepts is beyond the scope of this book. Several of the books devoted solely to critical path methods deal with at least some of these subjects, and the reader is encouraged to consult the bibliography for further information.

APPLYING NETWORK-BASED PROJECT CONTROL

From day 1 onward, those responsible for the management and control of a project have more information than did the planners who prepared the schedule for the project's guidance. Management should recognize this fact and should not be intimidated by the network, nor should they ignore it. The network is basically a graphical expression of the experience, judgment, intuition, decisions, and assumptions of the planner. If the planner is worth listening to, then his network deserves careful attention. But, since each day brings new information about the project, one must continuously be prepared to justify or question the thinking that went into the plan. If the network is indeed to be a viable control tool on the project, it must be subject to revision, change, and improvement. Planners can aid this process if, in defining activities and preparing the original network, they think ahead to the people who will be expected to use it. As a follow-up to the technical details of network diagramming and computations, we shall therefore suggest a few practical guidelines for this purpose.

Defining Activities

In defining the content and level of detail of a network, consider who will use it: Project management? Staff scheduling engineers? Foremen? What will they use it for, and how? What is the size of the project, and what is the scope of work under the intended users' control? Especially on larger projects, a hierarchy of schedules is necessary, with selectivity and subreporting. For example, a CPM network might serve for general and detailed project control by superintendents and the engineering staff. But it might be reduced to a summary network or even a bar chart or progress curve for higher management. At the other extreme, individual activities might be expanded to narrative work plans for foremen.

Who will maintain and update the schedule? Managers and supervisors on an intermittent basis? Or a trained staff of scheduling engineers? How will they do this? Manually or by computer? On site or in the home office? If the work is manual, does the network conveniently allow for selectively detailing portions of it, say activities on the critical path? Will the accuracy of logic and time estimates be affected by the level of detail? How?

If the planner considered alternatives and exceptions that are not shown on the network, or if there are assumptions that were made for diagrammatic or computational convenience rather than necessity, are these facts clearly cross-referenced from the diagram to a supporting set of notes? For example, with basic CPM networks it is difficult to distinguish in logic between technological constraints and resource constraints. To illustrate, Figure 12-44a shows two concrete pours that might be done in either order. However, the planner has assumed that only one crew will be available, and so, to make the project duration come out right, he has arbitrarily sequenced the activities as shown in Figure 12-44b. When the superintendent goes to do pour 1, he finds the subgrade is too wet, so he shifts the crew to pour 2 instead. When this was reported to the computer, it blew a fuse, and the network had to be revised and submitted again. This is but one example of where computational details peculiar to the technique can have little to do with practical field operations.

Will the feedback time of the information system (recall Figure 10-2) be compatible with the scope and duration of activities? In general, activity durations should be long

FIGURE 12-44
Arbitrary use of network logic for resource constraint.

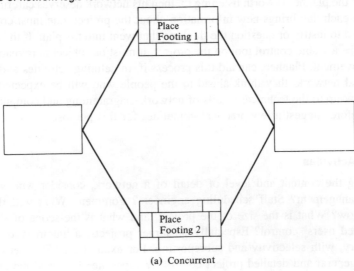

(a) Concurrent

(b) Sequential

enough so that management can take corrective action if schedule reports indicate this is necessary.

These are just a few ideas for the definition of activities. These decisions should be made with the overriding criteria being the needs, skills, and cooperation of the users.

Preparing a Network

Responsibilities for preparing CPM schedules should be clearly defined in company policies and procedures. As a general principle, however, it can be stated here that persons who will be responsible for carrying out a particular area of work in the project should substantially contribute to the corresponding part of the CPM schedule preparation. They may of course be assisted by a scheduler, and the scheduler may actually have the responsibility of converting the rough logic and preliminary network sketches into neatly drafted CPM diagrams.

Although the technical details of network preparation are beyond the scope of this book, it is worth reviewing the general procedure to further indicate where a good planner will include the needs of users. The steps listed here should also show that the actual mechanics of the process are fairly simple; they apply to large as well as small networks. Also, note that there are strong parallels between estimating and network planning; they are interdependent functions.

1 Begin by learning all you can about the project itself. Study the plans and specifications, include a site reconnaissance if possible, and above all, seek input from all key parties known to be involved in the planning or execution of the project. These can be the owner's representatives, the designer, subcontractors, major suppliers, labor organizations, regulators, and, of course, the professional construction manager's own staff designated for the project.

2 Make a preliminary listing of some key activities, keeping in mind the guidelines for defining activities.

3 Put a key activity on the diagram—first, last, or in between—but make a start. If one is intimidated by the prospect of constructing a large network, this first activity is often the hardest step. Steps 4, 5, and 6 follow fairly easily, and will produce an initial diagram.

4 Ask yourself the following questions:
   a What must be completed immediately before this activity can begin?
   b What activities can follow once this activity is complete?

5 Put these new activities on the diagram.

6 Repeat steps 4 and 5 until you have a reasonably comprehensive diagram of the project. Its organization at this stage will probably be chaotic, but tidiness can wait.

7 Reexamine the plans, specifications, and other sources of information to be sure that all parts of the project are covered and none is duplicated.

8 Check and double-check the logic and content for factors such as improper "dummies" (a common source of error only in arrow networks), possible recursive "loops," and for assurance that each activity has an intentionally defined start and finish point.

9 At this stage you have a rough diagram. Before drafting the final version, recheck it with those parties who were consulted for input in step 1 to be sure it does indeed represent their thinking. This step is particularly important: it is futile to impose an unworkable schedule on the people responsible for the execution of the project.

10 Repeat the earlier steps as needed to produce a satisfactory network. Only at this stage should one proceed to an elegant presentation. The next section will offer some guidelines for this purpose.

Level of Detail

The level of detail to which to subdivide a network's activities depends upon a number of factors, including the schedule's intended purpose, the type of project, whether computer or manual processing will be used, and the preferences of management. Schedules may be prepared at several levels of detail. A. T. Armstrong-Wright's example in *Critical Path Method* illustrates this idea:

The network for a board of directors of a power company considering a new power station might contain activities such as:

Raise capital
Obtain land
Plan project
Construct stage I

The project manager might take the above activity-plan project and in his or her network break it down further:

Determine required output
Survey site
Design plant
Order plant
Design building
Project plan to Managing Director

While even greater detail would be required by the Chief Electrical Design Engineer; he or she might break down the activity-design plan as follows:

Select generator size and type
Prepare preliminary layout
Design wiring diagram
Design transmission lines
Preliminary plans to Project Manager

Similarly, the construction manager will require a detailed breakdown of the broad activity Construct Stage I. Whether or not these detailed networks are then combined into one comprehensive network will depend on the method of control and liaison between departments [pp. 31–32].

Detailed planning is generally beneficial, but excessive detail in the network can overcomplicate the schedule and create unnecessary effort and cost in preparation and processing. For these reasons, selective detailing can be used to advantage. For example, different parts of a network might be prepared at different levels of detail. Starting with a fairly broad network, activities along the critical path, as well as other activities and areas that might be potential trouble spots, could be further subdivided for closer analysis. Another possibility is to take the general network and expand it into a progressive "wave" of detailed activities 1 to 3 months in advance of the commencement of the work in that interval.

In scope, activities should be defined and detailed so as to correspond to work items subject to cost control. Where possible, an activity should be contained entirely within one individual's responsibility. This requires a fairly high degree of homogeneity in resources—labor, materials, and equipment. Activity durations should be long enough to permit progress feedback to be reported in time for corrective action to be taken, if necessary.

Organizing the Network

The physical organization of activities on a network should proceed according to some rational pattern representing the needs and characteristics of the project. Some possible areas of organization that may be used alone or in combination include the following:

1 By responsibility: trades, type of work, supervisors, etc.

2 By geographic area and facility within the project, so there is some locational correlation between activities on the diagram and, perhaps, the plot plan of the site

3 By time scale

4 By project cost code (see Chapter 13)

These basic concepts of network organization were further amplified by A. T. Armstrong-Wright as follows:

In producing the fair diagrams of the network, careful attention must be given to the orderly arrangement of activities. The arrows should be drawn horizontally with only the ends turned to connect to events (nodes) so that descriptions can be easily written and read. The description of activities entered above each arrow should be clear and concise. Above all, the use of inexplicit abbreviations or codes should be avoided. Experience has shown that if constant and laborious reference to explanatory lists of abbreviations and codes is necessary, the network will soon fall into disuse and will be ignored from the start by all but the very enthusiastic.

By convention, networks are drawn with progress from left to right. This is, in fact, the normal method of indicating the passage of time on graphs, histograms and bar charts. There may be certain projects or processes where progress can best be indicated in some other direction, but these will generally be the exception. For example, the network for the construction of a skyscraper could be drawn with progress shown vertically upwards. One serious disadvantage of this is that arrows would be mainly vertical with descriptions either being written horizontally, thus taking much space, or written vertically, in which case the network would have to be read sideways. Inevitably the network would be turned on its side and hence any advantage of showing progress vertically would be lost.

The network should be divided into clear areas of responsibility so that it is not necessary to scan the whole network to locate activities or which one individual or works unit may be responsible. This is usually achieved by dividing the network into horizontal zones. This method is particularly suitable where the work of the individual units is for the most part independent, with only a few connecting activities.

Where there is much intermingling of the work of several teams or units, it may be better to divide the network into zones representing the location of the work rather than areas of responsibility. It is useful in this case to relate the network to a day plan [pp. 34–35].

Either type of organization described here is made possible by the use of "banding" in a computer system's plotter-drawn networks. Groupings could also be made by prime accounts or by other systems that would fit the needs of the project. In some computer systems, this grouping is specified by properly coding the nodes on the network diagram.

Managing Float

Both free float and total float can give management considerable flexibility in scheduling a project's activities. Judicious rescheduling of activities within their float ranges can effectively "level" resource usage and make a smoother, more efficient job. On the other hand, float is a valuable commodity that can be wasted if used indiscriminately. Total float, especially when it is reflected through a series of activities in a chain, must be regarded as "community property" among those activities. If a superintendent or supervisor carelessly lets an early activity in the chain slip to its late start and finish times, thinking that there is plenty of leeway, the remaining activities in the chain will become critical. Considerable managerial flexibility is thus lost. For this reason, companies sometimes do not reveal float values, late start dates, and late finish dates to their supervisor and subcontractors, giving them only scheduled dates determined by project management. Management does have all the information from the schedule. This of course takes much of the responsibility and flexibility away from line supervision. There is no fixed answer to this problem; it is a matter of policy that must be resolved by management on each project.

Reporting

Sources of information for schedule control are in part the same as those for cost engineering. They include labor and equipment time sheets, field quantity reports, and various kinds of trend reports. Also included are informal oral reports from field to office and conscientiously updated field diaries.

Particularly useful, however, are preprinted forms requesting actual and estimated time and resource information associated with specific activities. One good type of information form, and one that minimizes the amount of writing requested, will provide space for overwriting information on one of the duplicate copies of subreports sent to specific supervisors in the field. Used in this way, the subreport becomes a "turnaround" document. Figure 12-45 gives an example of a subreport concerning concrete activities that are either in progress or scheduled to start in the near future. The superintendent apparently believes that the report is accurate except for activity CPF, which is expected to start 2 days later than planned, and activity CFF, which is only 30 percent complete instead of the 60 percent scheduled. This document is then

Copy 2

Subreport Code: 03 Concrete
Report Date: 19 June 1978

SUNNYSTATE CONSTRUCTION COMPANY, INC.

Project: Mountaintown Warehouse
Location: Mountaintown, Westamerica

Copy 2

Return Copy 2 to Home Office Scheduling Department.

Note changes to schedule in the blank spaces provided.

Act. Label	Description	Planned Duration	Planned Start	Planned Finish	Early Start	Early Finish	Total Float	Percent Complete	Critical
CFR	Concrete Fabricate Resteel	8	2JUN78	14JUN78	2JUN78	14JUN78	0	100%	**
CFF	Concrete Fabricate Forms	6	13JUN73	21JUN78	2JUN78	12JUN78	8	30% 60%	
CPR	Concrete Place Resteel	10	14JUN73	28JUN78	14JUN78	28JUN78	0	20%	**
CPF	Concrete Place Forms	3 4	23 21JUN78	28 27JUN78	12JUN78	16JUN78	8	0%	
CPC	Place and Finish Concrete	2	28JUN78	30JUN78	28JUN78	28JUN78	0	0%	**

END SUBREPORT 03

FIGURE 12-45
Example schedule subreport "turnaround" document.

returned to the office where it is used as input for the next schedule update. Though normally used mainly in computer-based systems, this technique can also be applied in manual reporting systems.

Once data come in to be processed, numerous different reporting techniques may be employed to disseminate the updated information to the managers and supervisors who need it. The most notorious method is to use the computer as a sorting and printing device and to issue the same complete detail reports on the whole project to all parties, regardless of needs and interests. This approach has almost invariably failed. In keeping with the principles discussed in Chapter 10, computerized tabular reporting is most effective only when it has the capacity of selective subreporting, exception reporting, and summary reporting in order to get the right information to the right people at the right level of detail and in time for decisions and corrective action.

Other useful means of reporting include different kinds of graphical reports. Many projects make very effective use of color coding to update prominently located network diagrams to keep key people informed of project status. Computer-based plotters can be effectively employed to redraft updated schedules periodically. Some firms then use a Xerox-reduction process to issue these schedules, folded accordion-style, in a form that a supervisor can carry in his pocket. These graphical reports are particularly useful to field people.

Updating Considerations

There are several procedures for updating network-based schedules. With low-cost computer processing, it is feasible to carry out complete activity-by-activity computations on a periodic or demand basis. When larger networks are to be updated by hand, techniques such as those developed by John Fondahl enable schedulers to focus only on the parts of the network that are currently of interest, and thus greatly to reduce their computation burden.[5]

Regardless of the approach, in updating we wish to compare planned with actual progress and to calculate variances to highlight problem areas. Parameters of interest include the following:

1 Floats, especially in critical and near-critical activities
2 Changes in the critical path
3 Logic changes, including new and deleted activities
4 Resource usage, especially to predict constraints and reduce idleness
5 Changes in durations
6 Activities which were completed, and percentage complete on those in progress

One often hears debate among scheduling people about the best frequency with which to update schedules: daily? weekly? monthly? The debates, however, miss the point.

[5]John W. Fondahl, *A Non-Computer Approach to the Critical Path Method for the Construction Industry*, Technical Report No. 9, Stanford University, Dept. of Civil Engineering, The Construction Institute, Stanford, Calif., November 1961. Also, John W. Fondahl, *Methods for Extending the Range of Non-Computer Critical Path Applications*, Technical Report No. 47, ibid., 1964.

If we have gone 6 months and the project has essentially progressed as planned, fine! Apart from checking off completed activities and verifying this progress, no real updating is necessary. On the other hand, if we worked until late last night doing this month's revision, and then this morning we get word that a change order is brewing to add a whole new wing to our building project, we are back to base 1 already. The important thing is not how often a schedule is updated, but how accurately the schedule reflects the actual conduct of the work.

Corrective Action

Assuming that exception reports from our last schedule update show that some activities are deviating significantly from the plan, what do we do? Think before acting! Recall again that from day 1 onward we have more information than did the original planner who developed the schedule. It is quite possible that unforeseen favorable developments are simply causing the work to go better than expected. Certainly we should verify these trends and analyze the reasons behind them, but it will quite likely be the plan that needs corrective action, not the project.

In some cases, however, activities may be slipping well behind their scheduled start times, possibly because of a delayed material delivery. Here is where the "float" properties of the critical path method become a real asset. If the activities in question have sufficient free float, and if we can verify that the delayed deliveries will be made within this range, again it may be possible that no corrective action is necessary. This assumes, of course, that the resource allocation is not adversely affected.

But what if delays or resource overruns on critical activities appear to be seriously jeopardizing the completion date of the project? In the past, the response was too frequently to accelerate the whole job, without knowledge of which operations most affected duration. Here again, the critical path in CPM focuses management's attention on the activities that really count. Using the principles from CPMs "time-cost trade-off" extensions, a rational plan can be developed for most economically accelerating the project. Where feasible, new information and constraints might even make it economical to alter the logic of the schedule.

As mentioned in Chapter 10, one should not take corrective action merely for the sake of keeping to the original plan. That is, the schedule alone does not run the job. Rather, use it as a guide, but be prepared to take advantage of, and adapt to, new conditions as they arise. Where corrective action is warranted, the schedule can also help to focus thinking and set priorities for efficient solutions.

DOCUMENTATION FOR CHANGES, CLAIMS, AND DISPUTES

Changed conditions, change orders, delays, claims, and disputes occur in some measure on almost all projects of significant size. Occasionally there will be malicious intent or even dishonesty on behalf of one or more parties to a contract; the subject of this section, which deals with the documentation of facts, will be of no help to their cause, though it may help the other parties to defend themselves. But even when all parties to a contract are doing their level best to interpret its terms and conditions honestly and objectively, there are bound to be differences of opinion simply because of the specific interests of those involved in trying to get the job done as economically, quickly, and skillfully as possible. Thus, there has evolved a body of law and accepted practices which help to achieve just, equitable, and fair resolution of disputes. In contracts, these often take the form of clauses pertaining to changed conditions, change orders, delays, contract time, liquidated damages, disputes, and claims. But the process only starts here.

This section is by no means intended as an introduction to contracts, specifications, and their legal implications. Chapter 20 will have more to say on this subject. Rather, we assume some reader background in this area, and will illustrate a few cases where the planning, scheduling, and control tools discussed in this chapter can have a constructive impact. Briefly stated, five of the most important guidelines regarding matters of changes, delays, disputes, and claims in contract administration are as follows:

1 Documentation
2 Knowledge of contracts and the law
3 DOCUMENTATION
4 Good working relationships between all parties to the contract
5 D-O-C-U-M-E-N-T-A-T-I-O-N

Changed Conditions and Change Orders

Changed conditions occur when the nature of the work encountered on a project is significantly different from that described in the contract documents. Change orders, which are directives from the owner or his agent, and which usually result from negotiations with the contractor, can alter the terms and conditions of the contract, say, to add extra work, delete work, change the standards of work, etc. Change orders can thus provide an equitable means of dealing with changed conditions arising from unforeseen events, such as an unexpectedly bad foundation problem. Change orders can also be used, however, when an owner simply wishes to alter some part of the facility—say, add a wing to a building—after the contract has started.

The impact of changes differs depending on the nature of the contract. For example, if there is a quantity variation within the range of a unit-price contract, the nature of the price schedule automatically handles the changes. In a lump-sum contract, however, an overrun is likely to generate a claim. In a negotiated contract, a change in scope might be agreed upon, with adjustments in direct reimbursable costs and possibly an adjustment in the fee. In any of these types of contracts, however, one needs to be able to evaluate both the direct costs and the impact costs of the change and to determine how these are allocated among the parties to the contract.

The direct and impact costs of delays and time extensions will provide a simple illustration. Four situations will be considered and common settlements will be indicated.

Extra Work Requiring More Time   The document for this situation is a change order, and it normally justifies both a time extension and extra reimbursement for the contract. Either the time and cost are settled at the time of the request, or, to avoid default, the contractor can give written notice and proceed under protest.

Delay Caused by Owner or His Agent   If the owner or his agent causes a delay, say by late delivery of working drawings or tardiness in approving shop drawings, the contractor will normally be entitled to a time extension, and may also have a legitimate claim for extra compensation.

Excusable Third-Party Delays   Often there are delays caused by forces beyond either the owner's or the contractor's control. Examples that are normally unquestioned include fire, floods, earthquakes and other so-called "acts of God." Others that are sometimes subject to dispute include strikes, embargoes on freight, accidents, and reasonable delays in materials delivery. Excluded are conditions that existed at the time of bidding, and normal bad weather. Where agreed upon, these types of delays usually result in extensions of time but no additional compensation.

Contractor-Caused Delays   Such delays usually result in no extensions of time and no additional compensation. Indeed, in the extreme they can lead to breach of contract.

All these situations, even the fourth, if needed for the owner's defense, require accurate and equitable means of determining extensions to time and changes in compensation. We shall start by dissecting the direct and impact costs in more detail.

The Effects of Changes[6]

The effects of changes can be subdivided into three main categories:

1 Direct costs
2 Time extension
3 Impact costs

Even the first two can be difficult to assess, and impact costs are almost certain to provoke disagreement. It is worth exploring each in turn.

Direct Costs   All labor and all its overhead burdens, contractual and temporary materials, construction equipment, and even supervisory and staff time that can be clearly attributed to work associated with a change or delay constitute direct costs. *If* these costs are well documented, it is normally not too difficult to justify them in a claim. The main caution is to avoid settling the whole claim at this level without thoroughly analyzing the next two categories.

Time Extension   If a change can be shown to delay the completion date of a project, all parties to the contract will most probably incur additional expenses for the overhead associated with keeping the support staff and facilities for this extra time. Increasingly, this delay can also seriously increase the financing and escalation costs in a project. The problem is to verify the degree to which a delay in one or a few

---

[6]This section is based on an outstanding paper by Carroll J. Collins, "Impact—The Real Effect of Change Orders," *Transactions of the American Association of Cost Engineers*, San Francisco, June 21–24, 1970, pp. 188–191. His paper is strongly recommended reading for anyone involved in the construction process.

activities affects the project as a whole. In concept, the delay in an activity in a CPM network could be propagated through its successors to assess the effect. If there is ample float, there is no delay in the project. If not, then the amount of project delay can be directly computed. But this is often far too simple. For example, what if we had to complete the activity called "Divert river" before the spring runoff started? If successful, the project could continue as shown on the schedule. But a 1-month delay in this activity might result in overtopping a partially completed cofferdam and the loss of a full year's work. Nevertheless, networks and other types of scheduling tools can be useful in determining the effects of even these kinds of time extensions.

Impact Costs   The last example leads us indirectly to the area of impact costs. These costs are among the most difficult to define, let alone quantify, but they are very real and they can far exceed all others. Let us proceed with four increasingly accepted categories:

1 Acceleration
2 Job rhythm
3 Morale
4 Learning curve

Acceleration, or speeding up the project, is often a deliberate response when a delay in project completion cannot be tolerated. Methods here include (a) shift work, (b) overtime, and (c) increased crew sizes. None of these is as economical as the original plan, for reasons discussed in Chapters 10 and 11.

The impact on job rhythm is particularly severe on projects with a repetitive production cycle. For example, consider a high-rise reinforced-concrete building where flying forms are jumped on a 1-week cycle, and weekends are needed to satisfy the specified curing time. One day's loss can cause a week's delay, during which "morale" and the "learning curve" can also suffer. Similar situations occur on drill-and-blast tunnel jobs, where meal times and shift changes allow no-loss time to shoot and ventilate the fumes. Numerous other examples could be cited.

The relationship of morale to production is well understood in the armed services, but seems to make little headway as a factor in justifying claims in construction. But the fact is that construction workers and their supervisors sense pride in accomplishment just like anyone else—both in quantity of work in place and in the professional skill and efficiency that go into it. Changes requiring that existing work be extensively modified or torn out breed frustration, cynicism, and resentment. Doubts about the usefulness or permanence of one's work, whether conscious or unconscious, will most certainly reduce motivation, slow production, and drive up costs.

Learning curves were introduced in Chapter 10. The point here is that they do not happen automatically, but are a result of supervisors and workers who are determined to stay with a repetitive task and who strive to do it better. Recall from Figure 11-7d that there is an "unlearning curve" as well as a learning curve; interruptions in production also can only increase costs.

Assessing the Costs   This book is not really the proper forum for debating the relative merits of these various types of costs. There are volumes of legal precedents particularly in the areas of direct costs and time extension, and modern court decisions are increasingly recognizing some categories of impact costs. The problem is, however, that too frequently only the lawyers win in court; owners, engineers, and contractors—though they more and more frequently resort to litigation—are the losers. Ideally, as more commonly happened in the not too distant past, disputes should be settled among the parties to the contract. A reasonable, experienced resident engineer or a professional construction manager, trusted with sufficient authority by the owner, should be able to resolve most problems with a professional, experienced and reasonable contractor's project manager. But specific considerations for the new areas of cost, especially in the impact area, and the dissolution of authority have made this solution increasingly difficult. More and more emphasis is put on the importance of thorough documentation to arm the attorneys when they go to court. The fact is, however, that good documentation and analysis can short-circuit many such claims before they become a real issue. The following section describes a method that is valuable in either situation; its wider application could help bring back the times when the parties to the contract could "reason together" at the project level.

Factual Networks

Factual networks or their equivalent have been used for nearly two decades for the documentation of projects, but too often they have been prepared long after the fact—with the "facts" themselves muddled from poor memory and adversary objectives—for use in court. Ideally, they should be prepared as the project evolves, serving a purpose analogous to the "as-built" drawings. That is, they should document how things happen when they happen. Furthermore, they should be supplemented with good narrative documentation, and they should be cross-referenced to pertinent correspondence on file.

There are numerous special techniques for constructing such networks. Figure 12-46 shows a sample of the approach taken by Antill and Woodhead.[7] The networks can also be supplemented by other useful graphical methods. As an expert witness in litigation for a large underground powerhouse, John Fondahl of Stanford University developed a method for plotting accumulated delays versus time in a manner that clearly pinpointed causes and responsibilities for the delays. An example is shown in Figure 12-47.

SUMMARY

This chapter introduced several important methods and concepts for the planning, scheduling, and control of operations and resources on engineering and construction projects. The ideas discussed here are another major part of the overall system for the management of projects.

---

[7]James M. Antill, "Critical Path Evaluations of Construction Work Changes and Delays," *The Institution of Engineers, Australia Civil Engineering Transactions*, vol. 77, no. 1, April 1969, pp. 31–39. This paper also serves as the basis for chap. 11 in J. M. Antill and R. W. Woodhead, *Critical Path Methods in Construction Practice*, 2d ed., John Wiley & Sons, Inc., New York, 1970.

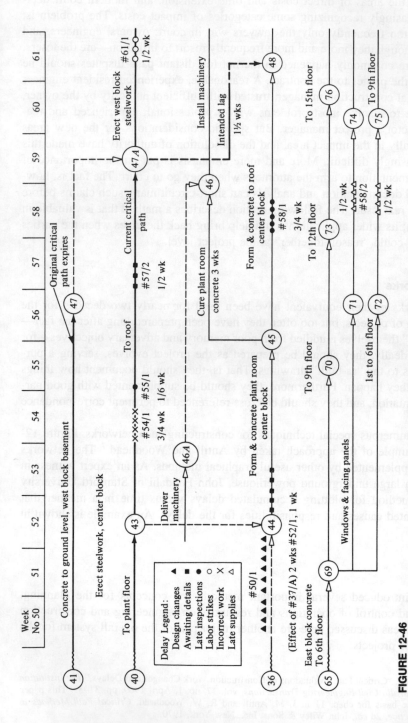

FIGURE 12-46
Portion of time-scaled factual network for building project. (From James M. Antill and Ronald W. Woodhead, *Critical Path Methods in Construction Practice.* John Wiley & Sons, Inc., New York, 1970, p. 282.)

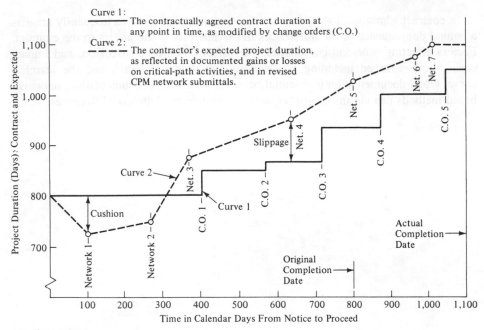

FIGURE 12-47
Plot of schedule delays versus time recognized.

Five basic scheduling tools include bar charts, progress curves, matrix schedules, linear balance charts, and critical path networks. Depending on the nature of the project and the needs and capabilities of its staff, each of these has advantages and disadvantages. Regardless of focus, any analytical tool for planning and control should be evaluated with respect to two main criteria. First, how well does it document the thinking of the planner; does it capture his intentions and the main constraints that influenced him? Second, how well does it communicate the planner's thinking to the people charged with the execution of the project? No matter how accurate and meticulously prepared, a plan is of no value if its language is foreign to its users.

Because of its generality and its power as an analytical tool, we focused in some depth on network-based project control. The success of critical path methods begins with clear, unambiguous definition of activities, involvement of users in preparation of the schedule, and the physical organization of the network itself. Once a good schedule-control standard has been developed, the main criterion for updating it should be how well it represents the actual operations on the project. When deviations from the schedule are detected, one should consider all main alternatives before applying corrective action. Where corrective action is warranted, the critical path concept can aid in its most effective application.

In contract administration, changes, delays, claims, and disputes frequently arise as a natural consequence of the differences in the interests of the parties to the contract. Factors affecting time and cost include the direct costs, time extension, and impact costs, the last-named including acceleration, job rhythm, morale, and the learning curve. Good documentation is essential here. As a means of documentation, network-based methods can aid in the just, equitable, and fair resolution of differences.

CHAPTER

CHAPTER **13**

COST ENGINEERING

Cost engineering, which is one of the key responsibilities of all members of the design-construction team, provides the analytical methods and procedures for monitoring, analyzing, forecasting, and, most important, controlling the costs on a construction project. Like estimating and schedule and resource control, it is but one of several highly interdependent parts of what is basically the same project planning and control system, often referred to as an integrated system. In scope, its concepts can be applied from conceptual planning, through engineering and design, to construction and start-up. In its effective application, all the principles of the feedback control system that were introduced in Chapter 10 are fundamental. These include consideration of nonlinearity in monitoring and measuring cost status, converting raw data to accurate information for analysis and reporting, the application and limitations of variances for exception reporting, forecasting and trending procedures, and the appropriate application or abstention from corrective action.

A good, definitive *control budget* is the basic document against which the cost engineer measures and compares actual progress. He also uses and thus must understand the project schedule and procurement documents. His reports to management must be accurate, current, and measured against a standard for them to have much value. In order to have effective cost control, the trends of the cost must be established as soon as possible, no matter how tenuous the indications, and be compared against both the planned and the actual progress so that management can take remedial action if required.

This chapter can only introduce some of the basic concepts of this large and diverse field. Recommendations for further study are given in the bibliography. Those who expect to be more deeply involved in this field should also seriously consider joining the American Association of Cost Engineers, and thus gain access to the publications, experience, and professional contacts available through that organization.

The topics to be introduced here include the development and application of a **work breakdown structure** (WBS), standard cost codes and project cost codes, conversion of a cost estimate to a control budget, guidelines for obtaining good input cost data, and the application of engineering economy in cost engineering.

COST CONTROL AND COST ENGINEERING

It is important that "cost engineering" not be confused with the financial accounting functions on a project. Certainly they are closely related, especially where cost engineering provides information for the general and cost ledgers and for payroll purposes, but there are major differences.

First, the word "engineering" in cost engineering is not mere window dressing. The reason for this is that in order to monitor and report costs properly, and especially to forecast trends, one must be able to read plans and specifications intelligently and must have a solid technical understanding of the work going on in the field.

A second major distinction is the emphasis on forecasting and trending in cost engineering. The accountant deals mainly in historical, documentable facts so that he can correctly pay the bills, make out invoices, prepare tax returns, compute the payroll, etc. Even the pennies count here. On the other hand, the cost engineer must often deal with and interpret some of the most tenuous information, including rumors and third-party knowledge to the effect that an item of materials might be late, or that there might be a jurisdictional dispute between crafts, in order to keep his forecasts as up to date as possible.

The cost engineer plays an important role in providing information and control systems that help toward the timely and profitable completion of a project. In this role, the engineer often represents the interface between all the other key figures in the project, including management, line supervisors, scheduling, procurement, accounting, and field engineers.

WORK BREAKDOWN STRUCTURE

A WBS describes the work elements of a project in a logical hierarchy which can be used for a number of related management control activities. Integrated management control systems are designed to reflect the interdependence upon cost, schedule, and other parameters. A common parameter between the cost and schedule portions of the system is reflected in worker hours. For planning purposes, the hours of effort reflected by the cost estimate are also identical to hours of effort indicated by the resource loading charts developed by the scheduler. The WBS is designed to set forth a common numbering system that can be applied at various levels to structure both cost and schedule planning and reporting. Some organizations also prepare the WBS to further indicate individual or departmental responsibility and accountability, to organize drawings and specifications and other applications. An important feature of the WBS is the ability to identify both cost and schedule parameters at various levels of detail. Figure 13-1 shows a typical WBS for a major project. Home office and owner management can follow cost and schedule performance at the higher levels

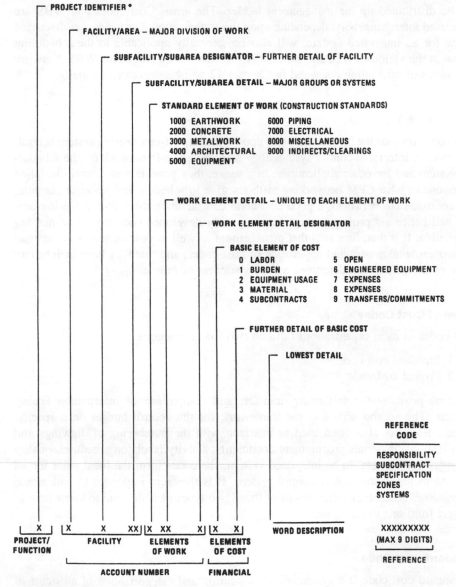

CLASSIFICATION OF ACCOUNTS
BASIC NUMBERING FORMAT

PROJECT IDENTIFIER *

FACILITY/AREA – MAJOR DIVISION OF WORK

SUBFACILITY/SUBAREA DESIGNATOR – FURTHER DETAIL OF FACILITY

SUBFACILITY/SUBAREA DETAIL – MAJOR GROUPS OR SYSTEMS

STANDARD ELEMENT OF WORK (CONSTRUCTION STANDARDS)

1000	EARTHWORK	6000	PIPING
2000	CONCRETE	7000	ELECTRICAL
3000	METALWORK	8000	MISCELLANEOUS
4000	ARCHITECTURAL	9000	INDIRECTS/CLEARINGS
5000	EQUIPMENT		

WORK ELEMENT DETAIL – UNIQUE TO EACH ELEMENT OF WORK

WORK ELEMENT DETAIL DESIGNATOR

BASIC ELEMENT OF COST

0	LABOR	5	OPEN
1	BURDEN	6	ENGINEERED EQUIPMENT
2	EQUIPMENT USAGE	7	EXPENSES
3	MATERIAL	8	EXPENSES
4	SUBCONTRACTS	9	TRANSFERS/COMMITMENTS

FURTHER DETAIL OF BASIC COST

LOWEST DETAIL

REFERENCE
CODE

RESPONSIBILITY
SUBCONTRACT
SPECIFICATION
ZONES
SYSTEMS

X	X	X	XX	X	XX	X	X	X	WORD DESCRIPTION	XXXXXXXX

PROJECT/
FUNCTION

FACILITY

ELEMENTS
OF WORK

ELEMENTS
OF COST

ACCOUNT NUMBER

FINANCIAL

(MAX 9 DIGITS)

REFERENCE

* PROJECT IDENTIFIER IS USED ONLY ON MULTIPROJECT
PROGRAMS AND IS THEREFORE OPTIONAL.

FIGURE 13-1
Work breakdown structure for a major project.

while individual area and craft superintendents can be appraised of detail performance in their sphere of responsibility at the lower levels. With the use of the computer, full detail can be made available at lower levels and fully consistent summary information can be distributed up the management ladder. The terms Cost codes and WBS are often used interchangeably, depending upon the organization. Cost codes, as discussed below for an integrated system, will also be generally applicable to the scheduling format at the various levels, and are usually referred to as a part of the WBS. Network and bar chart scheduling discussed in Chapter 12 will be structured similarly.

COST CODES

Cost codes provide the basic framework upon which a cost-engineering system is built. In a modern integrated control system, they also provide a framework for the scheduling system and for other applications. In a sense, they provide a structural discipline analogous to what CPM networking methods give to schedule and resource planning and control. Like networks, a good cost code can facilitate the cost-engineering process and hence aid project management; a poorly developed code can cause nothing but trouble. It is therefore vital that management as well as cost engineers understand the proper development and application of cost codes, and that they have insight into some of the details that determine a code's success or failure.

Types of Cost Codes

Cost codes in most organizations fall into two major categories:

1 Standard cost code
2 Project cost code

The first provides for uniformity, transfer, and comparison of information among projects. The second serves as the framework for the control budget on a specific project. Both are also often used to interface with the numbering of drawings and specifications, materials procurement documents, activity labels on schedules, quality assurance reports, etc. In an integrated system, the codes form the focal point for all these elements in the project control system. It is therefore important to understand the purpose, content, and differences of these two types of codes and to know how to convert from one to the other.

Standard Cost Code

A standard cost code is a systematic classification and categorization of all items of work or cost pertaining to a particular *type of work*. There may be different standard codes for different types of work, even within the same organization. In fact, one of the major problems in development of a cost code for a project results from forcing a standard cost code for one type of work to be used in another.

Some examples of different types of work that might each have its own standard cost code are the following:

General building construction (offices, schools, warehouses, etc.)
Thermal power plants (both nuclear and fossil-fueled)
Heavy engineering projects (dams, levees, hydroelectric schemes, etc.)
Process plants (oil refineries, petrochemical plants, etc.)

In some sectors of construction, there are more or less widely accepted industry standard cost codes. The best-known example is the *Masterformat* published by the Construction Specification Institute,[1] reproduced in Appendix D, which was developed through a joint effort of eight industry and professional associations.[2] It is primarily designed for building construction.

The *Masterformat* is actually a WBS. It also provides a standard for numbering and classifying sections of plans and specifications, a standard system for manufacturers to catalog literature about construction materials and equipment, and an office filing system for a contractor's own documents and correspondence on projects. It is thus possible to organize both estimate and schedule by the same numbering system which can then be used for the control budget and schedule, to look up technical data on materials for estimating or procurement, to locate sections of the plans and specifications needed by various specialty subcontractors, and to keep all correspondence with the architect, vendors, and subcontractors filed the same way. It is not a perfect system, but it is a good one for building construction and has thus been widely adopted. Unfortunately, there is little to compare with it in other sectors of the construction industry. The Nuclear Regulatory Commission imposed one system on the nuclear power industry, and there have been attempts at other industry standards, but none has been nearly as comprehensive or widely adopted as the *Masterformat*.

Each major design-construct company specializing in industrial work has developed its own WBS based principally upon the requirements peculiar to its type of work. These companies were the first to pioneer the integrated control system featuring a WBS that was integrated between costs, schedules and other parameters. Standards among different companies are similar, but each company has developed its own proprietary WBS and applicable cost codes.

Heavy construction companies have developed similar work breakdown structures designed to permit system output in accordance with the unit price bid schedule as well as to provide additional detail and summaries to help manage and control critical items in substantially more detail.

In developing a standard cost code, it is appropriate to create a relatively exhaustive checklist of all the items that might be found in its generic type of construction. For example, the *Masterformat* lists items as diverse as "Navigation Equipment," "Cementitious Decks and Toppings," "Stone Facing," "Shingles and Roofing Tiles," "Entrances and Storefronts," "Library Equipment," "Flagpoles," "Ecclesiastical Equipment," "Ice

---

[1]*Masterformat*, The Construction Specifications Institute, 601 Madison, Alexandria, VA 22314-1791, 1972.

[2]The American Institute of Architects, the Associated General Contractors of America, Inc., The Construction Specifications Institute, the Consulting Engineers Council of the United States, the Council of Mechanical Specialty Contracting Industries, Inc., The Professional Engineers in Private Practice/National Society of Professional Engineers, The Producers Council, Inc., and the Specification Writers Association of Canada.

Rinks," "Detention Equipment," "Fire Protection," and "Electrical Resistance Heating." No one building would require more than a few items even from this subset, but in constructing many buildings over a period of several years, a diversified contractor might encounter most of them. If they should crop up, the standard cost code provides a good checklist to help prevent such items from being overlooked in the estimate, budget, and materials procurement schedule.

Project Cost Code   A project cost code is a systematic classification and categorization of all items of work or cost pertaining to a particular *project*. There is normally a different project cost code for each project, but each should be derived from the standard cost code so that different projects can be compared, and especially so that meaningful information can be maintained for estimating purposes. It is also important that the project code be prepared as soon as possible after the project is authorized so that costs from the very beginning can be accurately distributed.

A project code, however, is adapted to incorporate the particular features and characteristics of the specific project. It thus contains some components not found in the standard code, and it deletes anything not required for the job at hand. In contrast with the exhaustive checklist desirable for the standard code, the project cost code is a day-by-day working document; for practical purposes it must be kept as concise and simple as possible in keeping with the objectives for planning, documentation, and control. A good rule of thumb in selecting items for the project code is, "If in doubt, delete it."

Major design-construction companies and many design engineers will first prepare an estimate and schedule at the conceptual level during the feasibility stage. With an orderly WBS, estimates, schedules and preliminary budgets can be developed conceptually at high levels. As additional detailed information becomes available, new estimates are prepared and directly compared with the preliminary budgets to facilitate management control in a manner as illustrated for the Mountaintown Warehouse Project. Figure 13-1 illustrates a WBS suitable for this type of management control from conceptual planning through actual construction.

Once developed, the project cost code may be used for referencing and documenting items such as the following:

Expenditures and commitments for:
   Labor
   Materials
   Equipment
   Subcontracts
   Indirect costs
Procurement documents, such as:
   Requisitions
   Purchase orders
   Receiving slips
   Invoices

The project cost code serves as an interface among all parties involved in the administration and supervision of the project, as indicated in Figure 13-2.

FIGURE 13-2
Project cost code: A means for communication.

Additional features of the project code will become more clear if we first show how it is derived from the standard cost code.

Deriving a Project Cost Code

The item code taken from the standard cost code becomes just one part of the code for a typical project code item. This part we will refer to as the "work-type code." In addition to this, it is common to add a "project number," an "area-facility code," and a "distribution code" to form the complete code for a particular item in the project code. An organization will normally have standards specifying how these additional elements are formed, but the actual codes, at least for the project number and area facilities, will be unique to the project at hand. Figure 13-3 illustrates how the conversion from standard code to project code takes place.

The following section shows how each of the four elements is derived.

Project Number The project number is chosen to identify the costs collected for this code specifically with the particular project from which they came. The number will often be a shorthand notation in itself that indicates such things as the type of project (e.g., H = heavy construction), the type of contract (e.g., L = lump sum), the year it was started, its sequence with other projects started that year, and possibly its location. In this way, if the home-office estimator is trying to figure out why one set of costs for, say, concrete footings differs so much from another set, the project

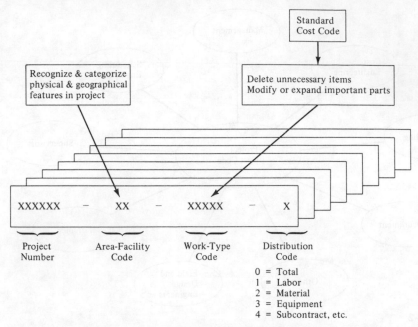

FIGURE 13-3
Developing project code from standard code.

numbers, which may show that one set was from an arctic project and the other from Southern California, help give the reason for the discrepancy.

The project number is normally implicit rather than explicit in the day-to-day project reports that use the cost code. For example, rather than include the extra digits for the project number on each line item reported, it is normally printed just in the heading of the report. Similarly, in computer files using magnetic media, the project number will normally appear only at the beginning of the file and not in each cost record.

Area-Facility Code Often in a project, there are certain distinct geographic and physical features that logically separate one part of a project from another. Often the management of a project is also structured according to these physical or technological features. For example, a hydroelectric project might logically be separated into "dam," "tunnel," "shaft," "penstock," "powerhouse," "transmission lines," and "support facilities." These major breakdowns we shall refer to as "areas." Within each area, there are normally logical subdivisions which we shall refer to as "facilities." For example, the powerhouse might have four separate turbine-generator units, each of which might be considered a "facility." Further breakdown into intake valves, scrollcase, turbine, drive shaft, generator, transformer, controls, etc., would normally be handled in the work-type code, which is based on the standard code, rather than in the area-facility code. These are characteristics common to almost all hydroelectric turbines, whereas the fact that there are four such units is unique to this project. It is important not to confuse the application of the area-facility code with the work-type code.

Once derived, the area-facility code helps keep track of costs in the different areas, and it can isolate the costs attributable to the different managers and supervisors on the project. The work-type code, especially where, like the *Masterformat*, it is categorized along recognized trade and subcontractor specialties, also helps in identifying costs with the parties responsible for them.

Work-Type Code   The work-type code is the part that is based on the standard cost code. However, as indicated in Figure 13-2, considerable thought should go into the conversion. Clearly, one should start by deleting all items that have no relationship to the project at hand. For example, all mortuary and ecclesiastical equipment would normally be deleted from a code for a public school building. But the process should go deeper than this. For example, is the level of *detail* in the standard code appropriate to all items in the project? To illustrate, if the only *Masterformat* code 5 metalwork on a precast concrete[3] building is a stair railing, one might simply collect this cost at the level of the major breakdown. On the other hand, the standard cost code may not have enough detail for the other parts of the project. For example, if this is a four-story precast-prestressed-concrete parking garage, code 03400 may need to be subdivided into more components to record accurately the costs of the elements involved. With these considerations, the work-type components of the project cost codes are developed.

Distribution Code   The *Masterformat*, like many standard codes, does not separately break out the resource components of the various types of work, such as separating out the labor, materials, equipment, and subcontract costs. In practice, however, such a breakdown is always important to contractors and is becoming increasingly important to owners, design-constructors and engineers to facilitate preparation of reliable cost estimates and schedules; such breakdowns are normally handled with an element of cost or "distribution code" such as that shown on Figures 13-1 and 13-3. In this way the costs associated with labor can interface with the payroll system, the materials costs can tie into the procurement schedule and the accounts payable, equipment costs can be internally accrued or paid to rental agencies, subcontractor costs can go into accounts payable, and all components can aid in the development of more accurate estimates to complete and cash-flow forecasts. Normally there will be a company standard for the distribution codes, so that, for example, "1" is always labor, "2" is material, "3" is equipment, "4" is subcontracts, etc.

Example   Figure 13-4 gives an example of a code for a particular item in a project cost code. Here the project code 88NB04 means that this is the fourth (04) negotiated (N) building construction (B) job started in 1988 (88). The area-facility breakdown, assuming this is a typical high-rise building, is chosen to give a separate "area" to each floor, with no further breakdown to "facilities" on the floor. The code "11" here

---

[3]Note that *Masterformat* reinforcing steel is in code 03.

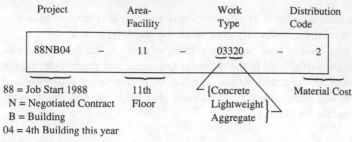

Project	Area-Facility	Work Type	Distribution Code
88NB04 –	11 –	03320 –	2

88 = Job Start 1988   11th   {Concrete   Material Cost
N = Negotiated Contract   Floor   Lightweight}
B = Building   Aggregate}
04 = 4th Building this year

FIGURE 13-4
Example project code.

refers to work on the eleventh floor. The work-type code, "03320," indicates that the costs are for placing lightweight-aggregate concrete. Distribution code "2" means that in this code we are collecting the material cost.

Account Hierarchy

Normally the desired level of detail in a cost code is reflected in the account hierarchy. This concept can apply in a standard cost code and in each of the four or more elements that make up a typical project cost code. Terminology commonly encountered here includes "prime account" and "subaccount." Figure 13-1 illustrates this concept through the use of eight levels of accumulation as is common for major industrial or heavy construction projects or programs.

Prime Account The "prime account" is the highest level of enumeration in a cost code. For example, in the *Masterformat*, the first two digits of each code form the prime account. "Concrete," "Wood and Plastics," and "Electrical" are represented by prime accounts "03," "06," and "16" respectively.

Subaccounts Subdivisions for greater detail within prime accounts are referred to as "subaccounts" and these in turn can have their own subhierarchies. For example, at the next level down within "03 Concrete," we have "031—Concrete Formwork," "032—Concrete Reinforcement," "033—Cast-in-place Concrete," "034—Precast Concrete," etc. These in turn can be subdivided to the desired level of detail.

Applications The account or WBS hierarchy has several important applications. First, as mentioned above, the level of detail of the code can be set at the level appropriate to the scope of work that is represented. For example, if concrete is a relatively minor part of a structure, consisting of a few miscellaneous footings for a large but temporary asphalt batch plant, one might appropriately collect all costs at the prime-account level, rather than go to the trouble and expense of breaking out the cement, aggregates, mixing, placing, labor, materials, etc.; it might be sufficient for future estimates to know that it cost $92 per cubic meter to construct the footings for this type of batch plant at this location in this year.

The account hierarchy is also useful for tuning the level of detail of reporting to the appropriate levels of management. For example, the concrete superintendent might want the full detail for the concrete accounts only. On the other hand, the project manager might, as a routine matter, look only to the major areas, the prime accounts, and the first level of subaccounts, but for all the prime-account divisions. If a particular item at that level of detail appeared to be deviating significantly from the planned budget, he or she could then request additional detail on that item.

These capabilities for summary reporting, exception reporting, and other techniques for getting the right information at the appropriate level of detail to the right person for decisions in time for corrective action, are made possible by a well-designed account hierarchy and structure.

Direct versus Indirect Costs

In general terms, *direct costs* are those which can be immediately associated in the field with work directly contributing to the physical completion of the permanent facility contracted for by the owner. Examples include (1) finishing labor for a concrete floor slab, (2) materials for a structural steel frame, (3) equipment for a foundation excavation, and (4) a subcontractor's charges for installing the air-conditioning system.

Conversely, *indirect costs* are those which necessarily contribute to the support of a project as a whole, but cannot be identified directly with specific work items in the permanent facility. Examples commonly include (1) job and office personnel salaries; (2) materials, supplies, and utilities for the temporary warehouse, field office building and change facilities; (3) staff vehicles; (4) safety and first-aid expenses; and (5) the portion of home-office support required for the project. The *Masterformat* puts most of these in prime accounts 00 and 01 with details geared to the designer's responsibilities during construction. Contractors, on the other hand, will develop their own individual cost accounts for indirect costs which can be as detailed as the complexities of the particular project require.

Although these are commonly accepted meanings of "direct" and "indirect" costs, the meanings are by no means uniformly interpreted. For example, in many negotiated contracts, including those involving the professional construction management approach, "direct" costs include those which are specifically billed to the owner, and those often comprise, for example, the salaries and benefits paid to the project manager and the field staff. "Indirects" in this context are those which can be recovered only in the negotiated general "overhead" figure, or in the profit for the contract.

Similarly, even in the traditional meanings as they were first introduced, there is no real uniformity in identifying exactly what is "direct" and what is "indirect." For example, although the wages of given trades might be identified specifically with items in the field, what about the "payroll burdens" such as union fringe benefits, payroll insurance, and social security payments? Similarly, if it would take excessive and meaningless effort to state specifically what an item of support equipment was doing at any given time, such as a hoist for a high-rise building that serves many different trades and subcontractors, should that item be directly charged? Or should it be in the indirect costs?

Actual practice varies widely from one company to another in questions of this type. Only two general guidelines can thus be offered here. First, such decisions should be thoroughly studied and deliberated, rather than be made haphazardly, and should recognize the special needs of the organization and its projects. Second, once these decisions are made, they should be documented and incorporated as standard policy, and they should be applied uniformly from one project to another. Exceptions should be thoroughly justified and should be approved by higher management. Otherwise, comparisons of costs and hourly production units from one project to another will be meaningless.

Practical Considerations

There are numerous practical considerations, such as those that were just mentioned, that can make all the difference in the success or failure of a cost code and the cost-engineering system that builds upon it. This section will further discuss these and add several more examples to illustrate the in-depth thinking that should go into the development of a cost code. This is by no means intended to be an exhaustive or balanced set of guidelines, but merely a set of representative examples. Some of the examples are fairly subtle, and to the novice reader some may at first even seem trivial. But in the authors' experience, each of these examples, at one time or another, has been an item of controversy that had potential adverse effects on one or more major projects. The experienced reader will no doubt appreciate this, and the novice will understand in time. Codes which arbitrarily ignore such considerations will be difficult if not impossible to live with in the field.

Tracking Bulk Materials   From the time they are identified in design or in the plans and specifications, through formal requisition, solicitation, purchase orders, shop-drawing or sample approval, vendor fabrication, shipping, delivery, inspection, storage, and installation in the permanent structure, it is necessary to keep track of all major and minor items of materials and equipment required on a project. Failure in this procurement process is one of the most common causes of delays and cost overruns in construction. The process can be greatly improved through a well thought and intelligently applied project code.

In some items it is indeed appropriate to account for costs at the level of the specific area-facility and detailed work-type codes in each of the above steps. For example, a large diesel-driven pump for an oil pipeline might have its final cost code attached to it when it is first requisitioned. This code will then be cross-referenced to equipment serial numbers in the factory and be tracked at this level until it is installed and tested on site. Similar guidelines will generally apply to items such as prefabricated structural steel and bent-and-tagged reinforcing steel.

In terms of bulk materials, however, such as raw stocks of lumber, unfabricated reinforcing steel, and small pipe, this level of detail is seldom desirable and generally is not even feasible until specific quantities are actually fabricated and installed in the structure in the final step. How, then, should one account for these items?

To illustrate, consider an extreme but too frequently attempted case. Globe valves for pipe are going to be required in 10 areas, in 4 work-type codes, and in 6 different sizes. Simple multiplication shows that the total number of detail accounts may potentially number 240. This would mean at least 240 arithmetic extensions and a minimum of 240 lines to check on purchase orders, invoices, receiving reports, etc. If this concept were carried through to all bulk items, the cost code would have created a bureaucratic nightmare requiring an army of clerks to handle the detail—or more probably, the system would be ignored altogether, with most items being classified as "miscellaneous."

These same concepts for materials procurement apply to the field labor associated with the materials until they are actually used. For example, it would be unrealistic to use area and detail accounts for labor required in unloading and warehousing materials when they are delivered to the site.

How, then, does one avoid this unnecessary detail? The usual procedure is to create a *suspense account*, also called a *holding account* or *clearing account*, to keep track of all costs associated with the materials up to the point where they are drawn for use in the project. In the example of the globe valves, there might only be one suspense account for them, or possibly one for each of the six sizes. All costs incurred during procurement, receiving, and storage on site would be accumulated at this level. Then, when a valve is actually drawn for installation in a particular area for a particular work type, the suspense account could be *relieved*, that is, decreased, to charge the area and detail account for the cost of the valve, and possibly for a pro rata share of the accumulated procurement and handling expenses incurred to that point. If all valves were used on the project, suspense accounts could feasibly be reduced to $0.

Control Costs Where Incurred   Problems sometimes arise when a cost code attempts to account for costs at some point where they are not really incurred. Consider, for example, prefabricated wooden forms. If all costs for prefabrication are charged directly to the area and detail accounts where the forms are actually used, but not made, one may be holding the concrete foreman responsible for costs over which he has no real control. The costs of fabrication were incurred in the carpentry shop, not at the point where the forms were erected for the pour.

The same idea applies to ready-mix concrete, on-site batch plants, and shops for fabrication of reinforcing steel, ventilation pipe, and like items. It is desirable to neutralize the impact that cost variances in these shops will have on the operations where their products are actually used. To do this, it is common to charge a constant rate per unit for the fabricated item at all points where such items are used, and not try to pass on daily, weekly, or monthly fluctuations in actual prefabrication costs to the point of use. If required, any net cost overruns or underruns can later be prorated to the points of use, once all related operations have been completed. Many firms periodically review the charge-out rate and increase or decrease the charges as construction proceeds to insure compatibility with client billings and to comply with accounting system requirements.

Alternatively, all costs of fabrication could be retained in the shop accounts, with no materials costs allocated to the point of use. This, however, would not put the responsibility for waste and damage where it belongs. In any case, attempts to pass on variations in prefabrication costs will not only create an unnecessarily complex documentation burden; it will also cause considerable frustration in line supervisory personnel if they are held accountable for costs over which they have no real control. Progressive firms will establish shop fabrication clearing accounts by element of cost to preserve element integrity and to facilitate shop control. Costs are cleared to the direct account at estimated charge-out rates through a material code.

Code for Cost Impact One common problem is to confuse engineering design requirements with factors that actually affect construction costs. For example, consider piping. Factors that do have a real impact on construction costs include size ($1\frac{1}{2}''$ and below, $2''$ to $4''$, etc.), whether the piping is on-site, or off-site, materials (carbon steel, alloy, etc.), and height of installation (below $15'$, $15'$ to $25'$ etc.). It serves no real purpose from the installation contractor's point of view[4] to segregate piping costs into "project," "process," "service," and "utility," for example. These are flowsheet and process definitions, not construction *cost* factors. However, such items can be very important to owners, engineers and design-constructors in the development of realistic cost estimates and budgets.

On the other hand contractor piping takeoffs by most mechanical contractors are usually performed by systems grouping on complex projects for ease in specification identification. Owners may request contractor quotations by system in order to compare to conceptual budget estimates which are normally prepared by individual system. Design-constructors usually estimate conceptually by process component, and feedback from actual projects helps to improve estimating accuracy and consistency. Utilization of a hierarchy of accounts as shown in Figure 13-1 permits both conceptual estimating and feedback at higher levels as well as detailed estimating, cost accumulation and full schedule integration at the detail direct work level element.

Prior to the widespread use of computers, industrial designers preferred to estimate by system while contractors preferred to accumulate costs by size and type of work, and one or the other method was chosen often to the detriment of project control. An intelligently designed WBS for computer utilization will identify both the process code and the size or material code in a single account number. The costs can then be identified by size and material specification for field installation control, and can also be identified by process for comparison to initial budgets and for schedule control where completion dates and milestones are expressed as system completion. Accumulated costs expressed in worker hours by system as well as by size and specification grouping are also important for continuing feedback for use in conceptual estimating for future projects.

Coding Direct Labor As a general rule, direct labor costs should be identified, to the extent possible, with the items where they are incurred. One can create real problems, however, when one tries to use the actual monetary costs associated with direct labor for control purposes. Hence the widespread use of installation costs expressed in unit hours for control purposes.

---

[4]Though it might to the owner. More on this later.

Especially on longer projects, estimates and budgets are prepared without exact knowledge as to what wages, fringe benefits, payroll-based insurance, and other related burdens will actually be when the item in question is eventually built. Wage rates often take several hikes during the course of a project. Social security and state disability insurance premiums can reach a cutoff point during each calendar year for a given worker who stays on the job long enough, and the next week's cost report can show reduced unit costs for no other reason than this. Workmen's compensation insurance premiums are subject to readjustment and rebates a year or more after the time they were actually incurred.

Most of these payroll details have little or nothing to do with controlling productivity on operations in progress. To impose them on a supervisor responsible for maintaining and improving production unnecessarily adds confusion to an already tough situation; yet many "cost-control" systems continue to operate this way.

Two means are commonly used to put the focus on production control. One is to budget, monitor, and control labor in terms of worker-hours, not monetary units. Increasingly, progressive firms will develop worker hourly rates in the estimate which include all labor related fringe benefits, payroll taxes, insurance etc. These rates are generally fully known or are subject to reasonable estimates in advance of bid time. Through the use of clearing accounts, a single charge-out rate can be modified to reflect actual conditions, on a periodic basis, as has been common practice on heavy construction for many years. As a part of estimate preparation, a computation of the anticipated escalation is also prepared and set up in the budget as a line item of indirect cost. As actual hourly rate increases are experienced, the cost engineer will estimate future costs to complete accordingly and reduce estimated future escalation costs in a similar manner. Therefore, any reduction in the escalation accounts will be offset by increases in the direct labor accounts. Actual comparison for control purposes and productivity calculations, of course, is based upon worker hours and all direct cost labor figures are based upon actual costs to track with payroll records. Escalation for materials and equipment costs can be handled in a similar manner.

Construction Equipment Costs As with labor, costs of construction equipment should theoretically be identified directly with operations on which they are incurred. In large, ongoing heavy construction operations this is often quite feasible and is the prevailing method of operation. Problems arise, however, on industrial and building work, especially when trying to charge off the ownership costs (depreciation, interest, taxes, insurance, etc.), and when equipment concurrently serves many different area and detail accounts.

Ownership costs are largely a function of annual utilization of equipment, its expected life, and the time value of money. Where the equipment can be identified with specific operations over long periods of time, or where the equipment is rented or leased, these ownership costs can be determined fairly accurately. Difficulties arise, however, when equipment is frequently moved from operation to operation and project to project. This is a large and complex field that can only be mentioned here. Those interested should refer to a book on this subject by James Douglas.[5]

---

[5]*Construction Equipment Policy*, McGraw-Hill Book Company, Inc., New York, 1975.

The second problem area, equipment that simultaneously serves several operations, occurs with machines, such as cranes and hoists, and central compressor and generating stations. For example, on a high-rise building, a crane might set a column on the top floor, deliver decking two floors down, assist in a concrete pour below that, help install some mechanical equipment still furthur down, etc. The same is often true of cranes on industrial construction projects. These situations can be handled by putting such equipment and its related nonlabor costs as separate cost centers in the indirect category. Other companies develop an all-cost rate for the crane utilizing a clearing account exactly as previously described to allocate crane nonlabor costs. With some exceptions, most progressive companies will code the crane operational labor costs directly to direct labor accounts in order to preserve the integrity of the account and associated worker hours for both cost and schedule control. Equipment usage (nonlabor costs) can be accumulated as a line item and may be distributed to individual accounts through a clearing account, if desired.

Avoid Imprecision and Ambiguity   Imprecision and ambiguity in the written descriptions for the codes are two of the most common sources of confusion and error in cost engineering. Paperwork as a rule is anathema to most field construction people, and they have little patience to figure out what was meant by a vague code description. Given a choice between two or more codes where an item might be classified, they are more likely to choose at random than to call for clarification. Along this line, many knowledgeable cost engineers strongly recommend against using codes for "miscellaneous" or "other." Experience has taught them that if field line supervisors are left to code their work, somewhere on the order of half the costs in the project is "miscellaneous," and the bulk of the remainder becomes "other"; something is lacking here for control or for future estimating!

Human Considerations   Several of the suggestions that have been made thus far relate to the needs and limitations of the human beings involved in the cost-engineering process. Two other examples, both from one of the largest heavy industrial engineering and construction companies, will further illustrate this need.

This company's cost-engineering department decided that one way to help their field superintendents accurately code the time sheets for work accomplished each day would be to print the code, plus a small plot plan showing the major areas and facilities, in a pocket-sized notebook. Unfortunately, the first edition failed in its purpose. The reason? The notebooks sent by the printer were approximately $\frac{1}{4}$-inch wider than the standard size pocket, and were therefore kept in the office.

Once the new, correctly sized notebooks arrived, the concept began to work as planned, and the accuracy of coding improved markedly. But, over time, another problem developed. On large projects of long duration, there is considerable turnover even of supervisory personnel. Also, as with drawings, specifications, and other works of engineers, there were successive revisions and new editions of the cost-code notebooks. The problem that arose was that in time, possession of older editions of the notebooks began to be associated with seniority and its related stature on the project. Consequently, several versions of the cost codes were being reported.

Each of these short examples may seem inconsequential in and of themselves. But in practice they are exactly the sorts of things that determine whether or not a system will be a success. In spite of twentieth-century computer automation, we must still make allowances for human idiosyncrasies, and we are better off for it.

Concluding Guidelines

Briefly stated, project cost codes must be simple, clean, concise, and easy to interpret. Use the standard cost code as a guide only. Do not incorporate its exhaustive detail. Ask only for essential and readily identifiable units of work. Area-facility breakdown should be logical and practical, and should never be changed in the middle of a project. Such a change destroys all chance of consistency in cost records, even though a new breakdown might later seem to suit the project better. Rarely is a project cost code made too simple. But far too often they are bogged down with responses to individual requests of, "Wouldn't it be nice if we had the costs of ...," or, "Although this isn't really needed on this project, someday it might be useful to have the costs of ...," etc. Like the travel itinerary planned to 20-foot intervals in Chapter 10, this type of code only gets in the way and causes problems once the project is underway. As a rule, if it doesn't add, it detracts. If an item cannot be coded easily and accurately by hard-pressed field personnel, forget it! Bad information is worse than none at all.

CONTROL BUDGETS

The control budget is the basic reference standard for monitoring and controlling cost status on a project. The structure for the control budget is the project cost code. Its standards for reference are derived from the cost estimate and include the quantities associated with each item of work in the code. In addition, the control budget usually makes provision for recording and reporting the following:

Actual performance, to-date and this period
Projections or forecasts to completion
Variances in absolute and/or relative terms
Reasons or conditions associated with excessive variances

Typical control budget report formats were shown in Figures 8-2, 8-3, 8-4 and 10-3. Figure 13-5 shows a simple format based on the example warehouse project. In creating such reports, there is always a trade-off between the amount of information to supply and the danger of confusing or inundating the user. Again, simplicity and consistency are the best guidelines.

For tight project control, it is not enough simply to take figures from the estimate and plug them into the cost codes. Some of the reasons for this were given in Chapter 11 in the section on converting an estimate to a control budget. First of all, for any number of good reasons, the estimate might not even have been done against the standard-cost-code categories. For example, where the contract bid documents specify a unit-cost proposal with a schedule of quantities, these may serve as the dominant breakdown, and will certainly have to be accommodated for payment

Division:	WestAmerica	Date: 8-1-8–	Project No. 7625
Facility	Dry Storage Warehouse		Prepared by R.A.G.
Location	Mountaintown, WestAmerica		Approved by G.L.R.

Description of project (attach separate schedules and drawings as required).

150,000 square foot Dry Storage Warehouse. Pallet and flow racks required. See attached drawings Nos. G-1, G-2, and G-3

	Allocation	Amount Requested	Total
1.	Architect/engineer		$ 260,000
2.	Site work	$ 798,000	
3.	Buildings and utilities	4,000,000	
4.	Repair and maintenance		
5.	Other owner's cost (CM)	372,000	
6.	Subtotal (items 2–5)	5,170,000	
7.	Contingency	430,000	
8.	Total building cost		5,600,000
9.	Operating and process equip.	360,000	
10.	Repair and maintenance		
11.	Other owner's cost	140,000	
12.	Subtotal equipment	500,000	
13.	Contingency	40,000	
14.	Total equipment cost		540,000
15.	Total land cost		100,000
16.	Total estimated cost		$6,500,000

FIGURE 13-5
Example project budget.

and often progress purposes in the course of a project. There must then at least be a conversion and redistribution process to get the costs into their appropriate project code categories. Similarly, owners, for purposes of capital depreciation and taxation, will often impose a code on a contractor that is of little value for construction cost control. Some government agencies, such as the Nuclear Regulatory Commission, also can impose a separate code on the contractor. This often requires the contractor to maintain project costs in two or three entirely different code classifications. With good computer programming, this is not as bad as it sounds, since the second and third distributions can be largely automated. Needless to say, however, without such automation it can be a huge paperwork burden.

Another important concept was implied in the discussion of learning curves in Chapter 10, and was amplified in Chapter 11. Estimates necessarily must be based on averages. However, if one controls only against averages, one might expect average results. Figure 11-14 gave a graphical expression of this idea. Some people believe it desirable to control against tighter standards, say the 20th percentile, assuming the concept is appropriately and intelligently applied, with an open and positive, rather than a punitive, managerial approach. Others believe that field supervision should be given the actual estimated unit work hours for control purposes and be encouraged to achieve superior performance through monetary or nonmonetary incentives.

SOURCES OF DATA FOR COST CONTROL

No control system, whether computerized or not, and regardless of the skill of its developers, is of any value without accurate, timely input data. In field cost control, this especially requires good data for materials, equipment, and labor. The last two are particularly important, since these are the resources whose productivity and costs can change most rapidly and are thus the ones over which a contractor has most control. As discussed in Chapter 10, regarding the "level of influence" concept, the die is largely cast for materials and permanent equipment costs by the time construction starts, and the feedback time is not nearly so critical.

The main sources of data for field cost control are (1) labor and equipment time sheets; (2) field surveys of quantities of work in place; (3) any other fragments of information that will assist in forecasting cost trends; and (4) data obtained from other parts of the project control system, including scheduling, procurement, and quality assurance. Each of these is important, and comparisons between these various measures are essential to evaluate project status satisfactorily.

The first two sources of data, time sheets and field quantity surveys, are the most basic sources of data for routine cost reports and will be discussed in more detail here. They are also important, of course, as sources of data for scheduling, procurement, and payroll accounting, but often they are the primary responsibility of the cost engineer.

Time Sheets

Labor and equipment time sheets are usually filled out daily and submitted either daily or weekly by foremen, operators, superintendents, or timekeepers, depending on company policy. They normally contain the following information:

Labor Time Sheets	*Equipment Time Sheets*
Employee name(s) and/or number(s)	Machine description(s) and/or number(s)
Date(s) worked	Date(s) worked
Craft or classification(s)	Type of work done
Hours worked (straight time (ST) and overtime (OT)	Hours worked (ST & OT)
Classification by cost code	Classification by cost code
Hourly rates (ST & OT)	Hourly rates
Total hours and dollars, by day and by code	Total hours and dollars, by day and by code
Special conditions (weather, etc.)	Special conditions (breakdowns, etc.)

The time sheets are generally preprinted on standard forms, and are organized so that they are partially self-checking through "crossfooting" (adding and extending both horizontally by rows and vertically by columns). An example of a daily time sheet for labor is shown in Figure 13-6. A weekly time sheet for a front-end loader is shown in Figure 13-7. Note that on daily sheets, it is common to list several employees or machines on the same sheet, since the two-dimensional matrix allows for both these and the cost codes in which they worked. Weekly sheets need one dimension for days of the week, and hence only one employee or machine is listed per card.

As a rule, to capture more accurate information it is desirable to use daily time sheets. Week-long memories can quickly fade in the last 15 minutes on Friday afternoon if one expects supervisors to recall everything at this time, even with the aid of their diaries. Also, several companies with computer systems and daily time-data input actually process their time-data information the night it is received and have critical

FIGURE 13-6
Daily labor time card. (Adapted from Richard Clough, *Construction Contracting*, 3rd ed., John Wiley & Sons, Inc., New York, 1975, p. 244.)

WESTAMERICA CONSTRUCTION COMPANY, INC.

DAILY LABOR TIME CARD

PROJECT: Mountaintown Warehouse   WEATHER: Clear, 90 F, light wind
DATE: August 23, 1980   PROJECT NO.: 83WH04   PREPARED BY: B. C. Paulson

Employee Number	Name	Craft	Str. Time Over Time	ST Wage Rate	F12- 03120	A27- 06181							Total Hours	Gross Amount
							Cost Code: Area & Work-Type							
622	J. Douglas	C	ST	29.28	8								8	$234.24
			OT											
714	J. Fondahl	C	ST	29.28		8							8	234.24
			OT	44.72		2							2	89.44
582	H. Parker	C	ST	29.28	4	4							8	234.24
			OT	44.72		2							2	89.44
529	M. Philips	C	ST	29.28		8							8	234.24
			OT	44.72		2							2	89.44
453	G. Roberts	L	ST	21.64	8								8	173.12
			OT											
642	G. Sears	L	ST	21.64		8							8	173.12
			OT	32.94		2							2	65.88
			ST											
			OT											
			ST											
			OT											
			ST											
			OT											
Total Hours			ST		20	28								
			OT			8								
Total Cost					$524.48	$1,092.92								$1,617.48 1,617.48

cost reports available on site the next morning. Some argue, however, that daily time cards are too time-consuming for field people. However, if you actually study the two examples, you will see that the total number of entries would be the same.

Both approaches can be greatly facilitated if a computer preprints all information on the cards except for the time entries and special condition notes. That is, a supervisor receives a sheet that already lists the names and numbers of his crew and equipment, plus the cost codes in which they are likely to work. There may be a few extra spaces each way for writing in additional names or codes if needed. Anything that can be done to minimize the time field people need to write will generally improve the accuracy and completeness of reporting.

Some also claim that field supervisors cannot be expected to code their work accurately. Their failure to do so usually results first of all from unnecessarily complex and ambiguous cost codes and, second, from lack of training in the use and importance of the cost system. Where problems are still experienced in obtaining accurate distributions from field supervisors, it may be necessary to have them submit their reports in narrative form and to let the timekeeper or cost engineer do the actual coding. However, the fact that many organizations do get good cost data coded directly by field supervisors is proof that this can be done, and it is the most desirable alternative where possible. Careful attention to the human factors in management can usually make for successful reporting.

FIGURE 13-7
Weekly equipment time card.

		Cost Code: Area & Work-Type					Total Hours					
PROJECT: Mountaintown Warehouse							**MACHINE:** Cat 977					
WEEK ENDING: Aug. 23, 198__ **PROJECT NO.:** 83WH04							**MACHINE NO.:** L-5		**RATE** ST: 84.80 /hr. Idle/Repair: 32.40 /hr. OT: 69.20 /hr.			

WESTAMERICA CONSTRUCTION COMPANY, INC.
WEEKLY EQUIPMENT TIME CARD

Day	Str. Time / Over Time	F10-02220	S01-02222			Str. Time	Over Time	Repair	Idle	Total Cost	Weather & Comments
Monday	ST	8				8				678.40	Clear, 85
	OT										
Tuesday	ST	6				6		2		573.60	Clear, 80 Hydraulic hose burst
	OT										
Wednesday	ST	8				8				678.40	Cloudy, 70
	OT										
Thursday	ST		8			8				678.40	75, light wind
	OT										
Friday	ST		8			8				816.80	80, clear
	OT		2					2			
Saturday Sunday	OT										
Total Hours		22	18			38	2	2			PREPARED BY:
Total Cost		1,865.60	1,495.20					64.80		3,425.60 3,425.60	B. C. Paulson

Measuring and Reporting Work Quantities

In order to associate labor and equipment costs with physical work achieved, it is necessary, on the same reporting time cycle (daily, weekly), to estimate or measure the quantities of the elementary work items that have been accomplished during that period. The same estimates will serve not only for cost reports on labor, materials, and equipment, but also for scheduling, procurement, and other parts of the control system.

Normally such quantity reports are a mixture of actual measurements and judgement estimates. For example, on major earthworks items where daily or weekly reports are desired, the actual surveying of cross sections and quantities will most likely be done only for monthly or semimonthly progress payments. In between, it will normally suffice to use projections based on past production rates, to count truck loads, or to use some similar approximation. These can then be readjusted if necessary when the results of the next physical survey are obtained. The point is that in all quantity reporting there is a trade-off between the cost of obtaining additional accuracy in information (field survey crews are expensive) and the value of that precision in reporting.

Once the quantity data are combined with the expenditure of labor, equipment, and material resources, one can compute unit costs, study learning-curve improvements, make projections of costs at completion, and apply corrective action to operations that are in trouble. Figure 13-8 gives an example summary report that brings two of these factors together to compare relative labor productivity against physical progress at the overall job level. See Figure 8-12 for a contractor's detail report showing productivity, physical progress and earned value for detail accounts.

ENGINEERING ECONOMY IN COST ENGINEERING

Especially in the planning and design stages but also in construction, cost engineers are often called upon for studies requiring a thorough understanding of the principles and methods of engineering economy. This section will not attempt to explain these methods, but it will mention some of the main concepts and give examples of where they apply. The bibliography lists a few basic texts that will provide the needed theory and procedures. It is worth emphasizing, however, that all engineers, whether involved in design, construction, or operations, should have a working knowledge of this vital subject.

Comparative Economic Studies

Cost engineers are often called upon for studies and recommendations requiring the comparative economic evaluation of alternatives in technology and procedures. The design of a building, for example, might require a comparison between two different systems for heating and air conditioning. The first, which incorporates some relatively unproven solar collectors, has a comparatively high first cost but, on the basis of expected energy costs, would have a lower annual operating cost than the alternative. The alternative has a lower first cost, uses reliable and proven technology, but relies

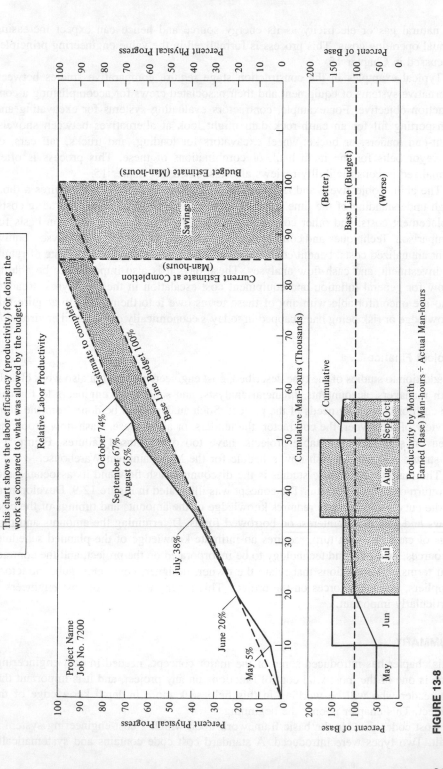

FIGURE 13-8
Relative labor productivity.

on natural gas or electricity as its energy source and hence can expect increasing annual operating costs. This process is formalized in the value engineering principles discussed in Chapter 15.

Typical examples in the construction stage require comparative studies between alternative systems of equipment and their associated crews for accomplishing a construction objective. For example, contractors evaluating systems for excavating and transporting fill for an earth-rock dam might look at alternatives between shovels, front-end loaders, or bucket-wheel excavators for loading, and trucks, rail cars, or conveyor belts for the main haul, or combinations of these. This process is often formalized in constructibility reviews also discussed in Chapter 15.

The ability objectively and rationally to make studies of this kind requires a thorough understanding of the time value of money so that capital costs, operating costs, replacement costs, and other costs can somehow be reduced to a common basis for comparison. Techniques and concepts involved include present-worth analyses, equivalent annualized costs, benefit/cost analyses, break-even or payout time, rate of return on investment, and cash-flow analyses. These days it is also important to be able to allow for general inflation and equipment cost escalation in these analyses. Readers who are uncomfortable with any of these terms owe it to themselves to acquire this knowledge or risk being handicapped in today's economically constrained enviroment.

Project Finance

In addition to studies of the type described, cost engineers must often also work closely with managers, accountants, financial analysts, and scheduling engineers to forecast the cash or borrowing needs of the project. Such analyses can be done from the point of view of the owner, the contractor, the lenders, or all of these. Cash-flow problems, even on seemingly profitable projects, have too often caused failures. Figure 5-3 illustrates a simplified cash-flow schedule for the Mountaintown Warehouse.

The main tool for these studies is the discounted cash flow and its associated rate of return on invested capital. The concept was illustrated in Figure 12-9. Development of the curves shown there requires knowledge of the amounts and timings of the cash flows and also of the interest on borrowed funds. Determining the amounts and timings of cash flows, in turn, requires an intimate knowledge of the planned schedule, resources, methods, and technology to be incorporated on the project, and the contractual terms and conditions that relate the owner, designer, contractor, subcontractors, suppliers, and labor forces on the project. This is why the input of cost engineers is particularly important.

SUMMARY

This chapter has introduced some of the major concepts needed in cost engineering. This is one of the most vital control functions on any project, and it is important that the reader who will be working in this field seek more in-depth knowledge of the subject. This chapter is only the beginning.

Cost codes provide the basic framework upon which a cost-engineering system is built. Two types were introduced. A standard cost code contains and systematically categorizes a relatively complete enumeration of all the types of work in a generic type of construction, such as general building construction. The *Masterformat* is one of the best-known examples. A project cost code, while derived from the standard code, is developed for a specific project and hence contains additional elements for the project number, major areas and facilities within it, and a means of distributing costs for resources, such as labor, materials, and equipment. Current nomenclature on major projects sets forth a work breakdown structure which is used to interrelate costs, schedules, responsibilities and other factors utilizing a common code. Use of the WBS is mandated upon many government projects as a part of a management information and control system and by design-constructors and others as a part of an integrated control system interrelating costs, schedules and other parameters.

In contrast with the exhaustive checklist sought in a standard cost code, a workable project cost code must be kept simple, clean, concise, and easy to interpret. Guidelines suggested for doing this include: (1) Use suspense accounts for procurement of bulk materials and other items as may be desirable; (2) control costs at the point where they are incurred; (3) code for cost, not design function, in the field. (4) do not attempt to use detailed payroll costs for control of labor productivity; (5) avoid unrealistically detailed equipment cost distribution; (6) avoid imprecision and ambiguity in code description; and (7) consider the human factors in cost engineering.

Control budgets are derived from estimates of costs and quantities, and usually provide a structure for recording actual performance, making forecasts, showing variances, and documenting reasons for unexpected problems. There is often need to recategorize the costs from the estimate to serve the needs of effective cost control or for owner or agency requirements.

Field construction cost control concentrates mainly on labor and equipment. The main sources of data here are labor, equipment time sheets, and estimates or surveys of quantities of work in place. Data from other parts of the control system, including scheduling, accounting, procurement, and quality assurance, are also important for consistency, comparisons, and interrelated factors.

Engineering economy supplies cost engineers with some of their most useful and powerful analytical tools. The cost engineer is often called upon to apply a thorough working knowledge of the theory and principles of engineering economy in making comparative economic evaluations of alternatives in design or construction and in developing cash flows for project finance.

As a key figure in project control, the cost engineer is often at the interface between management, line supervisors, scheduling, procurement, accounting, field engineers, and others. It can be an interesting and challenging job that takes the economic pulse of the whole project.

14

PROCUREMENT

In the broadest sense, procurement and procurement-related activities occur during all phases of a construction project. Periodically recurring scarcities of both manufactured goods and raw materials, both on a domestic and on a worldwide basis, will continue to offer a challenge to experienced industry professionals. Major procurement for a project may be handled almost entirely by one organization, as happens in a design-construct project, or it may be split between the owner, designer, general contractor, and subcontractors. On a professional construction management project, the manager often handles procurement of long-lead items in order to advance overall completion dates.

Methods and practices, of course, differ with individual firms and projects. A general contractor may receive bids from subcontractors, material suppliers, and equipment manufacturers who can refer to completed plans and specifications before bidding. On a traditional project he will have procurement practices that significantly differ from those of a design-constructor or turnkey firm normally engaged in a phased construction program. Nevertheless, certain basic principles are common to each general approach to construction procurement. This chapter will explain general industry practices and will show how procurement and procurement-related activities interface with other project activities and controls throughout the life cycle.

CONCEPTS OF PROCUREMENT

Procurement includes purchasing of equipment, materials, supplies, labor, and services required for construction and implementation of a project. It also includes related activities of tracking and expediting, routing and shipment, materials and equipment handling, accountability and warehousing, final acceptance documentation, and ultimate disposal of surplus items at job end.

Procurement is normally performed at several levels. Major equipment, such as a boiler on a coal-fired power plant, may be purchased by the owner long before the ground-breaking ceremony. A general contractor's home office may arrange for procurement of subcontractors and major materials and equipment to be installed by the contractor's own forces. The job-site field office will normally procure supplies, incidental rentals, and other requirements. On a professional construction management project, certain long-lead items may be procured by the owner, others by the manager, and less critical items by each of the contractors involved in the program.

Ethical standards of behavior in procurement-related activities are difficult to define. What is generally accepted practice in some industries, countries, and locations may be considered borderline or unethical in others. The conduct of owners, designers, general contractors, subcontractors, and professional construction managers is subject to varying degrees of economic risk that will also have an effect upon generally accepted standards. Therefore, a professional construction manager, general contractor, and subcontractor must have sufficient overall industry knowledge, as well as knowledge of practices in the area or country where they are doing business, to maintain an ethical and reasonable approach.

THE PROCUREMENT CYCLE

Procurement of materials ranges from simple purchases of supplies at the time they are needed, such as running down to the hardware store for a few more boards and a box of nails, to major design, manufacturing, and shipment procedures, such as those required for the fabricated structural steel for a 500-meter cantilever bridge. In general, however, the cycle involves all or some of the following steps:

1 Identification or recognition of the need during design or estimating

2 Determination of the design characteristics required to perform the desired function

3 Quantification of the elements needed, and preparation of procurement specifications

4 Issuance and processing of internal requisition

5 Solicitation of bids or price quotations

6 Receipt and evaluation of proposals

7 Issuance of purchase order, subcontract or lease

8 Vendor's or subcontractor's preparation and submission of shop drawings or samples

9 Review and approval of shop drawings by contractor and owner's architect/engineer

10 Fabrication by vendor or subcontractor

11 Tracking and expediting

12 Shipping and traffic

13 Delivery and inspection

14 Storage and handling on site prior to use

15 On-site fabrication operations

16 Installation and testing in constructed facility

17 Owner acceptance/rejection, warranties, corrections, and other follow-up

Not all these steps are required for all types of materials. For example, a contractor might already have an open purchase order with a ready-mix concrete supplier. The concrete superintendent simply phones in the specifications, quantity, and delivery requirements the afternoon before a day's pour, and the next morning the concrete is batched, trucked, and delivered and slump tests and cylinders are made, but there is no further on-site fabrication. The process is condensed to less than a day. On the other hand, for some procurement operations the above list is grossly oversimplified. Consider the procurement of bridge steel from Japan for erection in Australia, or the process of getting a nuclear steam supply system made in the United States delivered to a power plant in the Middle East. The preparation of shop drawings and fabrication are full-scale projects in themselves, months or years in duration, with hundreds or thousands of activities to be closely interfaced to the operations in the project for which the materials or equipment are intended. Shipping, especially of large or bulky items and particularly if multimode (truck, rail, barge, ship) carriers and international passage are needed, is a highly complex, time-consuming process that often baffles even the most knowledgeable and experienced professionals. In other words, procurement can involve far more than merely requesting the sterotyped "green-eye-shade" purchasing agent to phone around town for the lowest price on thin-wall conduit.

In today's projects, where design-construct and professional construction management contracts and performance specifications leave much to the judgment of the constructor or construction manager, procurement also requires up-to-date knowledge of the types of materials and equipment that can best meet the desired performance standards for the lowest capital and operating costs (life-cycle costs). If one material becomes unavailable or too costly, procurement specialists must know where they can get a substitute that will meet or exceed the original requirements.

DEFINITIONS OF PRINCIPAL DOCUMENTS

The following definitions will be helpful in understanding the principal documents needed in the purchasing process.

Prime Contract

A prime contract is one let by the owner to a contractor who is in turn responsible for performing the work according to the contract specifications. Under the single-contract traditional method, only one prime contract was generally awarded by the owner to a prime contractor. With multiple-contract and professional construction management approaches, the owner may award a number of prime contracts. Appendix C contains a sample prime contract.

Purchase Order

A purchase order, such as that shown in Figure 14-1, is a short form of contract normally issued for procurement of materials, permanent equipment, and supplies. It

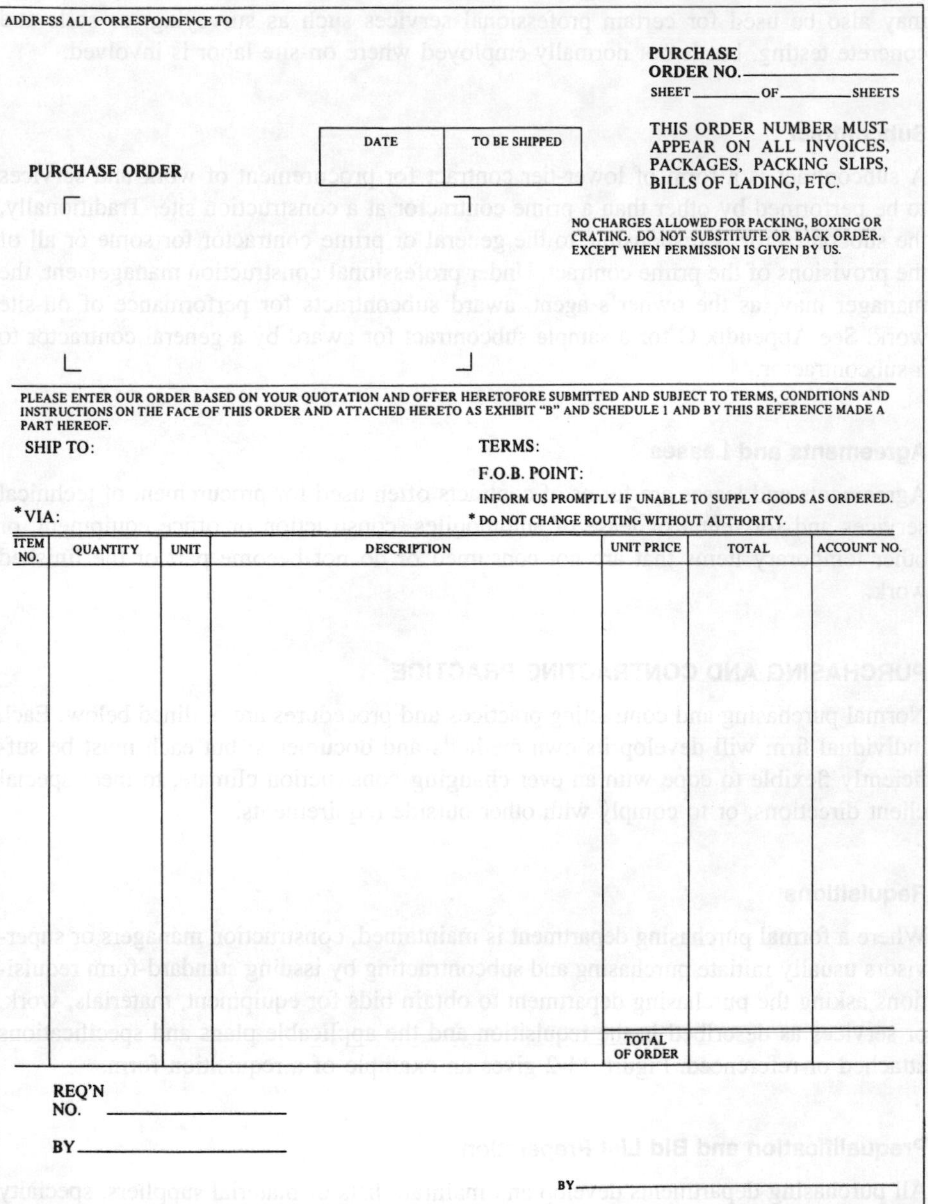

ADDRESS ALL CORRESPONDENCE TO

PURCHASE
ORDER NO._____
SHEET_____ OF_____ SHEETS

DATE	TO BE SHIPPED

PURCHASE ORDER

THIS ORDER NUMBER MUST
APPEAR ON ALL INVOICES,
PACKAGES, PACKING SLIPS,
BILLS OF LADING, ETC.

NO CHARGES ALLOWED FOR PACKING, BOXING OR
CRATING. DO NOT SUBSTITUTE OR BACK ORDER.
EXCEPT WHEN PERMISSION IS GIVEN BY US.

PLEASE ENTER OUR ORDER BASED ON YOUR QUOTATION AND OFFER HERETOFORE SUBMITTED AND SUBJECT TO TERMS, CONDITIONS AND
INSTRUCTIONS ON THE FACE OF THIS ORDER AND ATTACHED HERETO AS EXHIBIT "B" AND SCHEDULE 1 AND BY THIS REFERENCE MADE A
PART HEREOF.

SHIP TO:

TERMS:

F.O.B. POINT:

INFORM US PROMPTLY IF UNABLE TO SUPPLY GOODS AS ORDERED.

*VIA: * DO NOT CHANGE ROUTING WITHOUT AUTHORITY.

ITEM NO.	QUANTITY	UNIT	DESCRIPTION	UNIT PRICE	TOTAL	ACCOUNT NO.
					TOTAL OF ORDER	

REQ'N
NO. _____

BY _____

BY _____

FIGURE 14-1
Purchase order.

may also be used for certain professional services such as surveying, or soils and concrete testing, but is not normally employed where on-site labor is involved.

Subcontract

A subcontract is a form of lower-tier contract for procurement of work and services to be performed by other than a prime contractor at a construction site. Traditionally, the subcontractor is responsible to the general or prime contractor for some or all of the provisions of the prime contract. Under professional construction management, the manager may, as the owner's agent, award subcontracts for performance of on-site work. See Appendix C for a sample subcontract for award by a general contractor to a subcontractor.

Agreements and Leases

Agreements and leases are forms of contracts often used for procurement of technical services and for rental or lease of automobiles, construction or office equipment, or other temporary items that are not consumed or do not become part of the finished work.

PURCHASING AND CONTRACTING PRACTICE

Normal purchasing and contracting practices and procedures are outlined below. Each individual firm will develop its own methods and documents; but each must be sufficiently flexible to cope with an ever changing construction climate, to meet special client directions, or to comply with other outside requirements.

Requisitions

Where a formal purchasing department is maintained, construction managers or supervisors usually initiate purchasing and subcontracting by issuing standard-form requisitions asking the purchasing department to obtain bids for equipment, materials, work, or services as described in the requisition and the applicable plans and specifications attached or referenced. Figure 14-2 gives an example of a requisition form.

Prequalification and Bid List Preparation

All purchasing departments develop and maintain lists of material suppliers, specialty contractors, equipment suppliers, and other vendors. Prequalification of potential suppliers, manufacturers, contractors, and subcontractors is often appropriate on certain work, but may be less desirable in other situations. Where possible, prequalification to ensure that the award can always be made to the low responsive bidder can often improve job-site performance, particularly in the professional construction management method. On government and other selected work, use of minority contractors or subcontractors may be specified, and prequalification takes on a different aspect from the usual performance and financial requirements. On certain work, requiring a

REQUISITION

Date _____

For Purchase of: _____

End use: _____

In Accordance with Drawings and/or Specification _____

Rev. No. _____

Req. No. _____

Job No. _____

ITEM	QUANTITY	UNIT	DESCRIPTION	COST CODE & BUDGET EST.

Engineering Review Required. Yes _____ No _____

Date Required at Destination _____

Destination: _____

F. O. B. Point: Vendors _____ Destination _____

For Required Information Data and Conditions, See back of sheet.

Prepared By _____ Date _____

Approved-Project Engineer _____ Date _____

FIGURE 14-2a
Requisition (page 1).

PROCUREMENT CONDITIONS
(Please Check Items Required)

A. Information Required with Quotation
   1.___Specifications of proposed materials and workmanship.
   2.___General Dimension and Outline Drawings
   3.___General description of equipment operation.
   4.___Performance curves and data.
   5.___Vendor's recommended lists of spare parts. _____with prices.

B. Data Required After Purchase. In Sets, as follows:
   1.___General Arrangement Drawings.   8.___Wiring Diagrams.
   2.___Detail Drawings.   9.___Elementary Drawings.
   3.___Certified Dimension Prints.   10.___Spare Parts Lists.
   4.___Reproduced Tracings.   11.___Installation Instructions
   5.___Show Drawings.   12.___Operating Instructions.
   6.___Performance Curves.   13.___Maintenance Manuals.
   7.___Erection Drawings   14._____

   _____Sets of No._____ above for approval.
   _____Sets of No._____ above after approval.
      Above data to be directed to:

                       Attn:_____Project Engineer

C. Conditions and Services Required After Purchase
   15.___Inspection to be the responsibility of Vendor.
   16.___Inspection to be the responsibility of Purchaser.
   17.___Installation by Purchaser.
   18.___Installation supervision to be furnished by Vendor.
   19.___Guarantee to be Vendor's standard against defective workmanship and materials for a minimum
       period of one year from date of acceptance of material or equipment, with liability limited to
       replacement of parts.
   20.___Vendor to hold and save Purchaser harmless from patent liability of any nature or kind arising
       out of this equipment, its application, or its use.

Comments: _____

_____

FIGURE 14-2b
Requisition (page 2).

payment and performance bond results in prequalification of bidders by their ability to obtain a bond from a bonding company.

Preparation of bid lists varies with the different contractual methods. General contractors on competitively bid projects often advertise in trade papers that they will accept subbids up to a stated deadline. Other general contractors will make telephone calls or send out postcards to a predetermined list. When potential subcontractors cannot be prequalified, last-minute decisions regarding the qualifications of a low bidder can often become difficult just before bid time.

Design-constructors often keep card files and qualification data on a large number of potential vendors and subcontractors in different areas of the country. An area investigation will also normally turn up additional firms that may be more qualified in the local area than larger regional or nation wide suppliers and contractors. On the other hand, large international projects may include procurement of major engineered equipment from a prequalified list of firms from several different countries. On one major hydroelectric project, turbines were purchased from Germany, generators were purchased from Japan, and all the coordination required to ensure compatibility was handled from the California job site where the equipment was ultimately erected.

Prequalification and bid list preparation on professional construction management projects were discussed in Part 2, Professional Construction Management in Practice. Prequalification to ensure that only qualified firms are invited to bid, and restriction of the bid list to a reasonable number of firms so that all will be seriously interested, are extremely important in a professional construction management program.

Requests for Quotation

Requests for quotation can range from a simple advertisement in a local trade journal requesting subbids to an elaborate bid package with plans, specifications, and other contract documents fully delineating all aspects of the proposed purchase. Bid packages should be kept simple and consistent with individual requirements, but they might include some or all of the following documents:

- Specifications
- Drawings
- Scope of Work
- Bills of Materials
- Commercial Documents
- Notice to Bidders
- Proposal Form
- Contract Form
- Terms and Conditions
- Shipping Instructions
- Schedules
- Insurance Requirements
- Special Requirements
- Payment and Performance Bond Form

Bid Receipt and Evaluation

As in other procurement actions, good practice will vary with the nature of the project. Acceptance of last-minute telephone quotations on materials and subbids is common practice in many areas on competitively bid, single general contracts. Such a practice, however, would not be considered normal for an engineer-constructor who has sufficient time to accept written quotations on preprepared forms.

Government agencies generally hold public bid openings, whereas many private owners open all bids privately and never divulge the comparative figures to the bidders. In the professional construction management method, bids are often opened at a semiprivate bid opening to which all persons submitting a bid are invited. Other owners open bids in private but furnish all responsive bidders with a tabulation of actual evaluated results. Such a tabulation may show the component parts of each bid compared with the estimate (if available) and also note any omissions, inaccuracies, or other discrepancies. An evaluation to cover such discrepancies is often made so that the evaluated bid price may be higher than the actual bid.

If negotiations are required, they should normally be conducted with the low bidder or others equally in contention. It is in this sensitive area that ill will and misunderstandings often develop. The professional construction manager should endeavor to conduct himself like Caesar's wife, above even the hint of suspicion.

Recommendation for Purchase

Most medium-sized and large firms formalize recommendations for purchase with a document approved up through various levels of management; the level of approval required normally depends on the dollar value or potential risk associated with the transaction. In the case of complicated items of engineered equipment, an engineering review is usually also required as part of the preparation of the "Recommendation for Purchase." With reasonable care in preparation of bidding documents and in the selection of bidders, awards can almost always be made to the evaluated low bidder. See Figure 14-3 for a form combining the requisition and the recommendation for purchase.

Award and Preparation of Documents

Award of a purchase order, contract, or subcontract can be made verbally with completed documents to follow, or can be formalized by a "Notice of Award" advising the successful bidder that his quotation has been accepted. Figure 14-4 gives an example of such a notice. Purchase orders may be issued confirming a bidder's offer, and may not always require the vendor's signature. On the other hand, contracts and subcontracts are almost always formally signed by both parties. Bid bonds are often used by some owners, especially in the public sector, to ensure that the low bidder will in fact execute the contract.

Changes in Contract Documents

Changes in purchase orders, contracts, and subcontracts should conform where possible to general principles outlined in the original procurement action. Above all, one

REQUISITION & RECOMMENDATION TO PURCHASE

Required Delivery _____

Type of Quotation: Verbal □   Written □

Engineering Review Required □

Req. No. _____

Date _____

Material _____

Page _____ of _____

To be used for _____

Attention _____

			Coded by	Bid No.			Bid No.			Bid No.		
Quantity	Description Give Complete Data-Drawing Number, Etc.	Account Number	Unit	Unit Price	Amount		Unit Price	Amount		Unit Price	Amount
				Total							

Originator _____

Approved _____

P. O. No.	Purchase Order or Sub Contract Recommended To Be Placed With
Item No.	

F. O. B. _____

Terms _____

Shipping _____

Shipping Point _____

Price Policy _____

REASON SUPPLIER SELECTED

□ Lowest Price   □ Only Source Known

□ Required Design   □ Early Delivery

Remarks: _____

Prepared By _____

Purchasing Agent _____

APPROVED:

Field Engr. _____

General Supt. _____

Resident Mgr. _____

Owner _____

Distribution _____

FIGURE 14-3
Requisition and recommendation to purchase.

DOMESTIC
TELEGRAM
☐ Full Rate
☐ Day Letter
☐ Night Letter

WIRE MESSAGE

☐ *PREPAID*   ☐ *COLLECT*

INTERNATIONAL
CABLE
☐ Full Rate
☐ Night Letter

Job No._____ Authorized by_____ Sender's Extension No._____ Dept._____

Date ___9/24___

JENSEN EXCAVATORS
210 FIRST STREET
MOUNTAINTOWN, WEST AMERICA

RE: NOTICE OF AWARD
CONTRACT M1, EASYWAY WAREHOUSE,
MOUNTAINTOWN, WEST AMERICA

CONTRACT NO. M1, SITE EARTHWORK IN THE LUMP SUM AMOUNT OF TWO HUNDRED SIX

THOUSAND DOLLARS ($206,000) HAS BEEN AWARDED TO YOU THIS DATE. CONSIDER THIS

TELEGRAM AS NOTICE TO PROCEED WITH ALL WORK EXCEPT FIELD WORK. NOTICE TO PRO-

CEED WITH FIELD WORK WILL BE ISSUED BY O. HANSON, FIELD CONSTRUCTION MANAGER.

EASYWAY FOOD COMPANY

BY_____L. JAMES_____
PROJECT MANAGER

FIGURE 14-4
Notice of award.

should avoid verbal instructions, authorizations, or agreements; these may expedite initiation of the work, but can cause substantial honest disputes among the contracting parties when the bill is submitted. Figure 7-8 gave a sample change-order request form.

RELATION TO OTHER CONTROL SYSTEMS

Procurement is but one of many interdependent planning and control systems on a project, and as such it must fit in with the others. It must relate especially closely to the project schedule for operations and resources.

One approach would be simply to merge the procurement activities with the operations activities, say on a CPM diagram. If this were done, however, the detailed procurement activities would clearly dominate the overall schedule. Each of the operations activities could conceivably be preceded by a dozen or more steps from the list given early in this chapter.

The preferred approach, whether manually or by computer, is to develop modular, interrelated subsystems. For example, a field operation from the main schedule can define the desired latest availability date for one or more items of materials that will be needed in that operation. At most, that activity on the main schedule will be preceded by only one overall activity, such as "Procure steel doors," that will define the total time needed for all the steps in procuring those doors. Alternatively, if procurement is the time constraint, the cumulative times from the procurement steps will define the earliest time that the field operation will begin, or alert management to the need for expediting.

The details will be filled in on a detailed procurement schedule—possibly a network or bar chart, but more probably a two-dimensional matrix in paper or electronic form—which will list the item, details of its suppliers, its cost, and the start and end dates for each of the applicable procurement steps. The final column in the matrix will be the interface point to the project schedule, each referring, for example, to "Install doors in building 4." Figure 14-5 gives an example of such a procurement schedule and status report.

Reporting

Either through manual clerical attention, semimanual Cardex files, or full computer automation, the procurement schedule serves as a reference to assure that each item on the project is tracking through its required steps on time. Where items start falling behind, exception reports alert management that either field operations will be delayed from their scheduled start dates or, if the operations are critical, that procurement expediting will be necessary to bring them back on schedule. The principles of reporting and management by exception are basically the same as those described in Chapters 10 and 12, and will not be detailed again here. The concept of expediting, however, does deserve some explanation.

Procurement Schedule and Status Report   **JOB NO.**
REPORT: MASTER LIST BY COST-CODE

FAC. CMP. SHP EQ. ITEM	COST-CODE SPEC/REQ NO	P.O. CO AMOUNT	DESCRIPTION & QUANTITY VENDOR & ORIGIN	SCHED CLASS
46-0960-00-00-00	1501-5700	46001-00	STEAM METER #2	04
46-1010-00-00-00	1501-5700		HEATEXC FOR HOT WATER SYSTEM	06
46-1020-00-00-00	1501-5700		HEATEXC FOR HOT WATER SYSTEM	06
46-1030-00-00-00	1501-5700		HEATEXC FOR HOT WATER SYSTEM	06
46-1040-00-00-00	1501-5700		HEATEXC FOR HOT WATER SYSTEM	06
46-1060-00-00-00	1501-5700		DOMESTIC WATER HEATER	06
46-1080-00-00-00	1501-5700		DOMESTIC HOT WATER CIRCULAT-	03
46-1110-00-00-00	1501-5700		HOT WATER CIRCULATING PUMP	06
46-1120-00-00-00	1501-5700		HOT WATER CIRCULATING PUMP	06
46-1130-00-00-00	1501-5700		HOT WATER CIRCULATING PUMP	06
46-1300-00-00-00	1501-5700		SUMP PUMP FLOOR AREA 100 GPM	05
46-1310-00-00-00	1501-5700		SUMP PUMP FLOOR AREA 100 GPM	05
46-1350-00-00-00	1501-5700		BRINE PUMP	03
46-1360-00-00-00	1501-5700		BRINE PUMP	03
20-0001-00-01-00	2101-3100	101-00	PRIMARY CRUSHING PLANT	00
20-0001-00-02-00	2101-3200	101-00	PRIMARY CRUSHING PLANT	00
20-0001-00-03-00	2101-3200	101-00	PRIMARY CRUSHING PLANT	00
20-1100-00-00-00	2101-4400		VERTICAL LIFT DOORS MOTORIZED	06
20-1110-00-00-00	2101-4400		VERTICAL LIFT DOORS MOTORIZED	06
20-1140-00-00-00	2101-4400		VERTICAL LIFT DOORS MOTORIZED	06
20-1150-00-00-00	2101-4400		DUST SEAL SWING DOOR 44' W x	06
30-0550-00-00-00	2102-4800		MCC PRESSURIZING UNIT 6000	03
30-0600-00-00-00	2102-4800		VERTICAL LIFT DOORS MOTORIZED	06
30-0360-00-00-00	2102-5800		MONORAIL HOIST 3 T 10L 20'RUN	02

FIGURE 14-5
Procurement schedule and status report.

Expediting

The popular image of expediting is that it is a frantic, last-ditch effort to speed up the movement of materials that were not delivered when they were needed. In organizations that terminate the procurement function with the issuance of a purchase order, this is too often the case. With good project control, however, expediting involves monitoring all steps in the procurement cycle, with special focus on those involving the vendor or subcontractor, to assure reliable, economical, on-schedule delivery. The essence of professional expediting is in anticipating problems before they arise and in offering solutions before delays are encountered.

Expediting for major or critical items often requires periodic visits to vendors' shops and factories on a routine, not emergency, basis, plus frequent telephone follow-ups to check vendor progress, raw materials supplied, workload, completed product inventory, shop drawings, manuals, fabrication procedures, quality control, code certifications, and delivery status. An important underlying purpose in this is to be sure that the contractor retains high priority among other firms for whom the vendor may simultaneously be working.

PROJECT   **DATE**

                                                        PAGE

REQ-DTE BID-REC	KE-RECM CLNT-OK	AWARDED	ENGNRNG RELEASE	SHIP EX PLANT	ARRIVE AT JOB	EARLY START	DIFF CPM
A08FEB2		A30MAR2	02MAY2	31JUL2	15AUG2	01SEP2	17
A30MAR2	13MAY2	28MAY2	07JUL2	24NOV2	09DEC2	15SEP2	85-
A30MAR2	13MAY2	28MAY2	07JUL2	24NOV2	09DEC2	15SEP2	85-
A30MAR2	13MAY2	28MAY2	07JUL2	24NOV2	09DEC2	15SEP2	85-
A30MAR2	13MAY2	28MAY2	07JUL2	24NOV2	09DEC2	15SEP2	85-
F05JUL2	19AUG2	03SEP2	13OCT2	02MAR3	17MAR3	30APR3	44
F21JUN2	26JUL2	10AUG2	09SEP2	08NOV2	23NOV2	30NOV2	7
F21JUN2	26JUL2	10AUG2	09SEP2	08NOV2	23NOV2	30NOV2	7
A18JAN2		A01MAR2	F10FEB3	F01APR3	F15APR3	01MAY3	16
A18JAN2		A01MAR2	F10FEB3	F01APR3	F15APR3	01MAY3	16
A18JAN2		A01MAR2	F10FEB3	F01APR3	F15APR3	01MAY3	16
A 0MAR2	17MAY2	01JUN2	11JUL2	11NOV2	25NOV2	02NOV2	23-
A 0MAR2	17MAY2	01JUN2	11JUL2	11NOV2	25NOV2	02NOV2	23-
A 0MAR2	17MAY2	01JUN2	11JUL2	11NOV2	25NOV2	02NOV2	23-
F18SEP2	02NOV2	17NOV2	27DEC2	16MAY3	31MAY3	31MAY3	
F19MAY2	03JUL2	18JUL2	27AUG2	14JAN3	29JAN3	31MAR3	61
A30MAR2	17MAY2	01JUN2	11JUL2	28NOV2	13DEC2	31MAY3	169
F26MAY2	25JUN2	10JUL2	30JUL2	08SEP2	23SEP2	31JUL3	311

FIGURE 14-5
(continued).

Even with the best procedures, however, problems will arise that require corrective action. Sometimes, the contractor's own schedule changes require that procurement of critical items be accelerated ahead of dates originally agreed with suppliers or subcontractors. Other delays might arise from design, purchasing, vendor fabrication, shipping, etc. Figure 14-6 shows a typical form for reporting and seeking action when trouble arises. Table 14-1 provides a checklist for things to analyze in seeking both the sources of problems and solutions to them. It is largely the role of experienced and professional expediters to find the most effective means to keep procurement on schedule with the minimum adverse effect on cost. Alternatives may include seeking backup or secondary suppliers, switching to faster means of shipment, or merely identifying snarls in red tape. All too often, delays result simply because a procurement document—say a requisition, purchase order, or shipping request—gets lost in some clerk's IN file.

CONTROL OF MATERIALS PROCUREMENT

The control of materials procurement costs is somewhat different from the control of field labor and construction equipment costs. In the latter case, productivity is the

HOLMES & NARVER INC. _____ (Internal) EXPEDITING REPORT-TROUBLE _____
Date

H&N P.O. No. _____ Seller _____

Plant Location _____ Contact _____ Phone _____

Job No. _____ Client _____ Equipment _____

P.O. Item No.	1	2	3	4	5	6	7
Tag No.							
Shop Order No.							
Delivery Required							
Original Promise							
New Promise							

—PROBLEM—

—ACTION REQUIRED—

Signature of Expeditor

Distribution	—RESPONSE— (Optional)

Form 1540/18-175-1

Signature _____ Date _____

FIGURE 14-6
Expediting trouble report. (Courtesy Holmes & Narver, Inc., Anaheim, Calif.)

main criterion, and it requires continuous attention by management. Daily time sheets and overnight exception reporting can have an important impact.

In the case of materials, the main sources of information are requisitions, bids and quotations, purchase orders and subcontracts, shipping documents, receiving documents, and invoices. In most cases, these provide enough feedback for control. The

TABLE 14-1
EXPEDITING CHECKLIST. (COURTESY HOLMES & NARVER, INC., ANAHEIM CALIFORNIA.)

**HOLMES &
NARVER, INC.**
ENGINEERS · CONSTRUCTORS
A RESOURCE SCIENCES COMPANY

EXPEDITING CHECK LIST

Engineering Status	Material Status	Production Status	Analysis of Delays	Shipping Status
Is engineering required by the manufacturer? By purchaser?	Has the bill of material been issued?	When will the order be in production?	How many shifts working and number hours per shift.	Are the shipping instructions available and understood?
Obtain the engineering schedule. Is it being done on schedule?	Is the material in stock? Check inventory! Obtain a restricted allotment for H&N order where possible.	What is the estimated fabrication time? How does this compare with the time allowed in the shipping promise?	Will overtime improve delivery? How much? At what cost?	When will the invoice and packing lists be forwarded? Copies of all packing slips are to be issued to Expediting.
Do the drawings need approval?	Obtain the suborder numbers and schedules of the materials ordered.	What is the production and shipping schedule of the main unit and all auxiliary equipment?	Is there any other order interference? Give details.	On export orders report invoice number and date. Name of carrier and export agent.
When were drawings forwarded to whom?	Indicate those orders needing special expediting assistance. Give full details.	Are any production problems anticipated?	Can substitution be made? Obtain full details.	Is packing and crating material available?
Have the drawings been released to the shop?	Has advance placement of critical material been made?	If there are material shortages, take steps to get production started.	Press manufacturer to expedite critical sub-orders.	When will shipment go forward? What is the method and routing?
Can the order be released for fabrication?	If castings are required, what is the status of the patterns? Check the status of all accessories to be shipped with unit.	Are the inspection requirements understood on both the main unit and all auxiliary equipment?	Contact top management when necessary.	When shipment is extremely urgent, have manufacturer report full details. Pieces and weight. Name of carrier and Pro No. Rust Traffic will follow.

actual costs of the materials themselves are largely predetermined by the designer. Certainly the constructor can and should attempt to shop for the best price/performance characteristics he can obtain within the specifications.

Other important opportunities for control of materials costs on large projects are in:

1 Requisition procedures (specs for shipping, packaging, delivery, etc.)
2 Minimization of rehandling and shortages
3 Inventory procedures and policies

These are closely related to, and to a large extent reduce to, the problem of controlling the *timing* of the various steps in the procurement process. As long as everything runs smoothly and materials arrive on time and in good condition, their actual purchase costs are not of dominant concern in on-the-job project control. It is in its indirect impacts that the procurement process can cause the most concern at the working level. There is an old saying, "For the want of a shoe, the horse was lost." Analogously, the late arrival of a small embedded instrument can stop a large concrete pour. Similarly, materials arriving too soon, or in the wrong order, can also wreak havoc, particularly in a crowded work site, such as a downtown high-rise building or a nuclear power plant.

Comparison of actual receipts with invoiced quantities is also an important matter in project control. Figure 14-7 gives an example of a materials receiving report, which is a key document for this purpose.

INVENTORY THEORY

Regardless of the potential for their practical application in construction, the manufacturing and marketing industries' operations research techniques, collectively called "inventory theory," can provide a good understanding of the basic economics of materials procurement for construction projects. Their concepts will therefore be briefly introduced in this chapter and will be expressed in terms more applicable to construction.

Components of Costs

The general objective of quantitative inventory methods is to optimize trade-offs among the following four categories of costs in order to minimize total costs:

1 Purchase costs
2 Shipping costs
3 Holding costs
4 Shortage costs

Purchase Costs Purchase costs are those associated with (1) the actual overhead incurred in requisitioning, soliciting, and evaluating quotations, and in issuing purchase orders; (2) the actual materials prices obtained through effective negotiation, variations in unit costs with quantities, the amount of time allowed for fulfilling the order, etc.; and (3) costs associated with shipping materials to the site, which in turn are related to quantity, distance, and mode of transport.

On the administrative side, it is best to have policy guidelines for different procedures to use for different magnitudes of orders. Obviously, one should not spend $50 of administrative effort to order $10 worth of miscellaneous office supplies. On the other hand, one would not rely on a general foreman to place an order for elevators in a 20-story building. There should be clearly documented policies and procedures for these different procurement situations and for others in between.

Unit prices of materials vary not only with bargaining leverage, quantities, and delivery time, but also with design characteristics. Designers often unknowingly and

MATERIAL RECEIVED REPORT

M.R. No.:_____

PAGE____OF____

THIS ORDER IS:
COMPLETE ☐
NOT COMPLETE ☐
REC. DAMAGED ☐

P. O. NO.:_____

REQ. NO.:_____

DATE RECEIVED:_____

CARRIER		
FREIGHT BILL		NO.
EXPRESS BILL		WGT.
PARCEL POST		DATE
	PREPAID	AMT. $
	COLLECT	AMT. $
	RAIL LINE	
	CAR NO.	
	DOCK NO.	

ORDERED FROM. _____

_____

SHIPPED BY: _____

SHIPPED TO:_____ PACKING SLIP NO.:_____

VIA:_____ F.O.B. POINT:_____

P.O. ITEM NO.	QUANTITY	UNIT	DESCRIPTION	LOCATION	ACCOUNT NO.

DWG. NO._____

DELIVERED TO

B M NO._____

CHECKED BY

REQ. ORIGINATOR_____

WAREHOUSE SUPT.

FIGURE 14-7
Material received report.

unnecessarily increase prices by specifying odd sizes, say of windows or structural steel shapes, where a similar standard size could perform the same function equally well or better. Bargaining leverage often favors larger organizations whose long-term buying volume may cause vendors to give them lower prices on smaller orders also. Unit prices often vary inversely with quantity, much as they do in retail stores, but large quantities can often adversely affect holding costs, which are discussed below, and cause problems in cash flow. Sometimes it is possible to gain the best of both worlds, however, by placing a single large order with a low unit price and requesting incremental delivery dates. This is the usual approach for the bulk cement on a large concrete dam. Delivery time can adversely affect prices if the purchaser requests delivery sooner than the vendor's normal production schedule will allow. Some portion of the vendor's overtime and rescheduling costs will be passed on to the buyer.

Shipping Costs   The shipping component of purchase costs can be reduced on a unit basis as quantities become larger. For example, rail shipment by the carload, say of lumber, is much cheaper than less-than-carload (LCL) rates. Chartering an ocean freighter is usually cheaper and assures quicker delivery than shipping as just part of a cargo that is delivered to several different ports. The same applies to truck and air freight. Regarding variations in price with mode, water, rail, truck, and air shipping generally increase in price in that order, assuming convenient ports and terminals and ignoring transshipment; but shipping costs alone can sometimes be misleading. Clearly, on delicate, high value per weight items, such as computers and electronic controls, the better and quicker handling of air freight can often more than justify itself. In remote locations, also, costs can be deceptive. One of the authors once bid an arctic soil-cement runway project where the price of shipping bulk cement was about equal for ocean barge and air freight. Once the uncertainty of getting barges through the artic ice was considered, air freight became the obvious choice.

Holding Costs   Holding costs include storage space and warehouse overhead, deterioration and obsolescence, theft, misplacement, insurance and taxes, rehandling, and interest on funds invested in inventory.

Even outdoor storage space can be at a premium on crowded work sites. Renting a city block of prime commercial real estate near a building project can be absurdly expensive. Warehouse space even on remote, wide-open sites involves the capital and operations costs for construction, staff, security, and utilities. These costs can often exceed the "savings" obtained through purchasing large quantities of materials before they are needed.

Several types of construction materials are perishable and therefore cannot be delivered very far in advance of need. Portland cement is a common example. Similarly, changes in building codes and government health and safety regulations can make some items obsolete before they are installed as can the fact that new products can sometimes perform the jobs of their predecessors better and at less cost.

Theft is a real problem in some areas. Whole reels of copper cable, innumerable small tools and supplies, and even major items of equipment have been known to disappear from projects, sometimes at a cost of several million dollars on just one large job. Too often these losses have been attributable to well-organized criminal operations.

Even misplacement is a problem, especially where warehouse and storage yards are poorly organized. Materials cannot be found at the time they are needed on site, their purchases are duplicated, and they are found to be in surplus when the job is completed—or sometimes they have quite literally disappeared forever beneath the mud in the storage yard.

Insurance costs are higher when larger quantities of materials are on hand, and property tax assessors delight in finding surpluses for their coffers. In today's competitive climate, the cost of the money prematurely invested in inventory can also add an often unnecessary expense, either for the contractor or for the owner if the contract permits these costs to be passed on.

Perhaps the most important of all are the costs of rehandling resulting from premature deliveries. Ideally, there should be a perfectly coordinated flow of materials where trucks arrive on site and their burdens are hoisted directly to the point where they are needed. Even in the best-run operations, however, most materials are rehandled at least once. If management is in good control, the materials arrive, are unloaded, and are stored in logically preplanned and documented locations, either in the warehouse or in the yard, so that they can be readily and directly retrieved when they are needed. Too frequently, however, deliveries are unplanned, materials are stored at random locations, and when an item is finally needed in construction, there is first a frantic search to find it, and, once found, a whole load of reinforcing steel, two large pumps, and several other items must first be moved before workers can get to it. In this chaotic, unmanaged situation, materials are moved about like the pea in a shell game, are often rehandled a half-dozen or more times, and often are seriously damaged before their time comes for installation in the structure. This rehandling is done not for free, but by today's high-priced labor and equipment, and generally with poor supervision and inefficiency.

Shortage Costs The corollary of holding costs, of course, is the impact of shortages on project operations costs. It is often the fear of shortage costs that leads to the conservative delivery schedules that adversely affect holding costs.

In classic inventory theory, shortage costs are the "loss of sales," in the retail sense, from not having in stock the item that the customer wants—a toaster, TV, or whatever. The customer then walks down the street and buys from someone else, and the sale is lost. This accounts for the perennial inventory policy battle between the marketing side and the supply side of such organizations.

Shortage costs, however, can also result from interrupted production for lack of input materials. In construction, shortage costs are the direct and indirect costs of delayed and interrupted work (recall the interrupted learning curve in Figure 11-7), and they are especially catastrophic if the operations affected are critical to the completion schedule for the project.

These impact costs in turn motivate the incurrence of other types of shortage costs, those of expediting and special handling. To get things moving, management may be willing to pay more to get the materials from a different vendor, ship by air freight instead of rail, and complete some shop-fabrication operations at higher cost in the field.

Just-in-Time Inventory Management In recent years the just-in-time theory of inventory management has been popularized by the experience of the Japanese automobile manufacturers. These companies have realized considerable economy in scheduling sub-assemblies and parts to arrive at the assembly plant just in time to be incorporated into the automobile directly with minimum or no inventory requirements. Ready-mix concrete forms a good example of a basic construction material which by its nature cannot be inventoried and must be directly incorporated into the work. Applications in metropolitan areas where storage space is either nonexistent or very expensive have been developed to supply reinforcing steel, structural steel, piping assemblies and other basic materials in a similar manner. Taking advantage of just-in-time opportunities will become increasingly more important to the competitive contractor of the future.

Heavy construction operations on remote job sites have practiced a form of just-in-time inventory control through the use of contracts with vendors who maintain a stock of small tools, spare parts or other necessities in the vendor's warehouse or trailer at the job site or in a nearby location. The contractor then draws from these facilities as needed and is billed monthly.

Cement storage on a major dam or tunnel project is another example from heavy construction. Here relatively small cement silos are kept full by the cement manufacturer by a continuing delivery program designed to fit the concrete pouring schedule.

Applications of the just-in-time principles in other construction applications can be expected to minimize the holding costs resulting from materials being delivered too soon, in the wrong order, and in unpredictably large quantities.

Trade-offs It should be clear by this stage that purchase costs, holding costs, and shortage costs are highly interrelated, especially where risk and uncertainty are concerned, and in large measure they are in conflict with one another. A great deal of research has been done to show how their interrelationships can be modeled, and how, at least in theory, the expected sum of their costs can be optimized. Details of these techniques, however, are beyond the scope of this book. Some of the references in the bibliography will enable the interested reader to pursue this subject as deeply as desired.

SUMMARY

In an era combining global shortages of resources with a proliferation of new types of materials, procurement demands the increasing attention of professionals experienced in this phase of project planning and control. First of all, they must be knowledge-able of procedures and options in all parts of the procurement cycle, from recognition of need through requisition, prequalification of bidders, evaluation, selection, vendor fabrication, expediting, shipping, delivery, and inspection, to on-site storage and handling. The process is far more complex than just shopping around for the cheapest purchase price.

Purchase orders are a short form of contract generally used for the procurement of materials, equipment, and supplies. Contracts and subcontracts are normally used for the procurement of construction work involving on-site labor. Agreements and leases are often used for the rental of construction equipment and other items not incorporated into the finished work. Purchasing documents and steps will include requisitions, prequalification and bid list preparation, requests for quotation, bid receipt and evaluation, recommendation for purchase and award, and later finalization of change orders.

Materials control is different from the control of labor and equipment in several important ways. The importance of operations productivity, in particular, gives way to the importance of scheduling and timing and the indirect consequences that poor materials management can have on the whole project. It is particularly important that materials procurement systems be closely coordinated with the systems for the planning, scheduling, and control of operations and resources on the project. Reporting for materials control, in turn, emphasizes detailed tracking through all steps of the procurement cycle, and exception reporting can in turn lead to application of professional expediting procedures.

Quantitative analytical techniques, collectively known as "inventory theory," can provide real insight into the basic economics of materials management and can also find practical application in construction. In concept, these methods seek the optimum trade-off between procurement costs, holding costs, and shortage costs that will produce the lowest overall materials costs for a project.

15

VALUE ENGINEERING

Value engineering emerged during World War II when shortages of critical resources necessitated changes in methods, materials, and traditional designs; many of these changes resulted in superior performance at a lower cost. After the war, the General Electric Company pioneered in the development and implementation of an organized value analysis program for industry, and this technique was soon adopted by several other companies and government agencies. In 1962, value engineering became a mandatory requirement in the **Armed Services Procurement Regulations** (ASPR). This change in ASPR introduced value engineering to two of the largest construction agencies in the country, the U.S. Army Corps of Engineers and the U.S. Navy Bureau of Yards and Docks. During the 1960s and the 1970s, several other government agencies and jurisdictions adopted value engineering, including the Bureau of Reclamation, the **National Aeronautics and Space Administration** (NASA), the Department of Transportation, and the Public Buildings Service of the GSA.

Growth of construction applications of value engineering in the public sector has thus been fostered by legislation and regulation. However, there is little sign of an equally impressive growth in the private sector. As noted by James J. O'Brien in the preface to his book on the subject,[1] only about half the construction industry's designers and contractors are even aware of value engineering, and perhaps only about 1 percent are actively and successfully applying its techniques. If the approach is indeed useful, one wonders why its adoption is taking so long.

To help answer this question, this chapter will (1) summarize the state of the art in governmental and traditional applications; (2) briefly describe life-cycle costing and its specialized tools; (3) inquire into reasons slowing acceptance of value-engineering principles in the private sector; (4) suggest ways of overcoming this reluctance; and

---

[1] *Value Analysis in Design and Construction*, McGraw-Hill Book Company, New York, 1976.

(5) present documented results of a program that has received enthusiastic acceptance on projects in the private and public sector.

During the past decade, the term constructibility analysis has seen increasing use in the private sector and in some universities. Constructibility analysis is similar to value engineering in some or all respects. Some use the term to connote review of plans and specifications from the viewpoint of the constructor, which would include opportunities for prefabrication, preassembly, modularization, special construction methods and other considerations aimed at lessening the cost or improving the completion schedule. Others use the term to include both design and construction analysis in a manner similar to traditional value engineering but omitting the rigid certification and other requirements as recommended by the Society of American Value Engineers. For purposes of this chapter, value engineering, value analysis and constructibility analysis are used synonymously. Under any of these names, the technique represents a creative and organized approach whose objective is to optimize cost and/or performance of a construction facility or component.

POTENTIAL SAVINGS

L. D. Miles' book[2] on value analysis and engineering includes the following definition:

> Value analysis/engineering is an organized, creative approach which has for its purpose the effective identification of unnecessary costs, i.e., costs which provide neither quality nor use nor life nor appearance nor customer features.

But to whom do these savings accrue, and what are they really worth?

In 1974 the Army Corps of Engineers estimated that the total cumulative savings through value engineering was almost $234 million. The Public Buildings Service indicated that its value-engineering program had generated savings of $4.53 for every dollar spent, for total savings to GSA of $1.8 million in fiscal 1973. During fiscal year 1970, the Department of Defense estimated a saving of about $4.40 from contractor-sharing incentives for each $1 spent on the program. It further estimated an additional return of four times this amount resulting from in-house programs during the design phase and prior to contract award.

Alphonse Dell'Isola's book on value engineering[3] established potential savings guidelines as follows:

On total budget	1 to 3%
On large facilities	5 to 10%
On high-cost areas	15 to 25%

Realizing these potential savings requires a systematic and innovative approach. Generally accepted techniques include a job plan for value engineering which will have a number of phases:

---

[2] *Techniques of Value Analysis and Engineering*, 2d ed., McGraw-Hill Book Company, New York, 1972.
[3] *Value Engineering in the Construction Industry*, Construction Publishing Corp., Inc., New York, 1982. (With permission of Van Nostrand Reinhold Company.)

Develop information and requirements.
Speculate on alternatives.
Analyze and evaluate alternatives.
Develop the program.
Proceed with proposal, presentation, and selling.

This plan can be implemented over the life cycle of a construction project, and will have potential savings related to time and cost in varying degrees, depending upon the project development phase.

Conception
Development
Detail design
Construction
Start-up and use

The following section will discuss this approach in greater detail.

VALUE-ENGINEERING JOB PLAN

In his book, Dell'Isola designed a value-engineering job plan accomplished in four phases:[4]

1 *Information*: Get facts
2 *Speculative*: Brainstorm
3 *Analytical*: Investigate, evaluate
4 *Proposal*: Sell

These phases can be expanded and explained as follows:

Informative Phase

This phase includes these purposes:

1 To gather and tabulate data concerning the item as presently designed
2 To determine the item's function(s)
3 To evaluate the basic function(s)

During information gathering, certain questions must be answered:

1 What is the item?
2 What does it do?
3 What is the worth of the function?
4 What does it cost?
5 What are the needed requirements?
6 What is the cost/worth ratio?
7 What high-cost or poor-value areas are indicated?

---

[4]Ibid., pp. 12–58.

O'Brien says the question, "What does it do?" should be answered by two words (a verb and a noun) for each function; for example, a water pipe *transports water.*[5] Such definitions may require careful analysis. To illustrate, a door may provide access, limit access (as in a prison), provide security, exclude or contain fire, control traffic, provide visibility, or express prestige.

Considerable effort, ingenuity, and investigation are required to answer these questions. The value-engineering group must determine what criteria and constraints existed at the time of the original design and whether they still apply at the present time.

Other important questions may be:

1 How long has this design been used?

2 What alternative systems, materials, or methods were considered during the original concept?

3 What special problems were or are unique to this system?

4 What is the total use or repetitive use of this design each year?

Brainstorming

The purpose of this phase is to generate numerous alternatives for providing the item's *basic* function(s). By definition, a brainstorming session is a problem-solving conference wherein each participant's thinking is stimulated by others in the group. A team may consist of four to six people of different disciplines, sitting around a table and spontaneously generating ideas. Production of the maximum number of ideas is encouraged, and no idea is criticized.

The Gordon technique[6] has also been successful. With this technique only the group leader knows the exact nature of the problem, and he asks questions to generate ideas. This approach can stimulate freer thinking than brainstorming.

Analytical Phase

The purposes in this phase are:

1 To evaluate, criticize, and test the alternatives generated during the speculation phase

2 To estimate the dollar value of each alternative

3 To determine the alternatives which offer the greatest potential for cost savings

During this phase, also known as the evaluation and investigation phase, the group examines alternatives generated during the brainstorming and tries to develop lower-cost solutions.

The principal tasks are:

1 To evaluate

2 To refine

[5]O'Brien, *Value Analysis in Design and Construction*, pp. 12–14.
[6]William J. J. Gordon, *Synetics, The Development of Creative Capacity*, Harper & Row, New York, 1961.

3 To cost-analyze

4 To form a possible list of alternatives in order of descending savings potential

Aristotle said that worth can have economic, moral, aesthetic, social, political, religious, and judicial value. This is a valuable precept for value-engineering team members as well as for philosophers.

James J. O'Brien believes a value index such as "worth divided by cost" or "utility divided by cost" can be very beneficial during this phase.

The route of ideas is the following:

1 Eliminate ideas which do not meet environmental and operating conditions.

2 Set aside, for future discussion, ideas with potential but which are beyond present capability or technology.

3 Cost-analyze remaining ideas.

4 List ideas with useful savings, including their potential advantages and disadvantages.

5 Select ideas where advantages outweigh disadvantages and offer the greatest cost savings. (Often dollar values are not readily assignable and must be considered using statistical approaches.)

6 Finally, consider weighted constraints, such as aesthetics, durability, and salability, in order to produce a completed list.

Proposal Phase

Also called the program planning and reporting phase, this is the final portion of Alphonse Dell'Isola's plan. This phase must accomplish three things:

1 A thorough review of all alternate solutions must be prepared to assure that the highest value and significant savings are really being offered.

2 A sound proposal must be made to management.

3 The group must present a plan for implementing the proposal. If the proposal will not convince management to act, no savings will result.

U.S. GOVERNMENT VALUE-ENGINEERING JOB PLANS

A different perspective on this process can be achieved by considering other job plans. For example, the U.S. Army Corps of Engineers' job plan[7] involves five phases:

Information
Speculation
Analysis
Development
Presentation

This list differs from Dell'Isola's list in that it splits the proposal into two phases:

Development
Presentation

[7]U.S. Dept. of Defense, *Handbook*, 5010.8-4 (1963).

Development Phase

The purposes are:

1 To assess the technical feasibility of each surviving alternative
2 To obtain firm information concerning each surviving alternative
3 To develop written recommendations

Presentation Phase

The purposes are:

1 To present a value-engineering study report
2 To present the report to the decision maker(s)
3 To ensure that the report recommendations are implemented

In 1968 the **Department of Defense** (DOD) made two changes in this list. First, a new phase identified as "Orientation" was added at the top of the list. The orientation phase has three basic purposes:

1 Selection of appropriate areas to be studied
2 Selection of the appropriate team to accomplish the study
3 Determination of the policies needed to assist in the accomplishment of these determinations

Second, the last phase, "Presentation," was expanded to include follow-up. By 1972, the DOD job plan had again been expanded, this time by splitting the last phase, "Presentation and Follow-up," into separate phases and by adding the phase "Implementation" in between them. Figure 15-1 summarizes the job plan at this stage of development.

A composite list of value-engineering and job-plan categories is shown in Table 15-1. Though these plans emphasize various aspects of the process, they are basically similar in approach and sequencing. With a sound knowledge of both the purposes for each phase and the route of ideas, one can either select the job plan best suited to the project's needs or create a new one.

LIFE-CYCLE COSTING

Accurate cost measurement is one of the most important requirements of a successful value-engineering program. Most cost estimates and cost records used in the construction industry deal with capital costs from the viewpoint of the contractor or the ultimate user of the facilities. Yet the life of the building or facility will extend over 20 to 50 or more years. During this period the cost of maintaining and servicing the facility, including the cost of utilities such as fuel oil, electric power, or natural gas, will equal or exceed the capital cost. Value analysis from the viewpoint of the owner must therefore take into account both capital and future operation and maintenance costs if maximum value is to be achieved for minimum overall investment. In the final analysis, we are trying to find out how much additional capital expenditure is warranted today to achieve future cost benefits over the life of the facility.

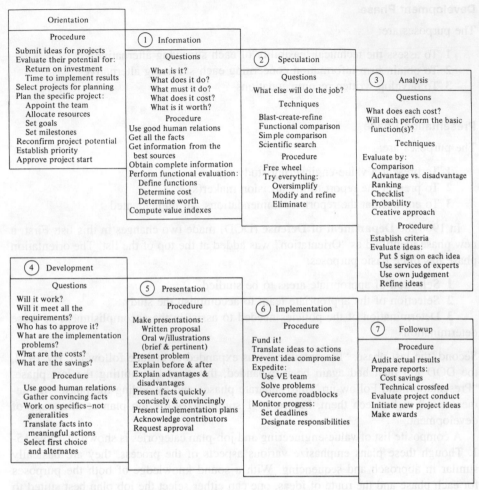

FIGURE 15-1
Major phases of a job plan. (Adapted from General Services Administration, Public Building Service, Manual No.P8000.1, *Value Engineering*, Washington, D.C.)

However, as we endeavor to estimate the cost of future events, our cost estimates for the life cycle become less reliable when compared with capital or construction cost estimates. To illustrate, consider some of the items that are important in analyzing future life-cycle costs for a project:

Maintenance and operating costs
Energy and utility costs
Value of money
Cost of insurance
Anticipated future income growth
Ease and timing of expansions

TABLE 15-1
JOB PLAN CATEGORY COMPARISON

Dell'Isola*	GSA-PBS P 8000.1 1972	L. D. Miles† 1961	L. D. Miles‡ 1972	E. D. Heller§ 1971	A. E. Mudge¶ 1971	PBS VM Workbook 1974
Information	Orientation	Orientation	Information	Information	Project selection	Information
	Information	Information			Information	Function
Speculation	Speculation	Speculation	Analysis	Creation	Function	Creative
Analysis	Analysis	Analysis	Creation	Evaluation	Creation	Judicial
Proposal	Development	Program planning	Judgment	Investigation	Evaluation	Development
	Presentation	Program execution	Development	Reporting	Investigation	Presentation
	Implementation	Summary and conclusion		Implementation	Recommendation	Implementation
	Follow-up					Follow-up

*A. J. Dell'Isola, *Value Engineering in the Construction Industry,* Construction Publishing Corp., Inc., N.Y. 1974.
†L. D. Miles, *Techniques of Value Analysis and Engineering,* 1st ed. McGraw-Hill, New York, 1961.
‡L. D. Miles, *Techniques of Value Analysis and Engineering,* 2nd ed., McGraw-Hill, New York, 1972.
§E. D. Heller, *Value Engineering and Cost Reduction:* Addison-Wesley, Reading, Mass., 1971.
¶Arthur E. Mudge, *Value Engineering.* McGraw-Hill, New York, 1971.

Fringe benefits difficult to analyze, including aesthetics, durability, and overall future image

Effect of the facilities on the productivity of operating, administrative, and maintenance personnel

Present and future trends in real estate and property taxes, income taxes, and investment credits

Location and operational costs based upon community growth, competitive patterns, and other factors

Each of the items includes choices among multiple alternatives, uncertain forecasts of future costs, and uncertain effects of future events. To help solve these general problems, a number of specialized tools is available; of these, the following are among the most important:

Present worth analysis
Sensitivity analysis
Break-even analysis
Discounted cash-flow and rate-of-return analysis

All these are useful economic tools for value engineering. However, it is beyond the scope of this book to examine them in detail. The reader who lacks knowledge in this area is strongly encouraged to consult the classic text on engineering economy by Grant, Ireson, and Leavenworth[8] or to read other related books listed in the bibliography.

VALUE ENGINEERING IN THE PRIVATE SECTOR

We have reviewed several accepted job plans that illustrate the state of the art of value analysis or value engineering and have emphasized the importance of life-cycle costing. Now consider some of the pitfalls of the traditional approaches in nongovernmental design and construction programs.

Major design-construct companies involved in large industrial projects have accounted for the greatest dollar value of construction work per individual firm in recent years. One might therefore expect that these firms would be the leaders in promoting and using value engineering since, in addition to offering considerable potential savings benefits to the clients, they control both design and construction operations. Yet value engineering has in general not been consistently adopted in any organized form by these firms or by many others in the private sector. In a magazine article, Dell'Isola appears to have identified some of the reasons for this.[9]

In this article, Al Dell, the value engineer, has convinced a reluctant Joe Weakley, design director, to try a value-engineering analysis. After the value-engineering team completed its study, a 15 percent overall savings developed. Even Joe Weakley was impressed by the action of the team. Later Dell inquired about the two engineers who

[8]Eugene L. Grant, W. Grant Ireson, and Richard S. Leavenworth, *Principles of Engineering Economy*, 6th ed., The Ronald Press Company, New York, 1976.

[9]Alphonse J. Dell'Isola, "A Value Engineering Case Study," *Heating, Piping and Air Conditioning*, June 1970, pp. 50–54.

had been responsible for the original design. "Those two?" replied Weakley, "I had them transferred. I won't tolerate people who can't design economical facilities." Note Joe Weakley's basic lack of understanding of value engineering: that it is a team effort and no one person can have all the answers.

In today's rapidly changing economy, the old traditional rules of thumb that designers relied on to produce economic designs and to choose between alternative solutions simply are not cost-effective. No designer can hope to master the current knowledge of the cost of individual items of labor and materials in the locality of the project. In fact, studies performed in a given locality often become out of date very quickly as certain materials and labor costs increase more rapidly than others or even become unattainable in reasonable delivery periods.

The theoretical value-engineering job-plan approach is technically sound and has been shown to be successful when given an opportunity, but it still has generally failed when it has been tried in the industrial and commercial areas where the greatest incentive for its use should be present. We shall further explore reasons for this in the context of the three main contractual approaches introduced in Chapter 2: those of the traditional general contractors, design-construct, and the professional construction manager.

Traditional Single General Contractor

The traditional method usually employs an architect or engineer who is often compensated on a fixed price, a percentage of construction cost, and/or a guaranteed maximum. If a multidiscipline value-engineering team reviews the design and comes up with suggested savings, a redesign is almost always involved. When confronted with major changes for which it is expected to pay additional design fees, the unsophisticated owner will generally feel that the designer should have done the work right the first time, even though estimates show that an overall saving may be achieved.

The designer responsible for the original design will generally resent any criticism of it, and the principals of the design firm will be concerned about the additional design hours for which they may not be compensated. This was illustrated in Chapter 10's discussion of the "level of influence." Other design principals or supervisors will criticize the original designer for not producing the right design the first time. Furthermore, when the fee is the traditional percentage of construction cost, the more money spent on engineering to lower the construction cost, the lower the profit the design firm will realize.

An additional difficulty is the accuracy of estimates. On competitive bids not involving a large percentage of subcontracts, bid spreads often range from 25 to 50 percent or more above the low bidder. Cost to the owner is influenced by how much the low bidder desires the work as well as its individual evaluation of the anticipated cost. This evaluation may be quite different from that of an estimator working for the value-engineering team, since it is not under such competitive pressures.

Design-Construct or Turnkey

Design-constructors have in-house all the disciplines associated with a value-engineering team. Why haven't they generally adopted value-engineering programs?

Again the method of compensation often creates the situation where spending more money on engineering may reduce the overall cost but may also reduce the amount of profit to the design-constructor. In other cases, the result may be an embarrassing overrun of the predetermined design budget .

Design-constructors also generally have highly structured design organizations. Internal design reviews are normally performed by discipline supervisors; outside review is generally interpreted as criticism, and it stimulates a natural reaction to justify the original design.

Estimating may generally be geared to the conceptual level rather than to the detailed level of the trade contractors. This approximate estimating may work well with broad trade-off studies and development of general concepts. However, it may prove to be misleading in the later verdict of the competitive marketplace.

Joe Weakley and his "we should have done it right the first time" philosophy continue to inhibit the use of value engineering by the design-construct firms who have all of the necessary internal skills to make the program work under a single management.

Professional Construction and Program Management

The emergence of professional construction and program management and the early involvement of the constructor in the design process have fostered the use of value engineering. The program and construction managers who have understood the attitude of Dell'Isola's Joe Weakley and who have appreciated the financial considerations of the design firms have generally had more success with value engineering in the private sector than have those firms that have tried the more traditional methods without modification. When a way can be found to support value engineering without outside independent design critique, and where each team member can be credited equally, in the eyes of the owner, for any savings, a healthy climate for implementation of suggestions exists.

One of the advantages of successful professional construction management is the absence of the adversary relationships often present between the architect/engineer and the general contractor. The architect who criticizes the professional construction manager to the owner quickly finds that he is equally vulnerable to criticism by the manager. Everyone benefits when discrepancies are discussed internally and straightened out before finalizing construction contracts. This same concept applies to the value-engineering effort.

PRACTICAL PRIVATE SECTOR VALUE-ENGINEERING PROGRAMS THAT WORK

A practical value-engineering program was developed in the mid to late seventies as a part of the senior author's professional construction management program. This program was created in an effort to eliminate the prejudices that have kept value engineering from realizing its potential in the private industrial and commercial sectors. Although the program is relatively unsophisticated, it has achieved significant cost savings for owners when compared with projects constructed along older, more traditional methods.

A Professional Construction Management Value-Engineering Program

In an effort to eliminate the objections of a traditional program, the following criteria were developed for the program, and all parties—owner, architect/engineer, and professional construction manager—were enlisted in carrying it out:

1 The professional construction manager, owner, and designer will educate all their personnel to be on the lookout for alternate methods and concepts at any stage of the project. Suggestions are actively encouraged, and they are to be submitted to the manager.

2 The professional construction manager is given the responsibility for documenting and investigating initial feasibility and for preparing estimates and working out a presentation for all suggestions deemed feasible.

3 The designer is given the responsibility for ruling on the engineering suitability of the suggestions and for making the determination as to whether the proposed suggestion is equal, superior, or inferior to the original concept.

4 Together with an analysis of effects on the schedule or other like factors, all suggestions which are deemed technically feasible by the designer and which result in cost savings are submitted to the owner for final approval. In the event that significant redesign is required, the owner will compensate the designer accordingly, and the design costs will be subtracted from the indicated savings.

5 If possible, each bid package should contain several alternates that are equally acceptable to the owner and designer. Thus the bidders themselves will participate in the program, and the owner will receive the benefit of bidder preference and current cost estimates.

6 Potential bidders should be encouraged to develop alternates, which, if technically equal to the base bid, will receive consideration in evaluating bids. Thus by its ingenuity a bidder who is not low under the base bid may become the low bidder. The owner receives a cost saving, and the ingenious bidder receives the contract.

7 The criterion for determining whether a suggestion should be classified as a value-engineering saving is whether or not the saving is a result of the three-party team concept. (Suggestions may be made by any party—owner, manager, designer, or even the construction contractors themselves.) If, in the final judgment of the owner, it is a product of the three-party team, it is classified as a value-engineering saving and is documented by the professional construction manager.

It is worth noting that the program, summarized in Table 15-2, was successful with five different owners and six different architects involved in these ten projects. Each enthusiastically accepted the program and lived up to its individual responsibilities. The key to this enthusiastic acceptance has been the team concept where all members receive credit for an implemented suggestion. Many of the suggestions were relatively simple, but would not have been incorporated on a project using the traditional methods.

TABLE 15-2
VALUE-ENGINEERING SAVINGS ACHIEVED ON PROJECTS

Description	No. of savings	Value of savings	Applicable building cost	% Savings
Distribution center	17	343,379	8,030,000	4.3
Meat fabrication plant	7	125,419	3,040,000	4.2
Bakery	7	63,240	4,020,000	1.6
Distribution center	10	215,979	3,600,000	6.0
Meat processing plant	11	164,219	8,000,000	2.0
Distribution center	12	279,043	8,000,000	3.5
Light manufacturing plant	16	484,950	6,600,000	7.3
Plating & reclaim facility	6	8,315	1,100,000	.8
Meat fabrication plant	18	196,525	6,000,000	3.3
Water treatment plant	7	56,047	4,400,000	1.3
Totals	111	1,937,116	52,790,000	3.7

Simplified Example

Savings on one of the projects will be outlined in more detail to show the types of things achieved. The following savings were achieved for a distribution center expansion with the joint cooperation of the owner, architect, and construction manager.

Value Engineering, Item 1 Metal roof deck for the produce warehouse in lieu of double-tee decking as initially called for. **Savings: $ 62,000**

Value Engineering, Item 2 Systems procurement package for the frozen foods warehouse which specified results rather than individual proprietary details and resulted in competition among all major manufacturers. Savings are based upon an evaluated life cycle of 20 years since maintenance, insurance, financing, and other costs varied depending upon the particular manufacturer's system. **Savings: $113,000**

Value Engineering, Item 3 Glu-laminated beams, wooden-metal span joists and plywood decking in lieu of an all-steel roof support system for the freezer room and canopy of the frozen foods warehouse additions. **Savings: $ 33,783**

Value Engineering, Item 4 Acceptance of general contractor's alternate bid for steel erection of the meat warehouse rather than of fabricator's price. **Savings: $ 6,130**

Value Engineering, Item 5 Acceptance of an alternate proposal to use a local painter for a specialty product rather than an out-of-town manufacturer's installer, while preserving original guarantees. **Savings: $ 12,500**

Value Engineering, Item 6 Proprietary cement product to increase joint spacing for the grocery warehouse's special floors, rather than the specified portland cement. **Savings: $ 10,800**

Value Engineering, Item 7   Forty-five concrete placings for special floors in the frozen foods warehouse in lieu of the ninety specified.   **Savings: $ 1,389**

Value Engineering, Item 8   Insulated structural-styrofoam panel walls along column line 2 of the meat warehouse in lieu of specified product.

Savings: $ 17,167

Value Engineering, Item 9   Glassboard facing over built-up insulation for meat warehouse columns rather than the prefabricated insulted panels as specified.

Savings: $ 6,300

Value Engineering, Item 10   Sloped concrete retaining walls in lieu of conventional walls around fuel-oil storage tanks.   **Savings: $ 6,105**

Value Engineering, Item 11   Call for alternate refrigeration-support-steel quotations incorporating modified design after initial bids appeared high.

Savings: $ 6,619

Value Engineering, Item 12   Credit allowance as a result of good working conditions and smooth job coordination during caisson installation in frozen foods section, which was performed on a negotiated guaranteed maximum-price, share-of-savings concept.   **Savings: $ 3,250**

Total value-engineering savings: **$279,043**

Demonstrated Results

Table 15-2 tabulates value-engineering results for 10 projects constructed over a 5-year period with the professional construction management approach. These projects were completed with the active innovation, cooperation, and participation of a number of leading owner and architectural and engineering firms.

Initial ideas for savings were contributed by the architect, the owner, the construction manager, contractor, and subcontractor. However, it is believed that the overall saving of 3.5 percent of the applicable building cost would not have been realized by the owner under traditional contracting programs; it is clearly a direct result of the value-engineering partnership program utilized on a professional construction management project.

PRIVATE SECTOR DEVELOPER PROGRAMS

Perhaps the most effective utilization of value analysis has been created by successful major developers. Using value-engineering techniques coupled with a sound knowledge of the marketplace, major developers have been able to dominate the hotel, high-rise office building, and major residential complexes. Working closely with a selected architect, successful developers have been able to substantially cut the design period from past norms while producing attractive salable or leasable space with a market-driven cost effectiveness that has allowed them to dominate in this type of project. Separating tenant improvement work from the basic building structure has enabled the developers to both improve schedules while minimizing use of "fast track" construction. Each area can be customized to fit the tenant's requirements without modification. Major developers have found that properly managed tenant improvement work can be the source of substantial increased profitability.

SUCCESSFUL PUBLIC SECTOR VALUE-ENGINEERING PROGRAMS

A Multibillion-Dollar Prison Construction Program

In the mid-1980s the Department of Corrections of the state of California embarked upon an unprecedented major prison construction program. The Department acted as overall project director employing a program management consultant and a number of construction management firms to manage individual prisons which were to be built using a phased construction schedule to minimize overall design-construction time. The program manager with the enthusiastic participation and backing of the Department introduced a structured value-engineering program which consisted of value-engineering studies held in parallel with scheduled design reviews for each design package along with designer participation. These meetings were held during the conceptual stage and periodically throughout design development and were chaired by the program manager.

In February 1989, the Auditor General of the state of California published a report which presented the results of a management audit covering the performance of six major prison construction projects which were completed by 1989. The report found, in part that: "The indicated final cost for the six new prisons is two percent less than the initial budgets prepared before detailed design and construction work began. This represents an outstanding performance for a public sector program pioneering a new conceptual approach designed to shorten design-construction schedules as well as to decrease costs." The report attributed much of this success to the value-engineering program developed by the program manager and embraced by the Department of Corrections which helped to effectively manage the ultimate construction costs throughout the design development and detail-design phases of each new prison. The program manager utilized a traveling team under the leadership of the project manager who visited each contract package designer to conduct a design review and value-engineering study at fixed points throughout the design period. In addition to the project manager, the team consisted of one each civil, mechanical and electrical engineer with estimating support. The team also included a correctional officer representing the Department who had the responsibility for the preservation or improvement of Department correctional and security requirements.

Table 15-3 summarizes the results of the value-engineering program for a 1700-bed facility located in Amador County, California. The project was completed in 21 months from start of construction utilizing twelve construction contractors under the

TABLE 15-3
VALUE-ENGINEERING SAVINGS, CALIFORNIA PRISON PROJECT

Description	Amount	Percent Savings
Total Cost at Completion	$98,994,000	
Design Phase VE Savings	$11,866,000	12.0
Construction Phase Savings	$1,172,000	1.2
Total Savings	$13,038,000	13.2
Breakdown		
Number of Contracts	12	
Number of Proposed Design Savings	137	
Number of Proposed Construction Savings	65	
Total Number of Proposed Savings	202	
Number of Accepted Design Savings	103	75
Number of Accepted Construction Savings	25	38
Total Number of Savings Proposed	128	
Total Number of Savings Accepted	128	63

direction of a construction management firm who reported directly to a project director employed by the Department.

A Major Secondary Treatment Plant

Initial cost estimates for the Metropolitan Seattle West Point Secondary Treatment Plant by two independent estimating groups prepared after completion of the 90 percent design development phase indicated a probable overrun of two hundred million dollars on this $534 million project. A 40-hour Value-Engineering Workshop carried out under traditional lines was able to document potential savings which are anticipated to enable the program to be completed within budgeted funds. While the formal workshop only took five days for a multidisciplinary team, analysis of the suggestions, completion of revised conceptual designs and reestimating the work took approximately four and one half months to establish design feasibility and more accurately estimate the proposed changes.

SUMMARY

This chapter reviewed value analysis methods utilized by a number of government agencies and summarized the job-plan concept. Numerous public agencies continue to report significant savings achieved through their formalized programs, but the approach has had less widespread acceptance in private work.

An important feature of value engineering is life-cycle costing; this allows for the effect of future operational, overhead, carrying, and maintenance costs as well as initial capital or construction costs. Such future operational phase costs may be equally important to the owner in determining the nature and cost of the initial investment.

The less-than-enthusiastic acceptance that value engineering has received in the private sector is due in a large measure to the nature of the marketplace, the inherent fear of criticism on the part of the designers, and the adversary relationships which are often present among designers, contractors, and owners. The professional construction management and program management approaches, however, offer an outstanding opportunity to realize the demonstrated benefits of value engineering by minimizing adversary relationships and involving the original designer as a member of the team with full authority to analyze the technical acceptability of proposed modifications. Practical effective programs created by major developers have proved economically sound in the competitive marketplace.

As our economy continues to mature, value-engineering techniques that reduce costs while preserving basic value will become even more important than they were during the decades of heavy growth following World War II.

CHAPTER 16

QUALITY ASSURANCE

Planning and controlling standards for quality are fundamental in both the design and construction phases of a project. This aspect of a project, while closely interrelated with costs, schedule, procurement, and value engineering, deserves its own amplification.

Quality assurance involves economic studies to select the types of materials and methods to be included in design, making certain that the design is in accordance with all applicable building codes and other regulations, and controlling the construction on the project to be sure that the work is performed according to the standards specified in the contract documents. Methodology here ranges from computerized documentation of accepted governmental, technical and professional criteria for design, to sampling and testing concrete, earthwork, welding, bolting, and structural dimensions in the field.

This chapter will begin with some basic definitions pertaining to different phases of quality on a project. It will then discuss important organizational factors on a project. Basic economic concepts related to quality in both the design and construction phases will be presented next, followed by a brief introduction of statistical methods for quality control in construction. Recent developments include the development by the American Society of Civil Engineers (ASCE) of a guide publication designed to help achieve improved quality in the constructed project and the increasing use of peer reviews both at the organizational and project level.

BASIC CONCEPTS AND DEFINITIONS

Quality criteria affect all phases of a project. This section will briefly introduce some basic concepts and define some common terms used in the industry.

Definitions

Three key aspects of quality relate to engineering, control, and assurance. Each will be briefly defined below.

Quality Engineering   This term often describes procedures used to ensure that the engineering and design for a structure proceed according to recommended and mandatory criteria set by related professional and trade associations, building code authorities, and federal, state, and local organizations such as the Environmental Protection Agency, the Nuclear Regulatory Commission, the Occupational Safety and Health Administration, and others. Many of these standards are required by law and several are revised frequently, so it is important for the architects and engineers to be both knowledgeable and up to date on all applicable standards. It is very expensive to correct mistakes once construction has begun.

Quality Control   This process includes (1) setting specific standards for construction performance, usually through the plans and specifications; (2) measuring variances from the standards; (3) taking action to correct or minimize adverse variances; and (4) planning for improvements in the standards themselves and in conformance with the standards. In other words, once the architects and engineers have set the criteria for construction, quality control ensures that the physical work conforms to those standards.

Quality Assurance   Although its definition is not well standardized, quality assurance is generally a broader, more nearly all-encompassing term for the application of standards and procedures to ensure that a product or a facility meets or exceeds desired performance criteria. It also usually includes the documentation necessary to verify that all steps in the procedures have been satisfactorily completed. The term transcends both quality engineering and quality control, where in the first phase it includes the design of a product whose quality is economical in terms of its end use, and in the second it includes the development and application of procedures which will, at economical levels, assure attainment of the designed quality.

Elements of Quality

Basic elements of quality include (1) quality *characteristics*, (2) quality of *design*, and (3) quality of *conformance*. Each will be explained in the following paragraphs.

Quality Characteristics   As with the East Indian legend of the blind men and the elephant, there are many ways we could evaluate and describe any given product. We use the term "quality characteristic" for the one or more properties that define the nature of a product for quality-control purposes. Quality characteristics include dimension, color, strength, temperature, etc.

To illustrate, consider concrete, a material common to all types of construction. Quality characteristics that commonly specify and control concrete quality include compressive strength after a fixed curing time (usually measured indirectly by failure-testing cylinders cast at the time concrete is placed), slump, size of aggregate, the ratio of water to cement, surface finish, and sometimes color. Clearly, these are not the only characteristics of concrete—one might judge its taste and smell, for example—but they are the characteristics considered relevant to quality for structural purposes.

Quality of Design Designers generally recognize that no human undertaking produces absolutely perfect results. Therefore they often specify not only the desired standard for the characteristics that define a product, such as a dimension or strength, but also tolerances or ranges for acceptable variations from the standard. For example, reinforcing-bar spacing might be specified as 12 cm ±0.5 cm, or a concrete specification requesting a compressive strength of 200 kg/cm^2 might further state that no more than 20 percent of the compressive-strength sample cylinder breaks can fall below this value. To specify a spacing tolerance of ±0.2 cm or an understrength limit of 10 percent would be setting a higher quality of design. In each case we are recognizing the statistical nature of work processes.

In design, we should further recognize the impact that higher standards of quality and tighter tolerance limits have on costs. One of the easiest ways to drive up the costs of a project unnecessarily is to design in standards of quality that are inappropriate to the intended function; this applies whether the standards are too high or too low. For example, to specify architectural concrete finish standards on buried footings or other unexposed surfaces is setting standards too high. The designer of one large sewage treatment plant specified almost unachievable tolerances of $\frac{1}{32}$ of an inch on submerged concrete surfaces; this specification required unnecessary and costly grinding on most of the concrete! On the other hand, specifying cheap and failure-prone materials may provide lower initial costs, but it will also drive up overall costs in the long run. It is the designer's responsibility, with input from the professional construction manager, to specify a quality of design that is most economical and functional for the overall project.

Quality of Conformance Once the quality of design has been specified, the quality of conformance is the degree to which the physical work produced conforms to this standard. For example, a journeyman welder producing pipe welds with a reject rate of only 3 per 100 has a much higher quality of conformance than an apprentice with a reject rate of 17 per 100.

As with quality of design, there is a close correlation between standards for conformance and the cost of achieving those standards. There are trade-offs to be considered between costs of the work methods and quality-control procedures and the costs of rejects. These economic considerations will be discussed further.

The Quality System

All the elements of quality that have been described thus far combine to determine the quality of the final product. The relationship between these elements is shown in Figure 16-1. The *owner's needs* are expressed in the *design criteria* that guide

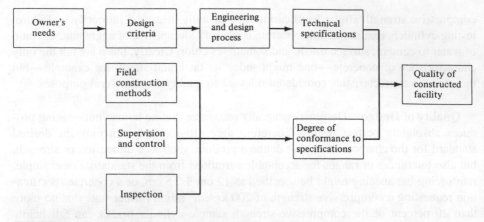

FIGURE 16-1
The elements of quality. (Adapted from Elwood G. Kirkpatrick, *Quality Control for Managers and Engineers*, John Wiley & Sons, Inc., New York, 1970, p. 5.)

the *engineering and design process* that produces the *technical specifications* for the project. This in effect sets the quality of design.

The quality of conformance is influenced by (1) the actual *field construction methods*, including the skill of the workers, the capabilities of their tools and equipment, and the quality of their raw materials; (2) the *supervision* they receive and the managerial *controls* that are applied to direct the workers in accordance with the plans and specifications; and (3) the *inspection* and quality-control procedures that are applied, including the knowledge and skill of the inspectors and the reliability of their methods and tools for measuring the quality characteristics specified by the designers. The last-mentioned is particularly important, because too often the problems are with factors such as poor statistical analysis or miscalibrated or inaccurate tools used for quality control. The combination of these three produces the *degree of conformance* to the design specifications.

Quality of design and quality of conformance, in turn, determine the *quality of the constructed facility*. In considerable measure, they also affect the cost. The economics of this process will be discussed in the next section.

ECONOMICS OF QUALITY

The economics of quality assurance must be considered in both design and construction. Economic concepts applicable to each will be introduced here.

Quality Economics in Design

In the simplest form, Figure 16-2 illustrates the relationship between the cost and value of quality. The horizontal axis relates to the quality of design as reflected by the quality characteristics chosen for the item concerned. As quality increases, the vertical axis shows that both the cost and the value of the quality increase as well, but in a different manner. The value curve is concave downward. As quality increases, the value increases, but at a decreasing rate. In other words, as quality increases, the marginal value, or the value of one additional unit of quality, becomes less.

On the other hand, the cost curve shows that as the quality of design increases, the marginal cost of each additional unit of quality increases more with each step. The last increment of quality costs far more than the first, until eventually it becomes too costly to specify higher standards.

In concept, the optimum level of quality occurs at the level where the marginal cost of one additional unit equals the marginal value. In Figure 16-2, this occurs where the slopes of the two curves are equal. Below this, an additional dollar of cost buys more than a dollar's worth of quality. Beyond it, the additional quality costs more than it is really worth for the functional objectives of the project.

The concept, of course, is simple compared with the reality of quantifying quality, cost and value, and determining the optimum point for design. Also, numerous other parameters affect the actual performance criteria. For example, as the quantity increases, it becomes more economical to specify higher levels of quality, since fixed costs can be written off over more units of production. To illustrate, one might justify high-quality steel forms for 100 precast concrete units, but these could not be justified for a half-dozen units. Nevertheless, by focusing on the trade-offs between cost and value, the concepts in Figure 16-2 are worth bearing in mind as design progresses.

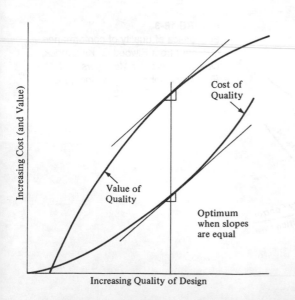

FIGURE 16-2
Economics of quality of design.
(Adapted from Elwood G. Kirkpatrick,
*Quality Control for Managers and
Engineers,* John Wiley & Sons, Inc.,
New York, 1970, p. 8.)

Cost of Quality

Value of Quality

Optimum when slopes are equal

Increasing Cost (and Value)

Increasing Quality of Design

Economics of Quality of Conformance

Simply stated, quality control costs money. There are two main ingredients to this cost: (1) the cost of the skilled labor, equipment, materials, methods, and supervision to produce quality output; and (2) the costs of monitoring and verifying the quality of output and of correcting or replacing defective work. Figure 16-3 illustrates the trade-offs between these two categories of costs.

Note that to achieve increasing quality of conformance directly from the resources and methods, one must invest more money in them, and hence the direct construction cost goes up. On the other hand, as the reliability of the methods and resources improves, less investment is required for monitoring their performance and for correcting and replacing defective work, so the costs of quality control go down. To optimize conformance costs for a given quality of design, one seeks to minimize the sum of the direct construction costs and the quality-control costs, as shown in the upper curve in Figure 16-3.

For example, a contractor might be faced with the need to prepaint metal components for a building. If the quantity is large, the contractor might invest in a semi-automated, on-site paint shop that will assure a quality paint job with few rejects. If the quantity is small, the production cost might more than offset the quality cost, so a more manual approach, with due allowance for touch-ups and rejects, might be better.

By considering quantity, this example has also illustrated that in reality there are more parameters than appear on the simplified model in Figure 16-3. Again, however, by focusing on just a few variables, the figure does emphasize the most important economic concepts in quality of conformance.

FIGURE 16-3
Economics of quality of conformance. (Adapted from Elwood G. Kirkpatrick, *Quality Control for Managers and Engineers*, John Wiley & Sons, Inc., New York, 1970, p. 10.)

ORGANIZATION FOR QUALITY ASSURANCE

The quality assurance objectives of the various parties associated with a construction project differ and often conflict. The owner wants to maximize the quality of characteristics associated with the intended function of a project, yet do so without undue costs. These functional characteristics might be aesthetic, production-oriented, or whatever. The designer wants a level of quality that will assure satisfactory performance of the structure and be a credit to his professional reputation, but again without undue cost overruns. A constructor on a fixed-price contract will be interested mainly in satisfying the specifications at minimum direct cost. And increasingly, external regulatory agencies are setting quality standards for characteristics that may not even be directly related to the primary function of the project, and often with little consideration for cost. Controls on emissions to the environment are among the best-known examples, and often are a primary area of conflict. One of the most important functions of a professional construction manager is to provide third-party objectivity in setting standards of quality that, first, satisfy all mandatory regulations, and second, provide the most economical quality-cost performance in keeping with the objectives of the project.

Typical Organizations

Resolving the conflicts in objectives between the various parties generally makes the organization and responsibilities of quality assurance somewhat different from those for other parts of the project planning and control system. Especially for conformance-oriented quality control, more attention is given to separating the responsibility for judging quality from those charged with carrying out the work. For example, in traditional competitively bid public works, quality control is one of the primary responsibilities of the resident engineer and his inspectors. The relationship is shown in Figure 16-4. Incentives for cost-and-time control are normally built into the contract so that the major responsibility remains with the constructors. In private works, also, the architect and his inspectors focus mainly on quality during the construction phase.

Even in design-construct work, owner or governmental requirements often mandate that quality assurance be separated from production operations. For example, Figure 16-5 shows a setup typical of a nuclear power plant. Although the quality assurance people are employed by the same company, they are organizationally separated not only from the project manager, but from the whole operations organization as well. The idea is to set up an autonomous, independent, and, it is hoped, objective group that is free to apply controls without fear of censure from higher levels in the operations organization. By and large, this approach has been adopted by the nuclear power industry, where the quality assurance organization does indeed put quality above cost considerations at the project level.

In professional construction management, the professional construction manager provides objective third-party input into determining the quality of design, as well as inspection services to control quality of conformance. In each case this objectivity provides a new dimension and a level of effectiveness not found in most alternative forms of contracts. Where design organizations are left to set their own quality of

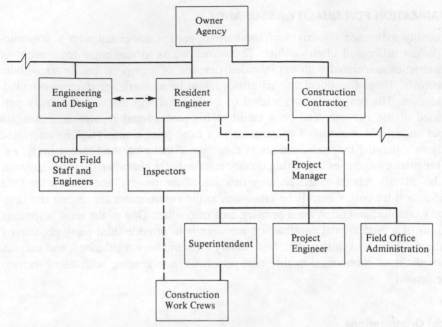

FIGURE 16-4
Quality control organization in a competitively bid public works project.

design, they commonly tend toward conservative overdesign, often out of ignorance of the implications for construction costs. Where the design organization also inspects the work of contractors, there are legitimate questions of fairness to the contractor in the resolution of disputes regarding conflicts between quality of design and quality of conformance. An experienced and knowledgeable professional construction manager can provide the necessary objectivity.

Tasks, Responsibilities, and Procedures[1]

Early in a project, tasks and functions must be identified and the methods of accomplishing them must be determined. Figure 16-6 uses a convenient matrix form to show the important relationships for quality assurance, and in effect summarizes the whole program. As with the detailed thinking required by CPM scheduling, the preparation of such a chart can spotlight areas of conflict and omission at the planning stage.

The left-hand column of the chart lists methods in five categories: documentation, tracking, inspection, testing, and administration. Along the top of the chart is a list of functions and tasks for a typical project. The reader is encouraged to study this

---

[1] The figures and concepts in this section are based on the paper titled "System for Control of Construction Quality," by Roland M. Parsons, *Journal of the Construction Division*, ASCE, vol. 98, no. CO1, March 1972, pp. 21–36. The paper contains a detailed narrative discussion of all four figures (16-6 through 16-9), and is strongly recommended to the reader for further study.

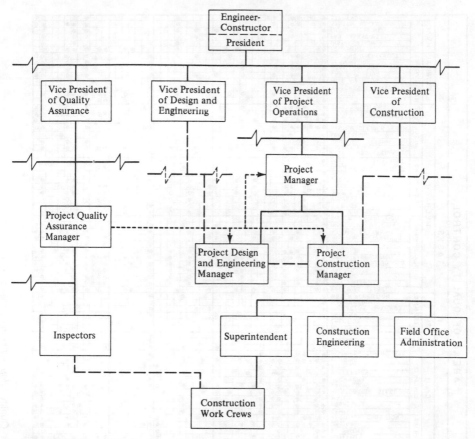

FIGURE 16-5
Quality control organization in a design-construct project.

figure for a moment to gain an appreciation for what is involved here and to see how quality assurance procedures differ for the different kinds of tasks. Note that critical or complex tasks are subject to more control procedures than are the simpler or less important ones. In preparing such a diagram, one tries to achieve a balanced program that will provide the most economical approach to quality assurance.

Delegating Responsibility and Authority Once the necessary tasks and methods have been identified, it is necessary to delegate responsibility and authority for accomplishing them to persons in the project organization. In addition to organization charts such as those given in Figures 16-4 and 16-5, a linear responsibility chart, such as that presented in Figure 16-7, clearly shows the responsibilities and relationships among the key parties. Tasks are listed in the left-hand column, positions are listed across the top, and symbols define responsibilities. Note that a single vertical column also provides the quality assurance portion of the listed individual's job description.

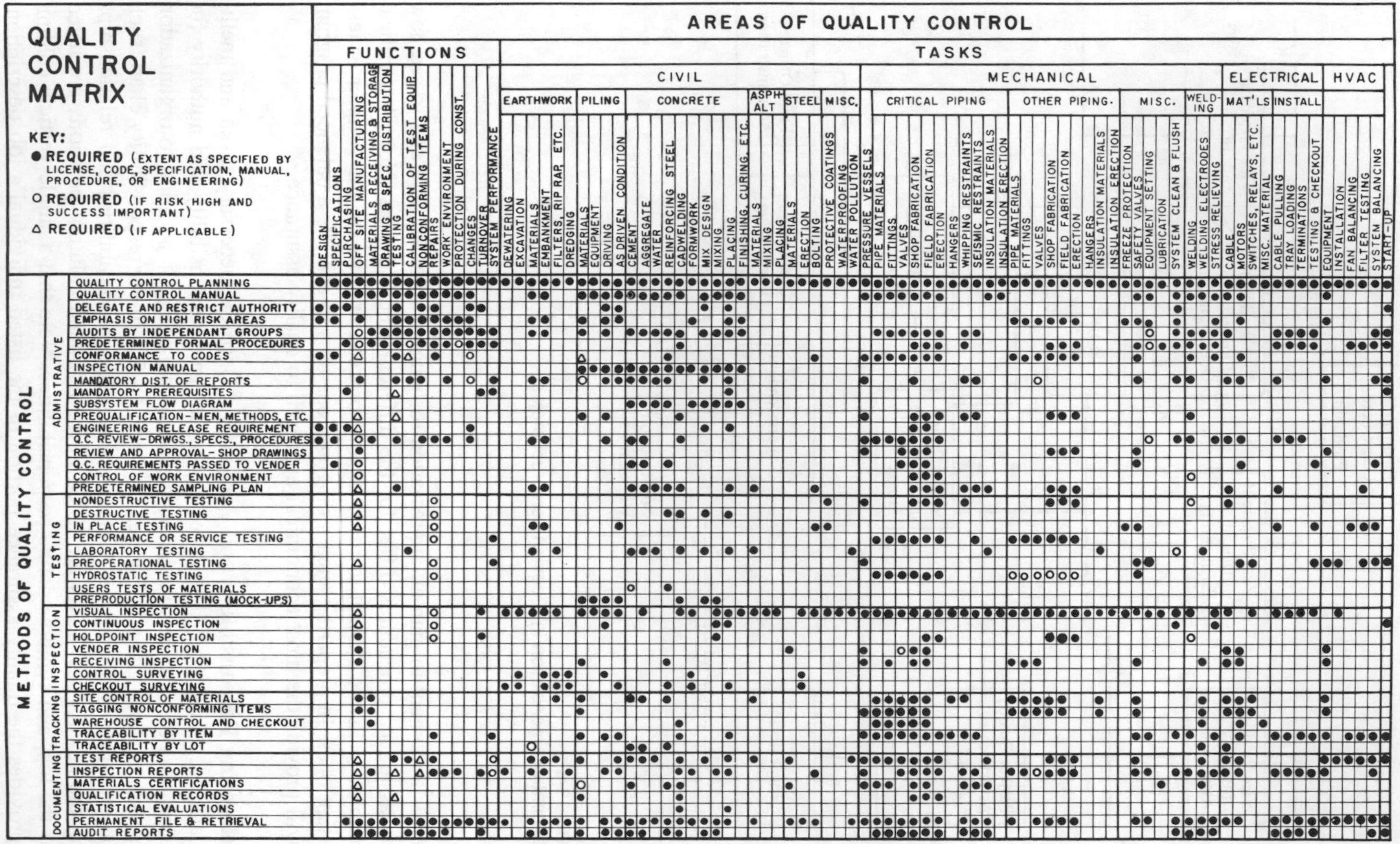

FIGURE 16-6
Quality control matrix. (From "System for Control of Construction Quality," *Journal of the Construction Division*, ASCE, vol. 98, no. CO1, March 1972, pp. 28–29.)

LINEAR RESPONSIBILITY CHART

• KEY •

●	PRIMARY RESPONSIBILITY
▲	JOINT RESPONSIBILITY
■	APPROVAL RESPONSIBILITY
○	MUST BE CONSULTED
△	MAY BE CONSULTED
□	AUDITS OR REVIEWS

	OWNER				ENGINEER				CONSTRUCTOR							
	Manager of New Construction	Chief Engineer	Board of Review	Site Representative	Project Manager	Project Engineer	Responsible Design Engineer	Materials Engineer	Project Superintendent	Construction Superintendent	Site Q.C. Supervisor	Site Q.A. Supervisor	Home Office Q.A. Chief	Purchasing Agent	Startup Engineer	Regulating Agency
Select quality objectives	▲				▲				▲							▲
Define activities affecting quality	■	△	□		●	△	△	△	○							□
Specify quality standards	■	△	□		■	●	△	△	○							□
Prepare quality control manual	■	△	□		■	○		△	○				●			□
Prepare quality control procedures		□			■						○	●	■			□
Prepare construction method procedures				□		□			■	●	○	□				□
Prepare welding and NDT procedures		□	□		■			●								□
Establish design criteria	■	●	□													□
Perform design	■	■	□		■	■	●									□
Define vendor quality control requirements	■				■	●	○	○						△		□
Prepare procurement documents	■	△	□		○								□	●		□
Evaluate vendor quality capability	■	△	□		■	○			○					●		□
Inspect off site manufacturing		□	□			□	△						●		○	□
Control distribution of plans and specifications			□		○				○	●						□
Specify sampling plans		□	□			□	△		○				●			□
Train and qualify craftsmen			□						○	●	○	□				□
Train and qualify inspection personnel			□	□								●	□			□
Direct construction operations									○	●						
Inspect work in progress			□	□		□	△			△	●	□				□
Accept work in progress			□	□	△	□	△				●	□				□
Stop work in progress			△			□	○		△		○					□
Inspect materials upon receipt				□			△	△				●	□			□
Monitor and evaluate quality trends				□	□	□						●				□
Maintain file of quality control documents					□	□						●				□
Determine disposition of nonconforming items	■	△	□	△	●	△	△	△	○		○	○				□
Investigate failures	○	□	□	□	○	○	△	△	△	△	△	△	△			□
Release systems and components to operations				□					□	●					■	□
Conduct flushing and cleaning operations					■										●	□
Conduct preoperational testing			□		■										●	□
Accept completed plant as to quality	●	○	△													■

FIGURE 16-7
Linear responsibility chart (LRC) for quality control organization. (From "System for Control of Construction Quality," *Journal of the Construction Division*, ASCE, vol. 98, no. CO1, March 1972, p. 31.)

Preparation of such a chart requires a thorough understanding of the project and the contractual relationships of the parties; it also clarifies ambiguities and identifies omissions and conflicts.

Implementation The two charts in Figures 16-6 and 16-7 fit into the overall system for quality assurance as shown in Figure 16-8. Note again that this follows the pattern of the feedback-control system introduced in Chapter 10, Figure 10-2.

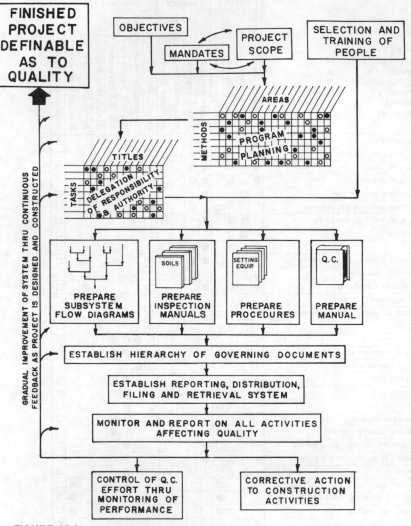

FIGURE 16-8
System for control of construction quality. (From "System for Control of Construction Quality," *Journal of the Construction Division*, ASCE, vol. 98, no. CO1, March 1972, p. 22.)

Figure 16-9 gives a detailed example of how the procedures operate in the case of quality control on concrete construction. Although there is more detail here than might be required on some simpler projects, a thorough study of the figure will deepen the reader's understanding of the quality assurance process.

METHODOLOGY

Most modern approaches to quality assurance, and more specifically to quality control, require an understanding of probability and statistics. It is beyond the scope of this

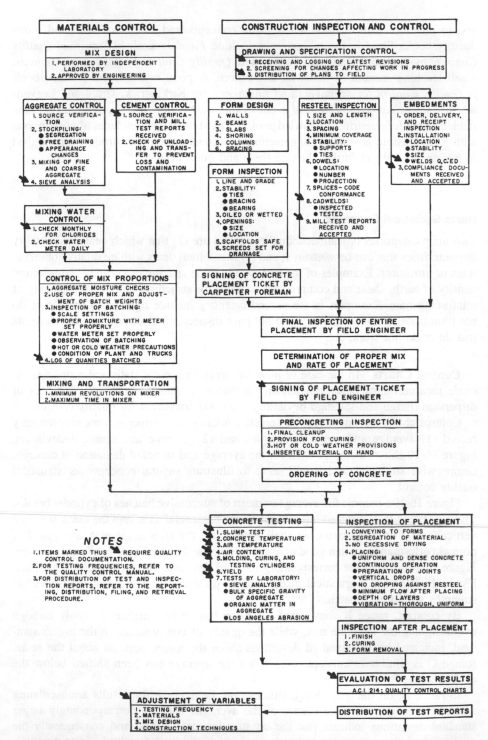

FIGURE 16-9
Flow diagram, quality control of concrete construction. (From "System for Control of Construction Quality," *Journal of the Construction Division*, ASCE, vol. 98, no. CO1, March 1972, p. 33.)

text to do more than mention a few basic concepts and methods. Two good introductory books listed in the bibliography include *Fundamentals of Statistical Quality Control*, by Samson, Hart, and Rubin, and *Quality Control by Statistical Methods*, a self-teaching manual by Knowler and others. A good management-oriented book is *Quality Control for Managers and Engineers*, by Kirkpatrick. Grant and Leavenworth's *Statistical Quality Control* is one of the most widely used in-depth texts in the field. The reader requiring a knowledge of this field—and this includes almost anyone with a responsible position in engineering and construction—is strongly encouraged to pursue this subject further through these and other books.

Basic Statistical Concepts

Two main categories of statistical quality control are (1) that which deals with quality characteristics that can be *measured*; and (2) that which deals with qualitative observations or *attributes*. Examples of the first type include spacing of columns, compaction density of earth, shear and compressive strength for structural lumber, and spacing of reinforcing bars. Examples in the second category include light bulbs that do or do not illuminate, welds which do or do not pass inspection, electronic control elements that do or do not work, etc.

Control Charts   In the case of measurements, important statistical properties include measurements of central tendencies (mean, median, mode) and measures of dispersion (range and standard deviation). Both are important for quality control.

Control charts documenting the central tendency and dispersion most commonly record (1) average and range of samples and (2) average and standard deviation. Figure 16-10 shows a control chart for the average and standard deviation of concrete compressive strengths, and it will serve to illustrate several concepts for statistical quality control.

Figure 16-10*a* records the average or mean of successive batches of cylinder breaks. The successive tests proceed left to right on the horizontal axis, and the mean is on the vertical axis. For comparison, we shall assume that the tests in range A are typical of normal performance, with a good quality of design and a good quality of conformance. Figure 16-10*b* shows the corresponding standard deviation for each batch of tests, and Figure 16-10*c* uses a graphical distribution to illustrate the pattern of the individual strengths within each batch.

By contrast, range B shows that the mean has shifted higher, probably through overdesigning the concrete mix, while the quality of conformance to the higher standard, indicated by the standard deviations about the mean, remains about the same. Range C is similar in concept, except that the average has been shifted below the desired strength.

In range D, on the other hand, although on the average the results are oscillating around the mean, the wide scatter of the averages and the correspondingly larger standard deviations indicate that the construction procedures, and consequently the quality of conformance, appear to have become badly out of control.

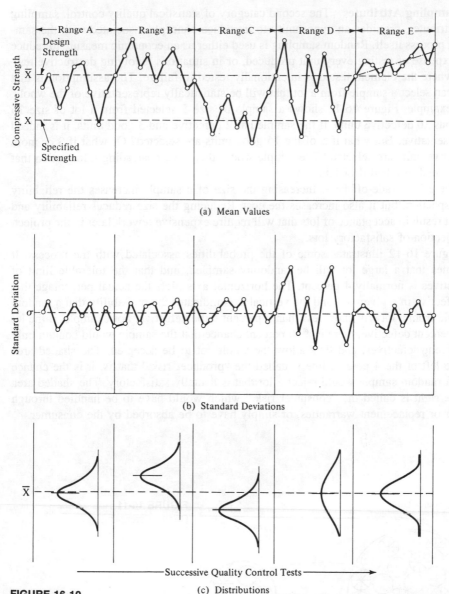

FIGURE 16-10
Control charts for mean and standard deviation.

Range E is included to show one of many other types of situations that can be revealed through control charts. The gradual uptrending pattern might indicate continuous wear on a key component in the process—possibly the mechanism that closes the gate that lets the cement into the concrete batch. These and other problems can be spotted by an experienced user of this type of quality-control tool.

Sampling Attributes The second category of statistical quality control, sampling by attributes, provides an opportunity to introduce concepts associated with the sampling process itself. Random sampling is used either as an economic measure to reduce the expense of testing every unit produced, or in situations involving destructive testing where the element tested is intentionally tested to failure. The assumption is that one can select a sample from a lot that will be statistically representative of the whole. For example, Figure 16-11 shows a sample of size 5 selected from a lot of size 25 that has 10 defective units. If the sample has 2 defective and 3 good units, it is indeed representative. But what if 5 of the 15 good units are selected? Or what if 3 or more defective units are selected? The sample would then give misleading information that could lead to a bad decision.

There is a trade-off here. Increasing the size of a sample increases the reliability of inspection, but it also increases the cost. Reducing the size reduces reliability and might result in acceptance of lots that will require expensive rework later in the project, or rejection of satisfactory lots.

Figure 16-12 illustrates some of the probabilities associated with the process. It assumes that a large lot will be randomly sampled, and that the tolerable limit of defectives is normally 4 percent. The horizontal axis plots the actual percentage that is defective in a given lot, and the vertical axis indicates the probability that a random sample from a lot will show less than 4 percent to be defective. For example, if the lot is 8 percent defective, there is a 50 percent chance that the sample would contain only 4 percent defectives, and thus allow the whole lot to be accepted. The shaded area to the left of the 4 percent line is called the "producer risk," that is, it is the chance that a random sample would reject a lot that is actually satisfactory. The shaded area to the right is called the "consumer risk," which would have to be handled through repair or replacement warranties, or simply have to be absorbed by the consumer.

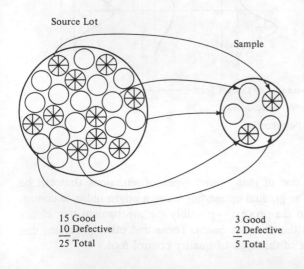

Source Lot

Sample

FIGURE 16-11
Random sampling.

15 Good
10 Defective
25 Total

3 Good
2 Defective
5 Total

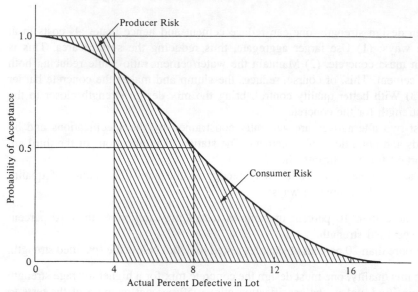

FIGURE 16-12
Risk and uncertainty in sampling. (Adapted from Charles Samson, Philip Hart, and Charles Rubin, *Fundamentals of Statistical Sampling*, Addison-Wesley Publishing Company, Reading, Mass., 1970, p. 87.)

Practical Considerations   At all times one must be on the lookout for extraneous factors affecting the quality assurance process. A real case history will illustrate this point. A supplier of ready-mix concrete for highway construction in the state of Oregon has reported that every summer the state-laboratory reports on the quality of his concrete start to go very bad about 28 days after the state university closes for the summer. His own testing indicates that his concrete really does not change at all in quality. Rather, what happens is that in the summer the state highway department temporarily hires college engineering students for the prime construction season. In their first few days on the job, the students are inexperienced in making the cylinders for concrete testing. The results are reflected in the tests after 28 days of curing. There are innumerable other examples of aberrations introduced by human factors, mechanical defects in test equipment, or environmental factors beyond the control of both constructors and testers.

Example: Concrete Quality Control

The design and control of concrete quality illustrate much of what we have discussed in this chapter. The designer, in quality engineering, prepares specifications on strength, size of aggregate, dimensional tolerances, slump, finish, etc. Assuming the designer has done so, the constructor still has considerable latitude in trading off quality control against methods to produce the required quality of conformance.

For the materials costs, cement is normally by far the most expensive component. The water/cement ratio is in turn a key factor in determining the strength of concrete.

For a given design strength, one can reduce cement and hence materials costs in at least three ways: (1) Use larger aggregate, thus reducing the surface area. This is common in mass concrete. (2) Maintain the water/cement ratio while reducing both water and cement. This, of course, reduces the slump and makes the concrete harder to place. (3) With better quality control, bring the mix design strength closer to the specified strength for the concrete.

The first two alternatives are generally constrained by the specifications and by the methods and conditions of placement. The statistical implications of the third are most important for our purposes here.

A typical concrete specification may recognize the statistical nature of quality control in one of the following ways:

1 No more than 10 percent of samples tested may fall more than 10 percent below the specified strength.

2 No more than 20 percent of samples tested may fall below the specified strength.

To achieve this quality, one must design the concrete mix for a higher average strength than that specified, unless the specification allows 50 percent or more of the tests to fall below the specified strength. This concept is illustrated in Figure 16-13, where X is the specified strength, and $\overline{X}$ is the mix design strength.

Knowing X, $\overline{X}$ can be derived as follows:

Assume X_i = observed strength for test i

σ = standard deviation

$$= \sqrt{\frac{\sum_{i=1}^{n}(X_i - X)^2}{n}} \qquad (16\text{-}1)$$

where n = number of cylinders tested

V = known coefficient of variation determined for the batch plant being used

$$= \frac{\sigma}{X} \times 100\% \qquad (16\text{-}2)$$

t = standard variable, the number of standard deviations from the mean (design) to the specified strength

$$= \frac{X - \overline{X}}{\sigma} \qquad (16\text{-}3)$$

Now by substitution of

$$\sigma = V\overline{X} \text{ for } \sigma$$

$$\therefore t = \frac{X - \overline{X}}{V\overline{X}} \qquad (16\text{-}4)$$

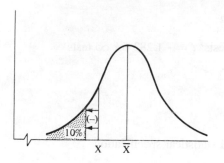

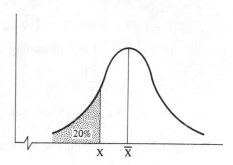

(a) No more than 1 in 10 less than 0.9X

(b) No more than 1 in 5 less than X

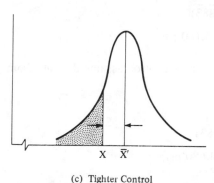

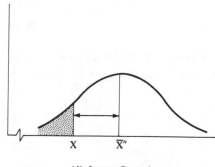

(c) Tighter Control

(d) Looser Control

FIGURE 16-13
Various levels of concrete quality control.

which gives

$$\overline{X} = \frac{X}{1 + Vt} \qquad (16\text{-}5)$$

This is the design strength that should satisfy the specification for the specified strength. The specifications define X. The standard variable t is from statistical tables for the normal distribution and corresponds to the owner's specification for the fraction of samples permitted to be substandard. V is developed from the contractor's or the ready-mix supplier's concrete quality control. If V can be kept low, one can design closer to the specified strength and thus keep the concrete material costs down. The trade-off is in the better equipment and supervision required to keep V under tight control.

Example 1 Determine the design strength required to satisfy the specified strength of 300 kg/cm² (4,270 psi) under the terms of the first specification above. Assume V = 20 percent.

$$X - 10\% = 300 - 30 = 270 \text{ kg/cm}^2$$

$$10\% \Rightarrow t = -1.383 \text{ (assuming 10 tests; } t = -1.282 \text{ for } \infty \text{ tests)}$$

$$\therefore \overline{X} = \frac{270}{1 + 0.2\,(-1.383)}$$

$$= 373 \text{ kg/cm}^2 \text{ (5,300 psi)}$$

Example 2 As above, but use the second statistical specification.

$$20\% \Rightarrow t = -0.383 \text{ (for 10-test average)}$$

$$\overline{X} = \frac{300}{1 + 0.2\,(-0.883)}$$

$$= 364 \text{ kg/cm}^2 \text{ (5,180 psi)}$$

Example 3 To illustrate the effect of better quality control, assume the data from example 2, but change V to 10 percent.

$$\overline{X} = \frac{300}{1 + 0.1\,(-0.883)}$$

$$= 329 \text{ kg/cm}^2 \text{ (4,680 psi)}$$

Example 4 To illustrate the effect of a specification that ignores statistical variations and in effect demands perfection by requiring that *no* cylinders fall below the specified strength, consider the following:

First, to have numbers to work with, approximate "no" cylinders to "no more than 3 standard deviations below" (i.e., 1 in 741). This gives a t of -3.000 standard deviations. Otherwise, assume X and V from example 2.

$$\overline{X} = \frac{300}{1 + 0.2\,(-3.0)}$$

$$= 750 \text{ kg/cm}^2 \text{ (10,670 psi)}$$

This is a very high design that would require extraordinarily expensive procedures. For this reason, almost all modern specifications do recognize the statistical implications of quality control.

QUALITY IN THE CONSTRUCTED PROJECT

The ASCE has developed a manual entitled *Quality in the Constructed Project*, Volume 1, issued in late 1990.[2] The manual was developed by a number of volunteer

---

[2]The concepts in this section are based upon *Quality in the Constructed Project: A Guide for Owners, Designers, and Constructors*, Volume 1 published by the ASCE, New York, New York, 1990.

authors and reviewers along with a number of individual critiques. The abstract from the manual describes its purpose as follows:

> This final version of the *Quality in the Constructed Project* (ASCE Manual No. 73) provides suggestions and recommendations to owners, design professionals, constructors and others on principles and procedures which have been effective in providing quality in the constructed project. It also provides guidance for establishing roles, responsibilities, relationships and limits of authority for project participants; and stresses the importance of concepts and practices that enhance the quality in the constructed project. Throughout the manual, various themes considered to be of particular importance are discussed. They include such concepts as: 1) the definition and assignment of responsibility; 2) the importance of teamwork; 3) the importance of concise contractual provisions; 4) the principles of good communication; 5) the owner's selection process for project team members; and 6) the procedures for design and construction.

Development of the publication began in 1985 and over 1000 professionals from ASCE and other design and construction industry organizations were involved. The 24-chapter manual discusses task requirements pertinent to quality and assigns prime responsibility, assisting or advising, or reviewing responsibilities to the owner, design professional or constructor.

PEER REVIEWS

Chapter 21 of *Quality in the Constructed Project* is devoted to peer reviews. Peer review is a technique that promotes quality in design organizations and in their services. It is a structured, independent review, by a team of peers external to the organization or project being reviewed. This section summarizes peer reviews in design organizations and for individual projects as outlined by the ASCE and by the *American Consulting Engineers Council* (ACEC).[3]

Types of Peer Reviews

The two main categories of peer review are organizational peer reviews and project peer reviews as illustrated in Figure 16-14. Organizational peer reviews look at the design organization as a whole, particularly its policies, procedures and practices. Project peer reviews consider individual projects and can include a project management peer review concentrating upon organizational aspects of the project and/or a project design peer review concentrating upon the technical results or recommendations as developed by the designers.

The peer review is a separate optional step in the design process conducted by one or more senior professionals who are separate and independent from the organization preparing the design. The scope of the review is defined in advance by the review team with the design professional and/or owner. It will include the functions to be reviewed, the process to be followed and the form of reporting.

---

[3] The figure and concepts in this section are substantially based upon Chapter 21, *Quality in the Constructed Project* op. cit and the *Peer Review Manual* published by the ACEC, Washington D.C., 3rd ed., 1987.

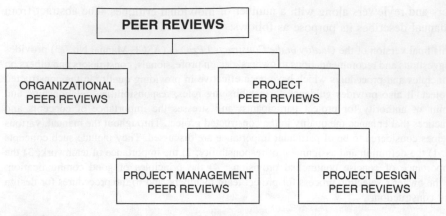

FIGURE 16-14
Categories of Peer Review.

The effectiveness of the team is greatly influenced by the independence, experience, skill and stature of the peer review team members. The size of the team depends upon the scope and complexity of the project, but in complex projects would normally involve several technical disciplines.

History of Peer Reviews

In the United States most building departments will perform plan checks or reviews before issuing building permits. However, scope and level of review can vary widely and in many instances are very rudimentary in nature. Budget restrictions and time constraints often limit such reviews to code requirements or major structural issues which could cause catastrophic failures. Other organizations, such as the city of Los Angeles Building Department, perform thorough reviews. The city of Boston developed peer review requirements shortly after a major structural collapse in 1971. The state of Connecticut requires independent engineering review on projects that exceed certain threshold limits. Costs of the Boston process have been estimated at 5 percent to 10 percent of the complete design.

In West Germany design reviews by independent "proof engineers" are mandatory for major structures and have been in use for many years. "Proof engineers" are federally licensed, independent peer consultants, retained by the municipality. Their responsibility is to confirm soil conditions, and architectural and structural integrity of the design. For complex structures, detailed checks of computations, drawings and temporary supports are also required. Cost of the reviews has been estimated to add from 0.6 percent to 1.0 percent to the construction cost.

In Belgium, the Bureau de Control pour la Sécurité de la Construction en Belgique (SECO) supervises all phases of design and construction. SECO is a nonprofit organization, similar to an engineering consulting firm, but which represents all Belgian insurance companies. Owners needing insurance submit the design to their insurance carrier who submits it to SECO for review prior to issuing a policy. A technical board arbitrates any differences in technical opinion between the designer and SECO. Failure to adopt SECO suggestions precludes obtaining insurance. France utilizes a similar system called SOCOTEC.

Large dams throughout the world have traditionally practiced a form of peer review. For each major dam or other similar project, the owner and/or the design professional appoints a consulting board of recognized experts who will meet periodically in order to review fundamental design criteria and to monitor actual development of the design as construction proceeds.

A number of detailed manuals on peer review have been published by the American Consulting Engineers Council (ACEC), including joint publications with other design organizations, and are referenced in the Bibliography.

SUMMARY

Quality assurance, which encompasses both quality engineering and quality control, involves (1) the application of standards and procedures to ensure that a product or facility meets or exceeds desired performance criteria, and (2) documentation to verify the results obtained. Quality engineering involves the application of procedures to ensure that design proceeds according to recommended and mandatory criteria set by professional associations, building code authorities, and the environment, while it produces a facility that most economically serves the owner's needs. Quality control includes the development and application of procedures that will, at economical levels, assure attainment of the designed quality when the project is constructed.

The elements of quality include quality characteristics, quality of design, and quality of conformance. Quality characteristics are those properties chosen to define the nature of a product for design and control purposes. The quality of design relates to the design tolerances set for the chosen characteristics that will enable a product to function at the desired level of reliability and economy. Quality of conformance is the degree to which the physical work produced conforms to the specified design. All these factors have cost implications on the project.

Quality economics in design involves trade-offs between the value of quality and the cost of obtaining it. In theory, the optimum design is at a level where the marginal value of an additional unit of quality equals its marginal cost. Economics of quality of conformance involves a minimum-cost balance between the costs of better but more costly construction methods to improve conformance and the costs of rejects and a greater control effort resulting from poor conformance. In reality, numerous other parameters, such as quantity, affect both the design and the conformance trade-offs, but the basic concepts must first be understood.

Organization for quality assurance differs from other phases of a project control system, and reflects the varying and often conflicting interests of the parties to the contract as well as the interests of society as imposed through governmental regulation. Briefly stated, there is often considerable emphasis on keeping the quality assurance function separate from project operations in order to ensure that third-party objectivity will prevail in decision-making. This chapter included four figures which detail the tasks and methods involved in quality assurance and show how responsibility and authority for these tasks can be delegated.

Methodology for most modern approaches to quality assurance requires at least some understanding of probability and statistics. There are two major categories of statistical quality control: one deals with quality characteristics that can be measured, and the other deals with qualitative observations or attributes. In both categories, there are various types of control charts and sampling procedures that can assure good quality control at economical levels.

A comprehensive manual, *Quality in the Constructed Project*, published by the ASCE, includes guidelines and recommendations for owners, design professionals and constructors on how to provide quality in construction projects. Peer reviews are increasingly being utilized on major projects and can provide significant quality assurance benefits.

17

SAFETY AND HEALTH
IN CONSTRUCTION

The state of safety and health in construction is reflected by the fact that of all the chapters in this book, this is the one that many readers might be most inclined to skip—or, with a slight nag from conscience, skim. Like "motherhood, the flag, and apple pie," safety is a subject to which most people are quite willing to pay lip service, but which too few are willing really to do something about. To put to rest any illusions of self-righteousness, the authors must admit to having been as remiss as many of their colleagues in this respect. For this very reason, however, the words in this chapter are more deeply felt than any others.

The facts show that construction is indeed a dangerous industry; U. S. Department of Labor and National Safety Council statistics indicate that although construction employees account for only about 6 percent of the total labor force, they incur 12 percent of all occupational injuries and illness (about 250,000 to 300,000 lost-time injuries per year in construction), and 19 percent of all work-related fatalities (about 3000 per year by National Safety Council estimates, and about 1000 according to the Occupational Safety and Health Administration). Related costs to the industry are estimated to run between $5 *billion* and $10 *billion* per year.[1] Research findings also show that these losses are far higher than they need to be. Many of the industry's most successful and profitable firms also have the best records in safety and health, as do many of the best and most productive workers, foremen, superintendents, and top managers. These findings are no coincidence, and they serve as a goal for the rest of the industry.

Safety and health are as much a part of effective project planning and control as are costs, schedules, procurement, and quality. Indeed, they are all closely interrelated.

[1]Raymond E. Levitt, *The Effect of Top Management on Safety in Construction*, Technical Report No. 196, Stanford University, Dept. of Civil Engineering, The Construction Institute, Stanford, Calif., July 1975, pp. 8–9.

Many of the same principles of management and engineering that have been described in earlier chapters also apply equally well here. In this chapter we shall describe the problem, then focus on constructive steps an organization can take to improve its occupational safety and health performance.

MOTIVATORS FOR IMPROVED PERFORMANCE

Safety and health are of concern to today's organizations on several levels. These include humanitarian concern, economic costs and benefits, legal and regulatory constraints, liability consequences, and organizational image. All are important, though changes in constraints and attitudes both within and imposed from outside the industry cause some factors to receive more emphasis than others.

Humanitarian Concern

On a purely humanitarian level, the purpose of improved occupational safety and health is to reduce the human pain and suffering, to workers' families as well as to themselves, that result from accidents and work-induced illness. It is difficult to quantify in economic terms, though the statistics quoted in the opening paragraphs tell at least part of the story. Three thousand deaths mean at least that many bereaved families: widows, children losing a parent, lost sons and daughters. Further suffering is inevitable for the 300,000 and their families who each year incur lost-time injuries. Even the strongest human beings are frail creatures when subjected to the forces of nature and the industrial hazards associated with the moving machinery, dust, explosives, heat, electricity, noise, potential for falling, and toxic substances that form the everyday environment of construction. The resulting injuries are often cruelly disfiguring and result in lifetime handicaps and disabilities. But because it is so difficult to quantify the humanitarian factors in preparing estimates and making operating decisions on projects, we must move beyond these into the economic costs of safety and health programs and their economic benefits.

Economics

Like national statistics on injuries, illness, and deaths in construction, on a personal or organizational level it is also difficult to relate to costs in the range of "$5 billion to $10 billion per year." More relevant, perhaps, is the fact that a high experience-modification factor on the workers' compensation insurance rate can put a contractor out of business in competitively bid construction work.

Standard manual rates for insurance depend on the performance of the whole industry; typical averages are about 18 percent of field labor costs in commercial building work. Electrical work averages about 6.5% and structural steel about 43.5%. But a contractor's individual premiums are adjusted up or down to reflect its performance relative to the industry, so a particular company may actually pay from less than 50 percent of the manual rate to over 150 percent. Furthermore, contractors with good safety records can also earn significant end-of-year "dividends" over and above the reduction in premiums reflected in the modification factor. Assuming general contractor labor to be 30 percent of project costs on a building construction job, the insurance premium will range from about 5 to 27 percent of the total direct costs, with the 22 percent difference being enough either to lose the bid or to wipe out the profit.

The impact of these differences is not limited to competitively bid work. In negotiating cost-plus-fee contracts, knowledgeable owners are increasingly taking a hard look at prospective contractors' insurance costs and safety records.

When the indirect as well as the direct costs of accidents and illness are considered, costs associated with insurance premiums, claims settlements, and the like are only a small portion of the whole. Even at the project level, it is often estimated that the indirect and impact costs resulting from interrupted production, reduced morale, lower productivity, and ripple effects on the schedule can be several times the direct costs associated with hospitalization, disability pay, spoiled materials, damaged equipment, and reconstruction. The indirect impact costs must normally be absorbed directly by the project.

Much the same analysis applies at the worker's level. Although today's headlines frequently report multimillion-dollar liability settlements, more typical settlements under workers' compensation are limited mainly to workers' medical expenses and a fraction of lost salaries. Furthermore, the indirect impact of the psychological and emotional disruption to the families is very real, but not easy to quantify. Where long-term or permanent disability results, there is an unknown lost potential for both earnings and future growth and development.

The industry as a whole also suffers from the loss of each worker. The cold fact is that skilled workers are a scarce and valuable resource; considerable time, money, and effort are invested in their training. A career interrupted at age 25 means that 40 additional years of skilled production have been lost. With today's shortages, it is a foolish, let alone callous, misjudgment to say there will be someone else to take the accident victim's place.

For these and other reasons, the more enlightened organizations in the engineering and construction industry—owners, constructors, designers, and agencies alike—have recognized that effective programs to improve performance in safety and health are not expenses, but investments. Some have estimated that each $1 invested in safety and health pays $4 to $8 in return. We need not be concerned so much, then, in justifying these investments as in optimizing them. Few other investment opportunities have such outstanding potential.

Legal and Regulatory Constraints

Like almost all laws, the increasing burden of legislation and regulation in occupational safety and health has developed in response to those individuals and organizations who have demonstrated irresponsible behavior when left to their own devices. Also like almost all other laws, the unfortunate side effect is further to curb the freedoms and even hamper the positive efforts of the responsible segments of the industry. The problem in construction is compounded by the intensely competitive nature of the industry, where short-run expediency in cost-cutting areas such as safety and health often seems attractive and even necessary for business survival. This not only runs counter to the productivity of long-term investments in safety described above, but this general attitude of expediency also accounts in large measure for the high rate of business failure in construction.

Regardless of the origins, the consequences of the irresponsible safety and health performance of some individuals and firms have brought on laws and regulations as a leveler for all. If these laws or their enforcers at times appear misguided and impractical, our recourse now is not to repeal the laws, but to redirect their application so that they do focus on real problem areas for the overall benefit of workers, the industry, and the economy. It is not our objective here to defend the current state of governmental regulation, but to recognize the reasons for it and to emphasize the importance of improving its effectiveness where it is really needed.

In construction, the main federal agency for regulation is the Occupational Safety and Health Administration, generally known as OSHA. The enabling legislation also authorizes state agencies to assume the functions of the federal OSHA, provided that the state agencies enforce standards at least as strict as those at the federal level. Some projects, however, will operate under different regulations. At the time of this writing, some construction projects on mine properties come under the Mining Enforcement and Safety Administration, known as MESA, which is within the Department of Interior's Bureau of Mines. Contracting agencies, such as the Army Corps of Engineers, will also sometimes prescribe regulations that go beyond those of OSHA.

For most construction, however, OSHA is the key agency. Established under the Occupational Safety and Health Act of 1970, it generally applies to all sectors of the private economy. In construction this includes designers, owners, and workers, as well as contractors. The law itself is a relatively concise document of 31 pages, but that is only the beginning. Hundreds of pages of regulations have subsequently been published in the *Federal Register* (starting with the issues of April 27 and May 29, 1971). These in turn incorporate by reference whole bookshelves full of published industry and government safety manuals in every conceivable field of endeavor. Employers have been made responsible for understanding and interpreting a seemingly impenetrable maze of material, a task which many agencies and committees are now striving to simplify. It is well beyond the scope of this chapter even to outline, let alone describe, those that apply just to construction.

In philosophy, however, the intent of the Act is quite straightforward. The essence is captured in Section 5, quoted here:

DUTIES

Sec. 5. (a) Each employer—

(1) shall furnish to each of his employees employment and a place of employment which are free from recognized hazards that are causing or are likely to cause death or serious physical harm to his employees;

(2) shall comply with occupational safety and health standards promulgated under this Act.

(b) Each employee shall comply with occupational safety and health standards and all rules, regulations, and orders issued pursuant to this Act which are applicable to his own action and conduct.

Like any other major piece of social or economic legislation, this law has been controversial, to say the least. First of all, to mandate that employers provide "employment and a place of employment which are free from recognized hazards" is a significant change in legal philosophy. Workers' compensation laws were considered a major step forward in the late nineteenth and early twentieth centuries when they relieved an employee's individual burden by denying employers the three old common law defenses of (1) the worker's assumption of risk in taking employment, (2) contributory negligence by the employee, and (3) negligence of fellow employees. But the underlying assumption remained that injury and death were inevitable expenses of production; what changed was that their burden shifted from individual workers to the industry as a whole. With OSHA, however, the assumption is that the "employment and place of employment" should be such that injury and death do not happen in the first place. This certainly is a noble goal, but the economy has only begun to comprehend the full implications of this shift in social and economic philosophy. Disputes over the details of its application and implementation will likely continue well into the twenty-first century. Suffice it to say that the law exists, it has "teeth," it is being enforced; all engineering and construction organizations must understand it and learn to operate within its constraints.

Liability Problems

A general trend in the courts to increasing frequency of lawsuits and increasingly large jury awards has also been causing problems for projects. The tentacles of liability have been reaching out to ensnare parties with only loose connections to the administration of a project, such as a bank that provides a mortgage loan. Therefore it is important that managers of engineering and construction organizations stay abreast of trends in liability suits, including those that at present appear frivolous, and take steps to minimize liability expense on projects.

In construction safety and health, this only begins with being sure that, to the extent humanly possible, all operations are carried out within recognized standards and regulations. Numerous other factors are also coming into play. For example, loss of hearing is now recognized as justification for a worker's compensation claim, but hearing loss is cumulative. Nevertheless, unless an employer can show that the claimant's hearing was poor before starting work on the project in question and that it suffered no further deterioration, the employer might be stuck with the full claim. To compound matters, medical testing has shown that the hearing of many young people these days is damaged by listening to overamplified music long before they seek their first job in construction. Some such music is played in the range of 120 to 130 decibels, which is roughly the unmuffled range of some of the noisiest construction operations, such as percussion drilling in a hard-rock tunnel. It appears that if liability claims for hearing loss become prevalent, it will be only prudent for employers to test employees' hearing when hired, retest when terminated, and document the results. The documentation should not backfire as long as the project is being run within current noise regulations. Already, however, union and legal restrictions are being imposed to prevent such testing.

Similar problems can be cited in other categories of occupational illness and disability. For example, spinal disorders appear to be a real consequence of operating some types of equipment, such as undampened scrapers with two-wheel tractors. Verified disabilities of this type can indeed be cause for legitimate claims, and manufacturers are working hard to correct the source of the problem. But as a generic problem, the so-called "backache" or "back trouble" also appears to some industry observers to have become a means for malingerers to collect unemployment and disability benefits. Their problem may actually be minor, or psychosomatic, or even pure fabrication, but when they go to a physician, who faces his own dire malpractice consequences if he is skeptical and ignores the "problem," it is not hard to get medical certification that the "problem" exists and that the employee should not work. This is a complex area, and a test-in/test-out procedure like the hearing test would be much more difficult to apply. The best present remedy is to include real or imagined backache complaints in past employee records, and to monitor such employees carefully. As in other liability and claims problems, good documentation is generally the best defense.

Organizational Image

Some companies and government agencies pride themselves on their good safety "image," both among their employees and among the public at large. Levitt cited this as one of three key motivators stated by top managers as justification for a safety program (the others being humanitarian concern, and workers' compensation and other accident costs).[2]

Especially in negotiated work, this image can be decisive in the award of contracts. Some owners, priding themselves on their own image, do not want a contractor coming onto their job who will do anything to tarnish that reputation. Others, simply looking at the balance sheet, want to minimize payment of excess insurance costs on the project and to avoid publicity-grabbing liability suits. In any case, the constructor with a proven good accident record and an effective no-nonsense safety program can also have a real competitive edge in negotiating contracts.

To other companies, a good safety image is both a matter of pride among industry peers and solid evidence of responsiveness to employee needs that can in turn engender higher morale and productivity and stronger employee loyalty. It is no coincidence that such companies can often better attract and keep skilled professionals in their ranks.

PROBLEMS IN SAFETY AND HEALTH

In the area of safety, it has long been recognized that many types of construction operations present serious hazards. Only recently, however, have occupational health problems in construction received much attention. This section will discuss each in turn.

[2]Raymond E. Levitt, op. cit., p. 5.

Safety

Safety hazards are those that pose imminent danger of causing injury or death to workers or damage to materials, equipment, or structures. They result not only from obvious physical dangers, but also from human factors such as lack of training, poor supervision, attitudes, poor planning, or even from workers who are so familiar with the work that they become oblivious to it.

Many of the physical aspects of construction safety result from the sheer scale of the work when compared with the frailty of the people performing it. In this environment workers are exposed to falls of hundreds of feet, yet are often injured or killed when falling off a stepladder. Even a hard hat is but an eggshell to a rock popping off the roof of a tunnel, a pipe wrench falling from the fortieth floor, or a 20-ton beam swinging out of control at the end of a cable. Fires in some types of building materials move quickly and generate intense heat. Lay people are generally astonished at the size of the 100-ton trucks, 20-yard shovels, and other large machines on an earthmoving job, but the operators and those working nearby too often become complacent about the amount of energy with which they are dealing. Similarly, electricians who admonish their children to be careful with 115-volt, 20-amp household wiring often need to heed their own warnings when taking shortcuts in making repairs to live 4000-volt circuits.

These forces are present on most large jobs, but they are not the real safety hazards. In most accidents, people are the problem. We shall examine this subject later in this chapter.

Health Problems in Construction

Until recently it was popularly assumed that construction provides rugged, out-in-the-fresh-air work and ideal summer training for athletes, and is healthy for anyone who can stand its pace. Although hard hats had gained some legitimacy as a safety device, health protectors such as earplugs, respirators, and shock absorbers were for "sissies"; no "real man" would be caught dead using them. Too many, unfortunately, are dead for lack of them, and others are handicapped for life.

Health hazards in construction include, among others, heat, radiation, noise, dust, shocks and vibrations, and toxic chemicals. Perhaps the main hazard here, however, is human optimism. Since the effects are not immediately felt, we say: "I can work in this dust from rock drilling for a few more hours. A hot, steamy shower will clear it out!" "I can go into the tunnel heading without ear plugs. The pain stops when I come out!" "I'll just ride these backbreaking scrapers until I'm 40, then I'll retire to a D-8!" "It sure is hot out, and I'm feeling dizzy and have a real headache coming on, but it's only an hour until quitting time. There's no point in stopping this truck for a drink of water now!" And, "I've been working with asbestos for 20 years and I'm not sick. What's this business about its causing cancer?"

Increasingly, it is being recognized that occupational diseases have indeed been a serious problem in construction. There are substantial direct costs for medical treatment and disability claims, and indirect costs through the premature loss of skilled workers. Many of the hazards are being not only identified but eliminated. Asbestos is but one of many recent examples. It is vitally important that all organizations involved in construction stay up to date with developments in occupational health and implement methods proven to reduce health hazards. If humanitarian concerns are insufficient, the liability implications should be more than enough reason. Two good starting points for information are two other McGraw-Hill Books in this series, *Productivity Improvement in Construction*, a book by Oglesby, Parker, and Howell,[3] and *Construction Safety Management*, by Raymond E. Levitt and Nancy M. Samelson.[4]

IMPLEMENTATION GUIDELINES

An effective construction safety and health program has many parallel functions. Oglesby, Parker, and Howell[5] broadly categorize these as follows:

Personal or behavioral factors

- Worker: his training, habits, beliefs, impressions, educational and cultural background, social attitudes, and physical characteristics
- Job environment: attitudes and policies of the employers and the managers, supervisors, foremen, and coworkers on the project

Physical factors

- Job conditions: dictated by hazards inherent in the work being performed, as well as by health hazards arising from methods and materials and the location of the job
- Mechanical hazard elimination: use of barriers, devices, and procedures to shield workers physically from hazardous areas or situations (trench shields, chain guards, etc.)
- Protection: use of such variables as hard hats, safety glasses, respirators, earplugs, seat belts, roll bars, and other devices to protect the individual's health and safety

All these factors are essential to a well-rounded safety program. Traditionally, major company safety expenditures as well as government regulatory programs have been aimed mainly at the physical factors. One senses this strong emphasis in most of the OSHA publications as well as in others, and OSHA inspections and fines also reflect this. Studies have shown, however, that roughly 80 percent of all industrial accidents result from unsafe acts in the accident chain, and not just from unsafe conditions. This finding implies that there should be much heavier emphasis on the personal and behavioral side rather than solely on the physical aspects. The disproportionate emphasis on the physical side partially accounts for the disappointing results of many safety programs, including those of OSHA.

Why the heavy emphasis on physical approaches to health and safety? For one thing, only recently have studies indicated the importance of the human side. For another, physical programs are much easier to visualize and implement, especially for people in technical or production-oriented industries. Finally, it is much more difficult

---

[3]Clarkson H. Oglesby, Henry W. Parker, and Gregory Howell, *Productivity Improvement in Construction*, McGraw-Hill Book Company, New York, 1989.

[4]Raymond E. Levitt and Nancy M. Samelson, *Construction Safety Management*, McGraw-Hill Book Company, New York, 1987.

[5]Oglesby, Parker, and Howell, op. cit.

to know how to approach the human side of safety and health, especially in a high-turnover and fast-changing industry like construction. Recent construction research studies, however, have been producing some clear and workable guidelines for the behavioral approach, and there is hope that in the near future we may see constructive changes and improved results in safety and health in construction.

Again, however, it is important to emphasize that both the behavioral and the physical sides must be developed simultaneously in an effective safety and health program. This chapter will thus present each in turn. The following section will summarize the findings of four research studies giving policy guidelines at levels from worker to top management. A subsequent section will present the physical aspects of construction safety. The reader is strongly encouraged to consult references cited in this chapter and in the bibliography to obtain detailed information in each area.

Behavioral Approaches to Safety and Health

Essentially all the findings and recommendations published in this section are the products of nearly a decade of research in Stanford University's Graduate Program in Construction Engineering and Management. This research was conducted by engineering faculty with years of experience in construction, a research social psychologist with some 30 years' experience in her field, and numerous graduate students.

In essence, the studies have been based on extensive survey work conducted in the field with the aid of construction companies, labor organizations, insurance companies, and their employees at all levels. Four separate but interrelated studies have focused on (1) top management, (2) superintendents and project management, (3) foremen, and (4) workers. The guidelines summarized below give only a glimpse of the depth of the research and the reported findings, but they at least will give the reader a point of departure. In effect, the source reports themselves contain a practical and workable program to enable an organization to increase its emphasis on the behavioral side of occupational safety and health in construction.

Guidelines for Top Managers The results below are quoted from the study of top managers by Dr. Raymond E. Levitt.[6] The study provided strong evidence that top managers can reduce accident costs significantly by:

1 Knowing the safety records of all field managers and using this knowledge in evaluating them for promotion or salary increases.
2 Communicating about safety on job visits, in the same way that they communicate about costs and schedules.
3 Using the cost accounting system to encourage safety by:
   • allocating safety costs to a company account.
   • allocating accident costs to projects.
4 Requiring detailed work planning to ensure that equipment or materials needed to perform work safely are at hand when required.
5 Insisting that newly hired employees receive training in safe work methods.

[6]Levitt, op. cit., p. 6.

6   Discriminating in the use of safety awards. The data suggest that:
   • Safety awards for workers, if used, should be incentives (awards of nominal monetary value), based on first-aid injuries rather than on lost-time accidents.
   • If correctly applied, safety awards for field managers should be bonuses (awards of substantial monetary value, made in private) based on lost-time accidents or insurance claims costs.
7   Making effective use of the expertise of safety departments, where these exist.

Guidelines for Superintendents and Project Managers   The middle-management study was conducted by Dr. Jimmie Hinze.[7] His findings, the summary of which is paraphrased here, showed that middle managers can reduce injuries significantly by:

1   Showing concern for and establishing rapport with foremen and workers. They can do so by making sure to:
   a   Orient new workers to the job and acquaint them with other job personnel. Particular attention should be given to the new workers in their first few days of employment.
   b   Be involved in worker-foremen conflicts, and in so doing, to recognize the worker's viewpoint. This is not to undermine the foreman's authority, but rather, to assure that the workers are fairly treated.
   c   Show respect for the ability of foremen, but also to accept the fact that foremen are not immune to error. This can be done by permitting foremen to select their own crew members (but not granting them the sole authority to terminate employees).
2   Keeping unnecessary pressures off the workers and foremen. Pressures to be avoided include:
   a   Stressing strict adherence to detailed cost estimates.
   b   Stressing adherence to detailed time estimates.
   c   Condoning or encouraging competition between crews on the job.
3   Actively supporting job safety policies, for example, by:
   a   Including safety as a part of job planning.
   b   Giving positive support to "toolbox" meetings.
4   Accepting responsibility for eliminating unsafe conditions and unsafe activities from the job.

Furthermore, Hinze found that top management can help supervisors reduce job accidents by:

   1   Personally stressing the importance of job safety through their informal and formal contacts with field supervisors.
   2   Stressing safety in meetings held at the company level. Because some of these findings indicate that pressures on cost, schedule, and competition should be reduced, it is worth pointing out that these studies also showed that safe supervisors, foremen, and workers can also be among the most productive. They in effect put to rest the myth that schedules and budgets must necessarily be traded off against safety and health. Indeed, by relaxing tensions, employees are in a better position for getting on with the job and doing

---

[7]Jimmie Hinze, *The Effect of Middle Management on Safety in Construction*, Technical Report No. 209, Stanford University, Dept. of Civil Engineering, The Construction Institute, Stanford, Calif., June 1976, pp. 4–6.

it well. Again, the reader is encouraged to consult the source reports for the documentation that supports these summary conclusions.

Guidelines for Foremen The foremen study was the primary focus of research social psychologist Dr. Nancy Morse Samelson.[8] In a preliminary outline, she sought answers to the question: How do highly effective (both safe and productive) construction foremen manage their crews? The answers form a set of probing guidelines in themselves:

1 *They handle the new worker differently*
   • They ask him more questions and less threatening ones.
   • They are watchful and keep a connection with the new worker rather than putting him right to work or putting him with an older worker and leaving it at that.
2 *They keep stresses off their crews*
   • They show high ability to "keep their cool" rather than show anger toward crew members.
   • Their reaction to lack of crew accomplishment is to analyze the problem rather than focus on changing the workers by telling them to work harder.
3 *Their approach to safety is different*
   • They integrate safety into the job with personal work rules rather than having a set of safety admonitions.
   • They are neither safety "nit-pickers" nor are they unaware of safety violations—they are in between.

Guidelines for Workers The study of construction workers was conducted by Lance deStwolinski.[9] His objective was more to identify characteristics of safe and unsafe workers than to prescribe a set of guidelines. These characteristics can then be used by management in selection and in tailoring supervision and assignments to recognize the needs and limitations of individuals. Oglesby, Parker, and Howell[10] summarized deStwolinski's findings as follows:

> Mere recognition of these relationships is not sufficient to solve the job-accident problem, however. A patterned program of evaluation is necessary. The following simple procedure offers one approach. The implication here is that the enlightened supervisor can work the odds to his benefit in safety if he takes the trouble to analyze his crews (especially those working under hazardous conditions) and gives extra attention to those individuals and situations that fit the following known critical characteristics that may lead to accident involvement:
>
> 1 The worker with abnormal time loss (absenteeism)
> 2 The worker whose time losses pattern into Mondays or days after payday as the days lost most frequently
> 3 The individual who requires the most supervision to produce normally

[8]Nancy M. Samelson, *The Effect of Foremen on Safety in Construction*, Technical Report No. 219, Stanford University, Dept. of Civil Engineering, The Construction Institute, Stanford, Calif., June 1977.

[9]Lance W. deStwolinski, *A Survey of the Safety Environment of the Construction Industry*, Technical Report No. 114, Stanford University, Dept. of Civil Engineering, The Construction Institute, Stanford, Calif., October 1969.

[10]Oglesby, Parker, and Howell, op. cit.

 4 The worker working in isolated areas

 5 The worker with problems from home, skirmishes with the law, and the like

 6 The individual who acts abnormally to attract attention (e.g., dress, hair style, hot rodders)

 7 The worker whose attitude changes with the time of day and the day of the week

 8 New workers with less than one year of service or those with more than 10 years' service

 9 Any individual whose name "crops up" frequently in any unfavorable light (e.g., absent, sick, frequently leaves job site)

 10 The worker whose personal appearance changes noticeably (watch for sudden change in gait, color, or actions)

When a new worker is hired, he should be assessed as to his possible susceptibility to accidents. The safety questionnaire (used by deStwolinski) indicated that answers to the questions listed below showed a high correlation to a man's accident record, indicating that attitudinal factors toward himself, his foreman, and job management may be significant in accident statistics.[11] Although these questions were designed for a study of a specific union in a specific section of the country, it is probable that other groups might have similar findings; however, such a correlation needs to be tested. Reevaluations of continuing employees is also necessary from time to time to determine those whose attitudes, home environment, or habits may have changed.

The significant questions as reported by deStwolinski have been reworded into statements and are listed here in the order of significance. "Yes" answers to these or similar questions indicate that the worker is in the accident-susceptible class.

 1 My job management does not know its job well.

 2 I have worked a relatively short time in my present job classification.

 3 I would like to have an opportunity for a good family life.

 4 My foreman is stubborn.

 5 My coworkers are boring.

 6 My job management does not praise good work.

 7 My coworkers are not safety minded.

 8 Risk taking is a part of the job.

It should be noted that most of these questions reflect worker-coworker-supervisor interactions rather than worker attitudes alone. In fact, recent studies which have attempted to isolate worker characteristics associated with accidents separately from other influences have been unsuccessful.

Other Behavioral Factors Several other results from deStwolinski's study are worth summarizing here. These relate to safety instruction, job safety meetings, and safety equipment.

Policies on Safety Instruction When new employees are hired, the introduction to the job and its surrounding conditions plays an important role. Safety instruction is an important aspect of this introduction. Thus, the response from 40 percent of the nonsupervisory operating engineers that instructions were not given at all, or if they

---

[11] deStwolinski, *A Survey of the Safety Environment*, op. cit., pp. 53–66.

were given, they were not to the workers' satisfaction, indicated that the situation is not good. Dissatisfaction with instruction probably results from poor-quality instruction, unsatisfactory instructional materials, or a lack of understanding on the part of the individual employee. The future trend may well be toward greater dissatisfaction with the quality of instruction. This results because of the rising education level of the younger operating engineers, an increasing number of whom have high school and some college education.

The most significant finding, related to safety instruction, is that there is a direct relationship between minor and lost-time accidents and the presence or absence of effective safety instruction at the time of initial employment. Those receiving instructions to their satisfaction have significantly better records than those who were not given, or did not understand, instruction, or who were not given effective instructions.

Job Safety Meetings   Job (toolbox) safety meetings have been used for some time to provide safety education on the job. In a number of states they are required by law. For example, in California it is required that job safety meetings be held at least every 10 working days.

A disturbing finding was that having or not having safety meetings seemed to have no effect on the lost-time or minor-accident rate. This is not to imply that all job safety meetings are ineffective. But it does say that the presence or absence of safety meetings had no marked effect on accident experience. A possible explanation was brought out through the survey. It was that, although those conducting the safety meetings were considered as knowledgeable, the meetings were often so dull and for the most part the "same old stuff" that they had no effect on worker attitude or behavior.

It seems that to make job safety meetings more effective, there is a need for more practical and current subject material given by a variety of qualified speakers. They might come from outside, from either the union or the company itself. Smaller meetings for specific crafts also may be appropriate, with more discussion dealing with immediate problems. For crews which have a variety of work assignments, discussion of the safety aspects of each new assignment might be held before the task is begun. The real point is that changes must be made if job safety meetings are to be effective and productive to the employee and employer.

Safety Equipment   Responses on this subject showed that, except for the requirement for hard hats (stipulated by 60 percent of the respondents), little else in the line of safety equipment was required by contractors. A good percentage used gloves (41 percent) and safety glasses or goggles (27 percent) on a voluntary basis. However, very few employers required that the workers wear earplugs or muffs, special clothing, steel-toed boots, gloves, or safety glasses. In light of the problems of accident potential and increased health hazards from dust, noise, and noxious agents, such items as these should be required on all jobs.

Physical Approaches to Safety and Health

The physical side of construction safety requires:

1 Education and training in correct methods and procedures.

2 Provision and proper utilization and application of good-quality, well-maintained tools and equipment, both for construction operations and for mechanical elimination of hazards. Examples of items currently being emphasized include roll-over protection on earthmoving equipment, and noise-level controls.

3 Enforced use of approved equipment for personal protection: hard hats, seat belts, earplugs, etc., as required by specific operations.

4 Good housekeeping on the job site.

5 Frequent and thorough job-site inspections by knowledgeable and objective professionals.

6 Incorporation of a safety review as a routine part of thorough preplanning for the actual methods and procedures to be carried out in field operations.

There are numerous excellent safety manuals that provide detailed elaboration on items 1, 2, 3, and 4 above. Many large construction companies have developed outstanding in-house manuals tailored to their own type of work. Good examples that are available to the public include the following:

California Construction Safety Orders, Dept. of Industrial Relations, Division of Industrial Safety, San Francisco. Also, *Tunnel Safety Orders*, from the same agency.

General Safety Requirements, Manual EM385-1-1, and Supplements 1 and 2, U.S. Army Corps of Engineers, Washington, D.C.

Safety Requirements for Construction by Contract, Dept. of the Interior, U.S. Bureau of Reclamation, Washington, D.C.

OSHA publications:

Compliance Operations Manual, OSHA 2006, U.S. Dept. of Labor, Occupational Safety and Health Administration, Washington, D.C., January 1972.

"Construction Safety and Health Regulations," U.S. Dept. of Labor, Occupational Safety and Health Administration; *Federal Register*, vol. 39, no. 122, June 24, 1974, Washington, D.C.

Construction Industry: OSHA *Safety & Health Standards*, OSHA 2207, Superintendent of Documents, U.S. Government Printing Office, Washington, D.C. Revised 1987.

Clearly, it is beyond the scope of this chapter to attempt the type of detailed elaboration contained in those manuals. Suffice it to say that detailed knowledge of this type is fundamental to an effective program in construction safety and health. We shall take a moment here, however, to comment on the fifth and sixth elements mentioned in discussing the physical factors: inspection and preplanning.

Inspection Good in-house inspection by personnel authorized to implement changes is becoming increasingly common these days, in part as a matter of self-defense against OSHA fines. Other companies are making good use of experienced inspectors provided by insurance companies. In some cases this approach either directly or indirectly affects insurance premiums. Regardless of the motivation, the trend toward objective and qualified inspection of work sites is a good one, and it has been a long time in coming. Inspection has always been an essential part of an effective safety and health program.

It is essential that the inspectors themselves be experienced in construction operations, and that they be objective, fair, and practical in their recommendations and directives to project managers and supervisors. Nothing can destroy a safety program more quickly than conspicuously ignorant and inexperienced inspectors who compensate with missionary zeal for what they lack in knowledge. However, given that we can have intelligent and objective inspectors, it is also important that they have the authority, either directly or through recommendations backed by higher management, to see that safety and health standards are maintained on job sites. Many companies are providing the necessary clout by making the inspector an independent entity on the job site, reporting directly to the project manager, or to the home office rather than to project management. There are advantages and disadvantages to this approach, but the organizational relationship is much like that for the quality control function shown in Figure 16-5.

The inspector's task is eased somewhat if he has a set of standards to use in his work. Figure 17-1 reproduces one of a complete series of checklists published by *Construction Methods and Equipment* magazine for this purpose; these are practical forms that have found wide acceptance in the construction industry. The example form, describing personal protective equipment, also illustrates typical factors that would be considered in items 1 through 4 of the list of physical factors at the beginning of this subsection.

Preplanning Thorough and conscientious preplanning is essential to economy, efficiency, and high productivity in almost all construction operations; safety and health considerations should be an integral part of this process. Safety professionals should participate in the development of standard procedures and should review job-operation plans for considerations such as the following:

  1 To verify that the method selected does indeed adhere to recommended and required standards and regulations for safety and health

  2 To be certain that the correct tools and equipment will be available for the work, including the necessary personal protection gear

  3 To express reservations about supervisors or workers who lack the skills that will be needed, and to suggest remedial training procedures where appropriate

  4 To anticipate hazards inherent in the work and recommend precautionary steps for dealing with them

Preplanning of this type, with good supervision to see that the plan is indeed executed, is one of the best methods to assure not only high levels of safety and health, but high production as well. Oglesby, Parker, and Howell's *Productivity Improvement in Construction* focuses on this subject in detail.[12]

[12]Oglesby, Parker, and Howell, op. cit.

Safety and Health On Worksites

CM&E's series of SHOW checklists are designed to make sure you're in full compliance with every aspect of the Construction Safety Act (Federal Register: April 17, 1971, Part II; May 29, 1971, Part II) and all the subsequent revisions, corrections, and amendments.....before things happen, not after.

No. 4

Project _____

Inspection area _____   Area supervisor _____

Inspected by _____   Date _____

Subject / Personal protective equipment	Yes	No	Action / comments
Head protection			
1. Do your employees wear protective helmets whenever they work in areas where there is the possible danger of a head injury from impact and penetration of falling and flying objects, or from electrical shock and burns?			
2. Do the helmets worn by your employees for the protection against impact and penetration of falling and flying objects meet the specifications in ANSI Z89.1-1969–Safety Requirements for Industrial Head Protection?			
3. Do the helmets worn by those of your employees exposed to high voltage electrical shock and burns meet the specifications in ANSI Z89.2-1970?			
Hearing protection			
1. Do you provide ear protective devices for your employees whenever it is not feasible to reduce noise levels or duration of exposure as specified?			

Hr duration per day	dBA level slow response	
8	90	
6	92	*a. Variations in noise level involving maxima at intervals of 1*
4	95	*sec or less are considered as continuous.*
3	97	*b. Daily noise exposure composed of two or more periods of noise*
2	100	*exposure should be considered in combination, rather than a indi-*
1½	102	*vidually.*
1	105	*c. Exposure to impulsive or impact noise should not exceed 140*
½	110	
¼ or less	115	*dBA peak sound pressure level.*

2. Do you make sure that your employees use the ear protective devices you provide for them?			
3. If these ear protective devices are inserted in the ear, are they fitted or determined individually for each employee by competent persons? *Important: Plain cotton is not an acceptable protective device.*			
Eye and face protection			
1. Do you provide your employees with eye and face protection equipment when machines or operators present potential eye or face injury from physical, chemical, or radiation agents?			
2. Does the eye and face protection that you provide your employees meet the requirements specified in ANSI Z87.1-1968–Practice for Occupational and Educational Eye and Face Protection?			
3. If any of your employees, whose vision requires the use of corrective lenses in spectacles, are required to wear eye protection, are the goggles or spectacles that you provide for their protection one of the approved types? *a. Spectacles whose protective lenses provide optical correction.* *b. Goggles that can be worn over corrective spectacles without disturbing the adjustments of the spectacles.* *c. Goggles that incorporate corrective lenses mounted behind the protective lenses.*			
4. Do you keep face and eye protection equipment clean and in good repair?			
5. Do you forbid the use of any of this type of equipment having structural or optical defects?			

continued on next page

FIGURE 17-1
Example safety checklist. (From *Construction Methods and Equipment*, vol. 54, no. 8, August 1972, pp. 23–24.)

Safety and Health On Worksites

No. 4

Project _____

Inspection area _____ Area supervisor _____

Inspected by _____ Date _____

Subject/Personal protective equipment	Yes	No	Action/comments

continued

6. Does the equipment that you provide your employees for face and eye protection conform to the standards for such protection from the hazards and operations as listed below?

Operation	Hazards	Recommended Protection
Acetylene–burning	Sparks, harmful rays	7,8,9
Acetylene–cutting	Molten metal	7,8,9
Acetylene–welding	Flying particles	7,8,9
Chemical handling	Splash, acid burns, and fumes	2,10 (for severe exposure add 10 over 2)
Chipping	Flying particles	1,3,4,5,6, 7A, 8A
Electric (arc) welding	Sparks, intense rays, and molten metal	9,11 (11 in combination with 4,5,6 in tinted lenses)
Furnace operations	Glare, heat, and molten metal	7,8,9 (for severe exposure add 10)
Grinding–light	Flying particles	1,3,4,5,6,10
Grinding–heavy	Flying particles	1,3,7A, 8A (for severe exposure add 10)
Laboratory	Chemical splash, and glass breakage	2 (10 when in combination with 4,5,6)
Machining	Flying particles	1,3,4,5,6,10
Molten Metals	Heat, glare, sparks, and splash	7,8 (10 in combination with 4,5,6 in tinted lenses)
Spot welding	Flying particles, sparks	1,3,4,5,6,10

Eye and face protector selection guide:
1. Goggles–Flexible fitting, regular ventilation
2. Goggles–Flexible fitting, hooded ventilation
3. Goggles–Cushioned fitting, rigid body
*4. Spectacles–Metal frame, with sideshields
*5. Spectacles–Plastic frame, with sideshields
*6. Spectacles–Metal-plastic frame, with sideshields
*7. Welding goggles–Eyecup type, tinted lenses
7A. Chipping goggles–Eyecup type, clear safety lenses
**8 Welding goggles–Coverspec type, tinted lenses
8A. Chipping goggles–Coverspec type, clear safety lenses
**9. Welding goggles–Coverspec type, tinted plate lens
10. Face shield–Plastic or mesh window
**11. Welding helmet

*—Non-side shield spectacles are available for limited hazard use requiring only frontal protection.
**—See following table for filter lens shade numbers.

Selection of shade numbers for welding filter

1. Do you provide your employees the proper filter lense or plate for protection against radiant energy in welding, as specified in the following table?

Welding operation	Shade number
Shielded metal-arc welding 1/16, 3/32, 1/8, 5/32-in.-dia electrodes	10
Gas-shielded arc welding (nonferrous) 1/16, 3/32, 1/8, 5/32-in.-dia electrodes	11
Gas-shielded arc welding (ferrous) 1/16, 3/32, 1/8, 5/32-in.-dia electrodes	12
Shielded metal-arc welding 3/16, 7/32, 1/4-in.-dia electrodes	12
5/16, 3/8-in.-dia electrodes	14
Atomic hydrogen welding	10-14
Carbon-arc welding	14
Soldering	2
Torch brazing	3 or 4
Light cutting, up to 1 in.	3 or 4
Medium cutting, 1 to 6 in.	4 or 5
Heavy cutting, over 6 in.	5 or 6
Gas welding (light) up to 1/8 in.	4 or 5
Gas welding (medium) 1/8 to 1/2 in.	5 or 6
Gas welding (heavy) over 1/2 in.	6 or 8

Laser protection

1. Do you furnish employees, whose occupation or assignment requires exposure to laser beams, suitable laser safety goggles which will protect them for the specific wavelength of the laser and be of optical density adequate for the energy involved as specified in the following table?

Intensity		Attenuation	
CW max power density (w/cm²)		Attenuation Optical density (O.D.)	Optical factor
10^{-2}		5	10^5
10^{-1}		6	10^6
1.0		7	10^7
10.0		8	10^8

(Output levels falling between lines in this table shall require the higher optical density.)

2. Do all your protective goggles bear a label identifying:
a. Laser wavelength for which use is intended?
b. Optical density of those wavelengths?
c. Visible light transmission?

FIGURE 17-1
(continued).

SUMMARY

Statistics show that safety and health are critical problems in construction. In the United States alone, construction deaths and lost-time injuries run up to 3000 and 300,000 per year respectively; annual direct and impact costs are estimated at $5 billion to $10 billion. Some companies, however, are bucking the trends; their outstanding safety and health records show that the industry as a whole can do much better. Basically, these industry leaders have discovered that safety and health are as much a part of effective project planning and control as are costs, schedules, procurement, and quality.

Motivations for improved performance include humanitarian concern, economic costs and benefits, legal and regulatory mandates, liability consequences, and organizational image. Humanitarian objectives boil down to reducing the human pain and suffering resulting from accidents and illness. Economic incentives include not only reducing insurance premiums, but minimizing the staggering indirect costs as well. Regulatory constraints have been imposed mainly by government agencies, and they serve as a lever in attempting to bring inept and irresponsible organizations into compliance with industry standards. Increasing liability problems, including high accident claims settlements for injuries and illness often taken for granted in the past, have led prudent employers to test and document key health factors of their employees. Finally, the positive image generated by good safety and health performance helps both in employee relations and in contract negotiations.

Safety hazards are those that pose imminent danger of causing injury or death to workers. They include falling from heights, fire, moving machinery and vehicles, explosives, electricity, and falling objects. Health hazards may or may not produce immediate symptoms, but they have been gaining wider recognition in construction. Among these dangers are heat, radiation, noise, dust, shocks and vibrations, and toxic chemicals.

Effective implementation programs should focus on both the physical and the behavioral sides of safety and health. Traditional approaches have concentrated on the physical or "hardware" aspects, yet roughly 80 percent of all industrial accidents involve unsafe acts and not just unsafe conditions. A balance between the different components of safety and health is therefore essential.

On the behavioral side, recent research has produced practical and workable guidelines aimed at the attitudes and actions of top management, project managers and superintendents, foremen, and workers. The intent is to identify the characteristics of managers and workers who are both safe and productive so that selection, assignments, supervision, and efforts for improvement can take these factors into account.

The physical side of safety involves: (1) education and training; (2) proper utilization and maintenance of correct tools and equipment; (3) equipment for personal protection; (4) good housekeeping; (5) frequent inspections by knowledgeable and objective professionals; and (6) integrating safety and health into thorough preplanning for field operations.

PART FOUR

BUSINESS METHODS IN MANAGING CONSTRUCTION

18

RISK MANAGEMENT, INSURANCE, BONDING, LIENS AND LICENSING[1]

Construction work is often a hazardous undertaking. Owners, contractors, construction managers, designers and others have a number of options to protect themselves from the hazards of the business. Risk management may be defined as an organized approach to identifying and dealing with potential exposures.

Insurance forms a major option in any risk management program to shift designated risks to a financially strong party who, for an agreed premium amount, is willing to assume some or all of the financial responsibility for the loss.

Payment and Performance bonds supply protection to owners against failure of the contractor to carry out his contractual obligations. The bond underwriter or surety agrees to indemnify the owner against default or other failure to perform upon the part of the contractor. Bonds are normally required on public works and are optional in the private sector. Bonding can also be included as a part of a risk management program.

A mechanic's lien is a statutory right which permits contractors, subcontractors and material suppliers to secure payment for improvements to real property. Architects and engineers are also eligible to file statutory liens in some states. Similar nonstatutory liens, called *stop notices*, also are prevalent in the public sector. Such liens provide contractors, material suppliers and others protection for non payment for completed work.

Licensing laws for contractors, architects/engineers and other design professionals are intended to help assure that reasonable qualifications are possessed by participants on a construction project in order to protect the general public.

[1]The authors are indebted to industry executive Henry Trainor, retired President and Chairman of Miller and Ames, San Francisco Division, for his review and helpful suggestions which have been incorporated into this chapter.

RISK MANAGEMENT PROGRAMS

Risk management in the construction industry includes an organized effort to identify and quantitatively evaluate potential exposures, along with an advance plan designed to eliminate or mitigate the consequences of the risks. A comprehensive risk management program throughout the life of the project continues the approach developed by the initial plan. A comprehensive safety program is an important component of a risk management program and is discussed in detail in Chapter 17, *Safety and Health in Construction.*

The owner, the architect/engineer, the general contractor, the developer or construction manager, and the subcontractors all have their individual needs and objectives along with individual risk tolerance. Some of these risks can be covered by insurance and bonding. Others can be eliminated by contractually transferring the risk to another party. Some construction methods may be more risky than others but offer potentially greater rewards. Evaluation of anticipated frequency and severity of the risk compared to the contemplated reward can assist in choosing the optimum method.

Risk Identification

Risks can be divided into internal and external, predictable and unpredictable, and technical and nontechnical. Some of the risks are best handled by the owner, others are best handled by the designer, and some are normally assumed by the contractor. Assigning risks best assumed by one of the parties to other parties not in a position to control or mitigate the risk has been the root cause of much of the adversarial relationships and litigation now experienced in the industry. Figure 18-1 lists a number of risks applicable to a construction project. Chapter 2, *Development and Organization of Projects*, lists advantages and disadvantages to both the owner and the contractor based upon risk assignment between the owner and contractor for various contract types.

CONSTRUCTION INSURANCE

This section will review a number of the more common and usual types of insurance which would be of general interest to owners, contractors, subcontractors, construction managers, developers and design professionals.

Worker's Compensation and Employer's Liability Insurance

Worker's compensation insurance is designed to provide the statutory benefits required by state law to an employee who is hurt or killed as a result of employment. Some states and the Canadian provinces have established mandatory funds which employers must utilize unless they can qualify as self-insured. Some states have established nonmandatory funds and in these jurisdictions the employer may choose between private insurance and the state program. In most states, however, the preponderance of worker's compensation insurance is provided by private insurance companies.

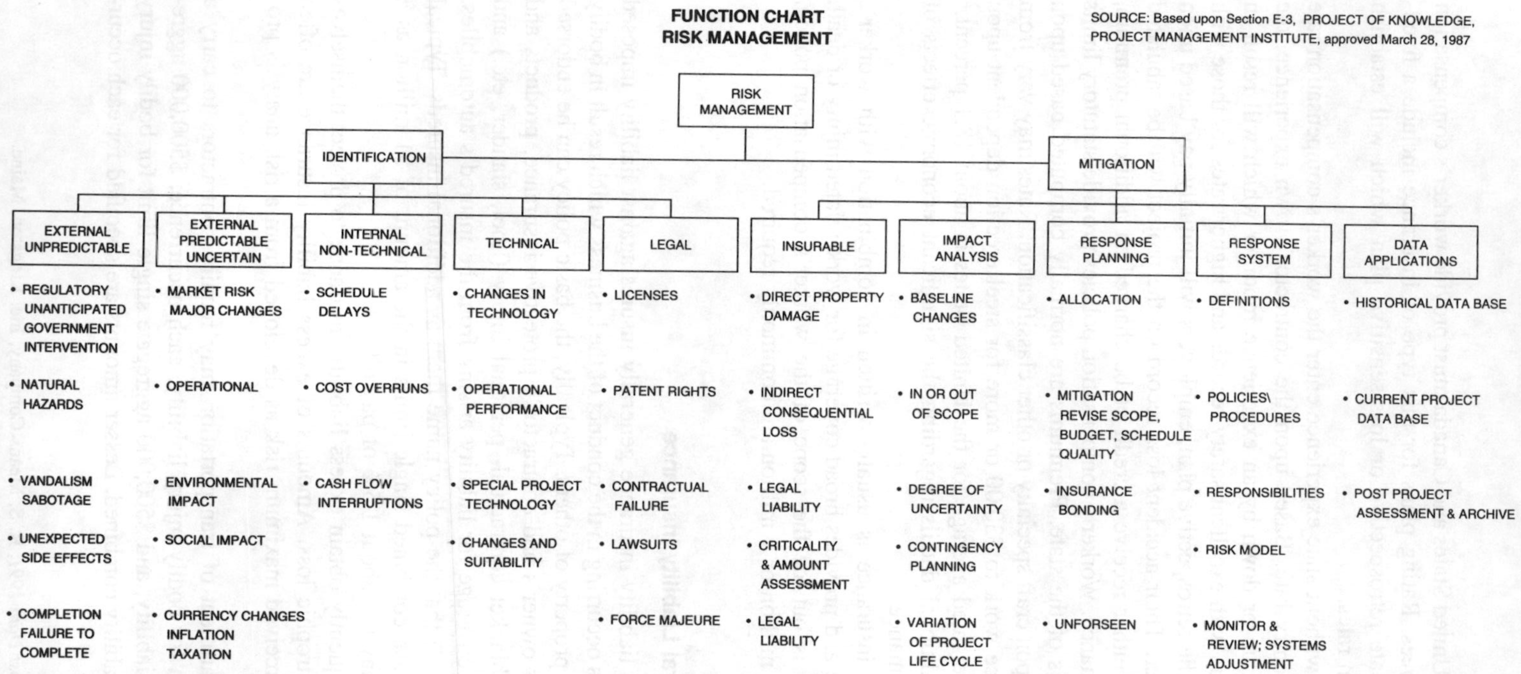

SOURCE: Based upon Section E-3, PROJECT OF KNOWLEDGE, PROJECT MANAGEMENT INSTITUTE, approved March 28, 1987

FUNCTION CHART
RISK MANAGEMENT

RISK MANAGEMENT

IDENTIFICATION

EXTERNAL UNPREDICTABLE
- REGULATORY UNANTICIPATED GOVERNMENT INTERVENTION
- NATURAL HAZARDS
- VANDALISM SABOTAGE
- UNEXPECTED SIDE EFFECTS
- COMPLETION FAILURE TO COMPLETE

EXTERNAL PREDICTABLE UNCERTAIN
- MARKET RISK MAJOR CHANGES
- OPERATIONAL
- ENVIRONMENTAL IMPACT
- SOCIAL IMPACT
- CURRENCY CHANGES INFLATION TAXATION

INTERNAL NON-TECHNICAL
- SCHEDULE DELAYS
- COST OVERRUNS
- CASH FLOW INTERRUPTIONS

TECHNICAL
- CHANGES IN TECHNOLOGY
- OPERATIONAL PERFORMANCE
- SPECIAL PROJECT TECHNOLOGY
- CHANGES AND SUITABILITY

LEGAL
- LICENSES
- PATENT RIGHTS
- CONTRACTUAL FAILURE
- LAWSUITS
- FORCE MAJEURE

MITIGATION

INSURABLE
- DIRECT PROPERTY DAMAGE
- INDIRECT CONSEQUENTIAL LOSS
- LEGAL LIABILITY
- CRITICALITY & AMOUNT ASSESSMENT
- LEGAL LIABILITY

IMPACT ANALYSIS
- BASELINE CHANGES
- IN OR OUT OF SCOPE
- DEGREE OF UNCERTAINTY
- CONTINGENCY PLANNING
- VARIATION OF PROJECT LIFE CYCLE

RESPONSE PLANNING
- ALLOCATION
- MITIGATION REVISE SCOPE, BUDGET, SCHEDULE QUALITY
- INSURANCE BONDING
- UNFORESEEN

RESPONSE SYSTEM
- DEFINITIONS
- POLICIES\PROCEDURES
- RESPONSIBILITIES
- RISK MODEL
- MONITOR & REVIEW: SYSTEMS ADJUSTMENT

DATA APPLICATIONS
- HISTORICAL DATA BASE
- CURRENT PROJECT DATA BASE
- POST PROJECT ASSESSMENT & ARCHIVE

FIGURE 18-1
Risk Management.

All employers in the United States and Canada must provide worker's compensation benefits to their employees. Rating plans for this type of insurance include a fixed-rate plan and an alternate retrospective or loss-sensitive plan which will establish maximum and minimum rates.

Initially, contractors without state experience enter the worker's compensation rate pool at the manual or entry rate. Based upon the contractor's own experience, the manual rate is adjusted up or down by an experience modifier which will result in lower rates for employers with excellent safety records and higher rates for those with poor safety records. In the retrospective plan, employers will initially pay based upon their experience modifier. Their accident loss record on the project will be tabulated and the contractor will either receive a credit or be charged an additional premium, depending upon performance. Worker's compensation policies provide statutory limits to comply with the laws of the state. Premiums are normally computed based upon $100 of payroll based upon craft specialty or other classification. Rates may vary from less than $1.00 for office work to $30.00 or more for steel erection dependent upon the state. An overall weighted average for the United States is about 18.1 percent.[2] See Chapter 17 for additional details regarding the significant economic effects of contractor safety performance.

Employer's liability insurance is usually written in combination with worker's compensation insurance and provides broad coverage for work-related injury or death of an employee which is outside the scope of the worker's compensation policy. Separate limits of $1/2 million or 1 million are commonly required.

Comprehensive General Liability Insurance

Comprehensive general liability insurance generally insures against liability imposed by law for negligent acts occurring in the conduct of the business which result in bodily injury or damage to the property of others. Typically, the basic policy can be endorsed to include coverage for owner's and contractor's protective insurance, products and completed operations, blanket contractual, personal injury (libel, slander, etc.) and will frequently include coverage for liability arising from the insured's automobiles. Owners frequently require that the policy name them as additional insureds. Typical contractors often purchase combined single limits in the amount of $1 million as a minimum. Premiums may run about 1.5% of payroll.

Contractors will frequently obtain excess liability insurance to protect themselves in the event of a catastrophic loss. Amounts of excess liability insurance are often dependent upon the perceived maximum risk as developed from a risk analysis program.

A typical state department of transportation may require contractors to carry a minimum of $1 million for bodily injury liability, each occurrence, $500,000 aggregate property damage liability, and $500,000 aggregate single limit for bodily injury and property damage liability combined. Lesser amounts are specified for each occurrence.

---

[2] *Building Construction Cost Data 1991*, R. S. Means Company, Inc. Kingston, Maine.

Contractual Liability Insurance

Contractual liability insurance protects the contractor when he assumes the legal liability of others, generally the owner, designer or other designated party. Most contracts will contain some form of indemnification clause or hold-harmless provision in which the contractor is required to assume the potential liabilities of others. Sample insurance and indemnification clauses are contained in the sample contracts reproduced in Appendix C.

Professional Liability Insurance

Professional liability coverage is designed to provide protection to architects and engineers from liability based upon professional errors or omissions in performing design, construction management or other services. A number of public and private owners require that architect/engineers, other design firms, construction managers and sometimes other consultants carry a specified minimum amount of this type of insurance.

The ASCE annual professional liability survey published in 1989 shows that consulting engineers paid 4.2 percent of their annual gross billings for professional liability insurance. Costs were down from a peak of just over 5 percent in 1987. Claims averaged about 45 per 100 firms.

Builder's Risk Insurance

Builder's risk insurance covers the cost of damage of a physical nature to a building or other component of a construction project. Coverage applies to material and equipment not yet incorporated into the work when located on site or in transit to the site.

Current policies generally include the interests of the owner, the general contractor and all subcontractors and material suppliers. While policies may be written upon a "named peril" form which insures against only those perils specifically named, such as fire, windstorm, vandalism, etc., the more common form is "all risk" which provides protection from all perils except those specifically excluded. Most common exclusions include war, nuclear, faulty design, faulty materials, faulty workmanship, earthquake and flood. Some exclusions can be eliminated for an additional premium. Deductibles are commonly used, the amount depending upon the size and type of the work.

Equipment Floater Policy

The equipment floater provides coverage for damage to mobile and stationary construction equipment which is not generally subject to vehicle registration. The floater does not normally provide for liability and property damage insurance for cars, trucks and other equipment subject to the motor vehicle licensing laws, which are covered under other liability policies. It usually provides coverage for damage to the equipment whether located at the job site, in transit, or at the contractor's yard. Premiums are normally computed periodically based upon the value of the equipment for contractor-owned items and based upon the rental rate for rented equipment. Newly acquired equipment is insured upon acquisition and reported for premium purposes in the forthcoming reporting period.

Payroll Taxes and Insurance

Other obligations of construction and other employers imposed by law include payroll taxes and insurance. Included is social security tax at 6.2 percent for both employers and employees for the first $53,400 of wages paid in 1991. The Federal unemployment (FUTA) tax rate is 6.2 percent on the first $7000 of wages. State disability insurance (SDI) in California is 1.0 percent to a taxable limit of $31,767. Other payroll based taxes include state unemployment tax, employment training tax, and other items, depending upon individual state requirements.

Wrap-Up Insurance

Particularly in the private sector, a number of owners have established wrap-up insurance programs to cover the owner, contractors, subcontractors and sometimes construction or program managers as well as the designers. Several major public programs including rapid transit, wastewater and other programs have utilized wrap-up insurance where state law does not prohibit their use. Basic policies can include comprehensive general liability insurance, excess general liability, builder's risk, worker's compensation, and occasionally errors and omissions insurance. Automobile liability insurance, equipment floaters and the deductible or self-insured retention normally continues to be the responsibility of the individual insured.

Considerable controversy exists within the industry regarding wrap-up programs. Certain large insurance brokers specialize in putting together wrap-up insurance programs for owners on major projects. Advantages claimed for the owner are lower insurance costs, control over the insurance program, standardization of risks and centralization of responsibility.

Contractors as a group are normally not in favor of the owner placing worker's compensation insurance and receiving the benefits of any savings on retrospective policies due to superior safety performance on the part of the contractor. Contractors with minimum experience in the state or those with poor safety records will pay more for this type of insurance than will experienced local contractors with excellent safety records. When this type of insurance is mandated by the owner to be excluded from contractor bids, the safe contractor with a favorable experience modification record will not normally benefit in his ability to bid lower than his less safe competition because of his favorable record.

The best-qualified contractors, as a group, generally prefer to place general liability insurance with their own broker, usually under continuing policies that automatically cover new projects as they are obtained. When the owner supplies this insurance, normal business practices can be disrupted and contractors must deal with a new set of administrators, insurance adjusters, safety inspectors and other representatives from several insurance companies. On the other hand, contractors often like to see the owner pick up builder's risk and excess liability insurance for the project providing subrogation rights are waived by the insurer.

The Los Angeles County Transportation Commission currently (1991) mandates use of a wrap-up program including worker's compensation on its rapid transit projects. However, the program includes a safety incentive clause in the contract which permits the contractor to earn an Incentive Value safety bonus of up to one percent of the contract price/(maximum $500,000) through the achievement of an Incidence Rate of up to two percent below an Incentive Value Incidence Rate goal (current goal 6.9 in 1991). The contractor can also be subject to additional costs of five percent of the Incidence Rate goal for each (0.1) increase above the Incentive Value up to two percent maximum. Such a program benefits safety-conscious contractors and should eliminate some or all of contractor objections toward wrap-up programs.

$$\text{Incidence Rate Calculation} = \frac{N \times 200,000}{MH},$$

where,

N = Number of lost time injuries and/or illnesses.

MH = Total hours worked by all on-site construction employees for this contract.

$200,000$ = Equivalent to the number of hours worked by 100 full-time employees at 40 hours per week/50 weeks per year.

Fatalities = 5 lost-time injuries and/or illnesses.

Glossary of Insurance Terms

Definitions for insurance and risk management terms follow:

Loss: The basis for a claim for damages under an insurance policy.

Primary insurance: Basic insurance coverage from first dollar to the stated insurance policy limit.

Excess insurance: Insurance coverage applying only to loss or damage in excess of a stated amount or of a primary insurance policy.

Self-insured retention: The amount of each loss for which the insured party agrees to be responsible before the primary insurance coverage begins to participate in the loss.

Loss-sensitive or retrospective policy: An insurance policy whose premiums are directly affected by the loss experience. Standard premiums can be increased or decreased based upon actual performance under the policy. This type of policy is most commonly used for worker's compensation insurance, where the frequency and severity of losses has a direct bearing upon the premium.

Subrogation: The right of subrogation preserves for insurers the right to sue to recover damages from the party who caused the loss, unless that party is a named insured under the policy, or the right has been waived by contract.

CONSTRUCTION SURETY BONDS

Surety bonds provide a third-party guarantee for the performance of construction contract obligations. This section will discuss the type of bonds most often used in public and private construction projects, including bid bonds, performance bonds and labor and material payment bonds.

Public Works

The Miller Act requires that, on federal construction projects greater than $25,000, the contractor shall furnish a performance bond for the full contract amount and a payment bond offering protection to subcontractors, material suppliers and others. Payment bonds are usually written using a sliding scale for 50 percent of the contract value or less, decreasing as contract value exceeds $1 million. Bid bonds, which guarantee that the contractor will sign the contract if awarded and furnish the required performance and payment bonds, are also normally required on federal and other governmental work.

Many states have established "Little Miller Acts" which include requirements similar to the federal statutes. A number of cities and other governmental agencies also require similar bonding requirements for publicly bid construction work.

Private Work

In private work, bonding requirements are at the owner's option. Some owners routinely require payment and performance bonds while other owners depend upon a program of prequalification to restrict bidding to financially sound contractors who have demonstrated that they have the required financial, resource, and experience qualifications for the proposed project. Bid bonds are also optional in the private sector. Some owners waive bid bonds and require that the contractor will furnish a payment and performance bond only if requested by the owner, in which case the owner will reimburse the contractor for the premium. Private owners often combine performance and payment bonds into a single instrument in the amount of 100 percent of the contract price.

Foreign Work

On most foreign work, performance and payment bonds are not acceptable to governmental clients and are replaced by an irrevocable letter of credit guaranteed by an acceptable financial institution. This letter of credit offers protection to the owner, who can draw on it in the event of default or nonperformance by the contractor.

Performance Bonds

The performance bond is not insurance. It does, however, guarantee performance of contract obligations to the owner as set forth in the contract. The amount of the performance bond represents 100 percent of the contract price, which is guaranteed to be made available by the surety to complete the contract in the event of default by the contractor. In the event of default, the surety is obligated to either complete the work or to make an arrangement with the owner to pay for the cost of completion of the work less the balance of the contract price previously unpaid.

The surety, of course, retains by agreement the right to recover costs from the contractor for his failure to complete. Therefore, contractors do not normally default on work unless they are financially unable to finish the work. In the event of default, the surety has several choices. It can provide additional funds to the contractor under an arrangement to permit the original contractor to complete the project. It can take bids from other contractors or he can negotiate with a favored contractor to complete the work. Alternately, it can arrange for the owner to take bids and to reimburse the owner for any additional costs up to the value of the bond. A number of construction contracts give the owner the right to take over the work, including the contractor's materials, plant and equipment, in the event of default and to complete the work as he chooses. The contractor and his surety would remain responsible for the additional costs incurred by the owner up to the value of the performance bond.

If the original contractor is capable of completing the work with financial assistance, bonding companies often find it less costly to provide prompt financial assistance rather than to assume the additional costs and delays resulting from shutting down and restarting the work. When this method is feasible, all parties, including the owner, may benefit.

Bid and Proposal Bonds

Bid and proposal bonds provide that the contractor's surety will guarantee that the contractor will enter into a contract and provide required bonding if selected. Public agencies and some private owners will require this type of bond. The contractor normally does not pay any premium for bid bonds. Surety remuneration comes from furnishing the performance and payment bonds if the bid or proposal is successful. Amounts for bid bonds usually are between 5 percent to 20 percent of the amount bid, expressed as a percentage, which means that the bond can be issued prior to the bid date, which helps to eliminate further complexities involving last-minute adjustments to the bid by the contractor.

Payment Bonds

Labor and material payment bonds provide for prompt payment to all those furnishing labor and/or materials to the work. This type of clause has been interpreted by courts to cover unpaid bills to a subcontractor even though payment has been made to the subcontractor by the general contractor. To help mitigate this situation, prudent contractors can:

prequalify and select subcontractors and materials suppliers with care;
require subcontractors to furnish evidence that they have paid for labor and materials furnished;
require subcontractors to furnish a performance and material payment bond; and
require lien releases prior to making final payment.

Additional liabilities may be incurred through mechanics lien laws and stop-notice requirements, which vary from state to state.

TABLE 18-1
TYPICAL BOND PREMIUM RATES PER $1000 OF CONTRACT
PRICE

Contract price	Highest priced bonding companies	Lowest priced bonding companies
First $500,000	$15.00	$9.00
Next 2,000,000	10.00	8.00
Next 2,500,000	7.50	6.50
Next 2,500,000	7.00	6.00
Over 7,000,000	6.50	5.75

Combination Payment and Performance Bonds

Private work often utilizes a combination payment and performance bond in the amount of 100 percent of the contract price. As the project proceeds to completion, the exposure of the surety to complete the work in the event of default normally decreases with time, except in unusual circumstances. However, the amount of the bond does not decrease with the amount of completed work, but remains constant throughout the life of the contract. Premiums for the combined bonds are normally the same as for separate performance and payment bonds. Table 18-1 shows a simplified tabulation of premium rates for performance and payment bonds applicable to fixed-price work (lump-sum or unit-price). Premiums for bonds for cost-reimbursable negotiated work where required are substantially smaller.

MECHANIC'S LIENS AND STOP NOTICES

Every state has mechanic's lien laws which convey a right to secure payment for work performed in the improvement of real property. Lien laws are subject to strict compliance by the courts and familiarity with such laws in each individual state is required. Mechanic's liens are recorded in a manner similar to a mortgage and represent a claim or lien against the real property. In California, for example, mechanic's liens must be recorded in the county where the work was performed within 90 days after completion of the work as a whole unless the owner records a notice of completion. In this case, contractors who contract directly with the owner have 60 days to record; others have 30 days. A mechanic's lien expires unless a foreclosure suit is filed within 90 days after the lien is recorded.

Stop notices are generally utilized on public property which is not subject to a mechanic's lien. A stop notice tells the public owner that bills have not been paid. State statutes require that the owner must withhold payment to the prime contractor until the obligation is paid. Some states exempt the owner from lien liability provided that the general contractor has furnished a payment bond. Architects, surveyors and engineers are given lien rights in some states, subject to certain conditions.

Where state law permits, construction contracts can require that the right to lien be waived by general contractors or others. Many construction contracts require that the contractor supply a signed release and waiver of lien prior to receiving final payment from the owner. In other cases, owners and contractors can require signed releases prior to receiving interim-progress payments.

SUMMARY

Risk management, insurance, bonding, liens and licensing considerations have become substantially more costly, complex and important for contractor survival during the past decade. The previous discussions represent a simplified outline of the subjects. Professional assistance in these areas is a necessity for today's contractors and engineers in a climate of increasing complexity. A helpful reference covering additional information can be found in *Construction Contracting* by Richard H. Clough.[3]

---

[3] *Construction Contracting*, 5th ed., John Wiley & Sons, New York, 1986.

CHAPTER 19

INDUSTRIAL RELATIONS

This chapter on industrial relations in the construction industry includes a brief discussion of labor laws in the United States, a discussion of organized labor summarizing the organization and control of the AFL-CIO building-trades unions and the advantages and disadvantages of union membership for both the contractor and the worker. The section on open shop similarly reviews the nonunion and coexistent positions. A comparison of the union and nonunion approaches sets forth the current situation in the industry.

Increasingly, disadvantaged business enterprises (DBE), including ethnic minorities and women, are becoming a sizable factor on public works, and compliance with the law and recognized affirmative action programs has become increasingly important in contractor operations. Also included is a description of contractor and owner organizations, designed to further the interests of each individual group and the industry as a whole. Finally, a description of the duties of an industrial relations department of a large contractor helps illustrate how contractors try to develop harmonious relationships at all levels.

LABOR LAW IN THE UNITED STATES

Over the years labor law has evolved to define the rights of employers, employees and unions or other labor organizations in the work place. Early practice in the construction industry was generally created in the courts based upon the principles of common law. Initially the interpretation of the law was generally weighted to favor employers until the advent of the Norris-La Guardia Act in 1932. Assisted by the National Labor Relations Act (Wagner Act) passed by Congress in 1935, the balance of power shifted to favor the labor unions until passage of the Labor Management Relations Act (Taft-Hartley Act) in 1947, which helped to equalize the situation. Later congressional action established restrictions on both employers and labor unions and defined unfair labor practices as set forth in the Landrum-Griffin Act in 1959. The Civil Rights Act of 1964 established certain individual rights and prohibited discrimination by both employers and unions. Arbitrary age discrimination was prohibited by the Age Discrimination in Employment Act of 1967.

Administration of the National Labor Relations Act as amended is handled by the National Labor Relations Board (NLRB). The NLRB has power over employers operating in interstate commerce and operates through regional offices located throughout major cities in the United States. Five board members and the general counsel are appointed by the President with the consent of the Senate. The board has primary responsibility to establish whether or not groups of employees wish to be affiliated with a particular union and to prevent and remedy unfair labor practices.

A summary of selected portions of major labor legislation is included below. For additional detail discussion of labor law and labor relations in the construction industry, see *Construction Contracting* by Richard D. Clough, Chapters 13 and 14.[1]

Norris-La Guardia Act

Provisions of the Norris-La Guardia Act of 1932 limit the power of federal courts to issue injunctions against labor unions. The act makes it difficult for employers to secure injunctions in federal courts against labor union activity during disputes. A number of states have established similar legislation covering state work. The act also prohibits employers from requiring a prospective employee to promise not to join or belong to a labor union during his employment ("yellow dog contract").

National Labor Relations Act (Wagner Act)

Enacted by Congress in 1935, the Wagner Act was designed to enhance organizing and collective bargaining abilities of the labor unions. Employers were forbidden to discriminate against their employees for union activity or to influence membership in a labor organization. Employers are also required to show good faith throughout the bargaining process.

The influence and power of organized labor increased strongly in the ensuing period resulting in considerable public resentment of increasing restrictive practice, strikes and illegal or criminal activity by certain labor leaders. To counter this growing power, a number of states enacted laws designed to develop a more balanced labor climate on state work.

Labor Management Relations Act (Taft-Hartley Act)

The Taft-Hartley Act enacted in 1947 amended the Wagner Act and imposed certain controls over the actions of the labor unions while adding additional provisions affecting both employers and labor unions. The act defined unfair labor practices for

---

[1]Richard D. Clough *Construction Contracting*, 5th Edition, John Wiley & Sons, Inc., New York, 1986.

both employers and labor unions and was principally designed to curb growing union power.

Landrum-Griffin Act

The Landrum-Griffin Act amended the Taft-Hartley Act in 1959. It added more detailed definitions of unfair labor practices, guaranteed certain rights to individual union members and set forth certain controls over internal union affairs. Nondiscriminatory hiring halls were made legal along with other provisions designed to help limit union power.

Unfair Labor Practices by Unions

The act spells out a number of unfair labor practices by unions which are prohibited:

1 Prohibits coercion by unions affecting employees' rights to refrain from union membership or to influence selection of union representatives.
2 Prohibits unions from attempting to influence employers to discriminate against individual employees for the purpose of fostering membership in a labor union. Nondiscriminatory pre-hire agreements are, however, allowed by the act.
3 Prohibits accredited labor unions from refusal to bargain in good faith with employers covering wages, work hours and other employment conditions.
4 Prohibits strikes, boycotts, and other threats designed to:
  • Force an employer to join a particular labor or employer organization or to enter into "hot-cargo" agreements under specified circumstances. A "hot cargo" agreement prohibits an employer from doing business with or handling products of a third party.
  • Create a secondary boycott by forcing a firm or individual to cease doing business with any other firm or individual. A secondary boycott occurs when a union endeavors to force a particular firm, producer or individual (Company A) from doing business with a third party (Company B) who is engaged in a primary dispute with the union. Primary boycotts, not illegal under the act, involve a union engaged in a dispute with an employer who endeavors to influence others such as the firm's customers to refrain from doing business with the employer.
  • Force an employer to recognize or bargain with a particular labor organization that has not been certified as an authorized representative.
  • Force or require an employer to assign work to a particular craft union or other labor organization unless conforming to a lawful NLRB order. This provision pertains to jurisdictional disputes where two or more unions claim entitlement to the work.
  • Force or require payment of excessive or discriminatory union membership fees by employees.
  • Force an employer to pay for services not performed or to unlawfully picket an employer in an effort to force collective bargaining except where permitted under the act.

Unfair Labor Practices by Employers

The act spells out a number of unfair labor practices which are prohibited by employers:

Interfere with, discharge, coerce, restrain or otherwise discriminate against employees exercising lawful rights.

Dominate or otherwise interfere with the formation or management of any labor organization. This provision prevents the formation of "Company Unions."

Discriminate against an employee in order to encourage or discourage membership in a labor union. Lawful compulsory union membership (union shop) is permitted pursuant to NLRB requirements.

Refusal to bargain in good faith with accredited labor unions covering wages, work hours and other conditions of employment.

Enter into "hot-cargo" (refrain from handling or using products of others) agreements with a labor organization.

Definitions of Labor Terms

1 Union shop: An employee need not be a union member at time of hire, but must join the union within a designated time period in order to retain his job.

2 Closed shop: An employee must be a union member at time of hire. This practice is illegal under current labor law.

3 Agency shop: An employee need not join a union but must pay dues, initiation fees and assessments equivalent to other union members.

4 Open shop: The firm does not have agreements with labor unions and can hire both employees and subcontractors irrespective of union membership. Union members, if employed, are not represented by the union in dealing with the employer.

5 Merit shop: A particular form of open shop practiced by members of the Associated Builders and Contractors of America, the leading open-shop employer organization.

6 Pre-hire agreement: Permitted only in construction, pre-hire agreements permit an employer to make an agreement with the union prior to an election by employees. If challenged, unions must show that they have a representative majority. The courts have held that employers are bound by such an agreement until repudiated. (See union shop for employee rights and duties.)

7 Right to work laws: States have the right to ban union-shop agreements permitted by the Taft-Hartley Act but must first pass a law making it illegal to require union membership as a condition of continued employment.

8 Hiring halls: Organizations established to refer workers to construction projects. The Landrum-Griffin Act permits nondiscriminatory union hiring halls to be required by agreement. Open-shop hiring halls and referral services give similar services to open-shop contractors.

9 Common-situs picketing: The NLRB and current labor law have placed restrictions upon the union's right to picket a primary employer in order to insure that other employers at the common site are not affected in order to avoid a secondary boycott. Sometimes called the separate gate doctrine, the NLRB holds that picketing a gate reserved for a particular contractor engaged in a primary labor dispute is legal. Picketing of other separate gates established for contractors or subcontractors who do not have a primary dispute with the union is illegal. Hence general contractors, construction managers and owners often utilize a number of separate gates on open-shop projects which involve both union and nonunion workers. Similar action can be taken on union projects in order to mitigate effects of a legal strike against contractors or subcontractors not directly involved in the dispute.

10 Jurisdictional disputes: Craft unions have historically fought to preserve their perceived jurisdiction over certain types of work. When work is claimed by two or more separate unions, a jurisdictional dispute exists. Many disputes were encountered in the industrial expansion after the second world war as new technology developed building materials and equipment which did not conform to past norms. Under today's conditions, jurisdictional disputes still occur on unionized work, but they are far less prevalent and disruptive than in prior years.

11 Davis-Bacon Act: This act is a prevailing wage law which establishes wages and fringe benefits as minimums to be paid by all contractors on federal and federally assisted construction projects. Many states have established similar laws. Administration of the act by the U.S. Department of Labor who initially established wage rates essentially equivalent to the union pay scales has received substantial challenge from open-shop contractor organizations. Recognition of nonunion pay scales and classifications in strong open-shop areas is now beginning to receive recognition in certain areas.

12 Union work rules: Internal work rules for union members established unilaterally by the union. Unions usually try to incorporate as many of their own work rules as possible in collective bargaining agreements with employers.

13 Collective bargaining agreements: A written contract between a labor union and an employer that spells out the terms and conditions of an agreement reached through collective bargaining. Agreements normally spell out wages, fringe benefits, working conditions, grievance procedure, and other provisions and are subject to renegotiation at stipulated intervals.

14 National agreements: A special form of collective bargaining agreement which covers all work in the United States and sometimes in Canada. National agreements usually adopt local wage scales, describe separate work rules and provide for a separate grievance procedure directly involving the national union and the employer.

15 Project agreements: A special form of collective bargaining agreement which applies to a large construction project or group of related projects. While the agreement often has a wage reopener clause, the agreement will usually terminate upon completion of the project. All project contractors are normally required to sign such agreements which sometimes provide for admission of open-shop contractors on stipulated terms.

16 Apprenticeship programs: Historically, apprenticeship programs have been a partnership between employer groups and building-trades unions at local or regional levels. Joint union-employer committees supervise and administer the programs which are geared to the training and development of journeyman craftsmen. The Bureau of

Apprenticeship and Training has recently approved national standards for programs developed by employers or employer associations for open-shop contractors. Sponsors of apprenticeship programs include the AGC, the Associated Builders and Contractors of America (ABC), and other national, regional or local employer organizations. Apprentice wage rates on Davis-Bacon projects can only be paid to apprentices who are participating in federally approved programs and the recent approval of employer only sponsored programs in the open-shop sector has been achieved after a considerable battle between union and open-shop organizations. Helper wage rates, not currently established under prevailing wage laws, continued to be proposed by open-shop proponents and resisted by the labor unions.

ORGANIZED LABOR[2]

The major force in organized labor within the construction industry is the American Federation of Labor (AFL), which began in 1886 under the leadership of Samuel Gompers. Gompers developed two policies that contributed to the successful growth of this building-trades union: (1) national unions were guaranteed trade autonomy; and (2) each union was granted exclusive jurisdiction. At the present time, continuing demands include higher wages, shorter hours, improved working conditions, job security, and the right to represent the work force. Of major concern is the growing penetration of the open shop and the minority position of unionized construction in many areas.

Table 19-1 lists the building-trades unions affiliated with the AFL-CIO. The Teamsters Union is now re-affiliated after a long period of suspension during which it usually continued to cooperate fully with the building trades at the local level.

Organization and Control

Organized labor is split into three levels, the federation, the national unions, and the local unions. The federation is the AFL and the Congress of International Organizations (CIO), which merged into the AFL-CIO in 1955 after a split of 17 years. Figure 19-1 shows the organization's chart. The federation is a coalition of 106 national and international (both United States and Canada) unions. The principal role of the AFL-CIO is political since it does not take part in local collective bargaining. The state-level organization is open to membership on a voluntary basis. Its structure and functions are similar to the national organization except that it is oriented to local and state activities.

The building-trades national and international unions are the basis of American union activities in the construction industry. The national union has its own exclusive jurisdiction, and although collective bargaining is performed at the local level, that power is delegated from the national union. Unions have their own officers and manage their own affairs. Usually a national union vice-president is given authority over local unions, districts, or other collective groups throughout the United States.

---

[2]This section is based partly upon a chapter from *Directions in Managing Construction*, Donald S. Barrie (ed.), prepared by Dr. John Borcherding, John Wiley & Sons, Inc., New York, 1981.

TABLE 19-1
AFL-CIO CONSTRUCTION UNIONS

1. Brotherhood of Painters and Allied Trades
2. International Association of Bridge, Structural and Ornamental Iron Workers
3. International Association of Heat and Frost Insulators and Asbestos Workers
4. International Association of Marble, Slate and Stone Polishers, Rubbers and Sawyers, Tile and Marble Setters Helpers and Terrazzo Helpers
5. International Brotherhood of Boiler Makers, Iron Ship Builders, Blacksmiths, Forgers and Helpers
6. International Union of Bricklayers and Allied Craftsmen
7. International Brotherhood of Electrical Workers
8. International Brotherhood of Teamsters, Chauffeurs, Warehousemen and Helpers
9. International Union of Elevator Constructors
10. International Brotherhood of Operating Engineers
11. Laborers International of North America
12. Operative Plasterers and Cement Masons' International Association of the United States and Canada
13. Sheet Metal Workers International Union
14. United Association of Journeymen and Apprentices of the Plumbing and Pipe Fitting Industry of the United States and Canada
15. United Brotherhood of Carpenters and Joiners of America
16. United Union of Roofers, Waterproofers and Allied Workers

FIGURE 19-1
Structural Organization of the AFL-CIO.

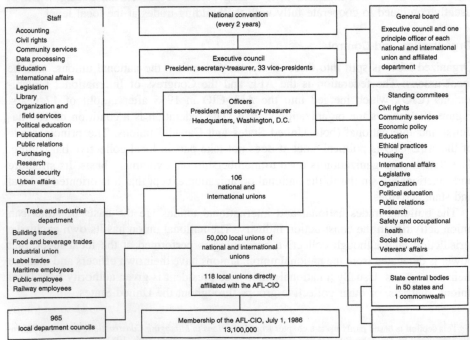

The third level in the union hierarchy is the local union. Locals are chartered by the national body and derive their power from the constitution of the national union. Local building-trades unions are fairly autonomous and responsible for collective bargaining and jurisdictional preservation for a specific area. The local union is responsible for all activities pertaining to the craft within its boundaries. Affiliated local unions often join together to form district councils to make rules or understandings for the district. Building-trades councils are made up of local unions from different construction trades and are instrumental in setting the political and collective bargaining pattern within their area.

Advantages and Disadvantages of Union Membership

Advantages of union membership are cited for both workers and construction company management.

Worker Advantages   Most workers are concerned with such economic and other issues as:

1 Hiring hall for job referral
2 Apprenticeship training
3 Right to strike, wages, job conditions, benefits, and job security
4 Power to act collectively as a group to enforce demands
5 Camaraderie through belonging to an organization

Management Advantages   Management can achieve a number of advantages through a healthy relationship with organized labor, including:

1 Source of a pool of skilled labor
2 Fixed wages and uniform conditions through negotiations that can have a stabilizing influence in the area and preclude unfair competition
3 Progressive unions may exercise a stabilizing influence among their own members and may help control the entry and actions of irresponsible or marginal contractors

Worker Disadvantages   Workers may also suffer disadvantages through union membership, including

1 Payment of initiation fees, dues, or assessments
2 No choice of employer, lack of merit promotions, equal pay regardless of performance, and general subordination of gain from outstanding individual performance
3 Restriction on utilizing certain methods, elimination of incentive programs, less flexibility in work assignments, and loss of work during jurisdictional disputes
4 Less number of hours of work during the year and less flexibility to make up for bad weather or other factors causing temporary layoff

Management Disadvantages   Management criticizes unions for

1 Restrictive work rules that decrease productivity, such as supervisor ratios, nonworking supervisors, guaranteed work week, prohibition against operating several pieces of equipment, and other featherbedding examples

2 No incentive upon the part of the individual worker to be innovative or especially productive since everyone gets the same pay

3 Little loyalty to the employer, which results in less management opportunities for innovation and development of a team spirit to improve both production and work-life quality

4 Jurisdictional disputes that can result in economic hardship through no fault of the contractor

THE OPEN SHOP[3]

Under United States labor law, construction firms have the right to decide whether they will operate an open or a union shop. The line between union and nonunion firms is not always clear. Many nonunion firms may sign formal project agreements or informal agreements that effectively make them union firms on a particular project. Nonunion firms often hire unemployed union members who "put their card in their shoe." Many open shop general contractors utilize union subcontractors, particularly in the mechanical and electrical trades.

Open shop has been steadily growing, both in its share of the overall construction market and in geographic penetration throughout the United States in more traditional union locations. Based upon the results of a study prepared for the Department of Housing and Urban Development by the Massachusetts Institute of Technology, a number of comparisons were developed between unions and nonunion firms that help clarify their interrelationship.

Organization and Control

To try to achieve some of the advantages that progressive labor unions can bring to both employees and employers, groups of open-shop contractors have bonded together to form the associations discussed in the "Employer Organizations" section. Open shop has achieved sharp growth in the past twenty years but may have peaked out or become stabilized in recent years. Union programs have recently been developed in an effort to halt or reverse this trend. See Figure 19-2 for the growth of the open-shop market share during the past two decades. Figure 19-3 shows the estimated percentage of open-shop construction in the United States in 1985 by region. Figure 19-2 is helpful in examining the growth trend in open-shop construction. The authors are not aware of any national study which endeavors to determine the market share of nonunion or union construction workers. Since many open-shop general contractors will utilize union subcontractors, the nonunion share of the market will be somewhat less than the open-shop market share.

Comparison of Union and Nonunion Construction

The Department of Housing and Urban Development (HUD) sponsored a major survey of the construction industry to compare many aspects of union and nonunion

---

[3]This section is prepared partly from a chapter in *Directions in Managing Construction*, prepared by Dr. Raymond L. Levitt and Donald S. Barrie (ed.), John Wiley & Sons, Inc., New York, 1981.

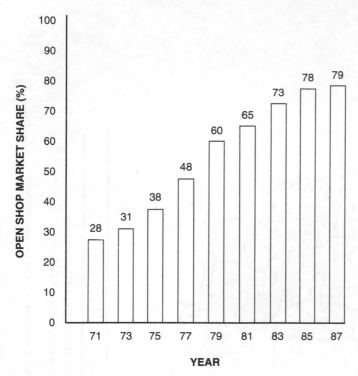

FIGURE 19-2
Open-shop market share. (From Associated Builders and Contractors, Inc., Washington, D.C.)

construction. Eight large metropolitan areas were chosen for the study, which was developed by a number of questionnaires and interviews. Findings from the 1978 study are summarized as follows:

Markets and Individual Firm Size In all eight areas, most of the union firms identified themselves as performing the majority of their work in commercial/industrial or heavy and highway construction. The open-shop firms were engaged primarily in residential or commercial/industrial work or both. Table 19-2 compares employment by product market. In all areas, the open-shop firms were smaller than their union counterparts. However, it was evident that there were a number of very large open-shop firms, both general and specialty contractors. Based upon the substantial continued growth of open-shop since the study, it is believed that this size differential has become less pronounced.

Occupational Structure and Overall Size Open-shop firms were found to be developing and utilizing important innovations in both occupational and skill structure. Open-shop firms often find it economical to use lead persons as working supervisors, helpers for unskilled work crossing trades lines, and new classifications relating pay to skill levels rather than to traditional craft lines. Figure 19-4 is a map of the indus-

Percentage of Open Shop Construction in the U.S.

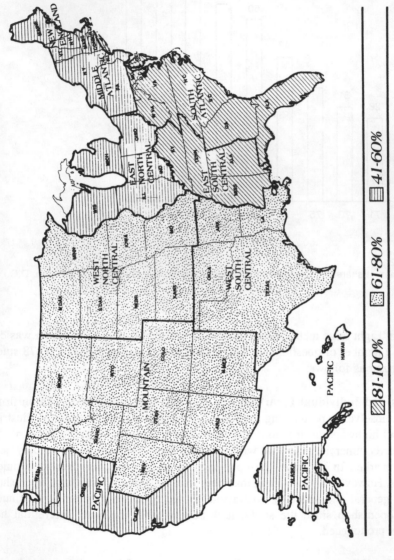

Total U.S. Construction Volume (1985): 221.5 billion*
Percentage of Open Shop Construction (average): 70% +

Legend: 81-100% | 61-80% | 41-60%

Sources: *F.W. Dodge Division, McGraw Hill Information Systems Co.
+ Open Shop Construction Revisited by Herbert R. Northrup, The Wharton School, Industrial Research Unit, University of Pennsylvania

FIGURE 19-3
Percentage of open-shop construction in the U.S.

TABLE 19-2
OPEN-SHOP VS UNION JOURNEYMAN WAGE RATES

Craft	Open shop wage	Fringes	Open shop total	Union wage	Fringes	Union total
Bricklayers	12.33	1.39	13.72	17.78	3.52	21.30
Carpenters	11.62	1.42	13.04	16.00	5.30	21.30
Cement masons	11.28	1.26	12.54	15.50	4.10	19.60
Electricians	11.76	1.56	13.32	19.40	4.00	23.40
Operators	12.44	1.46	13.90	18.05	3.90	21.95
Laborers	7.74	0.88	8.62	12.95	3.80	16.75
Plumbers	11.50	1.54	13.04	19.50	4.40	23.90
Ironworkers	11.29	1.31	12.60	17.60	5.50	23.10
1988 US Average	$11.25	$1.35	$12.60	$17.10	$4.32	$21.41

Notes:
1. Open-shop wages average 66% of union wages.
2. Open-shop fringes average 31% of union fringes.
3. Open-shop wages plus fringes average 59% of union total.
4. Open-shop journeyman wage data from ENR, June 29, 1989, p. 66.
5. Union journeyman wage data based upon *Richardson General Construction Estimating Standards*, October 1989.

try, with a rough assessment of the predominant classifications by size and type of construction.

Impact of Prevailing Wage Laws   There is considerable controversy about whether the Davis-Bacon Act and other prevailing wage laws significantly raise the cost of construction. The HUD study showed that there really was no prevailing wage in nonunion labor markets. The dispersion of nonunion wages for any trade in any given area automatically tends to favor the union rate, which was generally some 30 to 50 percent higher than the average nonunion rate. Whether increased productivity is achieved through higher wages was not addressed by the study and remains a controversial issue in many states with heavy union representation.

HUD Study Update

In 1978 it was estimated that the open-shop share of the construction market had grown to about 53 percent of the greater construction market. By 1988 the share had grown to about 79 percent although these figures have not been generally accepted. While no comprehensive updating of the HUD study has been performed, certain general trends have become apparent:

A survey by the National Association of Electrical Contractors indicated that in 1978 about 45 percent of the electrical craftsmen in the industry were unionized. By 1987 thirty percent remained organized. Total wages paid by union electrical contractors amounted to 57 percent of the total electrical industry payroll in 1978. By 1987 this share had declined to 36 percent reflecting a 5 percent drop in union craftsmen compared to an 80 percent increase in nonunion workmen.

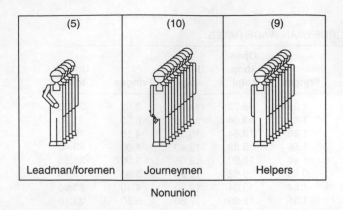

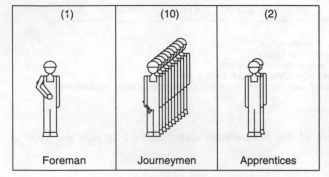

FIGURE 19-4
Comparison of crew structure in union and nonunion construction. From 1976 HUD/MIT study.

Residential Construction   Residential construction including remodeling and restoration work is probably approaching 95 percent open shop. The most recent gains have been in the major cities in the Midwest, Northeast and the West Coast. Most of these areas were 100 percent union in the early 1950s. By 1977 in *Open Shop Revisited*, Herbert R. Northrup cites that all areas of the country except for the east north-central and Pacific regions were probably more than 90 percent open shop based upon National Association of Home Builders data. By 1989 home building and remodel work in west coast cities had become predominantly open shop. A nondocumented estimate might put open-shop market share about 95 percent by 1991.

Building and Commercial Construction   No comprehensive studies regarding open-shop growth in building and commercial construction work have been identified. In the west as well as other metropolitan areas, the trend towards suburban shopping malls and other regional shopping centers has resulted in a shift to nonunion and open-shop contractors similar to home building. Union strengths remain in the public sector and in the larger projects such as hotels, hospitals, high rise office buildings and other facilities particularly in metropolitan areas.

Heavy and Highway Construction   Some modifications to federal and state Davis-Bacon laws have assisted open-shop penetration on government work. However, meeting apprenticeship training requirements, lack of helper classifications and other factors continue to favor union construction on major public works in most areas. Open-shop competition continues to increase on smaller projects and on nonfederal work.

Industrial Construction   A number of very large open-shop industrial builders such as Fluor Daniel Inc., CRSS, Brown & Root Inc., H. B. Zachary Co. and others continue to expand and grow in previously union-dominated markets. Some normally union firms have gone double-breasted or dual shop by establishing separately managed open-shop companies to complement their union operations. Organized labor, as can be expected, continues to fight this type of arrangement. However, heavy industrial construction continues to be one of the principal strongholds of organized labor and the union contractor. Increasingly, projects are utilizing open-shop general contractors or construction managers along with a mixture of union and nonunion subcontractors or contractors.

Current Wage Differentials   Comparing the true hourly cost of union vs nonunion construction is difficult on an industry-wide basis. Many nonunion contractors will market their internal wage scales compared to published union rates. In regions of heavy demand for open-shop construction, the nonunion rates must be competitive in order to secure sufficient manpower. An approximation is presented by Table 19-2, which compares typical nonunion rates with union rates including fringes on a broad basis by averaging regional rates. Table 19-3 compares regional rates for nonunion craftsmen which were utilized in computing averages shown in Table 19-2. The authors are not aware of any meaningful study of an effective comparison taking productivity into account and it is believed that the fragmentation of the industry would render such a study meaningless.

ETHNIC MINORITIES AND WOMEN [4]

Few subjects in construction have generated more controversy than the principle of affirmative action to increase participation of ethnic minorities and women in the construction work force. The Public Works Employment Act of 1977 made available $4 billion in federal subsidies for construction projects, provided that 10 percent or more of the work be awarded to ethnic minority companies, subcontractors, craftspeople, or suppliers. Subsequently, goals have been established to require that contractors use their best efforts to award subcontracts or sometimes other procurement actions to DBEs. A typical goal might be 15 percent of project value. The submission of a proposed program tabulating proposed DBE awards (or satisfactory documentation that best efforts have been utilized) is a prerequisite to award of the general contract. DBEs

---

[4]This section is based partly upon portions of a chapter from *Directions in Managing Construction*, Donald S. Barrie (ed.), prepared by Gerald Challenger and Audrey Barrie, John Wiley & Sons, Inc., New York, 1981.

TABLE 19-3
OPEN-SHOP JOURNEYMAN WAGE RATES

Location	Bricklayers		Carpenters		Cement masons		Electricians	
	Rate ($)	Fringe (%)	Rate ($)	Fringe (%)	Rate ($)	Fringe (%)	Rate ($)	Fringe (%)
New England	16.30	4.7	12.83	17.4	13.95	12.90	12.82	16.2
NY/NJ	14.26	12.4	12.73	13.4	12.91	9.34	12.61	15.7
Middle Atlantic	12.39	12.8	11.77	13.8	11.93	12.70	11.62	15.2
Southeast	11.90	9.3	10.60	10.7	10.93	10.30	10.97	15.0
Great Lakes	11.76	11.9	10.60	11.5	10.65	11.10	11.66	13.7
South Central	11.06	2.6	10.78	6.5	11.17	7.70	11.13	10.0
Central	11.67	13.0	10.23	9.4	10.17	6.30	10.57	11.0
Central Mountain	12.05	4.0	11.58	10.2	11.53	7.20	12.67	10.7
Mountain	14.75	7.3	14.50	10.3	13.04	6.10	14.85	11.8
Western	12.40	3.0	11.30	11.8	11.27	8.50	12.75	12.2
1988 U.S. average	12.33	11.3	11.62	12.2	11.28	11.20	11.76	13.3

Location	Operators		Laborers		Plumbers		Ironworkers	
	Rate ($)	Fringe (%)	Rate ($)	Fringe (%)	Rate ($)	Fringe (%)	Rate ($)	Fringe (%)
New England	13.25	13.8	8.83	14.4	12.14	20.20	11.09	14.5
NY/NJ	13.91	12.8	8.24	13.6	12.00	16.50	12.11	12.8
Middle Atlantic	12.44	19.0	7.69	13.4	11.64	14.40	10.95	8.1
Southeast	12.49	9.5	6.80	10.8	10.35	8.80	10.84	8.6
Great Lakes	11.82	10.8	7.55	10.9	11.56	14.20	10.83	14.7
South Central	11.76	6.8	6.91	7.5	10.77	10.00	11.83	7.5
Central	11.94	8.1	7.27	7.9	10.54	12.10	11.21	7.4
Central Mountain	13.15	10.5	8.77	11.0	12.71	9.70	11.65	9.4
Mountain	14.70	8.2	9.17	8.5	12.11	10.60	13.15	7.9
Western	13.08	11.3	8.59	13.1	12.77	11.60	13.02	10.0
1988 U.S. average	12.44	11.7	7.74	11.4	11.50	13.40	11.29	11.6

Notes:
1. Open-shop journeyman wage data based upon ENR, June 29, 1989, p. 66.
2. *Source:* Personnel Administrator Services, Inc.

include both women-owned businesses and ethnic minorities. Goals can be separately stated as owned by women and by ethnic minority or lumped together as DBEs.

Ethnic Minorities

A serious effort to resolve discrimination problems in the construction industry has been under way in the United States for some time. The Constitution has been interpreted to forbid discrimination in government and private employment. The Equal Employment Opportunity Commission was created in 1964 to investigate allegations of discrimination. Other acts forbid discrimination on the basis of age, sex, and national origin.

In many sections of the country, it was considered normal for blacks and other groups to occupy positions of leadership in local unions for laborers, cement finishers, and teamsters. In years gone by, there were often separate local unions for blacks in the southern states. The most significant areas of exclusion for blacks, hispanics, and sometimes others were in the metal trades. Until recently, black electricians, pipefitters, ironworkers, and other metal-trades personnel were largely nonexistent. The current accomplishments of black minorities have been largely tied to government-funded training or government-mandated affirmative action. In the final analysis, future affirmative action programs must be designed to increase productivity as well as to redress past wrongs through large-scale training and other incentives in a way that will benefit society, the individual, and the industry.

Minority firms play a major role in the overall development of minority participation in the construction industry. Ideally, the government-mandated project goals requiring that a designated percentage of work be performed by minority firms can be followed to both upgrade the utilization of minorities in the industry along with maintaining or improving present productivity standards. A support organization is the National Association of Minority Contractors, Washington D.C.

Women

The Civil Rights Act enacted in 1964 and amended in 1972 prohibits discrimination because of race, sex, or creed. Employment doors previously closed to women were suddenly open. The Department of Labor mandated that certain quotas of women be hired on construction projects. However, in spite of supportive legislation and an excellent pay scale, the women are clearly not yet beating a path to the crafts. Except for Rosie the Riveter, active in the shipyards in World War II, it was rare to find a woman who was in the building trades before the enactment of civil rights legislation. Society continues to perceive women as a protected, more delicate species, incapable of doing "man's work." Because of this perception, many women continue to limit themselves, or be limited by others, to employment deemed more traditional for their sex.

Professional and white collar women are making substantially better progress in the construction industry. Women are appearing in essentially all the salaried functional tasks required in the construction industry, including accounting, purchasing, labor relations, scheduling, estimating, design engineering, construction engineering, supervision, and project management. Of all the minority engineers being graduated in the United States, women are the fastest growing group.

Several support organizations have evolved to assist the upwardly mobile female construction professional. Women in Construction embraces all women in construction administrative, managerial, and other white collar positions. The National Association of Women in Construction (NAWIC) awards scholarships for outstanding achievement, sponsors a four-year home-study course leading to a Certified Construction Associate degree, and serves as a clearinghouse for job openings. Local NAWIC chapters hold periodic meetings and sponsor workshops and other educational events. NAWIC conducts similar undertakings on a national scale.

The Society of Women Engineers (SWE) is a professional, nonprofit, educational service organization of graduate engineers and others of equivalent engineering experience. Specific objectives of the society are to:

- Inform young women and other interested parties of qualifications, achievements, and opportunities open to women engineers
- Assist women engineers in readying themselves for a return to active work after temporary retirement or a leave of absence
- Serve as an information center for women in engineering
- Encourage women engineers to obtain higher levels of education and professional achievement

SWE administers several award, certificate, and scholarship programs. Sections are located in 40 states and in Puerto Rico. Student sections have been chartered in 92 colleges and universities. The Society has an international membership of over 5000 and about 60 corporate members, including many of the largest employers of engineers in the United States.

CONTRACTOR AND OWNER ORGANIZATIONS

Most contractors in the United States belong to one or more contractor or industry associations. Some of the organizations are purely national in scope; others are active at both the local and the national level; and others represent contractors only in a particular area. Table 19-4 lists a number of the major national construction contractor associations. In addition, the Business Roundtable, made up of a large number of major construction owners, has been very active in the construction area.

Union Contractor Organizations

Major associations of unionized contractors include the Associated General Contractors of America (also includes open shop), the National Constructors Association, the Mechanical Contractors Association of America (MCA), the National Electrical Contractors Association (NECA), and a number of associations for specialty trades.

National Constructors Association The National Constructors Association (NCA) was formed in 1947 to represent large national contractors engaged primarily in industrial construction and who generally also have engineering capability to perform design-build or design-manage projects. Engineer-contractor or contractor members of the NCA hold national or international agreements with the national building-trades unions that have been negotiated by the national office. The national agreements generally adopt the wage scale and fringe benefits that are negotiated by the local union in the job-site area. However, the national agreement provides for the adjudication of disputes first with the local union and then with the national union, bypassing the provisions of the local agreements. National agreements also often provide for the elimination of certain local practices that may conflict with the national agreement.

The NCA has negotiated several other agreements that provide more favorable terms to member union contractors competing against open shop in designated areas and in maintenance or "turnaround" work in operating plants. Project agreements are often specially negotiated for large projects by NCA members, which generally follow local wage rates but which may provide for different working conditions or

TABLE 19-4
LIST OF SEVERAL MAJOR NATIONAL CONSTRUCTION ASSOCIATIONS, 1991

Associated General Contractors of America 1957 E Street, N.W. Washington, D.C. 20006	National Constructors Association 1001 Fifteenth Street, N.W. Suite 1000 Washington, D.C. 20005
The Business Roundtable (owner association) 405 Lexington Avenue New York, New York 10014	National Electrical Contractors Association 7351 Wisconsin Avenue Washington, D.C. 20014
Mason Contractors Association of America 601 Fourteenth Street, #17W Oakbrook Terrace, Illinois 60181	National Insulation Contractors Association 1001 Connecticut Avenue, N.W. Suite 800 Washington, D.C. 20036
Mechanical Contractors Association of America 5530 Wisconsin Avenue, N.W. Washington, D.C. 20015	National Utility Contractors Association 815 Fifteenth Street, N.W. Washington, D.C. 20005
National Association of Plumbing-Heating-Cooling Contractors 1016 Twentieth Street, N.W. Washington, D.C. 20036	Painting and Decorating Contractors of America 7223 Lee Highway Falls Church, Virginia 22046
National Construction Employers Council 2033 K Street, N.W. Suite 200 Washington, D.C. 20006	Sheet Metal and Air Conditioning Contractors National Associations, Inc. 8224 Courthouse Road-Tysons Corner Vienna, Virginia 22180

other provisions. The NCA is generally the representative of the larger engineer-contractor who offers turnkey services to industrial owners on heavy industrial projects throughout the United States.

The Associated General Contractors of America   The Associated General Contractors of America (AGC) was formed in Chicago in 1918 and was the first major association of general contractors. The AGC has historically handled wages and contract negotiations with the local building trades throughout the United States. Members are usually signatory to local agreements through their membership in the AGC and can operate in other jurisdictions simply by joining the local group and becoming signatory to the agreements and grievance procedures. In contrast to the NCA, grievance procedures are generally set forth in the local agreement and are binding upon both the union and the contractor. The AGC generally negotiates for the five or six basic trades usually employed by a general contractor. Other specialty contractors are often negotiated by the National Electrical Contractors Association, the Mechanical Contractors Association of America, or other specialty or independent groups.

With the rise in open shop, a number of AGC chapters now include both union and open-shop contractors who cooperate in an effort to improve productivity and help mitigate unreasonable wage and work-condition demands by the unions. The AGC generally represents the local contractors and certain national contractors who are generally without the engineering capability to perform turnkey work. Most of the contracts are competitively bid, in contrast to the negotiated contract favored by members of the NCA. It is not surprising that relationships between the AGC and the NCA, while coexisting at the national level, have never been closely aligned at the local levels or mutually supportive nationally.

Specialty Contractor Associations Other specialty groups, including the NECA, MCA, Painting and Decorating Contractors of America, and Mason Contractors Association of America, have historically performed services for their members similar to those provided by the AGC, including the negotiation of local contracts covering wages and working conditions. These associations, like the AGC, operate both nationally and locally, but the major effort is in the local area, in contrast to the NCA, which negotiates agreements at the national level.

Open-shop Organizations

The Associated Builders and Contractors (ABC) is a national organization formed in 1950 by a group of Baltimore contractors who became disillusioned with accepting union control of the industry. They established what is called merit shop: open shop, but not necessarily anti-union. By concentrating in the suburban and outlying areas, ABC members and other open-shop groups began building up strength, until today they are clearly in the majority in their strong areas and are making solid gains in almost all areas of the United States.

In 1991 there are 80 actual ABC chapters in 40 states, representing all 50 states. The association has a national office in Washington, D.C. and publishes a monthly magazine, *Merit Shop Contractor*, which in format and content is not unlike the AGC's magazine, *Constructor*.

Advantages in ABC membership for open-shop contractors include the availability of insurance, employee benefit plans, referral agencies, craft training programs, and other programs that can be more economically obtained as a part of an organized group of employers.

A tribute to the success of the ABC is the emergence of the dual role of the AGC, which is increasingly being made up of both union and nonunion contractors. The AGC continues to negotiate wages and working conditions with local building-trades unions in areas with substantial open shop strength. The substantial pressure from the open-shop membership of the AGC has helped mitigate construction-wage increases in the unionized sector and is contributing to increased productivity in many areas through open competition in right-to-work states and elsewhere, where both union and nonunion contractors coexist peacefully on individual projects.

In other areas, local groups of open-shop or nonunion contractors have formed a local "builder's exchange" or other association that often supplies a central plan room, arranges for group insurance and other fringe benefits, and in general endeavors to promote the welfare of the building industry in their particular locale.

Owner Associations

Owner associations include such groups as the Electric Power Research Institute, American Public Works Association, American Association of State Highway Officials, American Water Works Association, and a number of others. These associations have largely been concerned with technical and engineering-oriented objectives rather than with those usually associated with work force or union relationships.

Certainly the major effort by owners to influence the industrial relations elements in construction has been through an association of many of the largest buyers of construction in the United States: the Business Roundtable. The Roundtable grew from the National Conference on Construction Problems sponsored by the U.S. Chamber of Commerce in 1968 to review the rapid rise in industrial construction costs. The conference called for the formation of an organization of major purchasers of construction to establish mutual cooperation between purchaser and contractors, particularly in construction industry labor relations. The construction User Anti-Inflation Roundtable was then established, with Roger Blough, Chief Executive Officer, U.S. Steel, as the first chairman.

The main thrust of "Roger's Roundtable" was to educate users about the detrimental effect of certain decisions, such as working through strikes or requiring excessive overtime, upon both the overall industry and their own jobs. The Roundtable also promoted the establishment of local user groups to help educate local member companies as well as other local business people. In the 1980s, the organization completed a massive four-phase project titled the Business Roundtable Construction Industry Cost Effectiveness (CICE) study, which explores a comprehensive list of areas in addition to labor relations that could increase productivity in the future. Initial results of the study along with recommendations to be placed in effect were outlined in 1982; publication of the complete study was completed in early 1983. A more recent spin-off of that effort was the establishment of the Construction Industry Institute (CII), located at the University of Texas in Austin. CII performs research and maintains task forces that address the major issues facing the construction industry.

Umbrella Industry Organizations

Over the past few decades a number of umbrella associations have made generally unsuccessful attempts to develop a single voice to discuss and influence broad construction industry issues and to help to reduce problems caused by industry fragmentation. Past efforts have included:

The Construction Industry Joint Conference (a joint labor-contractor body)
Council of Construction Employers
Contractor's Mutual Association
National Construction Employer's Council

Currently active organizations include the National Construction Industry Council (NCIC) and the American Construction Industry Forum (ACIF). The NCIC was formed in 1974 to include a federation of about 20 contractor associations in an effort to develop an industry position with regard to important issues. At peak some 30 national associations including contractors, engineering societies and bonding organizations held membership which has declined in recent years reportedly due to the inability of the group to reach consensus positions instead of backing individual positions. The ACIF includes the Associated Builders and Contractors, The Associated General Contractors, the American Subcontractors Association, the Sheet Metal and Air Conditioning Contractors National Association and the National Association of Plumbing-Heating-Cooling Contractors The ACIF reportedly will not necessarily strive for a consensus opinion but will discuss major industry issues frankly in a noncontroversial way while continuing on separate paths.

HIRING HALLS AND LABOR POOLS

One of the major advantages of the union contractor was the availability of craft hiring halls who were in position to refer skilled craftsmen to the contractor through a union operated program. Such hiring halls are required to be nondiscriminatory under the law. A relatively recent development has been the emergence of third party operated labor pools specially designed to service the construction industry. A typical organization will furnish skilled and semi-skilled construction workers to open-shop contractors. One organization in the San Francisco Bay Area will furnish carpenters, laborers, sheetrockers, electricians, tile setters, plumbers, painters and cement finishers. Workers remain employees of the referral firm who is responsible for its own payroll as well as for payroll taxes and insurance and health and welfare benefits. This type of service is especially helpful to the small firm operating in metropolitan areas and offers advantages to the open-shop contractor comparable to those of the union hiring hall.

Large open-shop contractors such as Fluor Daniel, BE & K, Brown and Root and H. B. Zachary have developed substantial internal records showing name, address, craft skills, work history and telephone numbers and addresses. In addition to hiring past employees, advertisements are often placed in low cost or depressed areas offering construction jobs in other locations throughout the country. Some form of relocation and subsistence allowances are often paid to the worker.

INDUSTRIAL RELATIONS DEPARTMENT

The industrial relations department of a major construction firm might include a number of functions, each headed by a separate manager in the case of a large company or shared in the case of a smaller firm. Functions are briefly described as follows.

Department Manager

The department manager is often responsible to the general manager for the integration of the personnel, safety, and security functions and the labor relations in the home office setting broad policies for the company. Field projects may be delegated some or all of the on-site responsibility for the various functions through one or more on-site representation. See Figure 19-5 for a simplified organization chart for an industrial relations division home-office organization.

Personnel

The personnel manager may be given overall responsibility for the development of policies and procedures applicable to the salaried work force. This will include development and administration of programs for employee benefits, such as vacations, holidays, medical insurance, life insurance, retirement plans, profit sharing plans, and others. Duties will also normally include responsibility for development of a uniform, competitive, and workable wage and salary program for the company, including position descriptions and pay grades. Recruiting, hiring, and development of employment policies round out the assignment.

Safety and Security

A manager for safety and security will often develop internal procedures for safety, security, and fire protection. Follow-up with periodic field visits and close liaison with the company's insurance carrier are equally important. Favorable workers' compensation rates and accident experience can give the farsighted contractor a substantial edge in today's competitive climate. Safety requirements are spelled out in considerable detail in Chapter 17, *Safety and Health in Construction.*

Labor Relations

The labor relations manager may be responsible for negotiating project labor agreements, may be a member of the AGC committee that negotiates craft agreements with the local building trades, and may be responsible for adjudicating grievances and developing a uniform and fair labor relations policy. In nonunion companies, the duties are similar but without the formal agreements and grievance procedures spelled out in the local contract. Legal representation, either through an in-house representative

FIGURE 19-5
Simplified organization chart for home-office industrial relations division.

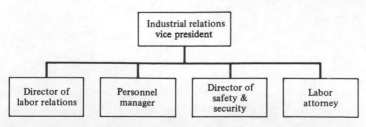

or through use of an outside law firm, is becoming increasingly important in today's complex environment. Ensuring equal employment opportunity and developing and administrating affirmative action plans in accordance with the law, the contract, or company policies and training individual employees continue to be increasingly important to the progressive firm.

In union construction companies, particularly those engaged in industrial construction, the handling of jurisdictional work assignments with the various trades is especially important. Often a pre-job conference is held prior to starting work, to delineate proposed jurisdictional assignments for work to be performed by the contractor's own forces sufficiently in advance to avoid work stoppages or other nonproductive attempts at solutions. The development of a complete file of area practice and a history of work assignments for similar work is an especially important task.

SUMMARY

Industrial relations is important to the progressive contractor in the union, nonunion, and open-shop areas of operation. This chapter briefly discussed the hierarchy of organized labor as represented by the building trades affiliated with the AFL-CIO. Open shop, and its leading proponent merit shop, continues to grow, and increasingly, both union and nonunion contractors are engaging in direct competition, to the overall competitive benefit of the industry. Affirmative action programs and equal employment opportunities for ethnic minorities and women have been mandated on federally funded projects and are offering significantly greater opportunities on private projects.

The industry has developed a number of contractor organizations that have proved very successful in certain areas. However, the local contractor operating through the AGC, the large engineer-contractor associated with the NCA, and the merit shop contractor associated with the ABC continue to go their separate ways, with little cooperation at the national level.

The responsibility for developing harmonious and beneficial relationships between the manual, salaried, and clerical work force in the construction industry is generally described as industrial relations. Its challenges in the future will be even more demanding than they are today.

CHAPTER 20

CLAIMS, LIABILITY AND DISPUTE RESOLUTION

Over the past few years, construction industry participants have experienced an increase in claims, liability exposure and disputes along with increasing difficulty in reaching reasonable settlements in an effective, economical, and timely manner. This chapter looks at current practice in claims and dispute resolution and suggests areas in which the participants themselves may help to prevent expensive litigation. The owner, the general contractor, the designer, subcontractor, and construction manager all can help to mitigate the consequences of disputes throughout the project life cycle.

CONTRACT TYPE AND CONTENT

Chapter 2 listed a number of advantages and disadvantages to both owner and contractor for the traditional fixed-price contract, the cost-plus-a-fee contract, the guaranteed maximum price (GMP) and the professional construction management approach. The choice of a contract method which supports the primary objective of the owner is important in minimizing or eliminating potential claims and disputes. Increasingly, some form of cost-reimbursement contract is being utilized in the public sector on important federal projects as well as in the private sector where earliest completion is a major objective. However, on most public projects, the use of the competitively bid fixed-price contract is mandatory. Private owners have a wide choice of contract types and management methods, as discussed in Chapter 2. Major owner objectives which may influence contract type will include:

  Shortest possible design-construct schedule
  The requirement for an advance price guarantee
  The flexibility to make changes during the construction period is desirable

Fixed-Price Contract

The fixed-price contract is utilized when a high quality advance price guarantee is required. A trade-off is that the design must be fully completed before construction starts so that the scope is fully known to the contractor at bid time, thus lengthening the design-construct period. For best results, a minimum number of changes during construction is important. The owner is protected from the risk of price increase in the absence of changes, changed conditions or attempts to accelerate the completion schedule.

Guaranteed Maximum Price Contract

Under the GMP contract, the owner takes the scope risk and the contractor takes the price risk over a guaranteed maximum or upset amount. Savings below the GMP are normally shared between contractor and owner on a negotiated basis. This type of contract permits work to begin prior to final design and, if the scope is well defined, can fix a maximum price for the work. As in the fixed-price contract, problems arise when scope changes are made by the owner or when changed conditions are encountered. Misunderstandings regarding scope are more prevalent when the GMP is negotiated prior to completion of detail design, a common practice.

When the owner, contractor and designer maintain non-adversarial relationships, this type of program can prove successful. If relationships deteriorate, many of the problems associated with changes to fixed-price contracts are encountered.

Construction Management Contracts

A professional construction management program featuring phased construction and competitively bid fixed-price contracts offers early completion advantages coupled with fixed-price contracts, but may not offer a high quality price guarantee since work begins prior to finalizing all individual contracts. Changes to the work after contracts have been awarded or nonperformance by individual contractors can result in major delays and disputes between owner, construction manager and project contractors.

Design-Construct Contracts

Design-construct contracts can utilize any of the three contract types discussed above. The preponderance of this type of contract on major projects has been on a cost-plus-a-fee or incentive-fee basis, with major firms specializing in this approach. Fixed-price and GMP contracts are also utilized. Since such contracts are negotiated prior to completion of design, scope misunderstandings may cause serious disputes between owner and the design-constructor. Fixed-price or GMP turnkey type contracts seem to work best where both owner and the design constructor have an ongoing relationship and are fully familiar with the type of work being constructed. Extensive contractor experience on similar type work can also improve performance.

CONTRACT DOCUMENT PREPARATION

Most domestic construction contracts are prepared by the owner or designer and the contractor has little opportunity to influence the fairness of the contract terms and conditions and the allocation of risk between the parties. On most public works projects, taking exception to contract terms and conditions is automatically cause for bid rejection. In the private sector, the willingness to negotiate in this area is determined by the owner and/or the engineer. On foreign work bid to contractors on the world market, negotiation of terms and conditions is commonly practiced. Domestically, owners and design professionals along with their lawyers have perceived that they could escape risk and liability by placing major responsibility upon the construction contractor through contract clauses that shift the preponderance of risk to the contractor and associated insurers while minimizing the risk to the owner, design professional, and their insurers.

Fairness in Allocating Risk

In the traditional construction process, the parties should have well defined duties and liabilities coupled with the ability to manage, carry out, and control the duties. The owner, designer, and contractor all have duties in a major construction project. Problems arise when the risk transfer provisions of the contract assign the responsibility and liability to a party who does not have the ability to manage and control the outcome or otherwise mitigate the risk.

An example is the site-of-the-work clause included in many state and private contracts providing that the contractor shall examine carefully the site of the proposed work including subsurface investigations. "Any interpretations or evaluation of the subsurface investigation record made by the bidder shall be at the sole risk of the bidder." The design professional may have spent months or years conducting geological studies, seismic evaluations, test borings and other investigations in order to develop the basis for a workable design. Obviously the contractors cannot duplicate or reasonably check such information during a short bid period, nor can they afford the expense on an individual basis. In this situation, the contractor has been given the duty and the liability for completion of the work under changed conditions without having the ability to anticipate such an event. (An encouraging trend has been the direction by the Federal Highway Administration (FHWA) and the Environmental Protection Agency to include a clause similar to the federal changed conditions clause on state work as a condition of federal revenue sharing.)

Another example is a CPM specification that provides for a sharing of float indicated in an approved schedule with the owner, thus impinging upon the contractor's right to manage the work in the most economical manner, a traditional right under historical fixed-price contracts. This practice also encourages the contractor to "play games" with the proposed schedule in an effort to minimize the owner's opportunity for claiming float or to help create opportunities for delay claims on owner furnished items.

The party bearing the risk without the ability to manage it will, of course, believe such provisions to be highly unfair. In these circumstances, creative lawyers have developed a number of innovative theories to support recovery through litigation or arbitration. In the end, a jury or arbitrator may well be influenced by the perceived unfairness of the provision and grant substantial redress. As more and more unfair restrictive or exculpatory provisions are included by owners, creative contractor lawyers continue to develop new legal theories to circumvent them. In many construction lawsuits, the sum of the parties' direct expense, court costs, expert witness and attorneys fees is often in excess of the final award and is not received until many years after construction completion. All parties to the litigation are often unhappy and the bulk of the expenditures go to third parties who prosecute the dispute for the principals.

Farseeing owners will understand that a contract which is fair to all parties is the first step in developing a cooperative and harmonious relationships. Such a contract can prove economically beneficial to both owner and contractor and help to create a climate of fairness in which both owner and contractor can directly settle most disputes in a timely and mutually beneficial manner.

Principal Risk Allocation and Payment Clauses

The general terms and conditions of any fixed-price contract contain a number of clauses which allocate or mitigate risk to the parties, define responsibility and liability and set forth the terms of payment. Review of these and similar clauses, including associated exculpatory language, requires careful analysis by contractors in order to identify contractual risks assigned by the owner.

The application of contingency to the bid price, and the development of a risk management program including insurance protection, represent some of the ways in which a contractor can help to mitigate potential exposure.

As an example, the following clauses which require careful analysis by contractors have been selected from a typical state highway standard specification. An examination of these and other similar clauses can give an excellent indication of the fairness of the contract from the contractor's standpoint which can be used to evaluate the bid/no-bid decision, to form the basis of contingency evaluation, and to develop a risk management and insurance protection program.

1 Definitions
2 Quantity interpretations and variations
3 Examination of work site
4 Subsurface exploration
5 Changes and alterations
6 Extra work
7 Authority of the engineer
8 Cooperation with others
9 Minimum wage rates
10 Responsibility for damage claims
11 Contract time for completion
12 Adjustment to contract time
13 Termination of contract
14 Failure to complete and liquidated damages

15 Right of way or access delays
16 Measurement of quantities
17 Compensation for changes and alterations
18 Claims for additional compensation
19 Notice requirements
20 Payment for extra and force account work
21 Progress payments and retention
22 Mobilization payment

CONTRACT CHANGES

Few construction contracts are completed without change. Changes or causes for change may be initiated by the designer, owner, regulatory agencies and others and the contractor. Changed conditions occur when the actual physical or other jobsite conditions prove different from those forseeable from the plans and specifications. Most minor changes are resolved on the job between the owner's representative and the contractor. A change order is then added to the contract by mutual agreement.

Constructive changes result from the failure of the owner or his representative to recognize contractual entitlement to a change order or other changed condition in a timely manner. This type of disputed change is among the most difficult to adjudicate. A claim occurs when the parties cannot reach a mutually agreeable solution leading to a change order. Litigation, arbitration or alternate dispute resolution processes conducted by third parties are then usually employed.

Owner or Designer Changes (Owner or Designer Initiated)

Common owner or designer changes may include the following

Numerous last-minute addenda during bid period
Delay in access to the site
Delay in furnishing approved for construction design drawings or clarifications
Delay in furnishing owner-furnished items
Defects in plans or specifications including errors and omissions
Major design changes
Many minor design changes
Scope additions
Scope deletions
Schedule improvement directives
Acceleration directives
Suspension of work
Interference by owner or his designated representative
Nonperformance by owner
Termination of contract
Equivocal or conflicting contract clauses
Slow or inadequate response to submittals and requests for information

Contractor Changes (Contractor Initiated)

Contractor changes usually involve some type of performance failure or installation of defective work.

Failure to start work as planned
Failure to supply a sufficient workforce
Contractor performance failure
Subcontractor performance failure
Supplier performance failure
Installation of defective work
Poor workmanship
Schedule delay
Subcontractor schedule delay

Changes Caused by Others

Other changes include acts or omissions by third parties, discovery of actual site conditions different than those represented, and other items not caused by owner, designer, or contractor:

Unforeseen changed physical site underground or other conditions
Other unforeseen site conditions
Unusual weather or other natural event
Regulatory agency change
Change in the law
Labor disputes
Third-party interference
Third-party nonperformance

MAJOR CLAIM CATEGORIES

For purposes of discussion in this chapter, claims are divided into four broad categories, including:

- Design and specification changes and additions
- Changed site conditions
- Delay claims
- Acceleration, impact and effect, and ripple effect of above delays and changes

A typical complex claim might involve numerous changes and additions to the work by the designer causing a schedule delay to the contractor, as well as additional direct costs to perform the changed or added work. The delay could cause a schedule extension which would involve extended overhead costs for job site and home office indirect costs. If the owner directed that the original completion date be maintained, the contractor would have an additional claim for acceleration which required overtime and larger work crews from initial plans. The overall impact and effect of the larger crews and possible overtime work could cause productivity losses to both the changed work and the unchanged work through a ripple effect which spread throughout the entire job site. Typically in this situation the contractor and owner might agree upon the direct costs of the work and the out-of-pocket costs for overtime premium. However, claimed labor inefficiency costs on all remaining work, which could run 50 to 100 percent higher than normal, could prove difficult or impossible for the owner to accept. The typical interim solution might be for the owner to pay the additional direct costs and to mutually agree to defer the acceleration and impact costs to the end of the project. All too often such disputes cannot be settled amicably, and a jury of individuals unfamiliar with the construction business will make the final decision many years and many additional costs later. The discussion on alternate dispute resolution process will review some of the ways in which such claims can be adjudicated in a less costly and more professional manner utilizing industry expertise in the adjudication process.

Definitions of Terms

This section will define some of the more common terms pertaining to claims and schedule analysis:

1 Change: A modification or impact to the project work which increases or decreases the original contract scope or which impacts the time or cost of completing the original scope. Prompt action by the parties to evaluate and to agree upon the monetary and time impacts of the change can often avoid increasingly complex and expensive adjudication methods at a later date.

2 Constructive Change: The contractor is entitled to recognition of a change but the owner refuses to allow additional time or monetary compensation in violation of the terms of the contract. For example, the owner insists upon completion by the originally specified date, even though the contractor is entitled to a time extension. The contractor maintains that he has been constructively directed to accelerate or compress the work even though he has not received a specific direction to accelerate.

3 Change Order: A formal document signed by both parties to compensate the contractor for changes, added work, delays or other impact by mutual agreement under the terms and conditions of the contract. In the absence of agreement some owners issue unilateral change orders based upon their own evaluation to acknowledge the change (but not the requested amount) and to permit interim progress payments pending future agreement or dispute resolution.

4 Claim: A disagreement that cannot be resolved by mutual agreement under the change-order process. Claims are often subject to a formal process as spelled out in the contract. Unresolved claims become disputes which may be adjudicated by arbitration, litigation or other alternate dispute resolution methods as set forth in the contract.

5 Entitlement: The right to receive remuneration, additional time or other benefit under the contract as a result of a change or other impact upon the work.

6 Excusable Delay: Delays which are excusable under the contract and which entitle the contractor to additional time for performance but may or may not give entitlement for additional remuneration.

7 Compensable Delays: Delays which are due to the owners failure to meet contractual obligations. The contractor is entitled to the reasonable additional costs attributable to the delay as well as an extension of the contract time. Excusable, noncompensable delays are normally due to factors outside the control of either owner or contractor such as unusual weather conditions, strikes or other labor disputes. The contractor is entitled to receive a time extension but is not entitled to receive additional compensation.

8 Compression: The result of requiring the contractor to perform more work during a given period of time than that which was reasonably initially contemplated.

9 Acceleration: The contractor is required by direction or constructive direction to complete the original work or a portion of the work in less than the originally scheduled time, to employ additional resources or to utilize overtime in order to make up for past delays or to complete additional or changed work without benefit of a time extension. Disputes arise when the owner directs nonreimbursable acceleration, believing that the contractor is behind schedule and the contractor believes that a time extension and/or reimbursement is in order.

10 Impact Costs (Impact and Effect): Impact costs result from lower productivity of workers or equipment which are the result of changes and changed conditions.

11 Ripple Effect: Ripple effects are impact cost increases resulting from the "ripple" of a change to one portion of the work spreading to unchanged work.

Design Changes, Additions and Deletions

Minor design changes are experienced upon almost every construction project. So long as changes form a small part of the contract, say up to about ten percent, the normal change order process incorporated into most contracts works reasonably well. The owner and contractor can agree upon a lump sum, utilize pre-determined unit prices if applicable, or perform the work upon a time and material (force account) basis. A change order can be issued possibly including additional time, if warranted. Progress payments can be made promptly and both parties can be satisfied with the results.

However, when changes exceed some 15 percent of the contract price, the effects of the changes can begin to impact schedule and cost for both the changed work and the remaining unchanged work, depending upon contractor management skills and conditions unique to each individual project. A large amount of minor changes can also impact the work due to the demands upon the contractor's staff in processing, estimating, scheduling and laying out the new work. While each project is of course unique, when changes exceed 20 percent or so, potential major impacts upon the performance of the changed work and upon the remaining original work are often experienced. Estimating the cost of the changed work and of the remaining unchanged work becomes very difficult under a climate of continual change, and the full extent and effects of the changes impacting schedule and craft productivity are difficult or impossible to evaluate until much later in the project.

The future impact and effect upon craft productivity under a climate of continual change and rework is not subject to quantification by either owner or contractor management, and wide differences of opinion will normally be found. When the owner endeavors to preserve initial completion dates in the face of major change, the impact and effect of acceleration through the use of additional resources, larger crews and overtime work can often result in a massive productivity decline dependent, of course, upon the management and supervisory skills of the contractor. Complex delay and productivity loss claims, including acceleration coupled with the impact and effect upon the overall jobsite, are difficult or impossible to settle during work performance by polarized parties whose primary objective remains to try to minimize the damage and to complete the project.

Changed Site Conditions

There is considerable difference of opinion among owners and design professionals regarding who should bear the responsibility for unforseen physical conditions encountered during construction. Many owners make no representation in their contract documents that conditions indicated by the bidding documents will actually be encountered; through restrictive and exculpatory language, all risk of initial interpretation and later change is assigned to the contractor. On the other hand, the FHWA has directed individual states without changed conditions clauses to include such a clause as a condition to obtain federal funding.

A typical differing site conditions clause adopted by the Florida Department of Transportation follows:

> The Contractor shall promptly, and before such conditions are disturbed, notify the Engineer in writing of (1) Sub-surface or latent physical conditions at the site differing materially from those indicated in this contract, or (2) unknown physical conditions at the site of an unusual nature, differing materially from those normally encountered and generally recognized as in work of the character provided for in this contract. The Engineer will promptly investigate the conditions, and if he finds that such conditions do materially so differ and cause an increase or decrease in the Contractor's cost of, or the time required for, performance of any part of the work under this contract, whether or not changed as a result of such conditions, an equitable adjustment will be made and the contract modified in writing accordingly.

The use of a Dispute Resolution Board can be used to evaluate changed conditions claims. The DRB is made up of knowledgeable industry experts who can be helpful in bringing the parties together by making nonbinding recommendations for settlement, as discussed in the section on Alternate Dispute Resolution.

Another progressive approach includes the development of a Geotechnical Design Summary Report[1] for complex underground or sub-surface projects in order to establish baseline geotechnical conditions which will become the basis for contractor bidding. The owner and the designer establish the geotechnical baseline for the anticipated conditions. The owner then accepts the risk for conditions more difficult than the baseline while the contractor accepts full price risk in the absence of material change.

Claims for changed site conditions can also lead to claims for delay, acceleration, impact and effect, including the ripple effect as discussed below.

---

[1] *Source: Avoiding and Resolving Disputes in Underground Construction*, ASCE, June, 1989.

Delay Claims

Contracts normally contemplate that there will be delays throughout the course of construction. Where minor delays affect a small part of the overall project, normal change and extra work provisions of the contract can usually be followed and a mutually agreeable change order negotiated. However, beyond a certain point major delays or numerous minor delays and other impacts to the work can disrupt normal construction activity and result in project delay, compression and acceleration causing cost overruns which lead to contractor claims and disputes.

Types of Schedule Impacts   Schedule impacts include compensable, excusable and noncompensable types. While each individual contract will normally define the type of delay, a general classification is set forth below:

Compensable impacts causing delays can include changes and additions, late delivery of owner furnished equipment, materials, drawings, and permits. Delays due to differing site conditions are compensable when covered by a changed conditions clause, owner representation, or other circumstance. Equitable adjustment for compensable impacts include both time extensions and additional compensation.

Excusable, noncompensable impacts may include strikes, other labor disputes, abnormally severe weather, epidemics, and other *force majeure* items. An equitable adjustment extends the time for performance but does not provide for additional compensation.

Noncompensable impacts include contractor impacts or delays which are not the responsibility of the owner, such as delays due to mobilization, lack of tools or equipment, faulty workmanship, and material delivery. Responsibility for such delays and additional costs remains with the contractor.

In complex claims, owners often develop a counter claim for damages due to actions or delays attributable to the contractor. Claims may be made up of interrelated compensable, noncompensable and excusable impacts and delays. Some of the impacts will affect the critical path while others may represent concurrent or noncritical delay. CPM scheduling techniques are normally utilized to analyze the effects of the delays and to assist in quantifying overall project delays, acceleration and the related responsibility.

CPM Schedule Analysis Methods   CPM analysis is a useful tool in sorting out overall impacts to the schedule. Normally three types of schedules are of interest:

The "As-Planned" CPM schedule may be based upon the approved project schedule or upon the contractors initial plan for completing the work.

The "Impacted Schedule" is a network model superimposing the various schedule impacts upon the As-Planned schedule including excusable, compensable, noncompensable and concurrent items. Delays to the critical path can be identified and a revised overall completion date for the project can be calculated. Input to the schedule will involve identifying and quantifying the individual delays and determining the responsibility and entitlement of the parties. This schedule adjusts the As-Planned schedule to reflect the overall effect of the impacts and delays. If this step can be performed early enough while construction is still underway, an agreed upon Impacted

Schedule can represent the basis for a change order extending the project completion date while becoming the adjusted approved project schedule. If this step is prepared after job completion, as is the norm, then contractor and owner diaries, records, correspondence and other records can turn up other impacts and delays which occured throughout the construction period to be included in the hypothetical As-Impacted schedule.

A construction project is a dynamic environment. No matter how well an initial CPM schedule is developed, modification throughout the program will be necessary in order to help achieve overall project objectives. Therefore, an "As-Built" schedule represents the third part of the analysis. Due to vast differences of opinion on responsibility for impacts, entitlement and other factors, most complex acceleration and impact and effect claims are not settled during the life cycle of the project. It is an unusual project that will keep the details of the original As-Planned CPM logic diagram fully up to date. However, most management control systems will routinely develop float and start and completion dates for the approved project schedule thus presenting a factual representation of the actual performance without necessarily documenting the many logic changes which were implemented during the construction period. The usefulness of this high-level record of actual performance can indicate actual delays, actual acceleration, if any, as well as other interim delays throughout the remaining construction period. Jobsite records and correspondence can be utilized to develop responsibility and entitlement for these and other impacts, delays, and effects which, if significant, can be included in the As-Impacted schedule to illustrate the overall effect of all impacts.

Since most CPM analyses on major claims involve litigation or potential litigation or arbitration, the final very complex and detailed product must be able to be simplified to present to an arbitration board or jury in an understandable manner. The summarization of the CPM notation through the use of bar charts is normally utilized by expert witnesses to illustrate major conclusions and to identify the magnitude of major delays and acceleration. A complex case will involve sophisticated conflicting CPM analysis developed and presented by experts for each side during trial or arbitration. Juries and many arbitration boards have great difficulty in following complex CPM logic diagrams. In many cases the verdict will depend upon the skills of the respective lawyers in interpreting and explaining the testimony of their experts and other witnesses while casting doubt upon the equally detailed presentation from the other side, rather than be decided upon the intrinsic merits of the case.

Figure 20-1, compares an As-Planned schedule to an As-Impacted schedule for a simple milestone as a portion of a project. Figure 20-2 indicates that by owner direction, the contractor accelerated the work to make up four days of the five-day delay through use of larger crews and overtime. Figure 20-3 presents bar charts summarizing conclusions developed by the other exhibits. Figure 20-4 presents an As-Built schedule with impacts and indicates the causes and effects of owner delays. Figure 20-5 illustrates an As-Built schedule with owner impacts removed. Figure 20-6 summarizes the previous exhibits in a different format.

Alternate dispute resolution methods, as discussed later, involve use of impartial and knowledgeable industry experts to assist the parties in resolving delay impacts,

SCHEDULE ANALYSIS

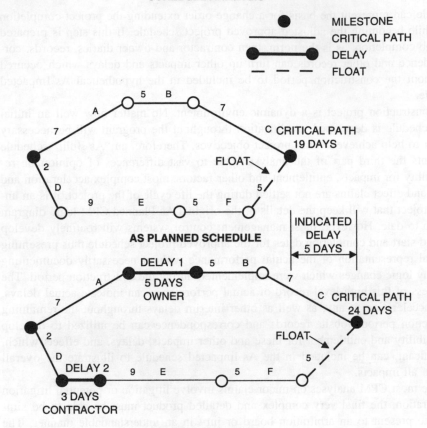

FIGURE 20-1
Schedule analysis, As-Planned vs As-Impacted.

DELAYS	DAYS	RESPONSIBILITY
DELAY 1 FURNISH EQUIPMENT	5	OWNER
DELAY TO START WORK	3	CONTRACTOR

CONCLUSION:
The Contractor would be entitled to 5 days compensable delay. The Contractor's own delay did not affect the critical path.

utilizing owner and contractor management and personnel during the construction program to avoid lengthy later adjudication by outside parties who may be unfamiliar with the construction industry.

Acceleration and Impact and Effect Claims

"An early example was the missile base program of the Air Force in the early 1960s. Contractors plagued by a combination of delays, scope changes, and government inde-

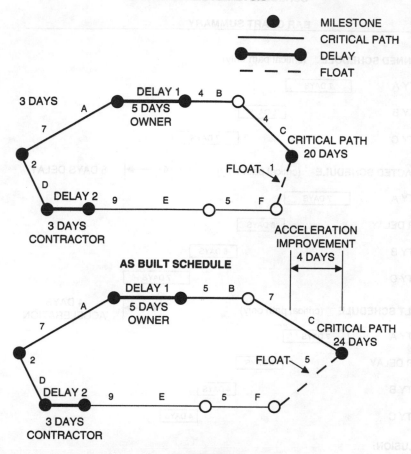

CONCLUSIONS:

Using additional manpower and overtime enabled Contractor to complete activities
B & C 4 days sooner than planned for a 4 day acceleration, thus decreasing the overall
delay to one day.

FIGURE 20-2
Acceleration analysis, As-Built vs As-Impacted.

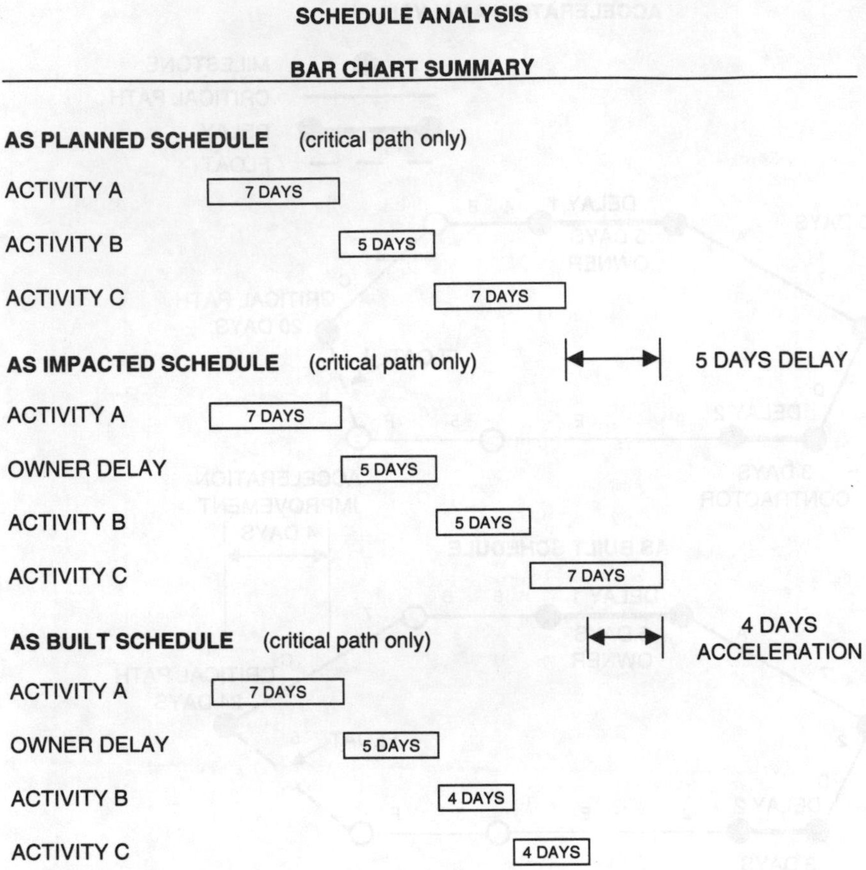

SCHEDULE ANALYSIS

BAR CHART SUMMARY

AS PLANNED SCHEDULE   (critical path only)

ACTIVITY A | 7 DAYS |

ACTIVITY B | 5 DAYS |

ACTIVITY C | 7 DAYS |

AS IMPACTED SCHEDULE   (critical path only) |◄——►| 5 DAYS DELAY

ACTIVITY A | 7 DAYS |

OWNER DELAY | 5 DAYS |

ACTIVITY B | 5 DAYS |

ACTIVITY C | 7 DAYS |

AS BUILT SCHEDULE   (critical path only) |◄——►| 4 DAYS ACCELERATION

ACTIVITY A | 7 DAYS |

OWNER DELAY | 5 DAYS |

ACTIVITY B | 4 DAYS |

ACTIVITY C | 4 DAYS |

CONCLUSION:

After an initial critical path delay of 5 days, the contractor worked additional personnel plus overtime and completed Activities B & C in 8 days compared to the originally planned 12 days, an acceleration of four days from the As Impacted Schedule. Overall completion took 20 days, a delay of 1 day from the As Planned Schedule. The Contractor would be entitled to recover all acceleration costs.

FIGURE 20-3
Schedule analysis, bar chart summary.

cision, found that the costs of their work under such conditions bore little relationship to what was originally contemplated.

"As a result, the government was ultimately required to pay not only the cost of additional work ordered during the performance period, but also for the additional cost of performing the original scope work. This additional cost resulted from the 'impact and effect' of the changes on each other and the original work. This in turn had a 'ripple effect' resulting in further increased costs. Productivity declined with fatigue and equipment utilization became less efficient. Additional supervisory recruitment and other overhead costs were incurred.

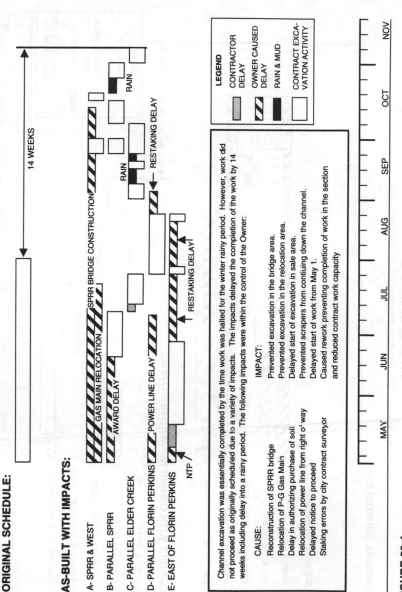

ABC CONSTRUCTION - EVERGREEN CREEK PROJECT
CHANNEL EXCAVATION
AS-BUILT SCHEDULE WITH IMPACTS

ORIGINAL SCHEDULE:

14 WEEKS

AS-BUILT WITH IMPACTS:

A- SPRR & WEST

GAS MAIN RELOCATION / SPRR BRIDGE CONSTRUCTION

B- PARALLEL SPRR

AWARD DELAY

RAIN

RAIN

C- PARALLEL ELDER CREEK

D- PARALLEL FLORIN PERKINS / POWER LINE DELAY

RESTAKING DELAY

E- EAST OF FLORIN PERKINS

RESTAKING DELAY

NTP

Channel excavation was essentially completed by the time work was halted for the winter rainy period. However, work did not proceed as originally scheduled due to a variety of impacts. The impacts delayed the completion of the work by 14 weeks including delay into a rainy period. The following impacts were within the control of the Owner:

CAUSE:	IMPACT:
Reconstruction of SPRR bridge	Prevented excavation in the bridge area.
Relocation of P-G Gas Main	Prevented excavation in the relocation area.
Delay in authorizing purchase of soil	Delayed start of excavation in sale area.
Relocation of power line from right of way	Prevented scrapers from contiuing down the channel.
Delayed notice to proceed	Delayed start of work from May 1.
Staking errors by city contract surveyor	Caused rework preventing completion of work in the section and reduced contract work capacity

LEGEND

CONTRACTOR DELAY

OWNER CAUSED DELAY

RAIN & MUD

CONTRACT EXCA-VATION ACTIVITY

MAY   JUN   JUL   AUG   SEP   OCT   NOV

FIGURE 20-4
As-Built schedule with impacts.

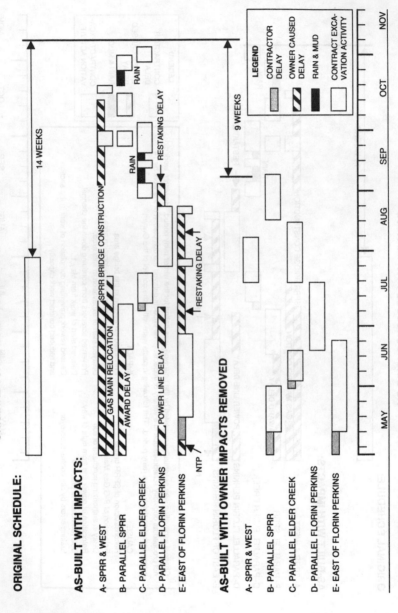

ABC CONSTRUCTION - EVERGREEN CREEK PROJECT
CHANNEL EXCAVATION
ORIGINAL VS. AS-BUILT SCHEDULES COMPARED

ORIGINAL SCHEDULE:

14 WEEKS

AS-BUILT WITH IMPACTS:

A- SPRR & WEST
B- PARALLEL SPRR
C- PARALLEL ELDER CREEK
D- PARALLEL FLORIN PERKINS
E- EAST OF FLORIN PERKINS

GAS MAIN RELOCATION
AWARD DELAY
POWER LINE DELAY
SPRR BRIDGE CONSTRUCTION
RAIN
RAIN
RESTAKING DELAY
RESTAKING DELAY
NTP

AS-BUILT WITH OWNER IMPACTS REMOVED

A- SPRR & WEST
B- PARALLEL SPRR
C- PARALLEL ELDER CREEK
D- PARALLEL FLORIN PERKINS
E- EAST OF FLORIN PERKINS

9 WEEKS

MAY   JUN   JUL   AUG   SEP   OCT   NOV

LEGEND

CONTRACTOR DELAY	
OWNER CAUSED DELAY	
RAIN & MUD	
CONTRACT EXCA-VATION ACTIVITY	

FIGURE 20-5
Original vs As-Built schedules compared.

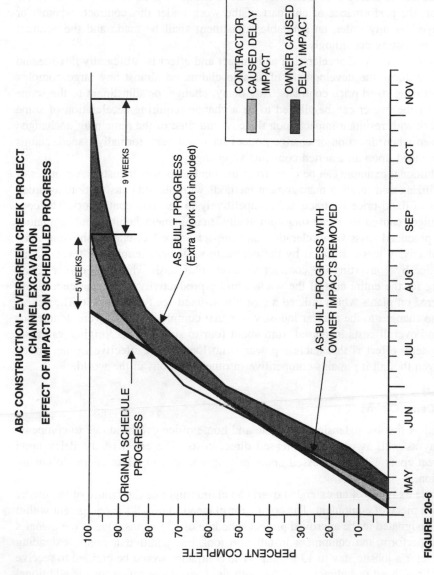

FIGURE 20-6
Effects of impacts on scheduled progress.

"Federal Procurement Regulations were changed to recognize the principles of acceleration providing:

"(a) The Contracting Officer may... at any time ... make any change in the work within the general scope of the contract, including but not limited to, changes: (4) Directing acceleration in the performance of the work.... (d) If the change under this clause causes an increase or decrease in the Contractor's cost of, or the time required for, the performance of any part of the work under this contract, whether or not changed by any order, an equitable adjustment shall be made and the contract modified in writing accordingly..."[2]

The current theory of acceleration and impact and effect is sufficiently flexible and accepted to allow the development of sizable claims on almost any large complex project utilizing fixed-price contracts. Any delay, change, or adjustment to the scope of work by the owner can be alleged to be a change requiring acceleration of some of the work with resultant impact upon the cost and time of the remaining unchanged work. Even when direction or change notices have not been formally issued, claims can be founded upon an asserted constructive change.

An additional example can be cited from the nuclear power industry. A number of plants utilizing construction management methods with phased construction, awarded a number of fixed-price contracts bid competitively to separate contractors. The constant changes caused by increasing complexity, management failures and regulatory changes produced massive acceleration and impact and effect claims showing massive productivity losses suffered by the contractors. A large number of these claims were settled by converting the contract to a cost-plus basis where the owner ended up paying for the entire cost of the work. Similar productivity loss experiences were encountered on plants which utilized a cost-plus-a-fixed-fee type contract. The net result of the changes to the nuclear industry was that completion times were doubled or tripled and overall costs increased from about four to six times the original estimate. The long-term effect is that nuclear power is no longer cost effective in the United States, even though it remains competitive in other locations in the world.

Pricing Delay Claims

Delay claims involve extension of jobsite and home office overhead due to compensable delay as well as possible increased direct costs. The effect of the delay upon other direct costs will be discussed under pricing impact and effect claims following this section.

A simple example of an extended overhead claim might be the failure of the owner to supply a piece of equipment. The contractor planned to install the equipment within the time designated in the approved project schedule of 12 months. Due to the owner's failure to perform, the equipment installation took one additional month, extending the contractor's jobsite stay to 13 months. The contractor would be entitled to receive additional overhead or indirect costs for both field and home office for the additional

---

[2]*Source:* This section is from *Directions in Managing Construction*, D. S. Barrie (ed.), Chapter 13, "Legal and Contractual Considerations," by S. R. McDonald and D. M. Bridges, John Wiley and Sons, 1981, New York, NY.

month plus any other costs directly increased by the delay, such as the additional cost of equipment rental during standby or delay periods. If the mechanical crafts received a wage increase which would not have been payable by the contractor, provided that work had been completed per the original schedule, the additional cost of the wage increase as well as the cost of required construction equipment for one month would also be compensable.

Even though the actual extent of the delay may be easily agreed upon by owner and contractor, it may be difficult to reach agreement on the actual pricing of the delay. In this simple case a contractor would normally claim reimbursement for all additional jobsite costs during the extended period less the estimated cost of the changed work if performed as originally planned. The contractor would also request reimbursement for any additional costs incurred during the original time period attributable to the owner's delay, such as stand-by equipment rental.

Home office costs during the delay period are often troublesome, since such costs are not normally identified with or charged to specific jobsites. Utilizing an agreed upon percent markup, including both home office costs and profit, in pricing change order work can help to avoid such disputes. The use of indirect cost allocation as developed in the Eichelay or other similar formulas has been accepted by certain courts; however, an analysis of individual circumstances in each case can often prove to be more realistic.

Percentage markups including both home, job and field office indirect costs are useful for normal change orders of small magnitude. However, additional indirect costs due to extended overhead often have little or no normal relationship to the volume of work performed, which may be quite small.

Complicated delay and extended overhead claims can usually benefit from a CPM analysis which separates delays into compensable, excusable and noncompensable delays. The contractor's reimbursement for any costs due to delays would be limited to compensable delays.

Pricing Acceleration, Impact and Effect, and Ripple Effect Claims

Acceleration, impact and effect and ripple effect claims represent the most difficult type of claims to adjudicate. Even if the delays can be agreed upon by the parties, the resultant effect of the delays upon the work can be subject to considerable disagreement. When the contractor alleges that the impact and effect due to owner changes and delay has spread to all other nonaffected work through the ripple effect, reaching agreement becomes even more difficult. Contractors preparing these types of claims may be able to establish the following conditions and seek a total cost or modified total cost plus profit recovery:

1 The work was subject to impacts caused by the owner, such as delays, changes, indecision, design deficiencies, changed conditions and other factors of such magnitude that it is impossible for the contractor to accurately identify and quantify the discrete effects of the impacts upon the individual work items.

2 The contractor's bid was reasonable and commensurate with normal industry standards for the type of work involved.

3 The contractor was not responsible for additional costs or delays through his own errors, inefficiencies or lack of resources.

4 The actual costs were reasonable under the changed conditions experienced at the jobsite.

Contractors preparing this type of claim will normally minimize or ignore the cause and effect of their own poor bidding, errors, omissions, management failures and lack of resources. Owners, on the other hand, will concentrate on contractor-caused problems and minimize or ignore their own deficiencies. Due to this polarization, there is little opportunity to reach a meeting of the minds through negotiation at a detail level. Often the full extent of the contractor cost overruns is not fully realized until near job completion, further complicating the opportunity for early settlement. Even when the job is complete or near complete, timely agreement on these types of claims usually involves a bargaining session in which the contractor has developed a minimum acceptable settlement figure and the owner has prepared a similar maximum figure. If strained relations and adversarial relationships do not interfere, an early settlement can be reached if an amount can be negotiated that is above the contractor's minimum and below the owner's maximum. As time passes and the heavy expense of pre-trial litigation and increasing polarization affects both parties, settlement by agreement often becomes more difficult. Experienced negotiators on both sides will estimate the potential costs of litigation or arbitration and recognize such costs in their offers and counter offers during the settlement bargaining session.

Impact and effect and related claims can be developed and priced utilizing several major claim theories:

Total Cost Method
Modified Total Cost Method
Base Line Method
Modified Base Line Method

Total Cost Method Total cost claims are the most difficult to understand, negotiate and adjudicate through third parties. The contractor claims that the owner should pay not only for the changes and impacts to changed work items, but also for the resultant impact and effect upon all or most of the remaining work. This impact was of such complexity and magnitude that it was not possible to measure the magnitude of the disruption and efficiency loss on individual work items, and the contractor is therefore entitled to be reimbursed for his total cost plus a fair profit.

The method assumes that the contractor's bid estimate was a proper measure of the cost of the original work; that the contractor was not responsible for any of the increased costs which were due to owner action or inaction; and that the increased costs were reasonable under the changed conditions created by the owner. This type of claim is often submitted when the contractor has not made a real effort to segregate costs and identify individual impacts upon discrete portions of the work. Owners are understandably reluctant to seriously attempt to settle such claims except in a bargaining situation where they may be prepared to pay more than their initial evaluation to avoid long delays, expensive discovery, pre-trial and litigation costs including substantial manager and executive time which would be better served in another capacity.

As the parties each prepare for arbitration or litigation, the presentation of the claims and the defense continue to improve. However, total cost claims are extremely difficult for juries and arbitrators to follow and final decisions may be heavily weighted in favor of the presentation skills of the parties.

Modified Total Cost Method The modified total cost method attempts to improve on the total cost presentation by validation and/or making adjustments to the contractor's bid estimate before evaluating productivity losses, correcting for contractor errors and removal of costs associated with excusable noncompensable delays, along with exclusion of portions of the work which were not affected by owner caused problems

The original estimate is often validated through macro comparison to bids received from other contractors, even though each individual bid may have wide internal variations not apparent from summary analysis. Actual costs are often validated through the use of impact factors developed in technical papers or published by contractor-based industry organizations such as the Mechanical Contractors Association of America, the National Electrical Contractors Association and others. Overtime studies showing the affect of extended overtime on productivity as published by the Business Roundtable are also normally cited when extended overtime is a factor.

The contractor association figures are usually highly speculative, present no documentation, and may bear little resemblance to actual jobsite conditions. Published overcrowding, excess manpower, and trade stacking curves usually plot productivity loss or inefficiency verses over-manning percent as developed from the contractors bid estimate and consider the overall project as a whole. If the contractor underbid a labor item by 50 percent, the different studies might show 50 percent over-manning with a consequent 30 to 50 percent productivity loss depending upon the selected curve. If the bidder's estimate is corrected to eliminate the under bidding, there would be no over-manning indicated since the original manpower would double, and there would be no productivity loss indicated by any of the studies. It is difficult or impossible to determine manning levels for individual bid components in the hectic pre-bid competitive environment on most fixed-price contracts. Except for macro manpower curves showing summary project manpower, individual task manpower is normally assigned and defined by the contractor after contract award. Manpower leveling is further modified and redefined by the superintendents as the normal changing priorities of the individual work items are planned by craft supervision.

Other modifications to the total cost method can exclude portions of the work which the contractor agrees were not affected by owner actions. Often work performed at a profit or subcontract work without claims is deemed not to have been affected by owner actions and is excluded from the claim.

Base Line Method Since total cost and modified total cost claims are difficult to prove, a baseline or "measured mile" approach has evolved. Pareto's principle, sometimes called the 80–20 rule, states that approximately 20 percent of individual items will account for approximately 80 percent of the cost. By selection of individual items with large cost impacts, the use of the total cost method can be avoided, with total claim amounts comparing favorably. In this method the contractor will often utilize his estimated cost as a base line for principal loss items, tabulate impacts and effects caused by the owner and conclude that the difference between the actual labor cost and the bid labor cost represents the productivity loss due to owner interferences to the original planned program.

Documentation of over-manning, stacking of trades, extended overtime and other impacts can be developed for individual work items of large magnitude in an effort to explain in a more rational and believable manner than when applied to the job as a whole. Combined with validation of the original estimate and schedule, this method can form the structure for an effective and explainable claim. However, this approach can be substantially more time consuming, complicated and expensive to the contractor to develop as well as to the owner to refute.

Modified Base Line Method Since the validation of the original estimate can be difficult and controversial at best, the preferred approach is to select actual costs from a portion of the claimed item which was not subject to disruption by the owner. This cost then forms a baseline independent of the validity of the original estimate and represents the actual cost of the work under unimpacted conditions. The difference between the actual impacted cost and the demonstrated unimpacted base line represents the productivity loss caused by the owner's or contractor's disruption to an orderly and efficient program. If the chosen base is truly representative of unimpacted conditions, the method will develop the productivity loss due to all impacts upon the work, including those caused by both owner and contractor. Documentation and explanation of cause and effect by both parties is, therefore, very important, especially where a third party is adjudicating the claim.

ALTERNATE DISPUTE RESOLUTION

The high cost and time delays associated with litigation and the increasing complexity of arbitration has convinced a number of industry leaders that certain alternate dispute resolution methods can help to avoid both the delays and high costs of traditional methods. Such alternate methods focus upon creation of a favorable climate to encourage settlement of the dispute by the parties themselves.

The Northern California Chapter of the PMI formed a task force to develop and recommend methods to help solve disputes. The results of this task force were presented at the PMI Seminar/Symposium held in San Francisco in September, 1988. Selected portions of the report are included below.

Dispute Review Boards are being successfully utilized on major projects to help avoid the delays, bitterness and excessive costs of litigation or arbitration. A summary of the use of DRBs condensed from an article in *Civil Engineering* is also included.

Step-By-Step Dispute Resolution Proposed by PMI[3]

"Because of the many claims (disputes) affecting nearly all projects and the severe impact of these disputes on project performance and insurance costs, the Northern

---

[3]This section is based upon excerpts from a paper entitled, "The Disputes Resolution Clause," presented by Marc Caspe, John Igoe and S. R. McDonald in the 1988 Proceedings Project Management Institute, September 1988, San Francisco, CA.

California Chapter of the Project Management Institute formed a Task Force on Disputes Resolution. This task force was chartered to devise and recommend specific methods and procedures for limiting the escalation of disputes. The task force realized that the most effective means to fulfill its purpose was to develop a concise contract clause that could function as a basis for enabling negotiators to customize their agreements for streamlining the disputes resolution process—and to get on with the project!

"The Disputes Resolution Clause (DRC) recommended by the task force involves a mandatory four-step process:

Step 1 Direct negotiation between the disputants involved (not binding).

Step 2 Mediation between disputants, with a third-party expert to facilitate early resolution (not binding).

Step 3 Mini-trial, with company officers cross-examining disputants in each other's presence (not binding).

Step 4 Adjudication either by private judging, litigation or arbitration (binding).

"The four-step sequence for resolving a dispute can best be defined by identifying the people involved in each step and their roles in trying to reach an optimized resolution between the contracted parties, without further escalating the dispute.

"During the first three steps, it is assumed that the parties are acting in good faith to try to resolve the dispute. Therefore, clear and accurate communication between all parties is encouraged in an informal (i.e., oral) manner. In the fourth step, procedures and responsibilities become more formally defined.

STEP 1: Negotiation

"The disputants themselves seek to directly negotiate resolution of the dispute without the participation of others (such as mediators, arbitrators, judges, attorneys or officers of their own companies). Attorneys and officers can advise their respective negotiators, but only outside of the negotiation sessions.

STEP 2: Mediation

"The disputants seek to directly negotiate resolution of the dispute using the assistance of a single mediator. The mediator is selected by the Disputes Resolution Panel (DRP) from its own ranks or is an outside person whom the DRP judges best qualified to mediate the dispute. The mediator's principal role is to facilitate the settlement process. Mediation is held directly with the disputants in an effort to find common ground for resolution, except that experts (such as those on the project's Board of Consultants) may be consulted. Attorneys may advise disputants but are prohibited from participating at the mediation sessions without the consent of all parties or a specific request from the mediator for guidance on legal matters. This is intended to provide the disputants with the greatest flexibility to communicate directly and reach an optimal resolution.

STEP 3: Mini-Trial

"A mini-trial panel—consisting of the Step 2 mediator and an officer of each disputing company—hears the positions of the various disputants, who are seeking to establish the reasonableness of their position. The mediator and officers are free to question all disputants and witnesses in order to best evaluate their respective positions and to determine an appropriate course of action. Attorneys are permitted to observe the proceedings but cannot participate without the consent of all parties or a specific request from the mediator about legal matters. The officers must then negotiate their companies' respective positions among themselves, with the assistance of the mediator, who emphasizes the desirability of agreement and the high cost/risk of continuing to Step 4. The officers should be at an appropriate level within their respective companies and should have suitable decision-making power (i.e., the legal authority to settle the dispute).

STEP 4: Adjudication

"Option A, Arbitration: A one- or three-person arbitration panel sits in judgment of the dispute. The mediator (from Steps 2 and 3) may not participate as an arbitrator but may participate in the selection of an arbitrator. The rules and regulations of arbitration shall be those required by the contract or, in the absence of specific contractual requirements, those of the American Arbitration Association.

"Option B, Litigation or Private Judging: A judge (or jury) shall sit in judgment of the dispute, in accordance with all rules of law for the jurisdiction identified in the contract. No limitations shall be placed on the appeals process. Where permitted by state law, private judging is encouraged to expedite a more timely resolution of the dispute. Private judges should be unanimously selected by the parties to the dispute. The ruling of the judge is subject to the standard provision of law as to judgment and appeal."

TEXT OF THE CONTRACT GUIDELINE FOR THE DISPUTES RESOLUTION CLAUSE (DRC)

"*NOTE: The text of the Disputes Resolution Clause included here is intended as a guideline to limit the escalation of disputes. It is intended to serve as a starting point and to enable contract negotiators to draft a brief contract clause that replaces the typical arbitration clause. . . . The agreed-upon contract clause should also be approved by the appropriate insurance carriers prior to contract signing.*

"The parties to this contract agree to use a compulsory four-step process to resolve all disputes between contracted parties. Should good-faith negotiations (Step 1) not result in a solution to the dispute, the remaining three steps of the process shall be invoked in sequence to resolve the dispute. On large projects (over $100 million), a three-member Disputes Resolution Panel (DRP) shall be selected at the start of the project. On projects in the $50-million to $100-million range, a single Disputes Resolution Person (DRP) shall be selected at the project start. On projects under $50 million, the DRP should be selected when a significant dispute arises.

"To better serve as a facilitator of the disputes resolution process, the DRP should also have a familiarity with construction. The DRP shall be kept current about the project by receiving copies of the minutes of all coordination meetings between the parties, and shall attend at least two project-wide coordination meetings per year.

"The four-step DRC sequence shall be as follows:

Step 1: NEGOTIATION: Negotiations shall be between the individuals directly involved in the dispute.

Step 2: MEDIATION: Mediation shall include the individuals directly involved in the dispute—excluding attorneys who are not principal disputants—and a mediator. The mediator shall be selected by the parties or by the DRP from its ranks, as best suited to address the issues involved.

Step 3: MINI-TRIAL: The mini-trial shall include all parties to the dispute, plus one senior officer for each of the companies involved in the dispute, excluding attorneys who are not principal disputants, except that uninvolved attorneys may attend the mini-trial as mute observers. The mini-trial shall be run as a multi-person panel of inquiry comprised of the mediator from Step 2, who will chair the panel, and one senior officer from each company involved. Each member of the panel shall be free to question the disputants and witnesses subject to limits set by the mediator.

Step 4: ADJUDICATION: If the dispute is not resolved in Step 3, all parties may unanimously agree to binding arbitration or, failing to so agree, the dispute will be litigated either by private judging or through the courts. When arbitration is chosen to adjudicate the dispute, the process shall include all parties to the dispute, with attorneys, and its results shall be binding on the parties.

"The rules and regulations to be applied in each step shall be mutually established by the parties, as tabulated in Appendix A,[4] which tabulation is hereby made part of the DRC. It is agreed by the parties that all subcontractors, vendors and suppliers will be bound by this clause, through its insertion into all contracts and purchase orders relating to the project.

"Any third-party legal actions shall be resolved internally by the appropriate contracting and subcontracting parties to the project—using the DRC's four-step process. All contracting and subcontracting parties agree not to enter cross-complaints against one another but instead to use the DRC to establish a joint position between all contracting parties in order to provide a unified response to third-party litigation.

"Nothing written or stated orally during the negotiation, mediation, or mini-trial proceedings may be used as evidence in later proceedings unless developed in accordance with standard methods of discovery."

Disputes Review Boards (DRB)[5]

On complex projects, often featuring underground work, the use of Disputes Review Boards has shown considerable promise in settling disputes during the project life

---

[4]Appendix A (not included here) outlines the resolution process.
[5]Based upon an article in *Civil Engineering*, December 1988, by E. M. Shanley entitled, "A Better Way."

cycle when the facts are readily available. Contract documents normally describe the functions of the DRB and it is organized and placed in operation before a dispute arises. The DRB normally has three members, one chosen by the owner, one by the contractor and a third by the two members themselves. The third or neutral member is the chairman of the board. The board meets at the site every three months or so to stay informed about the job. Each member is an experienced construction professional and all three members must be acceptable to both parties of the contract. The owner and contractor supply the board with progress reports and other information between meetings.

In the event of a dispute, the parties first try to settle it themselves. If not successful, they refer the dispute to the DRB which holds a hearing on-site. The owner and contractor present their cases and the board members ask questions. The board then meets privately and based upon the information presented, presents its recommendations for settlement to the owner and contractor. Acceptance of the recommendation is not mandatory.

While the parties can reject the board's decision, experience has shown that such recommendations have generally been favorably received. DRB recommendations and supporting facts are normally also available to a court or arbitration panel. The recommendation reached by experts at the time when facts are fresh and issues are clear is most generally the fairest to all parties. Also, the potential for prolonging the conflict by arbitration or litigation helps to promote a sincere effort by all parties to reach an equitable agreement. Figure 20-7 illustrates the Dispute Resolution Process. Selected projects which have benefited from DRB utilization are shown in Table 20-1. Utilization of a DRB can help to save time and money, helps keep the job on schedule and maintain good relationships on the job. All participants benefit from a smooth running job with minimum friction and, when it is finished, it is not usually subject to years of further litigation or arbitration.

SUMMARY

Where possible, the owner's choice of the appropriate contract provisions can help to place the risks with the party who has the ability to influence the outcome. Fairness in allocating risk is one of the major steps that an owner can take to help achieve a smooth-running construction project and to avoid costly and time consuming litigation and arbitration. Changes during construction are the major cause for disagreements and disputes. Changes can be initiated by the owner or designer, the contractor or by outside regulatory agencies or other parties.

Major claims occur when the contracting parties are unable to agree upon change orders for change or disruption. Types of claims include:

Design and specification changes and additions
Changed site conditions
Delay claim
Acceleration, impact and effect, and ripple effect

Impact and effect claims involving delay and/or acceleration are the most difficult type of dispute to settle by agreement during the life of the project. Such claims can

DRB PROCESS OF DISPUTES RESOLUTION

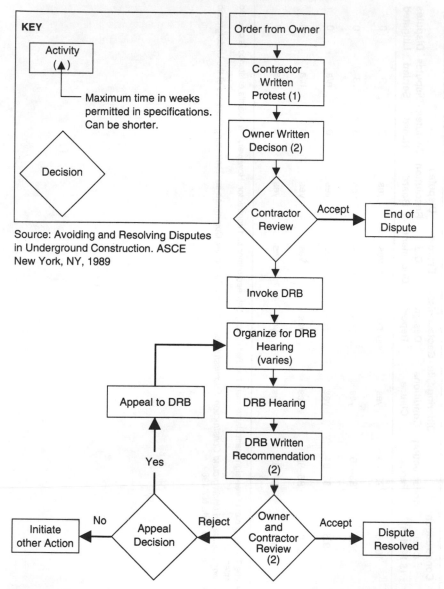

FIGURE 20-7
Dispute resolution process.

be priced under several major theories, including:

Total cost method
Modified total cost method
Base line method
Modified base line method

TABLE 20-1

USE OF SPECIAL CONTRACTING PRACTICES FOR UNDERGROUND CONSTRUCTION

Project	Construction Cost ($ Million)	Construction Period	Differing Site Conditions Clause	Geotechnical Design Report	Escrow Bid Documents	Disputes Resolution Board	Disputes Heard	Disputes Settled	Disputes Litigated
1. Eisenhower Tunnel Second Bore, I70, CO	106.0	'75–79	yes	no	yes	yes	3	3	0
2. Mt. Baker Ridge Tunnel, Seattle, WA	36.1	'83–86	yes	yes	yes	yes	3	3	0
3. Chambers Creek Interceptor, Tacoma, WA	9.6	'83–84	yes	yes	yes	yes	0	n/a	0
4. Downtown Seattle Transit, Seattle, WA	50.0	'86–88	yes	yes	yes	yes	0	n/a	0
5. Sewer Tunnel Relief Honolulu, HA	10.3	'87–89	yes	yes	yes	yes	0	n/a	0

Note: On the downtown Seattle Transit project three significant claims were handled directly between owner and contractor. However, the board's existence is credited with providing an incentive to settlement.

Source: Avoiding and Resolving Disputes in Underground Construction, Successful Practices and Guidelines, Committee on Contracting Practices of the Underground Technology Research Council, ASCE, New York, NY 1989.

Alternate dispute resolution methods show considerable promise in helping to avoid the high cost and delay of arbitration and litigation. Methods that involve the parties themselves, coupled with guidance from recognized impartial industry professionals, can include:

- Direct negotiation
- Mediation assisted negotiations
- Mini-trial
- Disputes Resolution Review Board

Developing fair contract documents, fostering timely jobsite communications in a non-adversarial climate and utilization of recognized construction industry professionals in an advisory capacity can be of great assistance in allowing the contracting parties to settle disputes in an economical and timely manner while avoiding the time delays and heavy expenses of adjudication by outside parties through arbitration and litigation.

CHAPTER 21

CURRENT STATE OF THE ART OF PROFESSIONAL CONSTRUCTION MANAGEMENT

Nonadversarial construction management by any of its names (CM, Professional Construction Management, Construction Management) is now mature and is generally accepted as a distinct method of construction. This chapter will review its evolution over the past two decades and its blending with the emerging program management concept. The chapter will also discuss the elimination of adversary relationships among team members and suggest guidelines for some of the more controversial areas where substantial differences of opinion still remain.

TWO DECADES OF CONSTRUCTION MANAGEMENT

Construction management has now been recognized as a distinct method of managing construction projects for over 20 years. In an April 16, 1981 editorial, the ENR stated: "For more than a dozen years, contracts increasingly have been for construction management (CM). And in recent years, with the emergence of super projects, program management has arrived as a construction service that might include no design, no direct construction responsibility—only overall management of a program of projects. ... CM's definition is hardly more agreed upon now than when ENR first attempted to pin it down in a May 4, 1972 cover story." Yet principles were generally agreed upon sufficiently to permit the ENR to tabulate a list of the largest CM firms and include a list of 20 of the largest annually beginning in its April 16, 1981 issue. In its June 15, 1989 issue ENR stated: "This year, for the first time, ENR is combining in one list all firms that perform construction management services for a fee—whether they are design firms, contractors or 'pure' construction managers. It is also treating CM offered as a professional service for a fee separately from CM where the company has total project responsibility and acts in effect, as a general contractor."

In the same issue, ENR published a list of the Top 100 CM firms based upon CM fees billed in 1988. Firms included are financially liable only for a professional fee, do not contract directly with prime contractors or subcontractors and do not assume the responsibility of a general contractor. The list includes design firms, contractors, engineer contractors and "pure" construction management firms.

If the GMP option developed by the AGC is excluded, the AGC, the AIA and the professional construction management concept outlined in this book differ only slightly if at all in their approach to construction management. Under the new classifications developed by ENR, the CM performance on at-risk contracts such as guaranteed maximum price is classified as CM at risk and also included in the general contractor rankings. CM under a fee type contract without general contractor risk is separately tabulated and also is in conformance with the definition of professional construction management as described in this book. The definition of PCM, first presented jointly by the authors at an ASCE convention in Denver, Colorado, in 1975, followed both the AGC definition (without the GMP option) and the *Study Committee Report on Construction Management* published by the Consulting Engineer Council in January, 1972. The architects' viewpoint was spelled out in *Professional Construction Management and Project Administration* in 1972. The AGC position was described in a handbook entitled *CM for the General Contractor*. An excellent article summarizing the overall concept entitled *How to Avoid Construction Headaches*, was published in the *Harvard Business Review* in 1973. Currently the Construction Management Association of America (CMAA) has also developed definitions, concepts and model agreements and forms for what they call Agency CM (equivalent to professional construction management) and GMP Type CM (equivalent to the AGC GMP option and the new ENR classification). Other organizations and books have developed similar definitions. However, when the GMP risk-taking option is excluded, it is clear that a generally accepted concept for what this book calls professional construction management has emerged irrespective of the label.

These organizations and their definitions generally reflect the three-party team concept of owner, architect-engineer, and construction manager. Increasingly, the principles of construction management are also being applied by design-constructors (engineer-contractors) using the design-manage form of contractual approach as a part of a turnkey project. Many architects as well as engineer-contractors have broadened their approach to perform both the design and the construction management for an individual project. Consulting engineers and project management firms have also entered the field. The definition of professional construction management and professional construction manager in this book remains middle-of-the-road, generally applicable to include somewhat more narrow definitions published by other, more specialized organizations. If the concept of professional construction management is interpreted to include design and construction departments of a design-constructor or knowledgeable architectural firm working in a nonadversarial relationship with the owner the concept and methods are equally applicable to the design-manage form of construction management, with one organization supplying both the design and construction expertise in a professional association with the owner, thus further complicating the issue.

Excluding the GMP option, all the above forms, including the new program management concept, are generally encompassed in the broad understanding of the term CM as used in the industry today.

GUIDELINES FOR SUCCESSFUL PROFESSIONAL CONSTRUCTION MANAGEMENT[1]

Professional construction management involves a three-party team of owner, designer and construction manager. Its success depends upon elimination of adversarial relationships among team members. Should one or more of the team members introduce concepts or policies detrimental to mutually satisfactory relationships, the concept deteriorates into an adversarial situation, with inevitable negative effects upon both the project and individual participants. This section reviews some of the still somewhat controversial and undefined areas of construction management and suggests guidelines to consider if the partnership atmosphere necessary for successful project performance is to be achieved.

Liability and Risk Implications

Many owners, designers and even some who profess to be construction managers do not fully appreciate the differences in liability and risk implications between PCM and the traditional fixed-price general-contractor method. With certain exceptions, PCM is much closer to the traditional negotiated cost-plus-a-fixed-fee general contractor who has settled for a smaller guaranteed profit as a trade-off to potential higher profits through the assumption of risks. For a fixed fee, the construction manager agrees to perform certain professional services for the client. Under this concept, the construction manager should certainly have some liability for actions under his control. However, under the PCM concept, he should not be assigned liability or risk for actions that are clearly outside his control.

It is not feasible to categorically define liabilities and risks that should or should not be assumed by the construction manager. Owners and CM firms are of varying degrees of size and financial strength, have varying degrees of insurance protection, and are willing to assume varying degrees of risk in any venture. Each project is unique and each agreement for services must contain a meeting of the minds regarding liabilities and risks of each party under the contract.

Generally, the following guidelines are suggested for PCM and program management projects:

1 The construction or program manager should be responsible for the actions of his key personnel, including reasonable and prudent skill and judgment. Many firms place a limit upon the amount of this liability in the contract.

2 The construction manager should be responsible for his own liability covering specific acts of the employees, including automobile accidents and other public liability and property damage exposure, which normally can be fully insured.

[1] This section is adapted in part from D. S. Barrie, "Guidelines for Successful Construction Management," *Journal of the Construction Division*, ASCE, vol. 106 no. CO3, September 1980, pp. 237–245.

3 CM firms normally try to eliminate any responsibility for consequential damages from the agreement with the owner since current fee structures do not permit assumption of risks that could have catastrophic consequences.

4 CM firms have a responsibility under the law regarding the Occupational Safety and Health Act. Whenever a number of contracts are let, there are bound to be overlapping safety requirements that must be handled by the construction manager. On the other hand, the project contractors must also be required to fully assume all their responsibilities.

5 By far the most difficult area of risk associated with the concept involves interfaces between the project contractors and the interfaces with the designer. In the traditional single-contract fixed-price approach, the general contractor took on the burden of adjudicating subcontractors' conflicting claims and other conflicts resulting from overlapping responsibilities, unclear specifications, or errors or misinterpretation in the bidding documents. Most owners were unaware of these conflicts since their single contract effectively insulated them from these internal squabbles. Architects and engineers were usually called in to adjudicate or interpret conflicting provisions, but in the absence of litigation, they were unaware of the underlying conflicts, many of which were a result of ambiguities or overlapping specifications. A professional construction manager has a higher duty than the general contractor. The professional construction manager, if the concept is to survive, must impartially adjudicate contractor claims regarding both interpretation of the design documents and the area of scope of the bidding documents for which the manager may have primary responsibility. The surviving firms will find a way to accomplish this without alienating the owners who employ them or the designers with whom they have an equally professional relationship.

6 Owners have a similar responsibility. If the concepts are successful, the owner will achieve the overall benefits. Minor errors or omissions by the designer or construction manager are sometimes inevitable in a fast-track program. A contingency of some 3 to 5 percent of the contract award values should take care of minor additions, minor omissions, or ambiguities in the drawings, specifications, or bid packages. This should be pointed out to the owner by both the designer and the construction manager and should be included in project budgets. If, in the judgment of the owner, managers or designers do not live up to these standards, the project is a failure and the owner will probably have the ultimate responsibility of accepting the consequences in the absence of litigation. On the other hand, if owners insist upon making an example of minor errors or omissions by team members, the professional nonadversarial concept will obviously not survive.

Labor Relations Considerations

The professional construction or program manager is not a general contractor. He must insulate himself from direct agreements with the labor unions, but it is extremely helpful for him to maintain a businesslike and cordial relationship with the local building trades and employer associations. On very large jobs, the owner and/or the construction manager may consider negotiation of a project agreement covering a specific project with the labor unions.

In right-to-work states and increasingly in other locations, there is often a mixture of union and open shop contractors. Local contractors have developed certain practices that tend to minimize or eliminate open conflicts or disruptive practices.

Labor relations considerations from the standpoint of the professional construction manager generally consist of making a careful study of the fundamental relationships between local contractors and the labor unions. Talking with both contractor and labor representatives is essential if an unbiased assessment of the facts is to be developed. The results of this type of investigation can be used in the development of bid packages that follow the normal pattern for the area.

The construction manager can help achieve harmonious labor relations by carefully studying the way business is normally done in the area and then taking this information into account when developing and choosing the contract-bid packages.

Compensation and Fee Structures

Fees for the construction manager are the result of negotiation between the parties and subject to considerable misunderstanding. Professional construction managers apply the skills of workwise general and specialty contractors to a three-party partnership program. Compensation should be comparable to that for the architect, engineer, or other professional. (See Chapter 9 for representative professional fees for home office costs and profit with all field costs being reimbursable.)

Professional construction management programs have been successfully performed under a variety of fee structures:

1 Cost reimbursement for home-office and field costs plus a fixed fee (sometimes award fee based upon performance)

2 Fixed fee for home-office services, including profit plus cost reimbursement for all job-site costs

3 Guaranteed maximum cost for CM services under a cost-reimbursement-plus-fixed-fee-type contract

4 Lump sum for CM services only

Under professional construction management and nonadversarial CM definitions, the construction manager does not guarantee the overall cost of the project or quote a lump sum for the entire project. His profits are therefore minus the risk component that must be included in the traditional lump-sum contract.

Forms of Contract

In the private sector, the professional construction management contract is the subject of negotiation between the owner and the construction manager. In the public sector, several agencies have developed a relatively standard form of contract. Normal forms of contract for PCM services parallel the compensation and fee structure previously described:

- Cost reimbursement for home-office and field costs plus a fixed fee
- Fixed fee for profit and home-office costs plus cost reimbursement for field costs

- Guaranteed maximum cost for professional services
- Lump sum for all professional services

Both the AGC and AIA have developed standard forms for CM services. The AGC form and the CMAA form both include a GMP alternate where the contractor guarantees the total cost in a manner similar to that of a general contractor. Many private and public owners, contractors and CM firms have developed their own standard contracts that are used with appropriate modification to fit the needs and negotiating skills of the parties. Selected sample contract forms from several organizations are included in Appendix C.

Organizational Concepts

Organization of a particular project will depend upon many variables. Planning, estimating, procurement, contract award, and other services may be performed in the home office. On the other hand, some or all of these services may be better performed at the job site, depending upon individual project considerations. Obviously, the choice of the field project manager and the staff may depend upon the organizational concepts involved.

The PCM concept is an extension of the old functional organization for traditional projects. The concept substitutes a CM firm for the general contractor under a nonadversary position. Thus the professional concept of a three-party team of owner-designer-construction manager has now been recognized as competition not only to the traditional single-contract concept but to the concept of design-construction, engineer-contractor, or turnkey.

Licensing Considerations

The question of licensing is being reviewed in almost every state. Should we have a separate licensing system for construction managers? Should construction managers hold licenses normally required for architects, engineers or general contractors who are licensed or registered in most states?

Since CM contracts are flexible, it is difficult to determine which licenses, if any, may be required. The prudent construction manager will obtain a general contractor's license when appropriate or unclear. He may also be wise to obtain an engineering or architectural license, depending upon his duties under the contract and the individual laws of the state.

In the absence of a universally accepted definition of construction management, separate licensing of construction managers is neither feasible nor desirable at present.

Owner Responsibilities

Some owners are never wrong. They hire an architect or an engineer, and a construction manager, and then place a number of constraints upon the authority of both parties. When things go wrong, either the construction manager or the designer must suffer the consequences.

Other owners are more tolerant and perceptive. They recognize that the PCM concept involves a three-party team in which the owner as well as the construction manager and designer have definite responsibilities. Even the owner can occasionally make a mistake. When the owner reserves certain functions for himself, he should be accountable to the team for performance.

No concept can prosper and grow in the absence of successful results. Designers and construction managers who cannot put together successful projects under the professional concept will not survive in the competitive environment. One of the strongest recommendations for a successful CM approach is reflected in repeat business from clients.

Designer Responsibilities

Association under a three-party team concept is a new experience for many architects and engineers. The professional construction manager has taken over some of the traditional duties previously carried out by the designer. On the other hand, the concept has released the designer from some of the responsibilities over which he had little or no control.

Many traditional designers (architects and engineers) are somewhat like general contractors. Some are so wrapped up in the traditional adversary position that they cannot see that the professional concepts offer an opportunity for both to compete with the turnkey or design-construction approach which has gained stature in the past several decades. However, the partnership program requires a mutually cooperative environment.

The construction manager must play fair with the designer. The designer must reciprocate and develop the ability to be cooperative and communicate with the construction manager in a manner that does not prejudice their individual responsibilities to the owner.

Designers who try to participate with construction managers in a three-party-team concept but who look upon the construction manager as an adversary will find that the concepts do not work. Construction managers who do not recognize the inherent professional design responsibilities of the architect or engineer or who criticize the designer to the owner will also find difficulty in achieving successful projects.

Construction Manager Qualifications and Responsibilities

The CM firm must live up to its responsibilities. First it must have the proper qualifications:

- An understanding of the workings of a design office, fundamental engineering principles, and appreciation of the role of the designer
- Skills of a successful general contractor, including the ability to estimate the cost of work and handle bid-package preparation, prequalification of contractors, bid evaluation, contract award and administration and overall management of the construction program

- Skills and estimating ability for electrical and mechanical work items and other specialties that old-line general contractors traditionally subcontracted
- Most importantly, the ability to visualize the overall objectives of the program and act as the leader of the three-party team in all matters relating to construction in an effort to achieve these objectives

The construction manager must carry out his responsibility to the other team members. In addition, he has a responsibility to the labor unions, to the overall industry, and to the general and specialty contractors operating in the area. He must faithfully try to achieve the owner's objectives while preserving a fair and businesslike relationship with project contractors, public agencies, and others.

Some general contractors continue to deprecate the CM firms, identifying them with "brokers" who have traditionally subcontracted all the work while engaging in "bid shopping" or other questionable practices. The professional construction manager understands this criticism and continues to conduct himself above even the appearance of suspicion.

SUMMARY

The construction management concept has had both successes and failures in the past twenty years. Now mature, it is generally accepted as another distinct option for completing construction projects. The ENR has separated construction managers into two groups: CMs who perform services for a professional fee (professional construction managers as defined by this book) and CMs who take the normal risks or responsibilities of a general contractor (construction managers who let subcontracts and or who perform services for a guaranteed maximum price).

Detail definitions and nomenclature differ among individual groups, but an overall consensus regarding the general principles of fee-based construction management has emerged.

The established professional construction manager and the emerging program manager must be prepared to offer a sizable list of services. Such services must be framed to take into account normal industry operating methods and practices in the area. Adversary relationships must be avoided if a successful professional program is to be achieved. Suggested guidelines for handling some of the still undefined and somewhat controversial aspects of construction management have been set forth, to encourage the partnership atmosphere necessary for successful project performance.

A

DESCRIPTION OF THE EXAMPLE PROJECT

The example project warehouse represents a simplified version of an actual project designed by Leo Rosenthal, A.I.A., a practicing Denver architect, and constructed using the professional construction management concept. Figure 4-2 in Chapter 4 shows photographs from a similar project which was featured in the *Kaiser Builder*. The following descriptive information is included in this section:

1. **Specification Summary**

   An outline of the specifications which are required for each bid package has been developed in summary format to illustrate the general nature of the technical specifications required for each bid package.

2. **Fair-Cost Estimates**

   Fair-cost estimates for each of the 10 bid packages have been prepared from the drawings and specifications.

3. **Engineering Drawings**

   The following drawings were required for the project:

   Drawing SI-1   Plot Plan and Utility Plan
   Drawing A-1   Foundation and Floor Plans, and Finish Schedule
   Drawing A-2   Roof Framing, Plan and Details
   Drawing A-3   Elevation and Roof Plan
   Drawing PH-1   Plumbing and Heating Plan
   Drawing E-1   Electrical Plan

   The architect has generously consented to make available full-size reproducible tracings or full-size prints to readers or educational institutions at prevailing reproduction and shipping prices. Inquiries should be addressed directly to the architect:

   Leo Rosenthal, A.I.A.
   311 S. Pennsylvania St.
   Denver, Colorado 80209

SPECIFICATION SUMMARY
CONSTRUCTION MANAGEMENT PROGRAM
MOUNTAINTOWN WAREHOUSE

1. Site Earthwork and Fencing
1.01 General
1.02 Scope
1.03 Related Work Not Included
1.04 Drawings and Site Examination
1.05 Removal of Vegetation and Top Soil
1.06 General Cut
1.07 Rejected Material
1.08 Scarifying and Recompaction
1.09 Fill
1.10 Grading at Railroad Tracks
1.11 Maintenance of Finish Grades
1.12 Dust Control
1.13 Inspection and Tests
1.14 Fencing and Gates

2. Structural and Yard Concrete
2.01 General
2.02 Scope
2.03 Related Work Not Included
2.04 Structural Excavation
2.05 Backfill and Fine Grading
2.06 Materials
2.07 Forms and Appurtenances
2.08 Metal Reinforcement
2.09 Embedded Items
2.10 Concrete Proportions and Consistency
2.11 Mixing and Placing of Concrete
2.12 Yard Concrete Paving
2.13 Placing and Finishing Yard Concrete
2.14 Expansion and Contraction Joints
2.15 Grouting and Dry Pack
2.16 Miscellaneous Work
2.17 Inspection and Tests

3. Interior Special Slabs
3.01 General
3.02 Scope
3.03 Related Work Not Included
3.04 Fine Grading
3.05 Materials
3.06 Forms and Appurtenances
3.07 Metal Reinforcement

3.08 Embedded Items
3.09 Concrete Proportions and Consistency
3.10 Mixing and Placing of Concrete
3.11 Interior Concrete Floors
3.12 Expansive Cement Alternate
3.13 Topping Installation
3.14 Curbs and Ramps
3.15 Miscellaneous Work
3.16 Inspection and Tests

4. Structural Steel and Deck
4.01 General
4.02 Scope
4.03 Related Work Not Included
4.04 Shop Drawings
4.05 Materials
4.06 Fabrication of Structural Steel
4.07 Design and Fabrication of Steel Joists
4.08 Welding
4.09 Shop Painting
4.10 Erection
4.11 Bearing Pads
4.12 Miscellaneous Work
4.13 Inspection and Tests

5. Precast Concrete Double-Tee Walls
5.01 General
5.02 Scope
5.03 Related Work Not Included
5.04 Shop Drawings
5.05 Materials
5.06 Prestressing Reinforcement
5.07 Concrete Proportions and Consistency
5.08 Construction
5.09 Construction
5.10 Erection
5.11 Finishing
5.12 Miscellaneous Work
5.13 Inspection and Tests

6. Plumbing, Heating, and Mechanical
6.01 General
6.02 Scope
6.03 Related Work Not Included

FAIR-COST ESTIMATE SUMMARY
MOUNTAINTOWN WAREHOUSE

Contract Package	Labor	Material	Markup	Total
1. Site earthwork	44,000	153,200	23,600	220,800
2. Structural & yard concrete	231,800	349,000	69,600	650,400
3. Special floors	219,800	271,500	59,000	550,300
4. Structural steel	113,000	686,000	95,800	894,800
5. Precast walls	79,200	418,200	59,600	557,000
6. Plumbing & HVAC	127,600	313,600	53,000	494,200
7. Fire protection	96,600	201,000	35,600	333,200
8. Electrical	104,000	250,000	42,000	396,000
9. Roofing	94,400	123,600	26,000	244,000
10. Building finish	255,000	311,300	68,000	634,300
Total	1,365,400	3,077,400	532,200	4,975,000

	Labor hours	Hours, sq. ft.	Cost, sq. ft.
1. Site earthwork	1,470	.0097	1.46
2. Structural & yard concrete	7,730	.0510	4.30
3. Special floors	7,330	.0484	3.63
4. Structural steel	3,770	.0249	5.90
5. Precast walls	2,640	.0174	3.68
6. Plumbing & HVAC	4,250	.0280	3.26
7. Fire protection	3,220	.0212	2.20
8. Electrical	3,470	.0229	2.62
9. Roofing	3,150	.0208	1.60
10. Building finish	8,470	.0559	4.18
Total	45,500	.3002	32.83

Job duration	8 months
Total manpower	45,500 hours
Total manpower	5,688 man-days
Total manpower	268 man-months
Average manpower	34 men
Estimated peak	68 men
Average labor cost	$30.00/hour (including taxes, ins., and fringes)
Square feet	151,600 sq. ft.

Title <u>Mountaintown Warehouse</u>
Client <u>Easyway</u> **Location** <u>M'tntown WA</u> **Date** <u>12–23–76</u>
Subject <u>Fair Cost Estimate Summary</u> **By** <u>E.P.M.</u>

Description	Quantity	Unit Cost Labor	Unit Cost Mat'l	Labor	Material	Total
Site Earthwork						
Site Grading	15,000 CY	1.00	2.40	15,000	36,000	
Compacted Building Fill	25,000 CY	.60	3.00	15,000	75,000	
Fencing	4,680 LF	3.00	9.00	14,000	42,200	
Direct Cost	1,470 HR			44,000	153,200	197,200
Overhead & Fee @ 12%						23,600
Total Construction Cost						220,800
Structural & Yard Concrete						
Structural Excav & Backfill	5,000 CY	8.00	6.00	40,000	30,000	
Hand Excavation	250 CY	20.00		5,000	–	
Fine Grade for Yard Paving	159,000 SF	.02	.04	3,180	6,360	
Concrete 3000-Yard Paving	3,850 CY	14.00	57.00	53,900	219,450	
-Bldg Fdns	440 CY	20.00	57.00	8,800	25,080	
-Misc	60 CY	50.00	57.00	3,000	3,420	
Forms-Edge for Paving	1,000 SF	2.00	.80	2,000	800	
-Bldg Fdns	11,300 SF	3.00	1.00	33,900	11,300	
-Misc	1,700 SF	4.00	1.00	6,800	1,700	
Reinforcing Steel-Bldg	30,000 LBS	.20	.40	6,000	12,000	
Grout at Base of Wall Panels	500 CF	4.00	4.00	2,000	2,000	
Set Anchor Bolts	432 EA	8.00	2.00	3,460	860	
Grout Base Plates	108 EA	20.00	6.00	2,160	660	
Paving-Expansion Joint "A"	2,500 LF	2.00	3.20	5,000	8,000	
-Construction Jt. "B"	3,440 LF	.40	.60	1,380	2,060	
-Contraction Jt. "C"	5,960 LF	1.00	.40	5,960	2,380	
-Edge Type "D"	1,280 LF	.40	.56	520	720	
Paving Slab Finish	159,000 SF	.30	.04	47,700	6,360	
6" φ Pipe Guards @ Islands	10 EA	80.00	120.00	800	1,200	
Embedded Rail in R.R. Bumper	30 LF	8.00	16.00	240	480	
Equipment Usage-Concrete	4,350 CY	–	3.00		13,050	
-Forms	14,000 SF	–	.08		1,120	
Direct Cost	7,730 HR			231,800	349,000	580,800
Overhead & Fee @ 12%						69,600
Total Construction Cost						650,400

Description	Quantity	Unit Cost Labor	Unit Cost Mat'l	Labor	Material	Total

Interior Special Slabs

Description	Quantity	Labor	Mat'l	Labor	Material	Total
Base Slab $5\frac{1}{4}''$ 3000#	2,500 CY	20.00	57.00	50,000	142,500	
Fill Blockouts at Columns	20 CY	30.00	57.00	600	1,140	
Ramp Slab	8 CY	30.00	57.00	240	460	
12″ Curbs at O.H. Doors	12 CY	60.00	57.00	720	680	
Forms for Blockouts	730 SF	4.00	1.00	2,920	720	
Reinforcing Steel	140,000 LBS	.20	.40	28,000	56,000	
Screed & Cure Base Slab	144,500 SF	.16	.04	23,120	5,780	
Float & Cure Ramp Slab	300 SF	.30	.04	100	20	
Topping on Base Slab $\frac{3}{4}''$	144,500 SF	.60	.20	86,700	28,900	
Const. & Contraction Joints	13,600 LF	.80	.20	10,880	2,720	
Greased Dowels #5	4,700 EA	.40	.50	1,880	2,360	
Premoulded Expansion Jt. $\frac{1}{2}'' \times 6''$	2,000 LF	.40	.40	800	800	
Curbs-4″ × 18″	1,000 LF	6.00	8.00	6,000	8,000	
-12″ × 12″	60 LF	6.00	10.00	360	600	
R.R. Dock Angle 3 × 3 × $\frac{5}{16}$	2,500 LF	.90	.50	2,240	1,260	
Channel at Dock Doors 10[15.3	4,000 LBS	.80	.50	3,200	2,000	
Steel Plate Curb Enclosures	2,900 LBS	.70	.80	2,040	2,320	
Equipment Usage	2,540 CY	–	6.00	–	15,240	
Direct Cost	7,330 HR			219,800	271,500	491,300
Overhead & Fee @ 12%						59,000
Total Construction Cost						550,300

Structural Steel & Deck

Description	Quantity	Labor	Mat'l	Labor	Material	Total
Structural Steel	280 Ton	20,000	1,350.00	56,000	378,000	
Roof Joists & Bridging	200 Ton	12,000	1,380.00	24,000	176,000	
Metal Deck $1\frac{1}{2}$; 22 Ga.	165,000 SF	.20	.80	33,000	132,000	
Direct Cost	3,770 HR			113,000	686,000	799,000
Overhead & Fee @ 12%						95,800
Total Construction Cost						894,800

Precast Double Tees

Description	Quantity	Labor	Mat'l	Labor	Material	Total
Prestressed TT's 22″ + 4″	44,300 SF	1.20	6.50	53,160	288,000	
Prestressed TT's 16″ + 5″	21,700 SF	1.20	6.00	26,040	130,200	
Direct Cost	2,640 HR			79,200	418,200	497,400
Overhead & Fee @ 12%						59,600
Total Construction Cost						557,000

Description	Quantity	Unit Cost Labor	Unit Cost Mat'l	Labor	Material	Total
Plumbing, Heating, Mechanical						
Plumbing						
Roof & Floor Drainage	165,000 SF	.12	.32	19,800	52,800	
Fixtures	30 EA	500.00	1,500.00	15,000	45,000	
Direct Cost	1,160 HR			34,800	97,800	132,600
Sheet Metal						
Cap Flashing	2,200 LF	2.00	3.00	4,400	6,600	
Surface Flashing & Reglet	600 LF	6.00	5.00	3,600	3,000	
Roof Expansion Joint	350 LF	20.00	20.00	7,000	7,000	
Roof Hatches	49 EA	200.00	1,000.00	9,800	49,000	
Misc. Flashing	Allow	–		5,000	10,000	
Direct Cost	990 HR			29,800	75,600	105,400
Heating						
Piping Av. Size = $2\frac{1}{2}''$	2,200 LF	12.00	20.00	26,400	44,000	
Unit Heaters-Gas	33 EA	100.00	920.00	3,300	30,360	
Unit Heater Supports	33 SFS	40.00	100.00	1,320	3,300	
Thermostats	33 EA	40.00	80.00	1,320	2,640	
Gas Meter Pit	1 EA	–	–	720	1,100	
Roof Vents	Allow	–	–	940	6,000	
Direct Cost	1,130 HR			34,000	87,400	121,400
Yard Underground						
Storm Sewers- 6″	500 LF	5.00	8.00	2,500	4,000	
8″	320 LF	6.00	9.00	1,920	2,880	
10″	780 LF	8.00	12.00	6,240	9,360	
12″	140 LF	9.00	14.00	1,260	1,960	
15″	200 LF	10.00	18.00	2,000	3,600	
18″	170 LF	12.00	24.00	2,040	4,080	
21″	130 LF	14.00	30.00	1,820	3,900	
Sanitary Sewers-4″	80 LF	8.00	8.00	640	640	
6″	370 LF	8.00	10.00	2,960	3,700	
Domestic Water-$\frac{3}{4}''$	160 LF	8.00	8.00	1,280	1,280	
2″	40 LF	12.00	12.00	480	480	
Manholes & Catch Basins	9 EA	650.00	1,880.00	5,860	16,920	
Direct Cost	970 HR			29,000	52,800	81,800
Total Direct Cost	4,250 HR			127,600	313,600	441,200
Overhead & Fee @ 12%						53,000
Total Construction Cost						494,200
Fire Protection						
Sprinkler System	150,000 SF	.50	1.10	75,000	165,000	
Fire Loop-10″ ϕ	1,800 LF	12.00	20.00	21,600	36,000	
Direct Cost	3,220 HR			96,600	201,000	297,600
Overhead & Fee @ 12%						35,600
Total Construction Cost						333,200

Description	Quantity	Unit Cost Labor	Mat'l	Labor	Material	Total
Electrical						
Site Work	Allow	–	–	7,000	9,000	
Distribution	Allow	–	–	17,000	50,000	
Lighting & Misc Power	Allow	–	–	66,000	166,000	
Motors & Controls	Allow	–	–	8,400	14,200	
Misc Systems	Allow	–	–	5,600	10,800	
Direct Cost	3470 HR			104,000	250,000	354,000
Overhead & Fee @ 12%						42,000
Total Construction Cost						396,000
Roofing						
Built-up Roofing	165,000 SF	.36	.44	59,400	72,600	
1″ Rigid Insulation	165,000 SF	.20	.30	33,000	49,500	
4″ Fibre Csnt.	4,000 LF	.50	.38	2,000	1,500	
Direct Cost	3,150 HR			94,400	123,600	218,000
Overhead & Fee @ 12%						26,000
Total Construction Cost						244,000
Building Finish						
Miscellaneous Metals						
Furnish only Angles	9,600	–	1.00	–	9,600	
Anchors	500	–	1.20	–	600	
Exp. Shields	50	–	4.00	–	200	
Tube Steel	500	–	2.40	–	1,200	
Pipe Rail	20 LF	–	30.00	–	600	
Nosings	100 LF	–	10.00	–	1,000	
Furn. & Install Angles	8,400	.80	1.00	6,720	8,400	
Pipe Guards	4,500	.80	1.00	3,600	4,500	
Plates	12,000	1.20	1.00	14,400	12,000	
Railing	16 LF	18.00	44.00	280	700	
Direct Cost	830 HR			25,000	38,800	63,800
Carpentry & Millwork						
Interior Structures Roof Fmg.	3,200 bm	1.00	1.00	3,200	3,200	
Plywood $\frac{3}{4}''$	2,000 SF	.40	1.20	800	2,400	
Roof Curbs, Nailers	10,000 bm	1.80	1.20	18,000	12,000	
Redwood Facia	600 bm	1.20	1.40	720	840	
Benches	200 bm	1.00	2.00	200	400	
Service Counters	2 EA	600.00	2,000.00	1,200	4,000	
Rough Hardware	Allow			80	760	
Direct Cost	810 HR			24,200	23,600	47,800
Doors & Windows						
Ind. Steel Doors, Fmes. Hdwe.						
Sing.	17 EA	200.00	500.00	3,400	8,500	
Dble.	1 PR	400.00	800.00	400	800	
Vert. Lift Steel Drs. 7′-0 × 7′-6	34 EA	200.00	1,400.00	6,800	47,600	
Rolling Stl. "B" Lab. N.O.						
17′-0 × 22′-6	1 EA	–	–	2,000	12,100	
Window Sash	500 SF	4.00	6.00	2,000	3,000	
Direct Cost	490 HR			14,600	72,000	86,600

Description	Quantity	Unit Cost Labor	Unit Cost Mat'l	Labor	Material	Total
Building Finish, cont'd						
Masonry						
Concrete Block-8″	5,500 EA	3.00	.80	16,500	4,400	
-6″	400 EA	2.80	.70	1,120	280	
-4″	1,100 EA	2.60	.60	2,860	660	
Scaffolding	4,200 SF	.60	.60	2,520	2,520	
Clean Down	8,100 SF	.24	.10	1,940	800	
Mortar	10 CY	–	60.00	–	600	
Durawal	2,000 LF	.20	.20	400	400	
Reinforcing Steel	4,200 LB	.30	.40	1,260	1,680	
Set Door Frames	13 EA	100.00	20.00	1,300	260	
Set Tubes	500 EA	1.20	–	600	–	
Supplies, Misc.	Allow	–	–	100	800	
Direct Cost	950 HR			28,600	12,400	41,000
Painting						
Structural Steel	480 Tons	100.00	40.00	48,000	19,200	
Doors	5,000 SF	.80	.60	4,000	3,000	
Office & Toilet Areas	3,700 SF	1.00	.60	3,700	2,220	
Pipe Coding	Allow	–	–	10,000	4,000	
Exterior Mechanical Units	Allow	–	–	12,000	4,100	
Misc. Iron not Galvanized	Allow	–	–	6,500	2,080	
Direct Cost	2,810 HR			84,200	34,600	118,800
Metal Partitions & Screens						
Toilet Partitions	13 EA	100.00	500.00	1,300	6,500	
Urinal Screens	7 EA	100.00	190.00	700	1,300	
Direct Cost	70 HR			2,000	7,800	9,800
Glass & Glazing						
Windows & Doors	600 SF	3.00	2.00	1,800	1,200	
Mirrors W/St./Stl. Fmes.	30 SF	6.60	20.00	200	800	
Pass Window Incl. Track Ass'y	1 EA	–	–	400	1,000	
Misc. Glazing Cleaning	Allow			1,400	1,000	
	130 HR			3,800	3,800	7,600
Building Finish, cont'd.						
Acoustical Treatment						
Gypsum Board $\frac{1}{2}″$	3,200 SF	.40	.60	1,280	1,920	
Ac. Tile on Gyp. Bd.	1,500 SF	.60	1.40	900	2,100	
Trim at Ceiling	400 LF	–	–	420	580	
Direct Cost	90 HR			2,600	4,600	7,200
Caulking						
Double II Panels–Both Sides	15,000 LF	2.00	1.40	30,000	21,000	
Doors & Windows in Ext. Walls	700 LF	2.00	2.00	1,400	1,400	
Direct Cost	1,050 HR			31,400	22,400	53,800

Description	Quantity	Unit Cost Labor	Mat'l	Labor	Material	Total
Building Finish, cont'd.						
Furnishings & Equipment						
Adjustable Dock Ramps	3 EA	1,000.00	4,400.00	3,000	13,200	
Permanent Dock Boards	31 EA	200.00	400.00	6,200	12,400	
Door Seals	34 EA	600.00	1,200.00	20,400	40,800	
Jib Crane	1 EA	–	–	1,000	10,000	
Double Post Lift	1 EA	–	–	600	3,000	
Direct Cost	1,040 HR			31,200	79,400	110,600
Bituminous Paving						
Base Course 4″ Gravel	80 CY	10.00	12.60	800	1,000	
Paving 2″	60 Ton	10.00	30.00	600	1,800	
Direct Cost	50 HR			1,400	2,800	4,200
Guard House	200 HR	–	–	6,000	9,100	15,100
Total Building Finish	8,470 HR			255,000	311,300	566,300
Overhead & Fill @ 12%						68,000
Total Construction Cost-Building Finish						634,300

Mountaintown Warehouse Milestone CPM

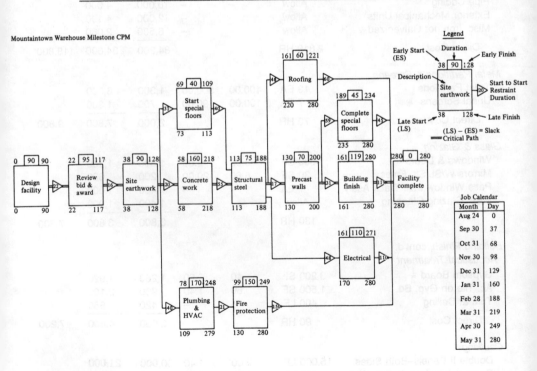

B

EXAMPLE BID PACKAGE

The example site earthwork bid package has been prepared, incorporating plans and specifications developed by the architect. The preparation of bid packages for a professional construction management program must be a joint effort involving all team members, owner, architect, and construction manager. One of the parties must be delegated the responsibility for preparation of the basic package. A careful review should be performed by the other parties. The example bid package contains the following information:

SITE EARTHWORK AND FENCING

I   Invitation to Bid
II Bid Form (including Bid Breakdown)
III Special Conditions (including Supplemental Provisions)
IV Simplified Specifications, Addendums, and Drawings
V   Owner-furnished Items
VI Construction Schedule

I INVITATION TO BID

September 10, 199_

EASYWAY FOOD COMPANY
DESIGN AND CONSTRUCTION DEPARTMENT
200 MADISON STREET
MOUNTAINTOWN, WEST AMERICA 99999

Gentlemen:

Re: INVITATION TO BID
NO. 16

We are pleased to enclose one set of plans and specifications covering Site Earthwork and Fencing for the Dry Storage Warehouse to be constructed in Mountaintown, West America.

Bids will be received until 2:00 P.M., September 22, 199_, at our main office at the above address. All bids shall be binding for 30 days thereafter. Bidders are invited to send one representative to a bid opening which will be conducted at the above specified time.

While we are asking all bidders to base their bids in accordance with the Owner's schedule, alternate completion dates will be given consideration based upon our evaluation of the advantages or disadvantages of the proposed schedule.

Attached hereto is the following information:
1. Three copies each of our Bid Form.
2. Two copies of the Construction Contract including Exhibit A and attachments listed therein.

Two copies of the executed Bid Form are to be submitted with your bid. All bids are requested to be itemized as noted on the Bid Form.

If you are the successful bidder, it will be necessary for you to furnish a 100% Payment and Performance Bond in accordance with "The Standard Form of Bond" latest edition, copyrighted by the American Institute of Architects and issued by a Surety Company satisfactory to Owner. You will also be expected to execute a contract in the form included with the attached documents.

Owner reserves the right to reject any and all proposals or to waive irregularities therein. Owner reserves the right to evaluate each and every proposal in his absolute discretion and to accept any particular proposal even though the price or completion date, or both, may not be as favorable as some other proposal.

Bidders requesting additional information are advised to address their inquiries to Construction Management and Control, Inc., Attention: Mr. O. Hanson, telephone number (000) 123-4567.

EASYWAY FOOD COMPANY

By Peter J. Cleaveland
Manager, Design and Construction
Department

II BID FORM (INCLUDING BID BREAKDOWN)

BID FORM

DATE _____ September 10 _____

INVITATION TO BID: 16
SITE EARTHWORK AND FENCING
DRY STORAGE WAREHOUSE
MOUNTAINTOWN, WEST AMERICA

TO: EASYWAY FOOD COMPANY
　　Design and Construction Department
　　200 Madison Street
　　Mountaintown, West America　99999

Gentlemen:
Having examined the plans and specifications including addendas Nos. _____
and having familiarized ourselves with the job and site conditions, the undersigned does hereby
tender the following bids for construction work at the Mountain Distribution Center, Mountaintown, West America, in accordance with invitation to Bid No. 16 and all attachments thereto.

Our firm price bid is in the amount of _____ Dollars ($_____)
which price includes the cost of the Performance bond. The undersigned further agrees to
complete the work in ____ calendar days after date specified in the Notice to Proceed with
work.

Attached hereto is a list of subcontractors whom we propose to use for those branches of work
exceeding $1,000 in contract cost.

We submit the following proposals for alternates and/or substitutions:

Alternate Bid A　　　　　　　　　　　$ Add _____　　Deduct _____

　Describe _____

_____

Alternate Bid B　　　　　　　　　　　$ Add _____　　Deduct _____

　Describe _____

_____

Our total firm price bid is itemized in component parts as follows:

　Site Grading　　　　　　　　　　　　　　　　　_____

　Imported Borrow　　　　　　　　　　　　　　　_____

　Fencing　　　　　　　　　　　　　　　　　　　_____

　Performance Bond　　　　　　　　　　　　　　_____

　　Total Firm Price Bid　　　　　　　　　　　　$_____

The undersigned agrees to execute the Construction Contract Form provided with the Invitation
to Bid. The undersigned certifies that he is:

1. A Corporation incorporated in the State of _____.

2. A Partnership consisting of the following partners _____

_____

_____.

3. An individual whose name is _____.

The foregoing offer shall be binding upon the undersigned until _____

_____.

Contractor

_____

Signed by   Date

_____

Title

_____

Address

_____

_____

Telephone Number

_____

Contractors License No.   State

EXHIBIT A

Attached to and a part of Agreement No. 16 are contract documents referred to in Article 6 the Contract.

In general the work consists of all site earthwork and fencing as described in the attached plans and specifications for the Dry Storage Warehouse located in Mountaintown, West America.

Drawings and specifications as set forth in Article 6 Contract Documents are enumerated and included as follows:

A. Invitation to Bid
B. Bid Form
C. General Conditions
D. Special Conditions
E. Technical Specifications
    1. Section E-1, Site Grading and Compaction
    2. Section E-2, Fencing and Gates
F. Drawings

Number	Rev. No.	Date
SI-1	0	9-1-9_
A-1	0	9-1-9_

G. Owner-Furnished items
H. Owner's Construction Schedule

III SPECIAL CONDITIONS

Article 1 Conflict with General Conditions

In case of conflict between these SPECIAL CONDITIONS and the GENERAL CONDITIONS, the Special Conditions shall govern.

Article 2 Superintendence and Supervision

The Contractor shall keep on the premises, during the progress of the work, a competent Supervisor, satisfactory to the Owner. The Supervisor shall not be changed except with the consent of the Owner unless the Supervisor proves to be unsatisfactory to the Contractor or Owner or ceases to be in the Contractor's employ. The Supervisor shall represent the Contractor in his absence and all instructions given to him shall be as binding as if given to the Contractor. The Contractor shall at all times enforce strict discipline and good order among his employees, and shall not employ on the work any unfit person or anyone not skilled in the work assigned to him.

Article 3 Compliance with Executive Order

Contractor agrees to be bound by, and to implement, all of the nondiscrimination requirements of Executive Order 11246 and any amendments thereof.

Article 4 Safety

Contractor shall comply with and shall require all subcontractors to comply with all applicable health and safety laws, rules and regulations, including without limitations, the Occupational Safety and Health Act of 1970 and the rules and regulations issued pursuant thereto.

Article 5 Surveys

Surveys furnished by the Owner as set forth in ARTICLE 5, GENERAL CONSITIONS, SURVEYS, PERMITS, & REGULATIONS shall include basic survey controls only consisting of two permanent benchmarks, monuments designating property corners, and basic horizontal control points for building corners. All construction layout shall be performed by Contractor who shall be responsible for the accuracy thereof.

Article 6 Construction Manager

The Owner has engaged Construction Management and Control, Inc., hereinafter referred to as Construction Manager, to act as Owner's designated representative in the administration of this Construction Contract and in the management and coordination of work performed on the project. The word "Architect" as set forth in Article 18 of the General Conditions shall be replaced by Construction Manager. The Architect, however, shall continue to exercise surveillance and observation over the work and will be responsible for interpretation of plans and specifications prepared under his direction.

IV SIMPLIFIED SPECIFICATIONS, ADDENDUMS, AND DRAWINGS

SECTION E-1 SITE GRADING AND COMPACTION

E-1:01 General

1. All this work shall be subject to the General Conditions, Special Conditions, and other Contract Documents.

E-1:02 Scope

The Contractor shall furnish all labor, material, tools and equipment required to complete all site grading as shown on the plans and herein specified. The principal items of work are the following:

1. Removal of vegetation and topsoil on the project area.
2. General cut, where called for on the plans, and as hereafter specified.
3. Scarifying of existing and cut surfaces and recompaction.
4. Furnishing and installation of compacted fill over areas of yards, and at building area and spur track.
5. Finish grading, swales, and slopes to swales.

E-1:03 Related work not included

1. Excavation and backfilling for footings and foundation walls of buildings.
2. All excavation and backfilling in connection with storm and sanitary sewer systems, utility lines, sprinkler mains and electrical service.
3. Topsoil for planting areas.
4. Aggregate base course under paving.

E-1:04 Drawings and site examination

1. The Contractor is referred to the site drawings, which show existing ground elevations as well as finish grades for building floors, paved yards, and unpaved areas. Where floors (concrete) or paving (concrete or asphaltic concrete) are indicated, the finish grades shown on the plans indicate the surface of such floors or paving. Where unpaved areas are indicated, the finish grades are finished ground surface.

   a *In Concrete Paved Areas*, grading work shall be to the elevation of the bottom of the aggregate base, or 18 in. below top of slab.

   b *In Asphaltic Concrete Paved Areas*, grading work shall be to elevation of bottom of the aggregate base, or 12 in. below top of slab.

2. It shall be the duty of the Contractor to visit the site and ascertain all existing conditions before submitting his bid. No allowance will subsequently be made on account of Contractor's neglect of this requirement.

E-1:05 Removal of vegetation, and topsoil

1. Over that portion of the site where project work is called for, areas where there is vegetation, loose earth, or trash, shall be cut sufficiently to remove such material. Remove brush and roots. Such cut material shall be removed from the site.
2. Filling work, as specified hereinafter, shall include allowance for this material removed, as part of the contract.

E-1:06 General cut

Contractor shall make general cut as required by existing ground surfaces to meet requirements outlined in Article E-1:04, paragraph 1, including allowance for compacted fill or rock where called for.

E-1:07 Rejected material

After excavation, and prior to scarifying and recompaction, the Soils Engineer shall inspect the exposed cut surface. Those areas of original in-place soil rejected by the Soils Engineer as being unsuitable as a base for construction of engineered fill, shall be excavated and removed from the site. This work shall be done only on written order from the Owner, and will be compensated for in accordance with an agreed price. Note that this does not include removal of vegetation, loose earth, and trash as specified above, or does not include correction of faulty work by the Contractor which has been rejected by the Soils Engineer as not meeting the specifications, or work under the contract which has become damaged due to rain, construction traffic, or failure to maintain proper drainage and dewatering, etc.

a Installation of fill to replace such rejected original material shall be done only on written order from the Owner, and will be compensated for in accordance with an agreed price. Use imported compacted fill, as hereafter specified.

E-1:08 Scarifying and recompaction

Before placing any fill or base course thereon, existing and cut surfaces shall be scarified to a minimum depth of 6", brought to, or just above, optimum moisture content and recompacted to a minimum relative compaction of 95%.

E-1:09 Fill

1. *Fill* for this project shall be as outlined below:
   a General fill material shall be on-site material, excavated from grading work, or imported fill. On-site material shall be placed first when both are required.
   b Imported granular earth, free of vegetation, debris, and other objectionable material, shall meet the following requirements:

Maximum particle size	3 inches
Passing #4 sieve	30–90%
Passing #200 sieve	0–20%
Maximum Plasticity Index	15

2. Organic soils, topsoils, and other material with sod or humus shall not be used for Compacted Fill. No frozen material, stumps, roots, all or parts of trees, brush, or other perishable material shall be placed. Clean sand shall not be used.

3. *Installation of Fill*
   a The Compacted Fill shall be constructed to provide adequate drainage at all times, and the surface kept uniformly graded and compacted. Each lift shall extend longitudinally or transversely over the entire area and be kept smooth. The Contractor shall route his equipment at all times, by equal distribution of travel, to prevent rutting. The equipment shall be operated so as to break up into small particles any cemented gravel or other soil particles so that they may be incorporated into the material of each lift or layer. Contractor shall be responsible for the stability of the fill, until the work is complete and accepted by the Owner, and shall replace any portion of which, in the opinion of the Soils engineer, has become displaced, unstable, or damaged.
   b Compacted fill shall be placed in 6" maximum layers of loose depth for the full width of fill, except that thick layers may be used providing the Contractor can satisfy the Engineer that requirement densities can be obtained. All earthfill shall be compacted to at least 95% of maximum density at the optimum moisture content as determined by the A.S.T.M. D-1577-66T method of testing. If the material is too wet or too dry, compaction work on all portions of the compacted fill affected shall be delayed until the material has been either dried or sprinkled, whichever is necessary, to provide compacted densities and moisture contents as specified. Compaction of fill shall be performed by Sheeps Foot rollers or Rubber Tired rollers. All equipment shall be approved by the Soils Engineer.

4. Toe of slope of building fill shall be as shown on Plot Plan Drawing SI-1 and slope shall extend on a 1:1 slope to underside of finish floor at elevation 99'6".

E-1:10 Grading at railroad tracks

Strip, scarify, and recompact as specified above, and furnish and install compacted imported fill as required for installation of railroad tracks at the location shown. Grading work shall be to the elevation of the bottom of ballast, shall meet fill requirements specified in E-1:09 above, and shall be to lines and grades approved by the Railroad Company.

E-1:11 Maintenance of finish grades

Finish grades, unless otherwise indicated, shall be uniform levels or slopes between points where elevations are indicated or between such ponts and existing grades at vicinity of boundary lines. The Contractor shall maintain and protect all grading, filling and excavation work, and drainage ditches, to the required elevations and slopes during the construction period. At completion of the work, finish surface of unpaved areas shall meet the above requirements to the satisfaction of the Owner.

a   The site of the work shall be left in a clean, orderly condition, free of all debris, trash, etc., resulting from said work.

E1:12 Dust control

Contractor shall prevent the blowing of dust and dirt, generated by the project work or vehicular traffic, over onto adjacent or nearby developed areas.

E1:13 Inspection and tests

Selection of fill material, placing, method of compaction, scarifying and recompaction of existing and cut surfaces, shall be performed under the supervision and control of a qualified Soils Engineer, appointed by and acting under the direction of the Architect. All costs in connection with this supervision and testing will be paid for by the Owner. The Soils Engineer will make "in-place" density and moisture tests, the Engineer may require the Contractor to make changes in his operations necessary to obtain the specified values for these items.

SECTION E-2   FENCING AND GATES

E-2:01 General

This work shall be subject to the General Conditions, Special Conditions, and other Contract Documents.

E-2:02 Scope

Furnish all labor, materials, and equipment to install complete all work required under this section, including, but not limited to, the following:

1. Demolition (paving, curbs, etc) at site entrances.
2. Fencing and gates.

E-2:03 Fencing and gates

1. Around site and parking, furish and install chain link fence and gates, as detailed. Fencing shall be chain link type as manufactured by Manufacturer A, B, or C. All materials shall be hot dip galvanized. Furnish shop drawings covering all fencing and gates to the Architect for approval.

   a   *Fabric*   shall be "galvanized after weaving" chain link, No 9 wire, woven in a 2″ mesh. Top and bottom selvages to have a twisted, barbed finish; barbing to be done by cutting wire on a bias, creating sharp points. Wire pickets of which fabric is made shall stand a tensile strength test of 90,000 psi, based on the cross-sectional area of the galvanized wire.

   b   *Line Posts*   H-column (2″ × 2¼″), weight 4.1 lbs. per lineal foot, or 2-3/8″ round, 3.65 lbs. No used, re-rolled or open seam material will be permitted in posts or rails.

   c   *Terminal & Gate Posts*   End, corner, and pull posts, small gate posts, 3″ O.D., 5.79 lbs. per lineal foot. Large Gate posts, 6″ O.D. Gate posts to have ball top.

   d   *Top and Bottom Rail*   1-5/9″ O.D., or H-section, weight 2.27 lbs. per lineal foot. Top rail to pass through base of line post tops and form a continuous brace from end to end of each stretch of fence. Rails shall be securely fastened to terminal posts by pressed steel connections.

   e   *Braces*   Same material as top rail. To be spaced midway between top rail and bottom rail and to extend from terminal post to first adjacent line post. Fasten braces securely to posts by means of suitable pressed steel connection, and truss from line post back to terminal post with a 3/8″ round rod.

f *Gates* Gate frames to be made of 2″ O.D. pipe, weight 2.72 lbs. per lineal foot. Corner fittings heavy pressed steel or malleable castings. Fabric to be same as fence fabric. Gates to be complete with malleable iron ball and socket hinges and center rest. Where so detailed, hinges shall permit gates to swing back against fence, 180° if required. Small gates to have catch stops, large gates hook posts and hooks. Provide sliding gates (manual) where shown.

2. *Erection* Install fencing and gates in locations shown on plans. Space posts in line of fence not farther apart than 10′ on centers. Posts in concrete paving and curbs shall be set in 2′ long pipe sleeves and grouted in place. Install gates, and turnstiles, and adjust for proper operation. Extend fence fabric across top of turnstiles. Fasten fabric to line posts with fabric bands spaced approximately 14″ apart, and to top and bottom rails with tie wires spaced approximately 24″ apart. Furnish sleeves and locations for setting in concrete (by others).

V OWNER-FURNISHED ITEMS

1 Temporary electrical supply at point designated by Owner.
2 Water in limited amounts for domestic and miscellaneous construction uses.
3 Areas for contractors' facilities, materials and equipment storage at locations determined by Owner.
4 Sanitary services (portable toilets) at locations determined by Owner.
5 Basic Survey Controls as set forth in the SPECIAL CONDITIONS.

VI CONSTRUCTION SCHEDULE

The attached Owner's Construction Schedule (Figure 4-4 in Chapter 4) is to be used as a guide for overall performance of the project. Actual dates to begin work will be transmitted by the NOTICE TO PROCEED. Actual completion time will be as set forth in the AGREEMENT.

Portions of the Example Bid Package have been adapted from bidding documents developed by E. A. Bonelli & Associates, San Francisco, California, and Leo Rosenthal, A.I.A., Denver, Colorado.

C

CONTRACT FORMS

This appendix contains the following example contract forms:

1. *Standard Form of Agreement Between Owner and Contractor, where the basis of payment is a Stipulated Sum*, AIA Document A101, 1987 Edition, American Institute of Architects, 1735 New York Avenue, Washington, D.C. 20006-5292. This document has been reproduced with permission under license number 91019. Further reproduction, in part or in whole, is not authorized. Please contact AIA directly to obtain additional copies.

2. *Standard Form of Agreement Between Owner and Construction Manager (Guaranteed Maximum Price Option)*, AGC Document 500, 1980 Edition, Associated General Contractors of America, 1957 E Street, N.W., Washington, D.C. 20006. This document has been reproduced with permission of the Associated General Contractors of America. Further reproduction, in part or in whole, is not authorized. Please contact AGC's Publications Department directly, at the above address, to obtain copies of current forms.

3. *Subcontract for Building Construction*, AGC Document 600, 1990 Edition, Associated General Contractors of America, 1957 E Street, N.W., Washington, D.C. 20006. This document has been reproduced with permission of the Associated General Contractors of America. Further reproduction, in part or in whole, is not authorized. Please contact AGC's Publications Department directly, at the above address, to obtain copies of current forms.

THE   AMERICAN   INSTITUTE   OF   ARCHITECTS

1. AIA copyrighted material has been reproduced with the permission of the American Institute of Architects under license number 91019. Permission expires February 28, 1992. FURTHER REPRODUCTION IS PROHIBITED.

2. Because AIA Documents are revised from time to time, users should ascertain from the AIA the current edition of this document.

3. Copies of the current edition of this AIA document may be purchased from The American Institute of Architects or its local distributors.

4. This document is intended for use as a "consumable" (consumables are further defined by Senate Report 94-473 on the Copyright Act of 1976). This document is not intended to be used as "model language" (language taken from an existing document and incorporated, without attribution, into a newly - created document.) Rather, it is a standard form which is intended to be modified by appending separate amendment sheets and/or fill in provided blank spaces.

AIA Document A101

Standard Form of Agreement Between Owner and Contractor

where the basis of payment is a

STIPULATED SUM

1987 EDITION

THIS DOCUMENT HAS IMPORTANT LEGAL CONSEQUENCES; CONSULTATION WITH AN ATTORNEY IS ENCOURAGED WITH RESPECT TO ITS COMPLETION OR MODIFICATION.

The 1987 Edition of AIA Document A201, General Conditions of the Contract for Construction, is adopted in this document by reference. Do not use with other general conditions unless this document is modified.

This document has been approved and endorsed by The Associated General Contractors of America.

AGREEMENT

made as of the   day of   in the year of
Nineteen Hundred and

BETWEEN the Owner:
(Name and address)

and the Contractor:
(Name and address)

The Project is:
(Name and location)

The Architect is:
(Name and address)

The Owner and Contractor agree as set forth below.

AIA DOCUMENT A101 • OWNER-CONTRACTOR AGREEMENT • TWELFTH EDITION • AIA® • ©1987
THE AMERICAN INSTITUTE OF ARCHITECTS, 1735 NEW YORK AVENUE, N.W., WASHINGTON, D.C. 20006   **A101-1987   1**

ARTICLE 1
THE CONTRACT DOCUMENTS

The Contract Documents consist of this Agreement, Conditions of the Contract (General, Supplementary and other Conditions), Drawings, Specifications, Addenda issued prior to execution of this Agreement, other documents listed in this Agreement and Modifications issued after execution of this Agreement; these form the Contract, and are as fully a part of the Contract as if attached to this Agreement or repeated herein. The Contract represents the entire and integrated agreement between the parties hereto and supersedes prior negotiations, representations or agreements, either written or oral. An enumeration of the Contract Documents, other than Modifications, appears in Article 9.

ARTICLE 2
THE WORK OF THIS CONTRACT

The Contractor shall execute the entire Work described in the Contract Documents, except to the extent specifically indicated in the Contract Documents to be the responsibility of others, or as follows:

ARTICLE 3
DATE OF COMMENCEMENT AND SUBSTANTIAL COMPLETION

3.1 The date of commencement is the date from which the Contract Time of Paragraph 3.2 is measured, and shall be the date of this Agreement, as first written above, unless a different date is stated below or provision is made for the date to be fixed in a notice to proceed issued by the Owner.
(Insert the date of commencement, if it differs from the date of this Agreement or, if applicable, state that the date will be fixed in a notice to proceed.)

Unless the date of commencement is established by a notice to proceed issued by the Owner, the Contractor shall notify the Owner in writing not less than five days before commencing the Work to permit the timely filing of mortgages, mechanic's liens and other security interests.

3.2 The Contractor shall achieve Substantial Completion of the entire Work not later than
(Insert the calendar date or number of calendar days after the date of commencement. Also insert any requirements for earlier Substantial Completion of certain portions of the Work, if not stated elsewhere in the Contract Documents.)

, subject to adjustments of this Contract Time as provided in the Contract Documents.
(Insert provisions, if any, for liquidated damages relating to failure to complete on time.)

AIA DOCUMENT A101 • OWNER-CONTRACTOR AGREEMENT • TWELFTH EDITION • AIA® • ©1987
THE AMERICAN INSTITUTE OF ARCHITECTS, 1735 NEW YORK AVENUE, N.W., WASHINGTON, D.C. 20006   **A101-1987   2**

ARTICLE 4
CONTRACT SUM

4.1 The Owner shall pay the Contractor in current funds for the Contractor's performance of the Contract the Contract Sum of
Dollars
($), subject to additions and deductions as provided in the Contract Documents.

4.2 The Contract Sum is based upon the following alternates, if any, which are described in the Contract Documents and are hereby accepted by the Owner:

(State the numbers or other identification of accepted alternates. If decisions on other alternates are to be made by the Owner subsequent to the execution of this Agreement, attach a schedule of such other alternates showing the amount for each and the date until which that amount is valid.)

4.3 Unit prices, if any, are as follows:

ARTICLE 5
PROGRESS PAYMENTS

5.1 Based upon Applications for Payment submitted to the Architect by the Contractor and Certificates for Payment issued by the Architect, the Owner shall make progress payments on account of the Contract Sum to the Contractor as provided below and elsewhere in the Contract Documents.

5.2 The period covered by each Application for Payment shall be one calendar month ending on the last day of the month, or as follows:

5.3 Provided an Application for Payment is received by the Architect not later than the
day of a month, the Owner shall make payment to the Contractor not later than
the   day of the   month. If an Application for Payment is received by the
Architect after the application date fixed above, payment shall be made by the Owner not later than
days after the Architect receives the Application for Payment.

5.4 Each Application for Payment shall be based upon the Schedule of Values submitted by the Contractor in accordance with the Contract Documents. The Schedule of Values shall allocate the entire Contract Sum among the various portions of the Work and be prepared in such form and supported by such data to substantiate its accuracy as the Architect may require. This Schedule, unless objected to by the Architect, shall be used as a basis for reviewing the Contractor's Applications for Payment.

5.5 Applications for Payment shall indicate the percentage of completion of each portion of the Work as of the end of the period covered by the Application for Payment.

5.6 Subject to the provisions of the Contract Documents, the amount of each progress payment shall be computed as follows:

5.6.1 Take that portion of the Contract Sum properly allocable to completed Work as determined by multiplying the percentage completion of each portion of the Work by the share of the total Contract Sum allocated to that portion of the Work in the Schedule of Values, less retainage of   percent
(%). Pending final determination of cost to the Owner of changes in the Work, amounts not in dispute may be included as provided in Subparagraph 7.3.7 of the General Conditions even though the Contract Sum has not yet been adjusted by Change Order;

5.6.2 Add that portion of the Contract Sum properly allocable to materials and equipment delivered and suitably stored at the site for subsequent incorporation in the completed construction (or, if approved in advance by the Owner, suitably stored off the site at a location agreed upon in writing), less retainage of
percent (%);

5.6.3 Subtract the aggregate of previous payments made by the Owner; and

5.6.4 Subtract amounts, if any, for which the Architect has withheld or nullified a Certificate for Payment as provided in Paragraph 9.5 of the General Conditions.

5.7 The progress payment amount determined in accordance with Paragraph 5.6 shall be further modified under the following circumstances:

5.7.1 Add, upon Substantial Completion of the Work, a sum sufficient to increase the total payments to
percent (%) of the Contract
Sum, less such amounts as the Architect shall determine for incomplete Work and unsettled claims; and

5.7.2 Add, if final completion of the Work is thereafter materially delayed through no fault of the Contractor, any additional amounts payable in accordance with Subparagraph 9.10.3 of the General Conditions.

5.8 Reduction or limitation of retainage, if any, shall be as follows:

(If it is intended, prior to Substantial Completion of the entire Work, to reduce or limit the retainage resulting from the percentages inserted in Subparagraphs 5.6.1 and 5.6.2 above, and this is not explained elsewhere in the Contract Documents, insert here provisions for such reduction or limitation.)

ARTICLE 6
FINAL PAYMENT

Final payment, constituting the entire unpaid balance of the Contract Sum, shall be made by the Owner to the Contractor when (1) the Contract has been fully performed by the Contractor except for the Contractor's responsibility to correct nonconforming Work as provided in Subparagraph 12.2.2 of the General Conditions and to satisfy other requirements, if any, which necessarily survive final payment; and (2) a final Certificate for Payment has been issued by the Architect; such final payment shall be made by the Owner not more than 30 days after the issuance of the Architect's final Certificate for Payment, or as follows:

ARTICLE 7
MISCELLANEOUS PROVISIONS

7.1 Where reference is made in this Agreement to a provision of the General Conditions or another Contract Document, the reference refers to that provision as amended or supplemented by other provisions of the Contract Documents.

7.2 Payments due and unpaid under the Contract shall bear interest from the date payment is due at the rate stated below, or in the absence thereof, at the legal rate prevailing from time to time at the place where the Project is located.

(Insert rate of interest agreed upon, if any.)

(Usury laws and requirements under the Federal Truth in Lending Act, similar state and local consumer credit laws and other regulations at the Owner's and Contractor's principal places of business, the location of the Project and elsewhere may affect the validity of this provision. Legal advice should be obtained with respect to deletions or modifications, and also regarding requirements such as written disclosures or waivers.)

7.3 Other provisions:

ARTICLE 8
TERMINATION OR SUSPENSION

8.1 The Contract may be terminated by the Owner or the Contractor as provided in Article 14 of the General Conditions.

8.2 The Work may be suspended by the Owner as provided in Article 14 of the General Conditions.

AIA DOCUMENT A101 • OWNER-CONTRACTOR AGREEMENT • TWELFTH EDITION • AIA® • ©1987
THE AMERICAN INSTITUTE OF ARCHITECTS, 1735 NEW YORK AVENUE, N.W., WASHINGTON, D.C. 20006

ARTICLE 9
ENUMERATION OF CONTRACT DOCUMENTS

9.1 The Contract Documents, except for Modifications issued after execution of this Agreement, are enumerated as follows:

9.1.1 The Agreement is this executed Standard Form of Agreement Between Owner and Contractor, AIA Document A101, 1987 Edition.

9.1.2 The General Conditions are the General Conditions of the Contract for Construction, AIA Document A201, 1987 Edition.

9.1.3 The Supplementary and other Conditions of the Contract are those contained in the Project Manual dated
and are as follows:

Document	**Title**	**Pages**

9.1.4 The Specifications are those contained in the Project Manual dated as in Subparagraph 9.1.3, and are as follows:
(Either list the Specifications here or refer to an exhibit attached to this Agreement.)

Section	**Title**	**Pages**

9.1.5 The Drawings are as follows, and are dated   unless a different date is shown below:
(Either list the Drawings here or refer to an exhibit attached to this Agreement.)

Number	Title	Date

9.1.6 The Addenda, if any, are as follows:

Number	Date	Pages

Portions of Addenda relating to bidding requirements are not part of the Contract Documents unless the bidding requirements are also enumerated in this Article 9.

AIA DOCUMENT A101 • OWNER-CONTRACTOR AGREEMENT • TWELFTH EDITION • AIA® • ©1987
THE AMERICAN INSTITUTE OF ARCHITECTS, 1735 NEW YORK AVENUE, N.W., WASHINGTON, D.C. 20006   **A101-1987**   **7**

9.1.7 Other documents, if any, forming part of the Contract Documents are as follows:

(List here any additional documents which are intended to form part of the Contract Documents. The General Conditions provide that bidding requirements such as advertisement or invitation to bid, Instructions to Bidders, sample forms and the Contractor's bid are not part of the Contract Documents unless enumerated in this Agreement. They should be listed here only if intended to be part of the Contract Documents.)

This Agreement is entered into as of the day and year first written above and is executed in at least three original copies of which one is to be delivered to the Contractor, one to the Architect for use in the administration of the Contract, and the remainder to the Owner.

OWNER   CONTRACTOR

_____   _____
(Signature)   *(Signature)*

_____   _____
(Printed name and title)   *(Printed name and title)*

AIA DOCUMENT A101 • OWNER-CONTRACTOR AGREEMENT • TWELFTH EDITION • AIA® • ©1987
THE AMERICAN INSTITUTE OF ARCHITECTS, 1735 NEW YORK AVENUE, N.W., WASHINGTON, D.C. 20006   **A101-1987   8**

THE ASSOCIATED GENERAL CONTRACTORS

STANDARD FORM OF AGREEMENT BETWEEN OWNER AND CONSTRUCTION MANAGER

(GUARANTEED MAXIMUM PRICE OPTION)

(See AGC Document No. 501 for Establishing the
Guaranteed Maximum Price)

This Document has important legal and insurance consequences; consultation with an attorney is encouraged with respect to its completion or modification.

AGREEMENT

Made this   day of   in the year of Nineteen Hundred and

BETWEEN

the Owner, and

the Construction Manager.

For services in connection with the following described Project: (Include complete Project location and scope)

The Architect/Engineer for the Project is

The Owner and the Construction Manager agree as set forth below:

TABLE OF CONTENTS

ARTICLE 1

The Construction Team and Extent of Agreement

The CONSTRUCTION MANAGER accepts the relationship of trust and confidence established between him and the Owner by this Agreement. He covenants with the Owner to furnish his best skill and judgment and to cooperate with the Architect/Engineer in furthering the interests of the Owner. He agrees to furnish efficient business administration and superintendence and to use his best efforts to complete the Project in an expeditious and economical manner consistent with the interest of the Owner.

1.1 *The Construction Team:* The Construction Manager, the Owner, and the Architect/Engineer called the "Construction Team" shall work from the beginning of design through construction completion. The Construction Manager shall provide leadership to the Construction Team on all matters relating to construction.

1.2 *Extent of Agreement:* This Agreement represents the entire agreement between the Owner and the Construction Manager and supersedes all prior negotiations, representations or agreements. When Drawings and Specifications are complete, they shall be identified by amendment to this Agreement. This Agreement shall not be superseded by any provisions of the documents for construction and may be amended only by written instrument signed by both the Owner and the Construction Manager.

1.3 *Definitions:* The Project is the total construction to be performed under this Agreement. The Work is that part of the construction that the Construction Manager is to perform with his own forces or that part of the construction that a particular Trade Contractor is to perform. The term day shall mean calendar day unless otherwise specifically designated.

ARTICLE 2

Construction Manager's Services

The Construction Manager will perform the following services under this Agreement in each of the two phases described below.

2.1 Design Phase

2.1.1 *Consultation During Project Development:* Schedule and attend regular meetings with the Architect/Engineer during the development of conceptual and preliminary design to advise on site use and improvements, selection of materials, building systems and equipment. Provide recommendations on construction feasibility, availability of materials and labor, time requirements for installation and construction, and factors related to cost including costs of alternative designs or materials, preliminary budgets, and possible economies.

2.1.2 *Scheduling:* Develop a Project Time Schedule that coordinates and integrates the Architect/Engineer's design efforts with construction schedules. Update the Project Time Schedule incorporating a detailed schedule for the construction operations of the Project, including realistic activity sequences and durations, allocation of labor and materials, processing of shop drawings and samples, and delivery of products requiring long lead-time procurement. Include the Owner's occupancy requirements showing portions of the Project having occupancy priority.

2.1.3 *Project Construction Budget:* Prepare a Project budget as soon as major Project requirements have been identified, and update periodically for the Owner's approval. Prepare an estimate based on a quantity survey of Drawings and Specifications at the end of the schematic design phase for approval by the Owner as the Project Construction Budget. Update and refine this estimate for the Owner's approval as the development of the Drawings and Specifications proceeds, and advise the Owner and the Architect/Engineer if it appears that the Project Construction Budget will not be met and make recommendations for corrective action.

2.1.4 *Coordination of Contract Documents:* Review the Drawings and Specifications as they are being prepared, recommending alternative solutions whenever design details affect construction feasibility or schedules without, however, assuming any of the Architect/Engineer's responsibilities for design.

2.1.5 *Construction Planning:* Recommend for purchase and expedite the procurement of long-lead items to ensure their delivery by the required dates.

2.1.5.1 Make recommendations to the Owner and the Architect/Engineer regarding the division of Work in the Drawings and Specifications to facilitate the bidding and awarding of Trade Contracts, allowing for phased construction taking into consideration such factors as time of performance, availability of labor, overlapping trade jurisdictions, and provisions for temporary facilities.

2.1.5.2 Review the Drawings and Specifications with the Architect/Engineer to eliminate areas of conflict and overlapping in the Work to be performed by the various Trade Contractors and prepare prequalification criteria for bidders.

2.1.5.3 Develop Trade Contractor interest in the Project and as working Drawings and Specifications are completed, take competitive bids on the Work of the various Trade Contractors. After analyzing the bids, either award contracts or recommend to the Owner that such contracts be awarded.

2.1.6 *Equal Employment Opportunity:* Determine applicable requirements for equal emloyment opportunity programs for inclusion in Project bidding documents.

2.2 Construction Phase

2.2.1 *Project Control:* Monitor the Work of the Trade Contractors and coordinate the Work with the activities and responsibilities of the Owner, Architect/Engineer and Construction Manager to complete the Project in accordance with the Owner's objectives of cost, time and quality.

2.2.1.1 Maintain a competent full-time staff at the Project site to coordinate and provide general direction of the Work and progress of the Trade Contractors on the Project.

2.2.1.2 Establish on-site organization and lines of authority in order to carry out the overall plans of the Construction Team.

2.2.1.3 Establish procedures for coordination among the Owner, Architect/Engineer, Trade Contractors and Construction Manager with respect to all aspects of the Project and implement such procedures.

2.2.1.4 Schedule and conduct progress meetings at which Trade Contractors, Owner, Architect/Engineer and Construction Manager can discuss jointly such matters as procedures, progress, problems and scheduling.

2.2.1.5 Provide regular monitoring of the schedule as construction progresses. Identify potential variances between scheduled and probable completion dates. Review schedule for Work not started or incomplete and recommend to the Owner and Trade Contractors adjustments in the schedule to meet the probable completion date. Provide summary reports of each monitoring and document all changes in schedule.

2.2.1.6 Determine the adequacy of the Trade Contractors' personnel and equipment and the availability of materials and supplies to meet the schedule. Recommend courses of action to the Owner when requirements of a Trade Contract are not being met.

2.2.2 *Physical Construction:* Provide all supervision, labor, materials, construction equipment, tools and subcontract items which are necessary for the completion of the Project which are not provided by either the Trade Contractors or the Owner. To the extent that the Construction Manager performs any Work with his own forces, he shall, with respect to such Work, perform in accordance with the Plans and Specifications and in accordance with the procedure applicable to the Project.

2.2.3 *Cost Control:* Develop and monitor an effective system of Project cost control. Revise and refine the initially approved Project Construction Budget, incorporate approved changes as they occur, and develop cash flow reports and forecasts as needed. Identify variances between actual and budgeted or estimated costs and advise Owner and Architect/Engineer whenever projected cost exceeds budgets or estimates.

AGC DOCUMENT NO. 500 • OWNER CONSTRUCTION MANAGER AGREEMENT JULY 1980

2.2.3.1 Maintain cost accounting records on authorized Work performed under unit costs, actual costs for labor and material, or other bases requiring accounting records. Afford the Owner access to these records and preserve them for a period of three (3) years after final payment.

2.2.4 *Change Orders:* Develop and implement a system for the preparation, review and processing of Change Orders. Recommend necessary or desirable change to the Owner and the Architect/Engineer, review requests for changes, submit recommendations to the Owner and the Architect/Engineer, and assist in negotiating Change Orders.

2.2.5 *Payments to Trade Contractors:* Develop and implement a procedure for the review, processing and payment of applications by Trade Contractors for progress and final payments.

2.2.6 *Permits and Fees:* Assist the Owner and Architect/Engineer in obtaining all building permits and special permits for permanent improvements, excluding permits for inspection or temporary facilities required to be obtained directly by the various Trade Contractors. Assist in obtaining approvals from all the authorities having jurisdiction.

2.2.7 *Owner's Consultants:* If required, assist the Owner in selecting and retaining professional services of a surveyor, testing laboratories and special consultants, and coordinate these services, without assuming any responsibility or liability of or for these consultants.

2.2.8 *Inspection:* Inspect the Work of Trade Contractors for defects and deficiencies in the Work without assuming any of the Architect/Engineer's responsibilities for inspection.

2.2.8.1 Review the safety programs of each of the Trade Contractors and make appropriate recommendations. In making such recommendations and carrying out such reviews, he shall not be required to make exhaustive or continuous inspections to check safety precautions and programs in connection with the Project. The performance of such services by the Construction Manager shall not relieve the Trade Contractors of their responsibilities for the safety of persons and property, and for compliance with all federal, state and local statutes, rules, regulations and orders applicable to the conduct of the Work.

2.2.9 *Document Interpretation:* Refer all questions for interpretation of the documents prepared by the Architect/Engineer to the Architect/Engineer.

2.2.10 *Shop Drawings and Samples:* In collaboration with the Architect/Engineer, establish and implement procedures for expediting the processing and approval of shop drawings and samples.

2.2.11 *Reports and Project Site Documents:* Record the progress of the Project. Submit written progress reports to the Owner and the Architect/Engineer including information on the Trade Contractors' Work, and the percentage of completion. Keep a daily log available to the Owner and the Architect/Engineer.

2.2.11.1 Maintain at the Project site, on a current basis: records of all necessary Contracts, Drawings, samples, purchases, materials, equipment, maintenance and operating manuals and instructions, and other construction related documents, including all revisions. Obtain data from Trade Contractors and maintain a current set of record Drawings, Specifications and operating manuals. At the completion of the Project, deliver all such records to the Owner.

2.2.12 *Substantial Completion:* Determine Substantial Completion of the Work or designated portions thereof and prepare for the Architect/Engineer a list of incomplete or unsatisfactory items and a schedule for their completion.

2.2.13 *Start-Up:* With the Owner's maintenance personnel, direct the checkout of utilities, operations systems and equipment for readiness and assist in their initial start-up and testing by the Trade Contractors.

2.2.14 *Final Completion:* Determine final completion and provide written notice to the Owner and Architect/Engineer that the Work is ready for final inspection. Secure and transmit to the Architect/Engineer required guarantees, affidavits, releases, bonds and waivers. Turn over to the Owner all keys, manuals, record drawings and maintenance stocks.

2.2.15 *Warranty:* Where any Work is performed by the Construction Manager's own forces or by Trade Contractors under contract with the Construction Manager, the Construction Manager shall warrant that all materials and equipment included in such Work will be new, unless otherwise specified, and that such Work will be of good quality, free from improper workmanship and defective materials and in conformance with the Drawings and Specifications. With respect to the same Work, the

Construction Manager further agrees to correct all Work defective in material and workmanship for a period of one year from the Date of Substantial Completion or for such longer periods of time as may be set forth with respect to specific warranties contained in the trade sections of the Specifications. The Construction Manager shall collect and deliver to the Owner any specific written warranties given by others.

2.3 Additional Services

2.3.1 At the request of the Owner the Construction Manager will provide the following additional services upon written agreement between the Owner and Construction Manager defining the extent of such additional services and the amount and manner in which the Construction Manager will be compensated for such additional services.

2.3.2 Services related to investigation, appraisals or valuations of existing conditions, facilities or equipment, or verifying the accuracy of existing drawings or other Owner-furnished information.

2.3.3 Services related to Owner-furnished equipment, furniture and furnishings which are not a part of this Agreement.

2.3.4 Services for tenant or rental spaces not a part of this Agreement.

2.3.5 Obtaining or training maintenance personnel or negotiating maintenance service contracts.

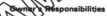

ARTICLE 3

Owner's Responsibilities

3.1 The Owner shall provide full information regarding his requirements for the Project.

3.2 The Owner shall designate a representative who shall be fully acquainted with the Project and has authority to issue and approve Project Construction Budgets, issue Change Orders, render decisions promptly and furnish information expeditiously.

3.3 The Owner shall retain an Architect/Engineer for design and to prepare construction documents for the Project. The Architect/Engineer's services, duties and responsibilities are described in the Agreement between the Owner and the Architect/Engineer, a copy of which will be furnished to the Construction Manager. The Agreement between the Owner and the Architect/Engineer shall not be modified without written notification to the Construction Manager.

3.4 The Owner shall furnish for the site of the Project all necessary surveys describing the physical characteristics, soil reports and subsurface investigations, legal limitations, utility locations, and a legal description.

3.5 The Owner shall secure and pay for necessary approvals, easements, assessments and charges required for the construction, use or occupancy of permanent structures or for permanent changes in existing facilities.

3.6 The Owner shall furnish such legal services as may be necessary for providing the items set forth in Paragraph 3.5, and such auditing services as he may require.

3.7 The Construction Manager will be furnished without charge all copies of Drawings and Specifications reasonably necessary for the execution of the Work.

3.8 The Owner shall provide the insurance for the Project as provided in Paragraph 12.4, and shall bear the cost of any bonds required.

3.9 The services, information, surveys and reports required by the above paragraphs or otherwise to be furnished by other consultants employed by the Owner, shall be furnished with reasonable promptness at the Owner's expense and the Construction Manager shall be entitled to rely upon the accuracy and completeness thereof.

3.10 If the Owner becomes aware of any fault or defect in the Project or non-conformance with the Drawings and Specifications, he shall give prompt written notice thereof to the Construction Manager.

3.11 The Owner shall furnish, prior to commencing work and at such future times as may be requested, reasonable evidence satisfactory to the Construction Manager that sufficient funds are available and committed for the entire cost of the Project. Unless such reasonable evidence is furnished, the Construction Manager is not required to commence or continue any Work, or may, if such evidence is not presented within a reasonable time, stop the Project upon 15 days notice to the Owner. The failure of the Construction Manager to insist upon the providing of this evidence at any one time shall not be a waiver of the Owner's obligation to make payments pursuant to this Agreement nor shall it be a waiver of the Construction Manager's right to request or insist that such evidence be provided at a later date.

3.12 The Owner shall communicate with the Trade Contractors only through the Construction Manager.

ARTICLE 4

Trade Contracts

4.1 All portions of the Project that the Construction Manager does not perform with his own forces shall be performed under Trade Contracts. The Construction Manager shall request and receive proposals from Trade Contractors and Trade Contracts will be awarded after the proposals are reviewed by the Architect/Engineer, Construction Manager and Owner.

4.2 If the Owner refuses to accept a Trade Contractor recommended by the Construction Manager, the Construction Manager shall recommend an acceptable substitute and the Guaranteed Maximum Price if applicable shall be increased or decreased by the difference in cost occasioned by such substitution and an appropriate Change Order shall be issued.

4.3 Unless otherwise directed by the Owner, Trade Contracts will be between the Construction Manager and the Trade Contractors. Whether the Trade Contracts are with the Construction Manager or the Owner, the form of the Trade Contracts including the General and Supplementary Conditions shall be satisfactory to the Construction Manager.

4.4 The Construction Manager shall be responsible to the Owner for the acts and omissions of his agents and employees, Trade Contractors performing Work under a contract with the Construction Manager, and such Trade Contractors' agents and employees.

ARTICLE 5

Schedule

5.1 The services to be provided under this Contract shall be in general accordance with the following schedule:

5.2 At the time a Guaranteed Maximum Price is established, as provided for in Article 6, a Date of Substantial Completion of the project shall also be established.

5.3 The Date of Substantial Completion of the Project or a designated portion thereof is the date when construction is sufficiently complete in accordance with the Drawings and Specifications so the Owner can occupy or utilize the Project or designated portion thereof for the use for which it is intended. Warranties called for by this Agreement or by the Drawings and Specifications shall commence on the Date of Substantial Completion of the Project or designated portion thereof.

5.4 If the Construction Manager is delayed at any time in the progress of the Project by any act or neglect of the Owner or the Architect/Engineer or by any employee of either, or by any separate contractor employed by the Owner, or by changes ordered in the Project, or by labor disputes, fire, unusual delay in transportation, adverse weather conditions not reasonably anticipatable, unavoidable casualties or any causes beyond the Construction Manager's control, or by delay authorized by the Owner pending arbitration, the Construction Completion Date shall be extended by Change Order for a reasonable length of time.

ARTICLE 6

Guaranteed Maximum Price

6.1 When the design, Drawings and Specifications are sufficiently complete, the Construction Manager will, if desired by the Owner, establish a Guaranteed Maximum Price, guaranteeing the maximum price to the Owner for the Cost of the Project and the Construction Manager's Fee. Such Guaranteed Maximum Price will be subject to modification for Changes in the Project as provided in Article 9, and for additional costs arising from delays caused by the Owner or the Architect/Engineer.

6.2 When the Construction Manager provides a Guaranteed Maximum Price, the Trade Contracts will either be with the Construction Manager or will contain the necessary provisions to allow the Construction Manager to control the performance of the Work. The Owner will also authorize the Construction Manager to take all steps necessary in the name of the Owner, including arbitration or litigation, to assure that the Trade Contractors perform their contracts in accordance with their terms.

6.3 The Guaranteed Maximum Price will only include those taxes in the Cost of the Project which are legally enacted at the time the Guaranteed Maximum Price is established.

ARTICLE 7

Construction Manager's Fee

7.1 In consideration of the performance of the Contract, the Owner agrees to pay the Construction Manager in current funds as compensation for his services a Construction Manager's Fee as set forth in Subparagraphs 7.1.1 and 7.1.2.

7.1.1 For the performance of the Design Phase services, a fee of
which shall be paid monthly, in equal proportions, based on the scheduled Design Phase time.

7.1.2 For work or services performed during the Construction Phase, a fee of
which shall be paid proportionately to the ratio the monthly payment for the Cost of the Project bears to the estimated cost. Any balance of this fee shall be paid at the time of final payment.

7.2 Adjustments in Fee shall be made as follows:

7.2.1 For Changes in the Project as provided in Article 9, the Construction Manager's Fee shall be adjusted as follows:

7.2.2 For delays in the Project not the responsibility of the Construction Manager, there will be an equitable adjustment in the fee to compensate the Construction Manager for his increased expenses.

7.2.3 The Construction Manager shall be paid an additional fee in the same proportion as set forth in 7.2.1 if the Construction Manager is placed in charge of the reconstruction of any insured or uninsured loss.

7.3 Included in the Construction Manager's Fee are the following:

7.3.1 Salaries or other compensation of the Construction Manager's employees at the principal office and branch offices, except employees listed in Subparagraph 8.2.2.

AGC DOCUMENT NO. 500 • OWNER CONSTRUCTION MANAGER AGREEMENT   JULY 1980

7.3.2 General operating expenses of the Construction Manager's principal and branch offices other than the field office.

7.3.3 Any part of the Construction Manager's capital expenses, including interest on the Construction Manager's capital employed for the project.

7.3.4 Overhead or general expenses of any kind, except as may be expressly included in Article 8.

7.3.5 Costs in excess of the Guaranteed Maximum Price.

ARTICLE 8

Cost of the Project

8.1 The term Cost of the Project shall mean costs necessarily incurred in the Project during either the Design or Construction Phase, and paid by the Construction Manager, or by the Owner if the Owner is directly paying Trade Contractors upon the Construction Manager's approval and direction. Such costs shall include the items set forth below in this Article.

8.1.1 The Owner agrees to pay the Construction Manager for the Cost of the Project as defined in Article 8. Such payment shall be in addition to the Construction Manager's Fee stipulated in Article 7.

8.2 Cost Items

8.2.1 Wages paid for labor in the direct employ of the Construction Manager in the performance of his Work under applicable collective bargaining agreements, or under a salary or wage schedule agreed upon by the Owner and Construction Manager, and including such welfare or other benefits, if any, as may be payable with respect thereto.

8.2.2 Salaries of the Construction Manager's employees when stationed at the field office, in whatever capacity employed, employees engaged on the road in expediting the production or transportation of materials and equipment, and employees in the main or branch office performing the functions listed below:

8.2.3 Cost of all employee benefits and taxes for such items as unemployment compensation and social security, insofar as such cost is based on wages, salaries, or other remuneration paid to employees of the Construction Manager and included in the Cost of the Project under Subparagraphs 8.2.1 and 8.2.2.

8.2.4 Reasonable transportation, traveling, moving, and hotel expenses of the Construction Manager or of his officers or employees incurred in discharge of duties connected with the Project.

8.2.5 Cost of all materials, supplies and equipment incorporated in the Project, including costs of transportation and storage thereof.

8.2.6 Payments made by the Construction Manager or Owner to Trade Contractors for their Work performed pursuant to contract under this Agreement.

8.2.7 Cost, including transportation and maintenance, of all materials, supplies, equipment, temporary facilities and hand tools not owned by the workmen, which are employed or consumed in the performance of the Work, and cost less salvage value on such items used but not consumed which remain the property of the Construciton Manager.

8.2.8 Rental charges of all necessary machinery and equipment, exclusive of hand tools, used at the site of the Project, whether rented from the Construction Manager or other, including installation, repairs and replacements, dismantling, removal, costs of lubrication, transportation and delivery costs thereof, at rental charges consistent with those prevailing in the area.

8.2.9   Cost of the premiums for all insurance which the Construction Manager is required to procure by this Agreement or is deemed necessary by the Construction Manager.

8.2.10   Sales, use, gross receipts or similar taxes related to the Project imposed by any governmental authority, and for which the Construction Manager is liable.

8.2.11   Permit fees, licenses, tests, royalties, damages for infringement of patents and costs of defending suits therefor, and deposits lost for causes other than the Construction Manager's negligence. If royalties or losses and damages, including costs of defense, are incurred which arise from a particular design, process, or the product of a particular manufacturer or manufacturers specified by the Owner or Architect/Engineer, and the Construction Manager has no reason to believe there will be infringement of patent rights, such royalties, losses and damages shall be paid by the Owner and not considered as within the Guaranteed Maximum Price.

8.2.12   Losses, expenses or damages to the extent not compensated by insurance or otherwise (including settlement made with the written approval of the Owner).

8.2.13   The cost of corrective work subject, however, to the Guaranteed Maximum Price.

8.2.14   Minor expenses such as telegrams, long-distance telephone calls, telephone service at the site, expressage, and similar petty cash items in connection with the Project.

8.2.15   Cost of removal of all debris.

8.2.16   Cost incurred due to an emergency affecting the safety of persons and property.

8.2.17   Cost of data processing services required in the performance of the services outlined in Article 2.

8.2.18   Legal costs reasonably and properly resulting from prosecution of the Project for the Owner.

8.2.19   All costs directly incurred in the performance of the Project and not included in the Construction Manager's Fee as set forth in Paragraph 7.3.

ARTICLE 9

Changes in the Project

9.1   The Owner, without invalidating this Agreement, may order Changes in the Project within the general scope of this Agreement consisting of additions, deletions or other revisions, the Guaranteed Maximum Price, if established, the Construction Manager's Fee and the Construction Completion Date being adjusted accordingly. All such Changes in the Project shall be authorized by Change Order.

9.1.1   A Change Order is a written order to the Construction Manager signed by the Owner or his authorized agent issued after the execution of this Agreement, authorizing a Change in the Project or the method or manner of performance and/or an adjustment in the Guaranteed Maximum Price, the Construction Manager's Fee, or the Construction Completion Date. Each adjustment in the Guaranteed Maximum Price resulting from a Change Order shall clearly separate the amount attributable to the Cost of the Project and the Construction Manager's Fee.

9.1.2   The increase or decrease in the Guaranteed Maximum Price resulting from a Change in the Project shall be determined in one or more of the following ways:

.1   by mutual acceptance of a lump sum properly itemized and supported by sufficient substantiating data to permit evaluation;

.2   by unit prices stated in the Agreement or subsequently agreed upon;

.3   by cost as defined in Article 8 and a mutually acceptable fixed or percentage fee; or

.4   by the method provided in Subparagraph 9.1.3.

9.1.3 If none of the methods set forth in Clauses 9.1.2.1 through 9.1.2.3 is agreed upon, the Construction Manager, provided he receives a written order signed by the Owner, shall promptly proceed with the Work involved. The cost of such Work shall then be determined on the basis of the reasonable expenditures and savings of those performing the Work attributed to the change, including, in the case of an increase in the Guaranteed Maximum Price, a reasonable increase in the Construction Manager's Fee. In such case, and also under Clauses 9.1.2.3 and 9.1.2.4 above, the Construction Manager shall keep and present, in such form as the Owner may prescribe, an itemized accounting together with appropriate supporting data of the increase in the Cost of the Project as outlined in Article 8. The amount of decrease in the Guaranteed Maximum Price to be allowed by the Construction Manager to the Owner for any deletion or change which results in a net decrease in cost will be the amount of the actual net decrease. When both additions and credits are involved in any one change, the increase in Fee shall be figured on the basis of net increase, if any.

9.1.4 If unit prices are stated in the Agreement or subsequently agreed upon, and if the quantities originally contemplated are so changed in a proposed Change Order or as a result of several Change Orders that application of the agreed unit prices to the quantities of Work proposed will cause substantial inequity to the Owner or the Construction Manager, the applicable unit prices and Guaranteed Maximum Price shall be equitably adjusted.

9.1.5 Should concealed conditions encountered in the performance of the Work below the surface of the ground or should concealed or unknown conditions in an existing structure be at variance with the conditions indicated by the Drawings, Specifications, or Owner-furnished information or should unknown physical conditions below the surface of the ground or should concealed or unknown conditions in an existing structure of an unusual nature, differing materially from those ordinarily encountered and generally recognized as inherent in work of the character provided for in this Agreement, be encountered, the Guaranteed Maximum Price and the Construction Completion Date shall be equitably adjusted by Change Order upon claim by either party made within a reasonable time after the first observance of the conditions.

9.2 Claims for Additional Cost or Time

9.2.1 If the Construction Manager wishes to make a claim for an increase in the Guaranteed Maximum Price, an increase in his fee, or an extension in the Construction Completion Date, he shall give the Owner written notice thereof within a reasonable time after the occurrence of the event giving rise to such claim. This notice shall be given by the Construction Manager before proceeding to execute any Work, except in an emergency endangering life or property in which case the Construction Manager shall act, at his discretion, to prevent threatened damage, injury or loss. Claims arising from delay shall be made within a reasonable time after the delay. No such claim shall be valid unless so made. If the Owner and the Construction Manager cannot agree on the amount of the adjustment in the Guaranteed Maximum Price, Construction Manager's Fee or Construction Completion Date, it shall be determined pursuant to the provisions of Article 16. Any change in the Guaranteed Maximum Price, Construction Manager's Fee or Construction Completion Date resulting from such claim shall be authorized by Change Order.

9.3. Minor Changes in the Project

9.3.1 The Architect/Engineer will have authority to order minor Changes in the Project not involving an adjustment in the Guaranteed Maximum Price or an extension of the Construction Completion Date and not inconsistent with the intent of the Drawings and Specifications. Such Changes may be effected by written order and shall be binding on the Owner and the Construction Manager.

9.4 Emergencies

9.4.1 In any emergency affecting the safety of persons or property, the Construction Manager shall act, at his discretion, to prevent threatened damage, injury or loss. Any increase in the Guaranteed Maximum Price or extension of time claimed by the Construction Manager on account of emergency work shall be determined as provided in this Article.

ARTICLE 10

Discounts

All discounts for prompt payment shall accrue to the Owner to the extent the Cost of the Project is paid directly by the

Owner or from a fund made available by the Owner to the Construction Manager for such payments. To the extent the Cost of the Project is paid with funds of the Construction Manager, all cash discounts shall accrue to the Construction Manager. All trade discounts, rebates and refunds, and all returns from sale of surplus materials and equipment, shall be credited to the Cost of the Project.

ARTICLE 11

Payments to the Construction Manager

11.1 The Construction Manager shall submit monthly to the Owner a statement, sworn to if required, showing in detail all moneys paid out, costs accumulated or costs incurred on account of the Cost of the Project during the previous month and the amount of the Construction Manager's Fee due as provided in Article 7. Payment by the Owner to the Construction Manager of the statement amount shall be made within ten (10) days after it is submitted.

11.2 Final payment constituting the unpaid balance of the Cost of the Project and the Construction Manager's Fee shall be due and payable when the Project is delivered to the Owner, ready for beneficial occupancy, or when the Owner occupies the Project, whichever event first occurs, provided that the Project be then substantially completed and this Agreement substantially performed. If there should remain minor items to be completed, the Construction Manager and Architect/Engineer shall list such items and the Construction Manager shall deliver, in writing, his unconditional promise to complete said items within a reasonable time thereafter. The Owner may retain a sum equal to 150% of the estimated cost of completing any unfinished items, provided that said unfinished items are listed separately and the estimated cost of completing any unfinished items likewise listed separately. Thereafter, Owner shall pay to Construction Manager, monthly, the amount retained for incomplete items as each of said items is completed.

11.3 The Construction Manager shall promptly pay all the amounts due Trade Contractors or other persons with whom he has a contract upon receipt of any payment from the Owner, the application for which includes amounts due such Trade Contractor or other persons. Before issuance of final payment, the Construction Manager shall submit satisfactory evidence that all payrolls, materials bills and other indebtedness connected with the Project have been paid or otherwise satisfied.

11.4 If the Owner should fail to pay the Construction Manager within seven (7) days after the time the payment of any amount becomes due, then the Construction Manager may, upon seven (7) additional days' written notice to the Owner and the Architect/Engineer, stop the Project until payment of the amount owing has been received.

11.5 Payments due but unpaid shall bear interest at the rate the Owner is paying on his construction loan or at the legal rate, whichever is higher.

ARTICLE 12

Insurance, Indemnity and Waiver of Subrogation

12.1 Indemnity

12.1.1 The Construction Manager agrees to indemnify and hold the Owner harmless from all claims for bodily injury and property damage (other than the Work itself and other property insured under Paragraph 12.4) that may arise from the Construction Manager's operations under this Agreement.

12.1.2 The Owner shall cause any other contractor who may have a contract with the Owner to perform construction or installation work in the areas where Work will be performed under this Agreement, to agree to indemnify the Owner and the Construction Manager and hold them harmless from all claims for bodily injury and property damage (other than property insured under Paragraph 12.4) that may arise from that contractor's operations. Such provisions shall be in a form satisfactory to the Construction Manager.

AGC DOCUMENT NO. 500 • OWNER CONSTRUCTION MANAGER AGREEMENT JULY 1980

12.2 Construction Manager's Liability Insurance

12.2.1 The Construction Manager shall purchase and maintain such insurance as will protect him from the claims set forth below which may arise out of or result from the Construction Manager's operations under this Agreement whether such operations be by himself or by any Trade Contractor or by anyone directly or indirectly employed by any of them, or by anyone for whose acts any of them may be liable:

12.2.1.1 Claims under workers' compensation, disability benefit and other similar employee benefit acts which are applicable to the Work to be performed.

12.2.1.2 Claims for damages because of bodily injury, occupational sickness or disease, or death of his employees under any applicable employer's liability law.

12.2.1.3 Claims for damages because of bodily injury, death of any person other than his employees.

12.2.1.4 Claims for damages insured by usual personal injury liability coverage which are sustained (1) by any person as a result of an offense directly or indirectly related to the employment of such person by the Construction Manager or (2) by any other person.

12.2.1.5 Claims for damages, other than to the Work itself, because of injury to or destruction of tangible property, including loss of use therefrom.

12.2.1.6 Claims for damages because of bodily injury or death of any person or property damage arising out of the ownership, maintenance or use of any motor vehicle.

12.2.2 The Construction Manager's Comprehensive General Liability Insurance shall include premises — operations (including explosion, collapse and underground coverage) elevators, independent contractors, completed operations, and blanket contractual liability on all written contracts, all including broad form property damage coverage.

12.2.3 The Construction Manager's Comprehensive General and Automobile Liability Insurance, as required by Subparagraphs 12.2.1 and 12.2.2 shall be written for not less than limits of liability as follows:

a. Comprehensive General Liability
   1. Personal Injury

   $_____ Each Occurrence

   $_____ Aggregate
   (Completed Operations)

   2. Property Damage

   $_____ Each Occurrence

   $_____ Aggregate

b. Comprehensive Automobile Liability
   1. Bodily Injury

   $_____ Each Person

   $_____ Each Occurrence

   2. Property Damage

   $_____ Each Occurrence

12.2.4 Comprehensive General Liability Insurance may be arranged under a single policy for the full limits required or by a combination of underlying policies with the balance provided by an Excess or Umbrella Liability policy.

12.2.5 The foregoing policies shall contain a provision that coverages afforded under the policies will not be cancelled or not renewed until at least sixty (60) days' prior written notice has been given to the Owner. Certificates of Insurance showing such coverages to be in Force shall be filed with the Owner prior to commencement of the Work.

12.3 Owner's Liability Insurance

12.3.1 The Owner shall be responsible for purchasing and maintaining his own liability insurance and, at his option, may purchase and maintain such insurance as will protect him against claims which may arise from operations under this Agreement.

12.4 Insurance to Protect Project

12.4.1 The Owner shall purchase and maintain property insurance in a form acceptable to the Construction Manager upon the entire Project for the full cost of replacement as of the time of any loss. This insurance shall include as named insureds the Owner, the Construction Manager, Trade Contractors and their Trade Subcontractors and shall insure against loss from the perils of Fire, Extended Coverage, and shall include "All Risk" insurance for physical loss or damage including, without duplication of coverage, at least theft, vandalism, malicious mischief, transit, collapse, flood, earthquake, testing, and damage resulting from defective design, workmanship or material. The Owner will increase limits of coverage, if necessary, to reflect estimated replacement cost. The Owner will be responsible for any co-insurance penalties or deductibles. If the Project covers an addition to or is adjacent to an existing building, the Construction Manager, Trade Contractors and their Trade Subcontractors shall be named as additional insureds under the Owner's Property Insurance covering such building and its contents.

12.4.1.1 If the Owner finds it necessary to occupy or use a portion or portions of the Project prior to Substantial Completion thereof, such occupancy shall not commence prior to a time mutually agreed to by the Owner and Construction Manager and to which the insurance company or companies providing the property insurance have consented by endorsement to the policy or policies. This insurance shall not be cancelled or lapsed on account of such partial occupancy. Consent of the Construction Manager and of the insurance company or companies to such occupancy or use shall not be unreasonably withheld.

12.4.2 The Owner shall purchase and maintain such boiler and machinery insurance as may be required or necessary. This insurance shall include the interests of the Owner, the Construction Manager, Trade Contractors and their Trade Subcontractors in the Work.

12.4.3 The Owner shall purchase and maintain such insurance as will protect the Owner and Construction Manager against loss of use of Owner's property due to those perils insured pursuant to Subparagraph 12.4.1. Such policy will provide coverage for expediting expenses of materials, continuing overhead of the Owner and Construction Manager, necessary labor expense including overtime, loss of income by the Owner and other determined exposures. Exposures of the Owner and the Construction Manager shall be determined by mutual agreement and separate limits of coverage fixed for each item.

12.4.4 The Owner shall file a copy of all policies with the Construction Manager before an exposure to loss may occur. Copies of any subsequent endorsements will be furnished to the Construction Manager. The Construction Manager will be given sixty (60) days notice of cancellation, non-renewal, or any endorsements restricting or reducing coverage. If the Owner does not intend to purchase such insurance, he shall inform the Construction Manager in writing prior to the commencement of the Work. The Construction Manager may then effect insurance which will protect the interest of himself, the Trade Contractors and their Trade Subcontractors in the Project, the cost of which shall be a Cost of the Project pursuant to Article 8, and the Guaranteed Maximum Price shall be increased by Change Order. If the Construction Manager is damaged by failure of the Owner to purchase or maintain such insurance or to so notify the Construction Manager, the Owner shall bear all reasonable costs properly attributable thereto.

12.5 Property Insurance Loss Adjustment

12.5.1 Any insured loss shall be adjusted with the Owner and the Construction Manager and made payable to the Owner and Construction Manager as trustees for the insureds, as their interests may appear, subject to any applicable mortgagee clause.

12.5.2 Upon the occurrence of an insured loss, monies received will be deposited in a separate account and the trustees shall make distribution in accordance with the agreement of the parties in interest, or in the absence of such agreement, in accordance with an arbitration award pursuant to Article 16. If the trustees are unable to agree on the settlement of the loss, such dispute shall also be submitted to arbitration pursuant to Article 16.

12.6 Waiver of Subrogation

12.6.1 The Owner and Construction Manager waive all rights against each other, the Architect/Engineer, Trade Contractors, and their Trade Subcontractors for damages caused by perils covered by insurance provided under Paragraph 12.4, except such rights as they may have to the proceeds of such insurance held by the Owner and Construction Manager as trustees. The Construction Manager shall require similar waivers from all Trade Contractors and their Trade Subcontractors.

12.6.2 The Owner and Construction Manager waive all rights against each other and the Architect/Engineer, Trade Contractors and their Trade Subcontractors for loss or damage to any equipment used in connection with the Project and covered by any property insurance. The Construction Manager shall require similar waivers from all Trade Contractors and their Trade Subcontractors.

12.6.3 The Owner waives subrogation against the Construction Manager, Architect/Engineer, Trade Contractors, and their Trade Subcontractors on all property and consequential loss policies carried by the Owner on adjacent properties and under property and consequential loss policies purchased for the Project after its completion.

12.6.4 If the policies of insurance referred to in this Paragraph require an endorsement to provide for continued coverage where there is a waiver of subrogation, the owners of such policies will cause them to be so endorsed.

ARTICLE 13

Termination of the Agreement and Owner's
Right to Perform Construction Manager's Obligations

13.1 Termination by the Construction Manager

13.1.1 If the Project, in whole or substantial part, is stopped for a period of thirty days under an order of any court or other public authority having jurisdiction, or as a result of an act of government, such as a declaration of a national emergency making materials unavailable, through no act or fault of the Construction Manager, or if the Project should be stopped for a period of thirty days by the Construction Manager for the Owner's failure to make payment thereon, then the Construction Manager may, upon seven days' written notice to the Owner and the Architect/Engineer, terminate this Agreement and recover from the Owner payment for all work executed, the Construction Manager's Fee earned to date, and for any proven loss sustained upon any materials, equipment, tools, construction equipment and machinery, cancellation charges on existing obligations of the Construction Manager, and a reasonable profit.

13.2 Owner's Right to Perform Construction Manager's Obligations and Termination by the Owner for Cause

13.2.1 If the Construction Manager fails to perform any of his obligations under this Agreement including any obligation he assumes to perform Work with his own forces, the Owner may, after seven days' written notice during which period the Construction Manager fails to perform such obligation, make good such deficiencies. The Guaranteed Maximum Price, if any, shall be reduced by the cost to the Owner of making good such deficiencies.

13.2.2 If the Construction Manager is adjudged a bankrupt, or if he makes a general assignment for the benefit of his creditors, or if a receiver is appointed on account of his insolvency, or if he persistently or repeatedly refuses or fails, except in cases for which extension of time is provided, to supply enough properly skilled workmen or proper materials, or if he fails to make proper payment to Trade Contractors or for materials or labor, or persistently disregards laws, ordinances, rules, regulations or orders of any public authority having jurisdiction, or otherwise is guilty of a substantial violation of a provision of the Agreement, then the Owner may, without prejudice to any right or remedy and after giving the Construction Manager and his surety, if any, seven days' written notice, during which period the Construction Manager fails to cure the violation, terminate the employment of the Construction Manager and take possession of the site and of all materials, equipment, tools, construction equipment and machinery thereon owned by the Construction Manager and may finish the Project by whatever reasonable method he may deem expedient. In such case, the Construction Manager shall not be entitled to receive any further payment until the Project is finished nor shall he be relieved from his obligations assumed under Article 6.

13.3 Termination by Owner Without Cause

13.3.1 If the Owner terminates this Agreement other than pursuant to Subparagraph 13.2.2 or Subparagraph 13.3.2, he shall reimburse the Construction Manager for any unpaid Cost of the Project due him under Article 8, plus (1) the unpaid balance of the Fee computed upon the Cost of the Project to the date of termination at the rate of the percentage named in Subparagraph 7.2.1 or if the Construction Manager's Fee be stated as a fixed sum, such an amount as will increase the payment on account of his fee to a sum which bears the same ratio to the said fixed sum as the Cost of the Project at the time of termination bears to the adjusted Guaranteed Maximum Price, if any, otherwise to a reasonable estimated Cost of the Project when completed. The Owner shall also pay to the Construction Manager fair compensation, either by purchase or rental at the election of the Owner, for any equipment retained. In case of such termination of the Agreement the Owner shall further assume and become liable for obligations, commitments and unsettled claims that the Construction Manager has previously undertaken or incurred in good faith in connection with said Project. The Construction Manager shall, as a condition of receiving the payments referred to in this Article 13, execute and deliver all such papers and take all such steps, including the legal assignment of his contractual rights, as the Owner may require for the purpose of fully vesting in him the rights and benefits of the Construction Manager under such obligations or commitments.

13.3.2 After the completion of the Design Phase, if the final cost estimates make the Project no longer feasible from the standpoint of the Owner, the Owner may terminate this Agreement and pay the Construction Manager his Fee in accordance with Subparagraph 7.1.1 plus any costs incurred pursuant to Article 9.

ARTICLE 14

Assignment and Governing Law

14.1 Neither the Owner nor the Construction Manager shall assign his interest in this Agreement without the written consent of the other except as to the assignment of proceeds.

14.2 This Agreement shall be governed by the law of the place where the Project is located.

ARTICLE 15

Miscellaneous Provisions

15.1 It is expressly understood that the Owner shall be directly retaining the services of an Architect/Engineer.

ARTICLE 16

Arbitration

16.1 All claims, disputes and other matters in questions arising out of, or relating to, this Agreement or the breach thereof, except with respect to the Architect/Engineer's decision on matters relating to artistic effect, and except for claims which have been waived by the making or acceptance of final payment shall be decided by arbitration in accordance with the Construction Industry Arbitration Rules of the American Arbitration Association then obtaining unless the parties mutually agree otherwise. This Agreement to arbitrate shall be specifically enforceable under the prevailing arbitration law.

16.2 Notice of the demand for arbitration shall be filed in writing with the other party to this Agreement and with the American Arbitration Association. The demand for arbitration shall be made within a reasonable time after the claim, dispute or other matter in question has arisen, and in no event shall it be made after the date when institution of legal or equitable proceedings based on such claim, dispute or other matter in question would be barred by the applicable statute of limitations.

16.3 The award rendered by the arbitrators shall be final and judgment may be entered upon it in accordance with applicable law in any court having jurisdiction thereof.

16.4 Unless otherwise agreed in writing, the Construction Manager shall carry on the Work and maintain the Contract Completion Date during any arbitration proceedings, and the Owner shall continue to make payments in accordance with this Agreement.

16.5 All claims which are related to or dependent upon each other, shall be heard by the same arbitrator or arbitrators even though the parties are not the same unless a specific contract prohibits such consolidation.

This Agreement executed the day and year first written above.

ATTEST: OWNER:

ATTEST: CONSTRUCTION MANAGER:

THE ASSOCIATED GENERAL CONTRACTORS OF AMERICA

SUBCONTRACT FOR BUILDING CONSTRUCTION

AGC DOCUMENT NO. 600 • SUBCONTRACT FOR BUILDING CONSTRUCTION
© 1990, The Associated General Contractors of America

TABLE OF CONTENTS

AGC DOCUMENT NO. 600 • SUBCONTRACT FOR BUILDING CONSTRUCTION
© 1990, The Associated General Contractors of America

SUBCONTRACT FOR BUILDING CONSTRUCTION

ARTICLE 1

AGREEMENT

This Agreement is made this _____ day of _____, 19 ___, by and between _____

_____

hereinafter called the Contractor and _____

hereinafter called the Subcontractor, to perform part of the Work on the following Project:

PROJECT:

OWNER:

ARCHITECT:

CONTRACTOR:

SUBCONTRACTOR:

CONTRACT PRICE:

(Here insert a lump sum, unit price or both. Bid schedules may be referenced.)

Notice to the parties shall be given at the above addresses.

3

ARTICLE 2

SCOPE OF WORK

2.1 SUBCONTRACTOR'S WORK. The Contractor contracts with the Subcontractor as an independent contractor, to perform the work described in Article 16. The Subcontractor shall perform such work (hereinafter called the "Subcontractor's Work") under the general direction of the Contractor and in accordance with this Agreement and the Contract Documents.

2.2 CONTRACT DOCUMENTS. The Contract Documents which are binding on the Subcontractor are as set forth in Paragraph 16.5. Upon the Subcontractor's request the Contractor shall furnish a copy of any part of these documents. Nothing in the Contract Documents shall be construed to create a contractual relationship between persons or entities other than the Contractor and Subcontractor.

2.3 CONFLICTS. In the event of a conflict between this Agreement and the Contract Documents, this Agreement shall govern.

ARTICLE 3

SCHEDULE OF WORK

3.1 TIME IS OF ESSENCE. Time is of the essence for both parties, and they mutually agree to see to the performance of their respective work and the work of their subcontractors so that the entire Project may be completed in accordance with the Contract Documents and the Schedule of Work. The Contractor shall prepare the Schedule of Work and revise such schedule as the Work progresses.

3.2 DUTY TO BE BOUND. Both the Contractor and the Subcontractor shall be bound by the Schedule of Work. The Subcontractor shall provide the Contractor with any requested scheduling information for the Subcontractor's Work. The Schedule of Work and all subsequent changes thereto shall be submitted to the Subcontractor in advance of the required performance.

3.3 SCHEDULE CHANGES. The Subcontractor recognizes that changes will be made in the Schedule of Work and agrees to comply with such changes.

3.4 PRIORITY OF WORK. The Contractor shall have the right to decide the time, order and priority in which the various portions of the Work shall be performed and all other matters relative to the timely and orderly conduct of the Subcontractor's Work. The Subcontractor shall commence its work within _____ days of notice to proceed from the Contractor and if such work is interrupted for any reason the Subcontractor shall resume such work within two working days from the Contractor's notice to do so.

ARTICLE 4

CONTRACT PRICE

The Contractor agrees to pay to the Subcontractor for the satisfactory performance of the Subcontractor's Work the amount stated in Article 1 subject to additions or deductions per Article 6.

ARTICLE 5

PAYMENT

5.1 GENERAL PROVISIONS

5.1.1 SCHEDULE OF VALUES. The Subcontractor shall provide a schedule of values satisfactory to the Contractor and the Owner no more than fifteen (15) days from the date of execution of this Agreement.

5.1.2 ARCHITECT VERIFICATION. Upon request the Contractor shall give the Subcontractor written authorization to obtain directly from the Architect the percentage of completion certified for the Subcontractor's Work.

5.1.3 PAYMENT USE RESTRICTION. Payment received by the Subcontractor shall be used to satisfy the indebtedness owed by the Subcontractor to any person furnishing labor or materials for use in performing the Subcontractor's work on this project before it is used in any other manner.

5.1.4 PAYMENT USE VERIFICATION. The Contractor shall have the right at all times to contact the Subcontractor's subcontractors and suppliers to ensure that the same are being paid promptly by the Subcontractor for labor or materials furnished for use in performing the Subcontractor's Work.

5.1.5 PARTIAL LIEN WAIVERS AND AFFIDAVITS. As a prerequisite for payment, the Subcontractor shall provide, in a form satisfactory to the Owner and the Contractor, partial lien or claim waivers and affidavits from the Subcontractor, and its subcontractors and suppliers for the completed Subcontractor's Work. Such waivers may be made conditional upon payment.

5.1.6 SUBCONTRACTOR PAYMENT FAILURE. Upon payment by the Contractor, Subcontractor shall promptly pay its lower-tier subcontractors and material suppliers the amounts to which they are entitled. In the event the Contractor has reason to believe that labor, material or other obligations incurred in the performance of the Subcontractor's Work are not being paid, the Contractor may give written notice of such claim or lien to the Subcontractor and may take any steps deemed necessary to assure that progress payments are utilized to pay such obligations including but not limited to the issuance of joint checks. If upon receipt of said notice, the Subcontractor does not (a) supply evidence to the satisfaction of the Contractor that the moneys owing to the claimant(s) have been paid; or (b) post a bond indemnifying the Owner, the Contractor, and the Contractor's surety, if any, and the premises from such claim or lien; then the Contractor shall have the right to withhold from any payments due or to become due to the Subcontractor a reasonable amount to protect the Contractor from any and all loss, damage or expense including attorney's fees arising out of or relating to any such claim or lien until the claim or lien has been satisfied by the Subcontractor.

5.1.7 SUBCONTRACTOR ASSIGNMENT OF PAYMENTS. The Subcontractor shall not assign any moneys due or to become due under this Contract, or under any Change Order thereto, without the written consent of Contractor, unless such assignment is intended to create a new security interest within the scope of Article 9 of the Uniform Com-

4

mercial Code. Should Subcontractor assign all or any part of any moneys due or to become due under this Contract, to create a new security interest or for any other purpose, the instrument of assignment shall contain a clause to the effect that the assignee's right in and to any money due or to become due to the Subcontractor shall be subject to the claims of all persons, firms and corporations for services rendered or materials supplied for the performance of the Work under this Subcontract and any Change Orders.

5.1.8 PAYMENT NOT ACCEPTANCE. Payment to the Subcontractor does not constitute or imply acceptance of any portion of the Subcontractor's Work.

5.2 PROGRESS PAYMENTS

5.2.1 APPLICATION. Subcontractor's application for payment shall be itemized and supported by substantiating data as required in the Contract Documents for the Contractor's payment application. Subcontractor's application shall be notarized if required. Subcontract payment applications may include payment requests on account of properly authorized Construction Change Directives. The Subcontractor's progress payment application for work performed in the preceding payment period shall be submitted to the Contractor per the terms of this Agreement and specifically Subparagraphs 5.1.1, 5.2.2, 5.2.3, and 5.2.4 for approval of the Contractor and _____

The Contractor shall forward, without delay, the approved value to the Owner for payment.

5.2.2 RETAINAGE/SECURITY. The rate of retainage shall be equal to the percentage retained from the Contractor's payment by the Owner for the Subcontractor's Work provided the Subcontractor furnishes a bond or other security to the satisfaction of the Contractor.

If the Subcontractor has furnished such bond or security, its work is satisfactory and the Contract Documents provide for reduction of retainage at a specified percentage of completion, the Subcontractor's retainage shall also be reduced when the Subcontractor's Work has attained the same percentage of completion and the Contractor's retainage for the Subcontractor's Work has been so reduced by the Owner. However if the Subcontractor does not provide such bond or security, the rate of retainage shall be _____ %.

5.2.3 TIME OF APPLICATION. The Subcontractor shall submit progress payment applications to the Contractor no later than the _____ day of each payment period for work performed up to and including the _____ day of the payment period indicating work completed and, to the extent allowed under Subparagraph 5.2.4, materials suitably stored during the preceding payment period.

5.2.4 STORED MATERIALS. Unless otherwise provided in the Contract Documents, and if approved in advance by the Owner, applications for payment may include materials and equipment not incorporated in the Subcontractor's Work but delivered to and suitably stored at the site or at some other location agreed upon in writing. Approval of payment applications for such stored items on or off the site shall be conditioned upon submission by the Subcontractor of bills of sale and applicable insurance or such other procedures satisfactory to the Owner and Contractor to establish the Owner's title to such materials and equipment or otherwise protect the Owner's and Contractor's interest therein, including transportation to the site.

5.2.5 TIME OF PAYMENT. Progress payments to the Subcontractor for satisfactory performance of the Subcontractor's Work shall be made no later than seven (7) days after receipt by the Contractor of payment from the Owner for the Subcontractor's Work.

5.2.6 PAYMENT DELAY. If for any reason not the fault of the Subcontractor, the Subcontractor does not receive a progress payment from the Contractor within seven (7) days after the date such payment is due, as defined in Subparagraph 5.2.5, then the Subcontractor, upon giving an additional seven (7) days written notice to the Contractor, and without prejudice to and in addition to any other legal remedies, may stop work until payment of the full amount owing to the Subcontractor has been received. To the extent obtained by the Contractor under the Contract Documents, the contract price shall be increased by the amount of the Subcontractor's reasonable cost of shutdown, delay, and start-up, which shall be effected by appropriate Change Order.

If the Subcontractor's Work has been stopped for thirty (30) days because the Subcontractor has not received progress payments as required hereunder, the Subcontractor may terminate this Agreement upon giving the Contractor an additional seven (7) days written notice.

5.3 FINAL PAYMENT

5.3.1 APPLICATION. Upon acceptance of the Subcontractor's Work by the Owner the Contractor, and if necessary, the Architect; and upon the Subcontractor furnishing evidence of fulfillment of the Subcontractor's obligations in accordance with the Contract Documents and Subparagraph 5.3.2, the Contractor shall forward the Subcontractor's application for final payment without delay.

5.3.2 REQUIREMENTS. Before the Contractor shall be required to forward the Subcontractor's application for final payment to the Owner, the Subcontractor shall submit to the Contractor:
(a) an affidavit that all payrolls, bills for materials and equipment, and other indebtedness connected with the Subcontractor's Work for which the Owner or its property or the Contractor or the Contractor's surety might in any way be liable, have been paid or otherwise satisfied;
(b) consent of surety to final payment, if required;
(c) satisfaction of required closeout procedures;
(d) certification that insurance required by the Contract Documents to remain in effect beyond final payment pursuant to Paragraph 13.4 is in effect and will not be cancelled or allowed to expire without at least thirty (30) days' written notice to the Contractor unless a longer period is stipulated in the Contract; and
(e) other data if required by the Contractor or Owner, such as receipts, releases, and waivers of liens to the extent and in such form as may be designated by the Contractor or Owner. Final payment shall constitute a waiver of all claims by the Subcontractor relating to the Subcontractor's Work, but shall in no way relieve the Subcontractor of liability for the

5

obligations assumed under Paragraph 9.10, or for faulty or defective work appearing after final payment.

5.3.3 TIME OF PAYMENT. Final payment of the balance due of the Contract Price shall be made to the Subcontractor:

(a) upon receipt of the Owner's waiver of all claims related to the Subcontractor's Work except for unsettled liens, unknown defective work, and noncompliance with the Contract Documents or warranties; and

(b) within seven (7) days after receipt by the Contractor of final payment from the Owner for such Subcontractor's Work.

5.3.4 FINAL PAYMENT DELAY. If the Owner or its designated agent does not issue a certificate for Final Payment or the Contractor does not receive such payment for any cause which is not the fault of the Subcontractor, the Contractor shall promptly inform the Subcontractor in writing. The Contractor shall also diligently pursue, with the assistance of the Subcontractor, the prompt release by the Owner of the final payment due for the Subcontractor's Work. At the Subcontractor's request and expense, to the extent agreed upon in writing, the Contractor shall institute reasonable legal remedies to mitigate the damages and pursue payment of the Subcontractor's final payment including interest thereon.

5.4 LATE PAYMENT INTEREST. To the extent obtained by the Contractor, under the Contract Documents, progress payments or final payment due and/unpaid under this Agreement shall bear interest from the date payment is due at the rate provided in the Contract Documents, or, in the absence thereof, at the legal rate prevailing at the place of the Project.

ARTICLE 6

CHANGES, CLAIMS AND DELAYS

6.1 CHANGES. When the Contractor orders in writing, the Subcontractor, without nullifying this Agreement, shall make any and all changes in the Work which are within the general scope of this Agreement. Adjustments in the Contract Price or contract time, if any, resulting from such changes shall be set forth in a Subcontract Change Order or a Subcontract Construction Change Directive pursuant to the Contract Documents. No such adjustments shall be made for any changes performed by the Subcontractor that have not been ordered by the Contractor. A Subcontract Change Order is a written instrument prepared by the Contractor and signed by the Subcontractor stating their agreement upon the change in the scope of the Work, adjustment in the Contract Price or Schedule of Work. A Subcontract Construction Change Directive is a written instrument prepared by the Contractor directing a change in the Work and stating a proposed adjustment, if any, in the Contract Price or Schedule of Work or both. A Subcontract Construction Change Directive shall be used in the absence of agreement on the terms of a Subcontract Change Order.

6.2 CLAIMS RELATING TO OWNER. The Subcontractor agrees to make all claims for which the Owner is or may be liable in the manner and within the time limits provided in the Contract Documents for like claims by the Contractor upon the Owner and in sufficient time for the Contractor to prosecute such claims against the Owner in accordance with the Contract Documents. The Contractor agrees to permit the Subcontractor to prosecute a claim in the name of the Contractor for the use and benefit of the Subcontractor in the manner provided in the Contract Documents for like claims by the Contractor upon the Owner.

6.3 CLAIMS RELATING TO CONTRACTOR. The Subcontractor shall give the Contractor written notice of all claims not included in Paragraph 6.2 within five (5) days of the occurrence of the event for which claim is made; otherwise, such claims shall be deemed waived. All unresolved claims, disputes and other matters in question between the Contractor and the Subcontractor not relating to claims included in Paragraph 6.2 shall be resolved in the manner provided in Article 14.

6.4 ADJUSTMENT IN CONTRACT PRICE. If a Subcontract Change Order or Construction Change Directive requires an adjustment in the Contract Price, the adjustment shall be established by one of the following methods:

1. mutual agreement on a lump sum with sufficient information to substantiate the amount;

2. unit prices already established in the Contract Documents or if not established by the Contract Documents then established by mutual agreement for this adjustment; or

3. a mutually determined cost plus a jointly acceptable allowance for overhead and profit.

6.5 SUBSTANTIATION OF ADJUSTMENT. If the Subcontractor does not respond promptly or disputes the method of adjustment, the method and the adjustment shall be determined by the Contractor on the basis of reasonable expenditures and savings of those performing the Work attributable to the change, including, in the case of an increase in the Contract Price, an allowance for overhead and profit of _____.

The Subcontractor shall maintain for the Contractor's review and approval an appropriately itemized and substantiated accounting of the following items attributable to the Subcontract Change Order or Subcontract Construction Change Directive:

1. labor costs, including Social Security, health, welfare, retirement and other fringe benefits as normally required, and state workers' compensation insurance;

2. costs of materials, supplies and equipment, whether incorporated in the Work or consumed, including transportation costs;

3. costs of renting, either from the Contractor or from others, of machinery and equipment other than hand tools;

4. costs of bond and insurance premiums, permit fees and taxes attributable to the change; and

5. costs of additional supervision and field office personnel services necessitated by the change.

6

6.6 DELAY. If the progress of the Subcontractor's Work is substantially delayed without the fault or responsibility of the Subcontractor, then the time for the Subcontractor's Work shall be extended by Subcontract Change Order or Subcontract Construction Change Directive to the extent obtained by the Contractor under the Contract Documents and the Schedule of Work shall be revised accordingly.

The Contractor shall not be liable to the Subcontractor for any damages or additional compensation as a consequence of delays caused by any person not a party to this Agreement unless the Contractor has first recovered the same on behalf of the Subcontractor from said person, it being understood and agreed by the Subcontractor that, apart from recovery from said person, the Subcontractor's sole and exclusive remedy for delay shall be an extension in the time for performance of the Subcontractor's Work.

6.7 LIQUIDATED DAMAGES. If the Contract Documents provide for liquidated or other damages for delay beyond the completion date set forth in the Contract Documents, and such damages are assessed, then the Contractor may assess same against the Subcontractor in proportion to the Subcontractor's share of the responsibility for such delay. However the amount of such assessment shall not exceed the amount assessed against the Contractor.

Nothing set forth herein shall limit the Subcontractor's liability to the Contractor for the Contractor's actual delay damages caused by the Subcontractor's delay. The Subcontractor shall be liable to the Contractor for the Contractor's actual damages caused by the Subcontractor's delay.

ARTICLE 7

CONTRACTOR'S OBLIGATIONS

7.1 CONTRACT DOCUMENTS. Prior to executing this Subcontract, the Contractor shall make available to the Subcontractor the Contract Documents which are binding on the Subcontractor and set forth in Paragraph 16.5.

7.2 AUTHORIZED REPRESENTATIVE. The Contractor shall designate one or more persons who shall be the Contractor's authorized representative(s) on-site and off-site. Such authorized representative(s) shall be the only person(s) the Subcontractor shall look to for instructions, orders and/or directions, except in an emergency.

7.3 STORAGE APPLICATIONS. The Contractor shall allocate adequate storage areas, if available, for the Subcontractor's materials and equipment during the course of the Subcontractor's Work.

7.4 TIMELY COMMUNICATIONS. The Contractor shall transmit, with reasonable promptness, all submittals, transmittals, and written approvals relating to the Subcontractor's Work.

7.5 NON-CONTRACTED SERVICES. The Contractor agrees, except as otherwise provided in this Agreement, that no claim for non-contracted construction services rendered or materials furnished shall be valid unless the Contractor provides the Subcontractor notice:
 (a) prior to furnishing of the services and materials, except in an emergency affecting the safety of persons or property;

 (b) in writing of such claim within three days of first furnishing such services or materials; and
 (c) the written charges for such services or materials no later than the fifteenth (15th) day of the calendar month following that in which the claim originated.

ARTICLE 8

SUBCONTRACTOR'S OBLIGATIONS

8.1 OBLIGATIONS DERIVATIVE. The Subcontractor binds itself to the Contractor under this Agreement in the same manner as the Contractor is bound to the Owner under the Contract Documents. The Subcontractor shall make available to its lower-tier subcontractors the Contract Documents which are binding on the lower-tier subcontractors.

8.2 RESPONSIBILITIES. The Subcontractor shall furnish all of the labor, materials, equipment, and services, including, but not limited to, competent supervision, shop drawings, samples, tools, and scaffolding as are necessary for the proper performance of the Subcontractor's Work in strict accordance with and reasonably inferable from the Contract Documents.

The Subcontractor shall provide a list of proposed subcontractors and suppliers, be responsible for taking field dimensions, providing tests, ordering of materials and all other actions as required to meet the Schedule of Work.

8.3 SHOP DRAWINGS. The Subcontractor shall be responsible to the Contractor for the accuracy and conformity with the Contract Documents and other submittals that pertain to its work in the same manner as the Contractor is responsible therefor to the Owner. Shop drawings, or their approval by the Contractor, shall not be deemed to authorize deviations or substitutions from the requirements of the Contract Documents.

8.4 TEMPORARY SERVICES. The Subcontractor shall furnish all temporary services and/or facilities necessary to perform its work, except as provided in Article 16. Said article also identifies those common temporary services, if any, which are to be furnished by the Subcontractor.

8.5 COORDINATION. The Subcontractor shall:
 (a) cooperate with the Contractor and all others whose work may interfere with the Subcontractor's Work;
 (b) specifically note and immediately advise the Contractor of any such interference with the Subcontractor's Work; and
 (c) participate in the preparation of coordination drawings and work schedules in areas of congestion.

8.6 AUTHORIZED REPRESENTATIVE. The Subcontractor shall designate one or more persons who shall be the authorized Subcontractor's representative(s) on-site and off-site. Such authorized representative(s) shall be the only person(s) to whom the Contractor shall issue instructions, orders or directions, except in an emergency.

8.7 PROVISION FOR INSPECTION. The Subcontractor shall notify the Contractor when portions of the Subcontractor's Work are ready for inspection. The Subcontractor shall at all times furnish the Contractor and its representatives adequate facilities for inspecting materials at the site or any place where materials under this Agreement may be

7

In the course of preparation, process, manufacture or treatment.

The Subcontractor shall furnish to the Contractor, in such detail and as often as required, full reports of the progress of the Subcontractor's Work irrespective of the location of such work.

8.8 CLEANUP. The Subcontractor shall follow the Contractor's cleanup and safety directions, and
(a) at all times keep the building and premises free from debris and unsafe conditions resulting from the Subcontractor's Work; and
(b) broom clean each work area prior to discontinuing work in the same.

If the Subcontractor fails to immediately commence compliance with cleanup duties within twenty-four (24) hours after written notification from the Contractor of non-compliance, the Contractor may implement such cleanup measures without further notice and deduct the cost thereof from any amounts due or to become due the Subcontractor.

8.9 SAFETY. The prevention of accidents on or in the vicinity of its Work is the Subcontractor's responsibility, even if Contractor establishes a safety program for the entire Project. Subcontractor shall establish a safety program implementing safety measures, policies and standards conforming to those required or recommended by governmental and quasi-governmental authorities having jurisdiction and by the Contractor and Owner, including, but not limited to, requirements imposed by the Contract Documents. Subcontractor shall comply with the reasonable recommendations of insurance companies having an interest in the Project, and shall stop any part of the Work which Contractor deems unsafe until corrective measures satisfactory to Contractor shall have been taken. Contractor's failure to stop Subcontractor's unsafe practices shall not relieve Subcontractor of the responsibility therefor. Subcontractor shall notify Contractor immediately following any accident and promptly confirm the notice in writing. A detailed written report shall be furnished if requested by the Contractor. Subcontractor shall indemnify Contractor for fines, damages or expenses incurred by the Contractor because of the Subcontractor's failure to comply with safety requirements.

8.10 PROTECTION OF THE WORK. The Subcontractor shall take necessary precautions to properly protect the Subcontractor's Work and the work of others from damage caused by the Subcontractor's operations. Should the Subcontractor cause damage to the Work or property of the Owner, the Contractor or others, the Subcontractor shall promptly remedy such damage to the satisfaction of the Contractor, or the Contractor may so remedy and deduct the cost thereof from any amounts due or to become due the Subcontractor.

8.11 PERMITS, FEES AND LICENSES. The Subcontractor shall give adequate notices to authorities pertaining to the Subcontractor's Work and secure and pay for all permits, fees, licenses, assessments, inspections and taxes necessary to complete the Subcontractor's Work in accordance with the Contract Documents.

To the extent obtained by the Contractor under the Contract Documents, the Subcontractor shall be compensated for additional costs resulting from laws, ordinances, rules, regulations and taxes enacted after the date of the Agreement.

8.12 SUBCONTRACTOR ASSIGNMENT OF WORK. The Subcontractor shall not assign the whole nor any part of the Subcontractor's Work without prior written approval of the Contractor. The Contractor's approval shall not be unreasonably withheld. Lower-tier subcontractors and suppliers previously approved by the Contractor may be listed at Paragraph 16.4.

8.13 NON-CONTRACTED SERVICES. The Subcontractor agrees, except as otherwise provided in this Agreement, that no claim for non-contracted construction services rendered or materials furnished shall be valid unless the Subcontractor provides the Contractor notice:
(a) prior to furnishing of the services or materials, except in an emergency affecting the safety of persons or property;
(b) in writing of such claim within three (3) days of first furnishing such services or materials; and
(c) the written charge for such services or materials no later than the fifteenth (15th) day of the calendar month following that in which the claim originated.

8.14 MATERIALS SAFETY. To the extent that the Contractor is not obligated by the Contract Documents or by law to perform work which involves pollutants, hazardous or toxic substances, hazardous waste, asbestos or PCB's, the Subcontractor likewise is not obligated. To the extent that the Contractor has obligations under the Contract Documents or by law regarding such materials within the scope of the Subcontractor's work, the Subcontractor likewise shall have these obligations.

ARTICLE 9

SUBCONTRACT PROVISIONS

9.1 LAYOUT RESPONSIBILITY AND LEVELS. The Contractor shall establish principal axis lines of the building and site whereupon the Subcontractor shall lay out and be strictly responsible for the accuracy of the Subcontractor's Work and for any loss or damage to the Contractor or others by reason of the Subcontractor's failure to set out or perform its work correctly. The Subcontractor shall exercise prudence so that the actual final conditions and details shall result in alignment of finish surfaces.

9.2 WORKMANSHIP. Every part of the Subcontractor's Work shall be executed in strict accordance with the Contract Documents in the most sound, workmanlike, and substantial manner. All workmanship shall be of the best of its several kinds, and all materials used in the Subcontractor's Work shall be furnished in ample quantities to facilitate the proper and expeditious execution of the work, and shall be new except such materials as may be expressly provided in the Contract Documents to be otherwise.

9.3 MATERIALS FURNISHED BY OTHERS. In the event the scope of the Subcontractor's Work includes installation of materials or equipment furnished by others, it shall be the responsibility of the Subcontractor to examine the items so provided and thereupon handle, store and install the items with such skill and care as to ensure a satisfactory and proper installation. Loss or damage due to acts of the Subcontractor shall be deducted from any amounts due or to become due the Subcontractor.

8

9.4 SUBSTITUTIONS. No substitutions shall be made in the Subcontractor's Work unless permitted in the Contract Documents and only then upon the Subcontractor first receiving all approvals required under the Contract Documents for substitutions. The Subcontractor shall indemnify the Contractor as a result of such substitutions, whether or not the Subcontractor has obtained approval thereof.

9.5 USE OF CONTRACTOR'S EQUIPMENT. The Subcontractor, its agents, employees, subcontractors or suppliers shall not use the Contractor's equipment without the express written permission of the Contractor's designated representative.

If the Subcontractor or any of its agents, employees, suppliers or lower-tier subcontractors utilize any machinery, equipment, tools, scaffolding, hoists, lifts or similar items owned, leased, or under the control of the Contractor, the Subcontractor shall defend, indemnify and be liable to the Contractor as provided in Article 12 for any loss or damage (including personal injury or death) which may arise from such use, except where such loss or damage shall be found to have been due solely to the negligence of the Contractor's employees operating such equipment.

9.6 CONTRACT BOND REVIEW. The Contractor's Payment Bond for the Project, if any, may be reviewed and copied by the Subcontractor.

9.7 OWNER ABILITY TO PAY. The Subcontractor shall have the right to receive from the Contractor such information as the Contractor has obtained relative to the Owner's financial ability to pay for the Work.

9.8 PRIVITY. Until final completion of the Project, the Subcontractor agrees not to perform any work directly for the Owner or any tenants thereof, or deal directly with the Owner's representatives in connection with the Project, unless otherwise directed in writing by the Contractor. All Work for this Project performed by the Subcontractor shall be processed and handled exclusively by the Contractor.

9.9 SUBCONTRACT BOND. If a Performance and Payment Bond is not required of the Subcontractor under Article 16, then within the duration of this Agreement, the Contractor may require such bonds before work is started and the Subcontractor shall provide the same.

Said bonds shall be in the full amount of this Agreement in a form and by a surety satisfactory to the Contractor.

The Subcontractor shall be reimbursed without retainage for cost of same simultaneously with the first progress payment hereunder.

The reimbursement amount for the bonds shall not exceed the manual rate for such subcontractor work.

In the event the Subcontractor shall fail to promptly provide such requested bonds, the Contractor may terminate this Agreement and re-let the work to another subcontractor and all Contractor costs and expenses incurred thereby shall be paid by the Subcontractor.

9.10 WARRANTY. The Subcontractor warrants its work against all deficiencies and defects in materials and/or workmanship and as called for in the Contract Documents.

The Subcontractor agrees to satisfy such warranty obligations which appear within the warranty period established in the Contract Documents without cost to the Owner or the Contractor.

If no warranty is required of the Contractor in the Contract Documents, then the Subcontractor shall warrant its work as described above for the period of one year from the date(s) of substantial completion of all or a designated portion of the Subcontractor's Work or acceptance or use by the Contractor or Owner of designated equipment, whichever is sooner.

The Subcontractor further agrees to execute any special warranties that shall be required for the Subcontractor's Work prior to final payment.

ARTICLE 10

RECOURSE BY CONTRACTOR

10.1 FAILURE OF PERFORMANCE

10.1.1 NOTICE TO CURE. If the Subcontractor refuses or fails to supply enough properly skilled workers, proper materials, or maintain the Schedule of Work, or it fails to make prompt payment for its workers, lower-tier subcontractors or suppliers, disregards laws, ordinances, rules, regulations or orders of any public authority having jurisdiction, or otherwise is guilty of a material breach of a provision of this Agreement, the Subcontractor shall be deemed in default of this Agreement. If the Subcontractor fails within three (3) working days after written notification to commence and continue satisfactory correction of such default with diligence and promptness, then the Contractor without prejudice to any rights or remedies, shall have the right to any or all of the following remedies:

(a) supply such number of workers and quantity of materials, equipment and other facilities as the Contractor deems necessary for the completion of the Subcontractor's Work; or any part thereof which the Subcontractor has failed to complete or perform after the aforesaid notice, and charge the cost thereof to the Subcontractor, who shall be liable for the payment of same including reasonable overhead, profit and attorney's fees;

(b) contract with one or more additional contractors, to perform such part of the Subcontractor's Work as the Contractor shall determine will provide the most expeditious completion of the total Work and charge the cost thereof to the Subcontractor;

(c) withhold payment of any moneys due the Subcontractor pending corrective action in amounts sufficient to cover losses and compel performance to the extent required by and to the satisfaction of the Contractor and _____ ; and

(d) in the event of an emergency affecting the safety of persons or property, the Contractor may proceed as above without notice.

9

10.1.2 TERMINATION BY CONTRACTOR. If the Subcontractor fails to commence and satisfactorily continue correction of a default within three (3) working days after written notification issued under Subparagraph 10.1.1, then the Contractor may, in lieu of or in addition to Subparagraph 10.1.1, issue a second written notification, to the Subcontractor and its surety, if any. Such notice shall state that if the Subcontractor fails to commence and continue correction of a default within seven (7) working days of the written notification, the Agreement will be deemed terminated and the Contractor may use any materials, implements, equipment, appliances or tools furnished by or belonging to the Subcontractor to complete the Subcontractor's Work.

The Contractor also may furnish those materials, equipment and/or employ such workers or subcontractors as the Contractor deems necessary to maintain the orderly progress of the Work.

All costs incurred by the Contractor in performing the Subcontractor's Work, including reasonable overhead, profit and attorney's fees, shall be deducted from any moneys due or to become due the Subcontractor. The Subcontractor shall be liable for the payment of any amount by which such expense may exceed the unpaid balance of the Contract Price.

10.1.3 USE OF SUBCONTRACTOR'S EQUIPMENT. If the Contractor performs work under this Article or sublets such work to be so performed, the Contractor and/or the persons to whom work has been sublet shall have the right to take and use any materials, implements, equipment, appliances or tools furnished by, belonging or delivered to the Subcontractor and located at the Project.

10.2 BANKRUPTCY

10.2.1 TERMINATION ABSENT CURE. If Subcontractor files a petition under the Bankruptcy Code, this Agreement shall terminate if the Subcontractor or the Subcontractor's trustee rejects the Agreement or, if there has been a default, the Subcontractor is unable to give adequate assurance that the Subcontractor will perform as required by this Agreement or otherwise is unable to comply with the requirements for assuming this Agreement under the applicable provisions of the Bankruptcy Code.

10.2.2 INTERIM REMEDIES. If the Subcontractor is not performing in accordance with the Schedule of Work at the time a petition in bankruptcy is filed, or at any subsequent time, the Contractor, while awaiting the decision of the Subcontractor or its trustee to reject or to assume this Agreement and provide adequate assurance of its ability to perform hereunder, may avail itself of such remedies under this Article as are reasonably necessary to maintain the Schedule of Work.

The Contractor may offset against any sums due or to become due the Subcontractor all costs incurred in pursuing any of the remedies provided hereunder, including, but not limited to, reasonable overhead, profit and attorney's fees.

The Subcontractor shall be liable for the payment of any amount by which such expense may exceed the unpaid balance of the Contract Price.

10.3 SUSPENSION BY OWNER. Should the Owner suspend its contract with the Contractor or any part which includes the Subcontractor's Work, the Contractor shall so notify

the Subcontractor in writing and upon written notification the Subcontractor shall immediately suspend the Subcontractor's Work.

In the event of such Owner suspension, the Contractor's liability to the Subcontractor is limited to the extent of the Contractor's recovery on the Subcontractor's behalf under the Contract Documents. The Contractor agrees to cooperate with the Subcontractor, at the Subcontractor's expense, in the prosecution of any Subcontractor claim arising out of an Owner suspension and to permit the Subcontractor to prosecute said claim, in the name of the Contractor, for the use and benefit of the Subcontractor.

10.4 TERMINATION BY OWNER. Should the Owner terminate its contract with the Contractor or any part which includes the Subcontractor's Work, the Contractor shall so notify the Subcontractor in writing and upon written notification, this Agreement shall be terminated and the Subcontractor shall immediately stop the Subcontractor's Work, follow all of Contractor's instructions, and mitigate all costs.

In the event of such Owner termination, the Contractor's liability to the Subcontractor is limited to the extent of the Contractor's recovery on the Subcontractor's behalf under the Contract Documents.

The Contractor agrees to cooperate with the Subcontractor, at the Subcontractor's expense, in the prosecution of any Subcontractor claim arising out of the Owner termination and to permit the Subcontractor to prosecute said claim, in the name of the Contractor, for the use and benefit of the Subcontractor, or assign the claim to the Subcontractor.

10.5 CONTINGENT ASSIGNMENT OF SUBCONTRACT. The Contractor's contingent assignment of the Subcontract to the Owner, if provided in the Contract Documents, is effective when the Owner (a) has terminated the Contract for cause and (b) has accepted the assignment by notifying the Subcontractor in writing. This contingent assignment is subject to the prior rights of a surety that may be obligated under the Contractor's bond, if any. Subcontractor hereby consents to such assignment and agrees to be bound to the assignee by the terms of this Subcontract.

10.6 SUSPENSION BY CONTRACTOR. The Contractor may order the Subcontractor in writing to suspend, delay, or interrupt all or any part of the Subcontractor's Work for such period of time as may be determined to be appropriate for the convenience of the Contractor. Phased or interrupted Work when required shall not be deemed a suspension of Work.

The Subcontractor shall notify the Contractor in writing within ten (10) working days after receipt of the Contractor's order of the effect of such order upon the Subcontractor's Work. To the extent allowed the Contractor under the Contract Documents, the Contract Price or contract time shall be adjusted by Subcontract Change Order for any increase in the time or cost of performance of this Agreement caused by such suspension, delay, or interruption.

No claim under this Article shall be allowed for any costs incurred more than ten (10) working days prior to the Subcontractor's notice to the Contractor.

Neither the Contract Price nor the contract time shall be adjusted under this Article for any suspension, delay or interruption to the extent that performance would have been so suspended, delayed, or interrupted by the fault

10

or negligence of the Subcontractor or by a cause for which Subcontractor would have been responsible.

The Contract Price shall not be adjusted under this Article for any suspension, delay or interruption to the extent that performance would have been suspended, delayed or interrupted by a cause for which the Subcontractor would have been entitled only to a time extension under this Agreement.

10.7 WRONGFUL EXERCISE. If the Contractor wrongfully exercises any option under this Article, the Contractor shall be liable to the Subcontractor solely for the reasonable value of work performed by the Subcontractor prior to the Contractor's wrongful action, including reasonable overhead and profit on the Work performed, less prior payments made, and attorney's fees.

ARTICLE 11

LABOR RELATIONS

(Insert here any conditions, obligations or requirements relative to labor relations and their effect on the project. Legal counsel is recommended.)

ARTICLE 12

INDEMNIFICATION

12.1 SUBCONTRACTOR'S PERFORMANCE. To the fullest extent permitted by law, the Subcontractor shall defend, indemnify and hold harmless, the Contractor (including the affiliates, parents and subsidiaries, their agents and employees) and other contractors and subcontractors and all of their agents and employees and when required of the Contractor by the Contract Documents, the Owner, the Architect, Architect's consultants, agents and employees from and against all claims, damages, loss and expenses, including but not limited to attorney's fees, arising out of or resulting from the performance of the Subcontract provided that:

(a) any such claim, damage, loss, or expense is attributable to bodily injury, sickness, disease, or death, or to injury to or destruction of tangible property (other than the Subcontractor's Work itself) including the loss of use resulting therefrom, to the extent caused or alleged to be caused in whole or in part by any negligent act or omission of the Subcontractor or anyone directly or indirectly employed by the Subcontractor or for anyone for whose acts the Subcontractor may be liable, regardless of whether it is caused in part by a party indemnified hereunder;

(b) such obligation shall not be construed to negate, or abridge, or otherwise reduce any other right or obligation of indemnity which would otherwise exist as to any party or person described in this Article 12.

12.2 NO LIMITATION UPON LIABILITY. In any and all claims against the Owner, the Architect, Architect's consultants, agents and employees, the Contractor (including its affiliates, parents and subsidiaries) and other contractors or subcontractors, or any of their agents or employees, by any employee of the Subcontractor, anyone directly or indirectly employed by the Subcontractor or anyone for whose acts the Subcontractor may be liable, the indemnification obligation under this Article 12 shall not be limited in any way by any limitation on the amount or type of damages, compensation or benefits payable by or for the Subcontractor under worker's or workmen's compensation acts, disability benefit acts or other employee benefit acts.

12.3 ARCHITECT EXCLUSION. Except as provided by the Contract Documents, the obligation of the Subcontractor under this Article 12 shall not extend to the liability of the Architect, the Architect's consultants, agents or employees of any of them, arising out of

(a) the preparation or approval of maps, drawings, opinions, reports, surveys, Change Orders, designs or specifications, or

(b) the giving of or the failure to give directions or instructions by the Architect, the Architect's Consultants, and agents or employees of any of them provided such giving or failure to give is the primary cause of the injury or damage.

12.4 COMPLIANCE WITH LAWS. The Subcontractor agrees to be bound by, and at its own cost, comply with all federal, state and local laws, ordinances and regulations (hereinafter collectively referred to as "laws") applicable to the Subcontractor's Work including, but not limited to, equal employment opportunity, minority business enterprise, women's business enterprise, disadvantaged business enterprise, safety and all other laws with which the Contractor must comply according to the Contract Documents.

The Subcontractor shall be liable to the Contractor and the Owner for all loss, cost and expense attributable to any acts of commission or omission by the Subcontractor, its employees and agents resulting from the failure to comply therewith, including, but not limited to, any fines, penalties or corrective measures.

12.5 PATENTS. Except as otherwise provided by the Contract Documents, the Subcontractor shall pay all royalties and license fees which may be due on the inclusion of any patented materials in the Subcontractor's Work. The Subcontractor shall defend all suits for claims for infringement of any patent rights arising out of the Subcontractor's Work, which may be brought against the Contractor or Owner, and shall be liable to the Contractor and Owner for all loss, including all costs, expenses, and attorney's fees.

ARTICLE 13

INSURANCE

13.1 SUBCONTRACTOR'S INSURANCE. Prior to start of the Subcontractor's Work, the Subcontractor shall procure for the Subcontractor's Work and maintain in force Workers' Compensation Insurance, Employer's Liability Insurance, Comprehensive or Commercial General Liability Insurance on an occurrence basis, and all insurance required of the Contractor under the Contract Documents.

The Contractor, Owner and other parties as designated in the Contract Documents shall be named as additional insureds on each of these policies except for Workers' Compensation.

11

This insurance shall include contractual liability insurance covering the Subcontractor's obligations under Article 12.

13.2 MINIMUM LIMITS OF LIABILITY. The Subcontractor's Comprehensive or Commercial General Liability Insurance and Comprehensive Automobile Liability Insurance, as required by Paragraph 13.1, shall be written with limits of liability not less than the following:

A. Comprehensive General Liability Insurance including completed operations

1. Combined Single Limit

   Bodily Injury and
   Property Damage $ _____
                           Each Occurrence

                           $ _____
                           Aggregate

            or

2. Bodily Injury $ _____
                           Each Occurrence

                           $ _____
                           Aggregate

3. Property Damage $ _____
                           Each Occurrence

                           $ _____
                           Aggregate

B. Commercial General Liability Insurance

1. Each Occurrence
   Limit $ _____

2. General Aggregate $ _____

3. Products/Completed
   Operations Aggregate $ _____

4. Personal and Adver-
   tising Injury Limit $ _____

C. Comprehensive Automobile Liability Insurance

1. Combined Single Limit

   Bodily Injury and
   Property Damage $ _____
                           Each Occurrence

            or

2. Bodily Injury $ _____
                           Each Person

                           $ _____
                           Each Occurrence

3. Property Damage $ _____
                           Each Occurrence

13.3 NUMBER OF POLICIES. Comprehensive or Commercial General Liability Insurance and other liability insurance may be arranged under a single policy for the full limits required or by a combination of underlying policies with the balance provided by an Excess or Umbrella Liability Policy.

13.4 CANCELLATION, RENEWAL OR MODIFICATION. The Subcontractor shall maintain in effect all insurance coverage required under this Agreement at the Subcontractor's sole expense and with insurance companies acceptable to the Contractor.

All insurance policies shall contain a provision that the coverages afforded thereunder shall not be cancelled or not renewed, nor restrictive modifications added, until at least thirty (30) days prior written notice has been given to the Contractor unless otherwise specifically required in the Contract Documents.

Certificates of Insurance, or certified copies of policies acceptable to the Contractor, shall be filed with the Contractor prior to the commencement of the Subcontractor's Work.

In the event the Subcontractor fails to obtain or maintain any insurance coverage required under this Agreement, the Contractor may purchase such coverage and charge the expense thereof to the Subcontractor, or terminate this Agreement.

The Subcontractor shall continue to carry completed operations liability insurance for at least two years after final payment. The Subcontractor shall furnish the Contractor evidence of such insurance at final payment and one year thereafter.

13.5 WAIVER OF RIGHTS. The Contractor and Subcontractor waive all rights against each other and the Owner, the Architect, the Architect's consultants and agents or employees of any of them, separate contractors, and all other subcontractors for loss or damage to the extent covered by Builder's Risk or any other property or equipment insurance, except such rights as they may have to the proceeds of such insurance; provided, however, that such waiver shall not extend to the acts of the Architect, the Architect's consultants, and the agents or employees of any of them listed in Paragraph 12.3.

Upon written request of the Subcontractor, the Contractor shall provide the Subcontractor with a copy of the Builder's Risk policy of insurance or any other equipment insurance in force for the Project and procured by the Contractor. The Subcontractor shall satisfy itself to the existence and extent of such insurance prior to commencement of the Subcontractor's Work.

If the Owner or Contractor have not purchased Builder's Risk insurance for the full insurable value of the Subcontractor's Work less a reasonable deductible, then the Subcontractor may procure such insurance as will protect the interests of the Subcontractor, its subcontractors and subcontractors in the Work, and, by appropriate Subcontractor Change Order, the cost of such additional insurance shall be reimbursed to the Subcontractor.

If not covered under the Builder's Risk policy of insurance or any other property or equipment insurance required by the Contract Documents, the Subcontractor shall procure and maintain at the Subcontractor's own expense property and equipment insurance for portions of the Subcontractor's Work stored off the site or in transit, when such portions of the Subcontractor's Work are to be included in an application for payment under Article 5.

AGC DOCUMENT NO. 600 • SUBCONTRACT FOR BUILDING CONSTRUCTION
© 1990, The Associated General Contractors of America

13.6 ENDORSEMENT. If the policies of insurance referred to in this Article require an endorsement to provide for continued coverage where there is a waiver of subrogation, the owners of such policies will cause them to be so endorsed.

ARTICLE 14

ARBITRATION

14.1 AGREEMENT TO ARBITRATE. All claims, disputes and matters in question arising out of, or relating to, this Agreement or the breach thereof, except for claims which have been waived by the making or acceptance of final payment, and the claims described in Paragraph 14.2, shall be decided by arbitration in accordance with the Construction Industry Arbitration Rules of the American Arbitration Association then in effect unless the parties mutually agree otherwise. Notwithstanding other provisions in the Agreement, this agreement to arbitrate shall be governed by the Federal Arbitration Act.

14.2 EXCEPTIONS. The agreement to arbitrate shall not apply to any claim:
(a) of contribution or indemnity asserted by one party to this Agreement against the other party and arising out of an action brought in a state or federal court or in arbitration by a person who is under no obligation to arbitrate the subject matter of such action with either of the parties hereto or does not consent to such arbitration; or
(b) asserted by the Subcontractor against the Contractor if the Contractor asserts said claim, either in whole or part against the Owner, or asserted by the Owner against the Contractor, when the contract between the Contractor and Owner does not provide for binding arbitration, or does so provide but the two arbitration proceedings are not consolidated, or the Contractor and Owner have not subsequently agreed to arbitrate said claim. In either case the parties hereto shall notify each other either before or after demand for arbitration is made.

In any dispute arising over the application of this Paragraph 14.2, the question of arbitrability shall be decided by the appropriate court and not by arbitration.

14.3 INITIAL DISPUTE RESOLUTION. If a dispute arises out of or relates to this Agreement, or the breach thereof, the parties may endeavor to settle the dispute first through direct discussions. If the dispute cannot be settled through direct discussions, the parties may endeavor to settle the dispute by mediation under the Construction Industry Mediation Rules of the American Arbitration Association before recourse to arbitration. Mediation will be commenced within the time limits for arbitration stipulated in the Contract Documents. The time limits for any subsequent arbitration will be extended for the duration of the mediation process plus ten (10) days or as otherwise provided in the Contract Documents. Issues to be mediated are subject to the exceptions in Paragraph 14.2 for arbitration. The location of the mediation shall be the same as the location for arbitration identified in Paragraph 14.4.

14.4 NOTICE OF DEMAND. Notice of the demand for arbitration shall be filed in writing with the other party to this Agreement and with the American Arbitration Association. The demand for arbitration shall be made as required in the Contract Documents or within a reasonable time after written notice of the claim, dispute or other matter in question has been given, but in no event shall it be made when institution of legal or equitable proceedings based on such claim, dispute or other matter in question would be barred by the applicable statute of limitation, whichever shall first occur. The location of the arbitration proceedings shall be the location of the Project.

14.5 AWARD. The award rendered by the arbitrator(s) shall be final and judgment may be entered upon it in accordance with applicable law in any court having jurisdiction.

14.6 WORK CONTINUATION AND PAYMENT. Unless otherwise agreed in writing, the Subcontractor shall carry on the Work and maintain the Schedule of Work pending arbitration. If the Subcontractor is continuing to perform, the Contractor shall continue to make payments in accordance with this Agreement.

14.7 NO LIMITATION OF RIGHTS OR REMEDIES. Nothing in this Article shall limit any rights or remedies not expressly waived by the Subcontractor which the Subcontractor may have under lien laws or payment bonds.

14.8 SAME ARBITRATORS. To the extent not prohibited by their contracts with others, the claims and disputes of the Owner, Contractor, Subcontractor and other subcontractors involving a common question of fact or law shall be heard by the same arbitrator(s) in a single proceeding.

ARTICLE 15

CONTRACT INTERPRETATION

15.1 INCONSISTENCIES AND OMISSIONS. Should inconsistencies or omissions appear in the Contract Documents, it shall be the duty of the Subcontractor to so notify the Contractor in writing within three (3) working days of the Subcontractor's discovery thereof. Upon receipt of said notice, the Contractor shall instruct the Subcontractor as to the measures to be taken and the Subcontractor shall comply with the Contractor's instructions.

15.2 LAW AND EFFECT. This Agreement shall be governed by the law of the State of _____.

15.3 SEVERABILITY AND WAIVER. The partial or complete invalidity of any one or more provisions of this Agreement shall not affect the validity or continuing force and effect of any other provision. The failure of either party hereto to insist, in any one or more instances, upon the performance of any of the terms, covenants or conditions of this Agreement, or to exercise any right herein, shall not be construed as a waiver or relinquishment of such term, covenant, condition or right as respects further performance.

15.4 ATTORNEY'S FEES. Should either party employ an attorney to institute suit or demand arbitration to enforce any of the provisions hereof, to protect its interest in any matter arising under this Agreement, to collect damages for the breach of the Agreement, or to recover on a surety bond given by a party under this Agreement, the prevailing party shall be entitled to recover reasonable attorney's fees, costs, charges, and expenses expended or incurred therein.

13

15.5 TITLES. The titles given to the Articles of this Agreement are for ease of reference only and shall not be relied upon or cited for any other purpose.

15.6 ENTIRE AGREEMENT. This Agreement is solely for the benefit of the signatories hereto and represents the entire and integrated agreement between the parties hereto and supersedes all prior negotiations, representations, or agreements, either written or oral.

ARTICLE 16

SPECIAL PROVISIONS

16.1 PRECEDENCE. It is understood the work to be performed under this Agreement, including the terms and conditions thereof, is as described in Articles 1 through 16 together with the following Special Provisions, which are intended to complement same. However, in the event of any inconsistency, these Special Provisions shall govern.

16.2 SCOPE OF WORK. All work necessary or incidental to complete the _____ Work for the Project in strict accordance with and reasonably inferable from the Contract Documents and as more particularly, though not exclusively, specified in _____ _____ with the following additions or deletions:

16.3 COMMON TEMPORARY SERVICES. The following "Project" common temporary services and/or facilities are for the use of all project personnel and shall be furnished as herein below noted:

By this Subcontractor;

By others;

16.4 OTHER SPECIAL PROVISIONS. (Insert here any special provisions required by this Agreement.)

16.5 CONTRACT DOCUMENTS. (List applicable Contract Documents including specifications, drawings, addenda, modifications and exercised alternates. Identify with general description, sheet numbers and latest date including revisions.)

IN WITNESS WHEREOF, the parties hereto have executed this Agreement under seal, the day and year first above written.

_____   _____
Subcontractor   Contractor

By_____   By_____
   (Title)   (Title)

14

AGC DOCUMENT NO. 600 • SUBCONTRACT FOR BUILDING CONSTRUCTION
© 1990, The Associated General Contractors of America

APPENDIX

D

CSI MASTERFORMAT— BROADSCOPE SECTION TITLES

This appendix contains the *Masterformat* for construction classifications, as published by the Construction Specifications Institute (CSI), 601 Madison Street, Alexandria, Virginia 22314-1791, in conjunction with Construction Specifications Canada (CSC). It is applicable to organizing specifications sections, drawings, cost codes, materials information, office corespondence and other aspects of design, procurement and construction. This document has been reproduced with permission from CSI.

BIDDING REQUIREMENTS, CONTRACT FORMS, AND CONDITIONS OF THE CONTRACT

00010 PRE-BID INFORMATION
00100 INSTRUCTIONS TO BIDDERS
00200 INFORMATION AVAILABLE TO BIDDERS
00300 BID FORMS
00400 SUPPLEMENTS TO BID FORMS
00500 AGREEMENT FORMS
00600 BONDS AND CERTIFICATES
00700 GENERAL CONDITIONS
00800 SUPPLEMENTARY CONDITIONS
00900 ADDENDA

Note: The items listed above are not specification sections and are referred to as "Documents" rather than "Sections" in the Master List of Section Titles, Numbers, and Broadscope Section Explanations.

SPECIFICATIONS

DIVISION 1 – GENERAL REQUIREMENTS

01010 SUMMARY OF WORK
01020 ALLOWANCES
01025 MEASUREMENT AND PAYMENT
01030 ALTERNATES/ALTERNATIVES
01035 MODIFICATION PROCEDURES
01040 COORDINATION
01050 FIELD ENGINEERING
01060 REGULATORY REQUIREMENTS
01070 IDENTIFICATION SYSTEMS
01090 REFERENCES
01100 SPECIAL PROJECT PROCEDURES
01200 PROJECT MEETINGS
01300 SUBMITTALS
01400 QUALITY CONTROL
01500 CONSTRUCTION FACILITIES AND TEMPORARY CONTROLS
01600 MATERIAL AND EQUIPMENT
01650 FACILITY STARTUP/COMMISSIONING
01700 CONTRACT CLOSEOUT
01800 MAINTENANCE

DIVISION 2 – SITEWORK

02010 SUBSURFACE INVESTIGATION
02050 DEMOLITION
02100 SITE PREPARATION
02140 DEWATERING
02150 SHORING AND UNDERPINNING
02160 EXCAVATION SUPPORT SYSTEMS
02170 COFFERDAMS
02200 EARTHWORK
02300 TUNNELING
02350 PILES AND CAISSONS
02450 RAILROAD WORK
02480 MARINE WORK
02500 PAVING AND SURFACING
02600 UTILITY PIPING MATERIALS
02660 WATER DISTRIBUTION
02680 FUEL AND STEAM DISTRIBUTION
02700 SEWERAGE AND DRAINAGE
02760 RESTORATION OF UNDERGROUND PIPE
02770 PONDS AND RESERVOIRS
02780 POWER AND COMMUNICATIONS
02800 SITE IMPROVEMENTS
02900 LANDSCAPING

DIVISION 3 – CONCRETE

03100 CONCRETE FORMWORK
03200 CONCRETE REINFORCEMENT
03250 CONCRETE ACCESSORIES
03300 CAST-IN-PLACE CONCRETE
03370 CONCRETE CURING
03400 PRECAST CONCRETE
03500 CEMENTITIOUS DECKS AND TOPPINGS
03600 GROUT
03700 CONCRETE RESTORATION AND CLEANING
03800 MASS CONCRETE

DIVISION 4 – MASONRY

04100 MORTAR AND MASONRY GROUT
04150 MASONRY ACCESSORIES
04200 UNIT MASONRY
04400 STONE
04500 MASONRY RESTORATION AND CLEANING
04550 REFRACTORIES
04600 CORROSION RESISTANT MASONRY
04700 SIMULATED MASONRY

DIVISION 5 – METALS

05010 METAL MATERIALS
05030 METAL COATINGS
05050 METAL FASTENING
05100 STRUCTURAL METAL FRAMING
05200 METAL JOISTS
05300 METAL DECKING
05400 COLD FORMED METAL FRAMING
05500 METAL FABRICATIONS
05580 SHEET METAL FABRICATIONS
05700 ORNAMENTAL METAL
05800 EXPANSION CONTROL
05900 HYDRAULIC STRUCTURES

DIVISION 6 – WOOD AND PLASTICS

06050 FASTENERS AND ADHESIVES
06100 ROUGH CARPENTRY
06130 HEAVY TIMBER CONSTRUCTION
06150 WOOD AND METAL SYSTEMS
06170 PREFABRICATED STRUCTURAL WOOD
06200 FINISH CARPENTRY
06300 WOOD TREATMENT
06400 ARCHITECTURAL WOODWORK
06500 STRUCTURAL PLASTICS
06600 PLASTIC FABRICATIONS
06650 SOLID POLYMER FABRICATIONS

DIVISION 7 – THERMAL AND MOISTURE PROTECTION

07100 WATERPROOFING
07150 DAMPPROOFING
07180 WATER REPELLENTS
07190 VAPOR RETARDERS
07195 AIR BARRIERS
07200 INSULATION
07240 EXTERIOR INSULATION AND FINISH SYSTEMS
07250 FIREPROOFING
07270 FIRESTOPPING
07300 SHINGLES AND ROOFING TILES
07400 MANUFACTURED ROOFING AND SIDING
07480 EXTERIOR WALL ASSEMBLIES
07500 MEMBRANE ROOFING
07570 TRAFFIC COATINGS
07600 FLASHING AND SHEET METAL
07700 ROOF SPECIALTIES AND ACCESSORIES
07800 SKYLIGHTS
07900 JOINT SEALERS

DIVISION 8 – DOORS AND WINDOWS

08100 METAL DOORS AND FRAMES
08200 WOOD AND PLASTIC DOORS
08250 DOOR OPENING ASSEMBLIES
08300 SPECIAL DOORS
08400 ENTRANCES AND STOREFRONTS
08500 METAL WINDOWS
08600 WOOD AND PLASTIC WINDOWS
08650 SPECIAL WINDOWS
08700 HARDWARE
08800 GLAZING
08900 GLAZED CURTAIN WALLS

DIVISION 9 – FINISHES

09100 METAL SUPPORT SYSTEMS
09200 LATH AND PLASTER
09250 GYPSUM BOARD
09300 TILE
09400 TERRAZZO
09450 STONE FACING
09500 ACOUSTICAL TREATMENT
09540 SPECIAL WALL SURFACES
09545 SPECIAL CEILING SURFACES
09550 WOOD FLOORING
09600 STONE FLOORING
09630 UNIT MASONRY FLOORING
09650 RESILIENT FLOORING
09680 CARPET
09700 SPECIAL FLOORING
09780 FLOOR TREATMENT
09800 SPECIAL COATINGS
09900 PAINTING
09950 WALL COVERINGS

DIVISION 10 – SPECIALTIES

10100 VISUAL DISPLAY BOARDS
10150 COMPARTMENTS AND CUBICLES
10200 LOUVERS AND VENTS
10240 GRILLES AND SCREENS
10250 SERVICE WALL SYSTEMS
10260 WALL AND CORNER GUARDS
10270 ACCESS FLOORING
10290 PEST CONTROL
10300 FIREPLACES AND STOVES
10340 MANUFACTURED EXTERIOR SPECIALTIES
10350 FLAGPOLES
10400 IDENTIFYING DEVICES
10450 PEDESTRIAN CONTROL DEVICES
10500 LOCKERS
10520 FIRE PROTECTION SPECIALTIES
10530 PROTECTIVE COVERS
10550 POSTAL SPECIALTIES
10600 PARTITIONS
10650 OPERABLE PARTITIONS
10670 STORAGE SHELVING
10700 EXTERIOR PROTECTION DEVICES FOR OPENINGS
10750 TELEPHONE SPECIALTIES
10800 TOILET AND BATH ACCESSORIES
10880 SCALES
10900 WARDROBE AND CLOSET SPECIALTIES

DIVISION 11 – EQUIPMENT

11010 MAINTENANCE EQUIPMENT
11020 SECURITY AND VAULT EQUIPMENT
11030 TELLER AND SERVICE EQUIPMENT
11040 ECCLESIASTICAL EQUIPMENT
11050 LIBRARY EQUIPMENT
11060 THEATER AND STAGE EQUIPMENT
11070 INSTRUMENTAL EQUIPMENT
11080 REGISTRATION EQUIPMENT
11090 CHECKROOM EQUIPMENT
11100 MERCANTILE EQUIPMENT
11110 COMMERCIAL LAUNDRY AND DRY CLEANING EQUIPMENT
11120 VENDING EQUIPMENT
11130 AUDIO-VISUAL EQUIPMENT
11140 VEHICLE SERVICE EQUIPMENT
11150 PARKING CONTROL EQUIPMENT
11160 LOADING DOCK EQUIPMENT
11170 SOLID WASTE HANDLING EQUIPMENT
11190 DETENTION EQUIPMENT
11200 WATER SUPPLY AND TREATMENT EQUIPMENT
11280 HYDRAULIC GATES AND VALVES
11300 FLUID WASTE TREATMENT AND DISPOSAL EQUIPMENT
11400 FOOD SERVICE EQUIPMENT
11450 RESIDENTIAL EQUIPMENT
11460 UNIT KITCHENS
11470 DARKROOM EQUIPMENT
11480 ATHLETIC, RECREATIONAL, AND THERAPEUTIC EQUIPMENT
11500 INDUSTRIAL AND PROCESS EQUIPMENT
11600 LABORATORY EQUIPMENT
11650 PLANETARIUM EQUIPMENT
11660 OBSERVATORY EQUIPMENT
11680 OFFICE EQUIPMENT
11700 MEDICAL EQUIPMENT
11780 MORTUARY EQUIPMENT
11850 NAVIGATION EQUIPMENT
11870 AGRICULTURAL EQUIPMENT

DIVISION 12 – FURNISHINGS

12050 FABRICS
12100 ARTWORK
12300 MANUFACTURED CASEWORK
12500 WINDOW TREATMENT
12600 FURNITURE AND ACCESSORIES
12670 RUGS AND MATS
12700 MULTIPLE SEATING
12800 INTERIOR PLANTS AND PLANTERS

DIVISION 13 -- SPECIAL CONSTRUCTION

13010 AIR SUPPORTED STRUCTURES
13020 INTEGRATED ASSEMBLIES
13030 SPECIAL PURPOSE ROOMS
13080 SOUND, VIBRATION, AND SEISMIC CONTROL
13090 RADIATION PROTECTION
13100 NUCLEAR REACTORS
13120 PRE-ENGINEERED STRUCTURES
13150 AQUATIC FACILITIES
13175 ICE RINKS
13180 SITE CONSTRUCTED INCINERATORS
13185 KENNELS AND ANIMAL SHELTERS
13200 LIQUID AND GAS STORAGE TANKS
13220 FILTER UNDERDRAINS AND MEDIA
13230 DIGESTER COVERS AND APPURTENANCES
13240 OXYGENATION SYSTEMS
13260 SLUDGE CONDITIONING SYSTEMS
13300 UTILITY CONTROL SYSTEMS
13400 INDUSTRIAL AND PROCESS CONTROL SYSTEMS
13500 RECORDING INSTRUMENTATION
13550 TRANSPORTATION CONTROL INSTRUMENTATION
13600 SOLAR ENERGY SYSTEMS
13700 WIND ENERGY SYSTEMS
13750 COGENERATION SYSTEMS
13800 BUILDING AUTOMATION SYSTEMS
13900 FIRE SUPPRESSION AND SUPERVISORY SYSTEMS
13950 SPECIAL SECURITY CONSTRUCTION

DIVISION 14 -- CONVEYING SYSTEMS

14100 DUMBWAITERS
14200 ELEVATORS
14300 ESCALATORS AND MOVING WALKS
14400 LIFTS
14500 MATERIAL HANDLING SYSTEMS
14600 HOISTS AND CRANES
14700 TURNTABLES
14800 SCAFFOLDING
14900 TRANSPORTATION SYSTEMS

DIVISION 15 --MECHANICAL

15050 BASIC MECHANICAL MATERIALS AND METHODS
15250 MECHANICAL INSULATION
15300 FIRE PROTECTION
15400 PLUMBING
15500 HEATING, VENTILATING, AND AIR CONDITIONING
15550 HEAT GENERATION
15650 REFRIGERATION
15750 HEAT TRANSFER
15850 AIR HANDLING
15880 AIR DISTRIBUTION
15950 CONTROLS
15990 TESTING, ADJUSTING, AND BALANCING

DIVISION 16 -- ELECTRICAL

16050 BASIC ELECTRICAL MATERIALS AND METHODS
16200 POWER GENERATION - BUILT-UP SYSTEMS
16300 MEDIUM VOLTAGE DISTRIBUTION
16400 SERVICE AND DISTRIBUTION
16500 LIGHTING
16600 SPECIAL SYSTEMS
16700 COMMUNICATIONS
16850 ELECTRIC RESISTANCE HEATING
16900 CONTROLS
16950 TESTING

BIBLIOGRAPHY

The references in this bibliography have been grouped into general categories corresponding to some of the main topics in this book. The list is by no means exhaustive, but rather, is intended to provide supplementary reading for those who wish to study particular subjects in greater detail. Many of the references have also been chosen because they, in turn, have good bibliographies of their own.

General Construction Administration and Management

Barrie, Donald S., ed., *Directions in Managing Construction*, Wiley, New York, 1981.

Bonny, John B., and Joseph P. Frein, *Handbook of Construction Management and Organization*, 2d ed., Van Nostrand Reinhold Co., New York, 1980.

Building for Tomorrow: Global Enterprise and the U.S. Construction Industry, National Academy Press, Washington, D.C., 1988.

Business Roundtable, *More Construction for the Money*, 2 Park Avenue, New York, N.Y., January, 1983.

Clough, Richard H., *Construction Contracting*, 5th ed., Wiley, New York, 1986.

Del Re, Robert, "The Resident Engineer: Intermediary Between Owner and Contractor," *Journal of The Construction Division*, ASCE, vol. 108, no. CO3, September 1982, pp. 375–378.

Fisk, Edward R., *Construction Project Administration*, Wiley, New York, 1978.

Gerwick, Ben C., Jr., and John C. Wollery, *Construction and Engineering Marketing for Major Project Services*, Wiley, New York, 1983.

Halpin, Daniel W., and Woodhead, Ronald W., *Construction Management*, Wiley, New York, 1980.

Havers, John A., and Frank W. Stubbs, Jr., *Handbook of Heavy Construction*, 2d ed., McGraw-Hill, Inc., New York, 1971.

Hendrickson, Chris, and Tung Au, *Project Management for Construction*, Prentice-Hall, Englewood Cliffs, N.J., 1989.

Manual of The Associated General Contractors of America, Washington, D.C. (Contains a compilation of forms, contracts, procedures, etc.)

Nunnally, S. W., *Construction Methods and Management*, 2d. ed., Prentice-Hall, Englewood Cliffs, N.J., 1987.

Paulson, Boyd C., Jr.,"Education and Research in Construction," *Journal of the Construction Division*, ASCE, vol. 102, no. CO3, September 1976, pp. 479–495.

Reiner, Lawrence E., *Handbook of Construction Management*, Prentice-Hall, Englewood Cliffs, N.J., 1972.

Rossow, Janet A. K., and Fred Moavenzadeh, "Management Issues in the U.S. Construction Industry," *Journal of the Construction Division*, ASCE, vol. 102, no. C02, June 1976, pp. 277–294.

Tatum, Clyde B., "Organizing Large Projects: How Managers Decide," *Journal of Construction Engineering and Management*, ASCE, vol. 110, no. 3, September 1984, pp. 346–358.

——"Designing Project Organizations: An Expanded Process," *Journal of Construction Engineering and Management*, ASCE, vol. 112, no. 2, June 1986, pp. 259–272.

——and R. P. Fawcett, "Organizational Alternatives for Large Projects," *Journal of Construction Engineering and Management*, ASCE, vol. 112, no. 1, March 1986, pp. 49–61.

Wass, Alonzo, *Construction Management and Contracting*, Prentice-Hall, Englewood Cliffs, N.J., 1972.

Construction Management

Barrie, Donald S., "CM as Seen by an Engineer-Contractor," *Plant Engineering*, July 13, 1972, p. 85.

——"Guidelines for Successful Construction Management," *Journal of the Construction Division*, ASCE, vol. 106, no. CO3, September, 1980, pp. 237–245.

——and Boyd C. Paulson, Jr., "Professional Construction Management," *Journal of the Construction Division*, ASCE, vol. 102, no. CO3, September 1976, pp. 425–436.

Bush, Vincent G., *Construction Management*, Reston Publishing Co., Reston, Va., 1973.

CM for the General Contractor: A Guide Manual for Construction Management, The Associated General Contractors of America, Washington, D.C., 1975.

Construction Contracting Systems: A Report on the Systems Used by PBS and Other Organizations, General Services Administration, Public Buildings Service, Washington, D.C., March 1970.

"Construction Management Responsibilities During Design," Committee on Construction Management, *Journal of Construction Engineering and Management*, ASCE, vol. 113, no. 1, March 1987, pp. 90–98.

Dressler, Joachim, "Construction Management in West Germany," *Journal of the Construction Division*, ASCE, vol. 106, no. CO4, December, 1980, pp. 477–487.

Foxall, William B., *Professional Construction Management and Project Administration*, Architectural Record and The American Institute of Architects, New York, 1972.

The GSA System for Construction Management, General Services Administration, Public Buildings Service, Washington, D.C., April 1975.

Goldhaber, Stanley, Chandra K. Jha, and Manuel C. Macedu, Jr., *Construction Management Principles and Practices*, Wiley, New York, 1977.

Heery, George T., *Time, Cost and Architecture*, McGraw-Hill, Inc., New York, 1975.

Jordan, Mark H., and Robert I. Carr, "Education for the Professional Construction Manager," *Journal of the Construction Division*, ASCE, vol. 102, no. CO3, September 1976, pp. 511–519.

Kettle, Kenath A., "Project Delivery Systems for Construction Projects," *Journal of the Construction Division*, ASCE, vol. 102, no. CO4, December 1976, pp. 575–585.

——— "Proposed Construction Management Specification," *Journal of the Construction Division*, ASCE, vol. 105, no. CO4, December, 1979, pp. 367–380.

Murray, L. William, et al., "Marketing Construction Management Services," *Journal of the Construction Division*, ASCE, vol. 107, no. CO4, December, 1981, pp. 665–677.

"Professional Construction Management Services," Subcommittee Report, *Journal of the Construction Division*, ASCE, vol. 105, no. CO2, June, 1979, pp. 139–156.

"Qualification and Selection of Construction Managers with Suggested Guidelines for Selection Process," Committee on Construction Management, *Journal of Construction Engineering and Management*, ASCE, vol. 113, no. 1, March 1987, pp. 51–89.

"Standard Form of Agreement Between Owner and Construction Manager," Document B-801, The American Institute of Architects, Washington, D.C., 1980.

"Standard Form of Agreement Between Owner and Construction Manager," Document No. 8, The Associated General Contractors of America, Washington, D.C., 1980.

"Study Committee Report on Construction Management," Consulting Engineers Council, Washington, D.C., January 1972.

Tatum, Clyde B., "Evaluating PCM Firm Potential and Performance," *Journal of the Construction Division*, ASCE, vol. 105, no. CO3, September, 1979, pp. 239–251.

Project Planning and Control, and Related Methodologies

General Background

Clough, Richard H., and Glenn A. Sears. *Construction Project Management*, 2d ed. Wiley, New York, 1979.

Collier, C. A., and D. A. Halperin, *Construction Funding: Where the Money Comes From*, 2d. ed., Wiley, New York, 1984.

Coxe, Weld, *Marketing Architectural and Engineering Services*, Van Nostrand Reinhold, New York, 1971.

Dabbas, Majed A. A., and Daniel W. Halpin, "Integrated Project and Process Management," *Journal of The Construction Division*, ASCE, vol. 108, no. CO3, September 1982, pp. 361–374.

Douglas, James, *Construction Equipment Policy*, McGraw-Hill, Inc., New York, 1975.

Grant, Eugene, L.W. Grant Ireson, and Richard S. Leavenworth, *Principles of Engineering Economy*, 7th ed., The Ronal Press Company, New York, 1982.

Halpin, Daniel W., *Financial and Cost Concepts for Construction Management*, Wiley, New York, 1985.

——— and R. W. Woodhead, *Design of Construction and Process Operations*, Wiley, New York, 1976.

Hillier, Frederick S., and Gerald J. Lieberman, *Introduction to Operations Research*, 2d ed., Holden-Day, San Francisco, 1974.

Hollander, G. L., "Integrated Project Control," Part I, *Project Management Quarterly*, vol. 4, no. 1, April 1973, pp. 6–13; Part II, vol. 4, no. 2, June 1973, pp. 6–14.

Jones, G. L., *How to Market Professional Design Services*, McGraw-Hill, Inc., New York, 1973.

Kalk, Anthony, *INSIGHT–Interactive Simulation of Construction Operations Using Graphical Techniques*, Technical Report 238, The Construction Institute, Dept. of Civil Engineering, Stanford University, Stanford, Calif., July, 1980.

MASTERFORMAT–Master List of Section Titles and Numbers, The Construction Specifications Institute, Alexandria, Va., 1983.

Neil, James, *Project Control for Construction*, Publication No. 6-4, Construction Industry Institute, Austin, Tex., September 1987.

———— *Work Packaging for Project Control*, Publication No. 6-5, Construction Industry Institute, November 1988.

Oglesby, Clarkson H, Henry W. Parker, and Gregory Howell, *Productivity Improvement in Construction*, McGraw-Hill, Inc., New York, 1989.

Paulson, Boyd C., Jr., "Concepts of Project Planning and Control," *Journal of the Construction Division*, ASCE, vol. 102, no. C01, March 1976, pp. 67–80.

———— "Designing to Reduce Construction Costs," *Journal of the Construction Division*, ASCE, vol. 102, no. C04, December 1976, pp. 587–592.

Proceedings of the Annual Seminar/Symposiums, The Project Management Institute, Drexel Hill, Pa. Published annually since 1969.

Project Management Journal, The Project Management Institute, Drexel Hill, Pa. Published since 1969.

Quality in the Constructed Project, American Society of Civil Engineers, New York, 1989.

Rasdorf, William J., and Mark J. Herbert, "Bar Coding in Construction Engineering," *Journal of Construction Engineering and Management*, ASCE, vol. 116, no. 2, June 1990, pp. 261–280.

Russell, Alan D., and Emmanuel Triassi, "General Contractor Project Control Practices and MIS," *Journal of The Construction Division*, ASCE, vol. 108, no. C03, September 1982, pp. 419–437.

Sanvido, Victor, "Conceptual Construction Process Model," *Journal of Construction Engineering and Management*, ASCE, vol. 114, no. 2, June 1988, pp. 294–312.

———— and Deborah J. Medeiros, "Applying Computer-Integrated Manufacturing Concepts to Construction," *Journal of Construction Engineering and Management*, ASCE, vol. 116, no. 2, June 1990.

Schrader, Charles R., "Motivation of Construction Craftsman," *Journal of the Construction Division*, ASCE, vol. 98, no. C02, September 1972, pp. 257–273.

———— "Boosting Construction Worker Productivity," *Civil Engineering*, vol. 42, no. 10, October 1972, pp. 61–63.

Wagner, Harvey M., *Principles of Operations Research*, 2d ed., Prentice-Hall, Englewood Cliffs, N.J., 1975.

Wilson, A., *The Marketing of Professional Services*, McGraw-Hill, Inc., New York, 1972.

Wilson, Woodrow W., "Model Form of 'Instructions to Bidders'," *Journal of the Construction Division*, ASCE, vol. 100, no. C01, March 1974, pp. 27–31.

———— "Model Form of 'Notice to Bidders'," *Journal of the Construction Division*, ASCE, vol. 100, no. C03, September 1974, pp. 373–375.

Estimating

Ahuja, H. N., and W. J. Campbell, *Estimating: From Concept to Completion*, Prentice-Hall, Englewood Cliffs, N.J., 1987.

Bacarreza, Ricardo, *The Construction Project Markup Decision Under Conditions of Uncertainty*, Technical Report 176, The Construction Institute, Department of Civil Engineering, Stanford University, Stanford, Calif., June, 1973.

Behrens, H. J., "The Learning Curve," in F. C. Jelen (ed.), *Cost and Optimization Engineering*, Chap. 9, McGraw-Hill, Inc., New York, 1970, pp. 170–184.

Building Construction Cost Data, Robert Snow Means Co., Duxbury, Mass. Published annually.

Building Estimator's Reference Book, The Frank R. Walker Co., Chicago. Updated periodically.

Construction Users Anti-Inflation Roundtable, "Effect of Scheduled Overtime on Construction Projects," *AACE Bulletin*, vol. 15, no. 5, October 1973, pp. 155–160.

Contractors Equipment Manual, 7th ed., The Associated General Contractors of America, Washington, D.C., 1974.

Cooper, George, and Stanley Badzinski, Jr., *Building Construction Estimating*, 3d ed., McGraw-Hill, Inc., New York, 1971.

Deatherage, George E., *Construction Estimating and Job Preplanning*, McGraw-Hill, Inc., New York, 1965.

Diekmann, James E., "Probabilistic Estimating: Mathematics and Applications," *Journal of Construction Engineering and Management*, ASCE, vol. 109, no. 3, September 1983, pp. 297–308.

Dodge Manual for Construction Pricing and Scheduling, McGraw-Hill Information Systems Co., New York. Published annually.

Erikson, Carl A., and Leroy T. Boyer, "Estimating-State-of-the-Art," *Journal of the Construction Division*, ASCE, vol. 102, no. C03, September 1976, pp. 455–464.

Fondahl, John W., and Ricardo R. Bacarreza, *Construction Contract Markup Related to Forecasted Cash Flow*, Technical Report No. 161, Stanford University, Dept. of Civil Engineering, The Construction Institute, Stanford, Calif., November 1972.

Foster, Norman I., *Construction Estimates from Take-Off to Bid*, 2d ed., McGraw-Hill, Inc., New York, 1972.

Guthrie, Kenneth M., *Process Plant Estimating and Control*, Craftsman Book Company of America, Solana Beach, Calif., 1974.

Lichtenberg, Steen, "The Successive Principle–Procedures for a Minimum Degree of Detailing," *Proceedings of the Sixth Annual Seminar/Symposium of the Project Management Institute*, Washington, D.C., September 1974, pp. 570–578.

Lowell, E. D., "Estimating Building Construction Costs," in Frederick S. Merritt (ed.), *Building Construction Handbook*, 3d ed., section 25, McGraw-Hill, Inc., New York, 1975.

McGlaun, Weldon, "Overtime in Construction," *AACE Bulletin*, vol. 15, no. 5, October 1973, pp. 141–143.

National Construction Estimator, G. Moselle (ed.), Craftsman Book Company, Los Angeles, 1975.

Neil, James M., *Construction Cost Estimating for Project Control*, Prentice-Hall, Englewood Cliffs, N.J., 1982.

Parker, Albert D., Donald S. Barrie, and Robert N. Snyder, *Planning and Estimating Heavy Construction*, McGraw-Hill, Inc., New York, 1984.

Paulson, Boyd C., Jr., "Estimating and Control of Construction Labor Costs," *Journal of the Construction Division*, ASCE, vol. 101, no. C03, September 1975, pp. 623–633.

Peurifoy, Robert L., *Estimating Construction Costs*, 3d ed., McGraw-Hill, Inc., New York, 1975.

Ringwald, Richard C., "General Overhead Distribution to Project Costs," *Journal of Construction Engineering and Management*, ASCE, vol. 112, no. 1, March 1986, pp. 83–89.

Saylor, Lee, *Current Construction Costs*, Lee Saylor, Walnut Creek, Calif. Published annually.

Zimmerman, O. T., "Capital Investment Cost Estimating," in F. C. Jelen (ed.), *Cost and Optimization Engineering*, Chap. 15, McGraw-Hill, Inc., New York, 1970, pp. 301–337.

Planning and Scheduling

Antill, James M., "Critical Path Evaluations of Construction Work Changes and Delays," *Australia Institution of Engineers, Civil Engineering Transactions*, vol. 77, no. 1, April 1969, pp. 31–39.

────── and Ronald W. Woodhead, *Critical Path Methods in Construction Practice*, 2d ed., Wiley, New York, 1970.

Armstrong-Wright, A. T., *Critical Path Method*, Longman Group Ltd., London, 1969.

Battersby, A., *Network Analysis*, 3d. ed., The Macmillan Co., New York, 1970.

Burman, Peter J., *Precedence Networks for Project Planning and Control*, McGraw-Hill, Inc., London, 1972.

Carr, Robert I., and Walter L. Meyer, "Planning Construction of Repetitive Building Units," *Journal of the Construction Division*, ASCE, vol. 100, no. C03, September 1974, pp. 403–412.

Dressler, Joachim A., "Stochastic Scheduling of Linear Construction Sites," *Journal of the Construction Division*, ASCE, vol. 100, no. C04, December 1974, pp. 571–587.

────── *Development of an Interactive Computer Program for Resource Allocation*, Technical Report 189, The Construction Institute, Department of Civil Engineering, Stanford University, Stanford, Calif., December, 1974.

Fondahl, John W., *A Non-Computer Approach to the Critical Path Method for the Construction Industry*, Technical Report No. 9, Stanford University, Dept. of Civil Engineering, The Construction Institute, Stanford, Calif., November 1961.

────── *Methods for Extending the Range of Non-Computer Critical Path Applications*, Technical Report No. 47, Stanford University, Dept. of Civil Engineering, The Construction Institute, Stanford, Calif., 1964.

────── *Some Problem Areas in Current Network Planning Practices and Related Comments on Legal Applications*, Technical Report 193, Stanford University, Dept. of Civil Engineering. The Construction Institute, Stanford, Calif., April 1975.

Harris, Robert B., *Precedence and Arrow Networking Techniques for Construction*, Wiley, New York, 1978.

Moder, Joseph J., Cecil R. Phillips, and Edward Davis, *Project Management with CPM, PERT, and Precedence Diagramming*, 3d ed., Van Nostrand Reinhold Co., New York, 1983.

Naaman, Antoine E., "Networking Methods for Project Planning and Control," *Journal of the Construction Division*, ASCE, vol. 100, no. C03, September 1974, pp. 357–372.

Neil, James, *The Impact of Changes on Construction Cost and Schedule*, Publication No. 6-10, Construction Industry Institute, April 1990.

O'Brien, James J. (ed.), *Scheduling Handbook*, McGraw-Hill, Inc., New York, 1969.

────── *CPM in Construction Management*, 2d ed., McGraw-Hill, Inc., New York, 1971.

────── "VPM Scheduling for High-Rise Buildings," *Journal of the Construction Division*, ASCE, vol. 101, no. C04, December 1975, pp. 895–905.

Paulson, Boyd C., Jr., "Man-Computer Concepts for Planning and Scheduling," *Journal of the Construction Division*, ASCE, vol. 98, no. C02, September 1972, pp. 275–286.

Peer, Shlomo, "Network Analysis and Construction Planning," *Journal of the Construction Division*, ASCE, vol. 100, no. C03, September 1974, pp. 203–210.

Priluck, H. M., and P. R. Hourihan, *Practical CPM for Construction*, Robert S. Means Co., Duxbury, Mass., 1968.

Sears, Glenn A., *A CPM-Based Cost Control System*, Technical Report No. 199, Stanford University, Dept. of Civil Engineering, The Construction Institute, Stanford, Calif., August 1975.

Wiest, Jerome D., and Ferdinand K. Levy, *A Management Guide to PERT/CPM*, Prentice-Hall, Englewood Cliffs, N.J., 1969.

Willis, E. M., *Scheduling Construction Projects*, Wiley, New York, 1986.

Cost Engineering

AACE Bulletin, American Association of Cost Engineers, Morgantown, W. Va. Published bimonthly.

Ahuja, Hira N., *Successful Construction Cost Control*, Wiley, New York, 1980.

Au, Tung, and Chris Hendrickson, "Profit Measures for Construction Projects," *Journal of Construction Engineering and Management*, ASCE, vol. 112, no. 2, June 1986, pp. 273–286.

Construction Cost Control, ASCE Manuals and Reports of Engineering Practice No. 65, rev. ed., American Society of Civil Engineers, New York, 1985.

Coombs, W. E., and W. J. Palmer, *Construction Accounting and Financial Management*, McGraw-Hill, Inc., New York, 1977.

Horngren, Charles T., *Cost Accounting: A Managerial Emphasis*, 3d ed., Prentice-Hall, Englewood Cliffs, N.J., 1972.

Humphrey, Kenneth K. (ed.), *Project and Cost Engineers' Handbook* (sponsored by the American Association of Cost Engineers), 2d ed., Marcel Decker, New York, 1984.

Jelen, F. C. and James H. Black, *Cost and Optimization Engineering*, 2d ed., McGraw-Hill, Inc., New York, 1983.

Kharbanda, O. P., E. A. Stallworthy, and L. F. Williams, Revised by James T. Stoms, *Project Cost Control in Action*, Prentice-Hall, Englewood Cliffs, N.J., 1981.

Mueller, F. W., *Integrated Cost and Schedule Control for Construction Projects*, Van Nostrand Reinhold Co., New York, 1986.

Park, William R., *Cost Engineering Analysis*, Wiley, New York, 1973.

Teicholz, Paul, "Requirements of a Construction Company Cost System," *Journal of the Construction Division*, ASCE, vol. 100, no. C03, September 1974, pp. 255–263.

——"Labor Cost Control," *Journal of the Construction Division*, ASCE, vol. 100, no. C04, December 1974, pp. 561–570.

Transactions, American Association of Cost Engineers, Morgantown, W. Va. Published annually since 1956.

Procurement

Ali, A. M., "Inventory Problems," in F. C. Jelen (ed.), *Cost and Optimization Engineering*, Chap. 10, McGraw-Hill, Inc., New York, 1970, pp. 185–206.

Bell, Lansford C., and George Stukhart, "Attributes of Materials Management Systems," *Journal of Construction Engineering and Management*, ASCE, vol. 112, no. 1, March 1986, pp. 14–21.

——"Costs and Benefits of Materials Management Systems," *Journal of Construction Engineering and Management*, ASCE, vol. 113, no. 2, June 1987, pp. 222–234.

Fabrycky, W. J., and J. Banks, *Procurement and Inventory Systems*, Reinhold Publishing Corporation, New York, 1967.

Ibbs, C. William, Jr., "Product Specifications Practices and Problems," *Journal of Construction Engineering and Management*, ASCE, vol. 111, no. 2, June 1985.

——" 'Brand Name or Equal' Product Specifications," *Journal of Construction Engineering and Management*, ASCE, vol. 112, no. 1, March 1986, pp. 1–13.

Kumar, A., and H. Leng, "A Material Control System for Large Construction Projects," in *Proceedings of the Fifth International Seminar/Symposium of the Project Management Institute*, Toronto, Canada, October 1973, pp. 601–623.

Lee, Lamar, Jr., and Donald W. Dobler, *Purchasing and Materials Management*, 2d ed., McGraw-Hill, Inc., New York, 1971.

Project Materials Management Primer, Publication No. 7-2, Construction Industry Institute, November 1988.

Value Engineering

Barrie, Donald S., and Gordon L. Mulch, "The Professional Construction Management Team Discovers Value Engineering," *Journal of the Construction Division*, ASCE, vol. 103, no. C02, Sept., 1977.

Dell'Isola, Alphonse J., "A Value Engineering Case Study," *Heating, Piping and Air Conditioning*, June 1970, pp. 50–54.

———— *Value Engineering in the Construction Industry*, Construction Publishing Company, New York, 1974. (now Van Nostrand Reinhold Company.)

DOD Handbook (Value Engineering), 5010.8-H, U.S. Government Printing Office, Superintendent of Documents, Washington, D.C., Sept. 12, 1968.

Miles, L. D., *Techniques of Value Analysis and Engineering*, 2d ed., McGraw-Hill, Inc., New York, 1961.

O'Brien, James J., *Value Analysis in Design and Construction*, McGraw-Hill, Inc., New York, 1976.

O'Connor, James T., Stephen E. Rusch, and Martin J. Schulz, "Constructability Concepts for Engineering and Procurement," *Journal of Construction Engineering and Management*, ASCE, vol. 113, no. 2, June 1987, pp. 235–248.

Value Engineering (Handbook), PBS P 8000.1 (Jan. 12, 1972) and Change 0.1 (March 2, 1973), U.S. General Services Administration, Washington, D.C.

Value Engineering in Federal Construction Agencies, Symposium-Workshop Report No. 4, National Academy of Sciences, Federal Construction Council, Building Research Advisory Board, Washington, D.C., May 1969.

Quality Assurance

American Society of Quality Control (ASQC) publications: *Annual Technical Conference Transactions* (theory and applications); *Journal of Quality Technology* (theory and methodology); *Quality Progress* (applications and trade articles).

Burati, James, *Cost of Quality Deviations in Design and Construction*, Publication No. 10-1, Construction Industry Institute, February 1989.

Cohen, Norman J., "Statistical Theory in Materials Sampling," *Journal of the Construction Division*, ASCE, vol. 97, no. C01, March 1971, pp. 95–111.

Goldbloom, Joseph, "Recommended Standards for the Responsibility, Authority, and Behavior of the Inspector," *Journal of the Construction Division*, ASCE, vol. 101, no. C02, June 1975, pp. 359–364.

Grant, Eugene L., and Richard S. Leavenworth, *Statistical Quality Control*, 4th ed., McGraw-Hill, Inc., New York, 1972.

Hester, Weston T., "Alternative Construction Quality Assurance Programs," *Journal of the Construction Division*, ASCE, vol. 105, no. C03, September, 1979, pp. 187–199.

Juran, Joseph M., Frank M. Gryna, and Richard S. Burgham (eds.), *Quality Control Handbook*, 3d ed., McGraw-Hill, Inc., New York, 1975.

Kirkpatrick, Elwood G., *Quality Control for Managers and Engineers*, Wiley, New York, 1970.

Knowler, Lloyd K., and others, *Quality Control by Statistical Methods*, McGraw-Hill, Inc., New York, 1969.

Ledbetter, Bill, *The Quality Performance Management System*, Publication No. 10-3, Construction Industry Institute, February 1990.

O'Brien, James J., *Construction Inspection Handbook*, Van Nostrand Reinhold Company, New York, 1974.

Parsons, Roland M., "System for Control of Construction Quality," *Journal of the Construction Division*, ASCE, vol. 98, no. C01, March 1972, pp. 21–36.

Quality in the Constructed Project: A Guide for Owners, Designers and Constructors, Volume 1, Manual No. 73, American Society of Civil Engineers, New York, 1990.

Samson, Charles, Philip Hart, and Charles Rubin, *Fundamentals of Statistical Quality Control*, Addison-Wesley Publishing Company, Reading, Mass., 1970.

Simmons, David A., *Practical Quality Control*, Addison-Wesley Publishing Company, Reading, Mass., 1970.

Willenbrock, Jack H., and Scott Shepard, "Construction QA/QC Systems: Comparative Analysis," *Journal of the Construction Division*, ASCE, vol. 106, no. CO3, September, 1980, pp. 371–387.

Safety and Health

Business Roundtable, *Improving Construction Safety Performance*, Construction Industry Cost Effectiveness Project Report A-3, New York, January, 1982.

California Construction Safety Orders, Dept. of Industrial Relations, Division of Industrial Safety, San Francisco.

Construction Industry: OSHA Safety & Health Standards Digest, OSHA 2202, U.S. Government Printing Office, Superintendent of Documents, Washington, D.C. Revised June 1975.

Construction Safety and Health Regulations: Part 1926, OSHA 2207, U.S. Government Printing Office, Superintendent of Documents, Washington, D.C., June 1974.

Construction Safety and Health Training, General Services Administration, National Audiovisual Center, Washington, D.C. (Manuals and slides for 30-hour course.)

deStwolinski, Lance W., *Occupational Health in the Construction Industry*, Technical Report No. 105, Stanford University, Dept. of Civil Engineering, The Construction Institute, Stanford, Calif., May 1969.

——*A Survey of the Safety Environment of the Construction Industry*, Technical Report No. 114, Stanford University, Dept. of Civil Engineering, The Construction Institute, Stanford, Calif., October 1969.

Gans, George M., Jr., "The Construction Manager and Safety," *Journal of the Construction Division*, ASCE, vol. 107, no. CO2, June, 1981, pp. 219–226.

General Safety Requirements, Manual EM 385-1-1, and Supplements 1 and 2, U.S. Army Corps of Engineers, Washington, D.C.

Hinze, Jimmie, *The Effect of Middle Management on Safety in Construction*, Technical Report No. 209, Stanford University, Dept. of Civil Engineering, The Construction Institute, Stanford, Calif., June 1976.

——"Human Aspects of Construction Safety," *Journal of the Construction Division*, ASCE, vol. 107, no. CO1, March, 1981, pp. 61–72.

——and Lori A. Figone, *Subcontractor Safety as Influenced by General Contractors on Small and Medium Sized Projects*, Publication No. 38, Construction Industry Institute, October 1988.

Knox, H., "Construction Safety as it Relates to Insurance Costs," *AACE Bulletin*, vol. 16, no. 3, June 1974, pp. 71–73.

Koehn, Enno, and Kurt Musser, "OSHA Regulations Effects on Construction," *Journal of Construction Engineering and Management*, ASCE, vol. 109, no. 2, June 1983, pp. 233–244.

Laufer, Alexander, and William B. Ledbetter, "Assessment of Safety Performance Measures at Construction Sites," *Journal of Construction Engineering and Management*, ASCE, vol. 112, no. 4, December 1986, pp. 530–542.

Levitt, Raymond E., *The Effect of Top Management on Safety in Construction*, Technical Report No. 196, Stanford University, Dept. of Civil Engineering, The Construction Institute, Stanford, Calif., July 1975.

————— and Henry W. Parker, "Reducing Construction Accidents–Top Management's Role," *Journal of the Construction Division*, ASCE, vol. 102, no. C03, September 1976, pp. 465–478.

————— Henry W. Parker, and Nancy M. Samelson, *Improving Construction Safety Performance: The User's Role*, Technical Report 260, The Construction Institute, Department of Civil Engineering, Stanford University, Stanford, Calif., August, 1981.

————— Raymond E., and Nancy M. Sanderson, *Construction Safety Management*, McGraw-Hill, Inc., New York, 1987.

Manual of Accident Prevention in Construction, 6th ed., The Associated General Contractors of America, Washington, D.C., 1971.

"The Occupational Safety and Health Act of 1970," P.L. 91–596 (OSHA 2001), U.S. Government Printing Office, Superintendent of Documents, Washington, D.C., December 1970.

OSHA Safety and Health Training Guidelines for Construction, (PB-239 312/AS), U.S. Dept. of Commerce, National Technical Information Service, Springfield, Va.

Robinson, M.R., *Accident Cost Accounting as a Means of Improving Construction Safety*, Technical Report 242, The Construction Institute, Department of Civil Engineering, Stanford University, Stanford, Calif., August, 1979.

Safety Requirements for Construction by Contract, U.S. Dept. of the Interior, Bureau of Reclamation, Washington, D.C.

Samelson, Nancy Morse, *The Effect of Foremen on Safety in Construction*, Technical Report No. 219, Stanford University, Dept. of Civil Engineering, The Construction Institute, Stanford, Calif., June, 1977.

Stanton, William A., and Jack H. Willenbrock, "Conceptual Framework for Computer-Based Safety Control," *Journal of Construction Engineering and Management*, ASCE, vol. 116, no. 3, pp. 383–398.

Business Methods in Managing Construction

Risk Management

Ashley, David B., *Construction Project Risk Sharing*, Technical Report 220, The Construction Institute, Department of Civil Engineering, Stanford University, Stanford, Calif., June, 1977.

————— Kazuyoshi Uehara, and Burke E. Thompson, "Critical Decision Making During Construction," *Journal of Construction Engineering and Management*, ASCE, vol. 109, no. 2, June 1983, pp. 146–162.

Derk, Walter T., *Insurance for Contractors*, 4th ed., Fred S. James & Co., Chicago, 1974.

Levitt, R. E., R. D. Logcher, and N. H. Quaddumi, "Impact of Owner-Engineer Risk Sharing on Design Conservatism," *Journal of Professional Issues in Engineering*, ASCE, vol. 110, 1984, pp. 157–167.

Industrial Relations

Anderson, Howard J., *Primer of Labor Relations*, Bureau of National Affairs, Washington, D.C., 1975.

Bonny, J. B. and J. P. Frein, eds., *Handbook of Construction Management and Organization*, 2d, ed., Chap. 24, L. E. Knack, "Labor Relations and their Effect on Employment Procedures," Van Nostrand Reinhold, 1980.

Borcherding, John D., "Construction Labor Unions in the United States," Chap. 10 in: *Directions in Managing Construction*, by Donald S. Barrie, ed., Wiley, New York, 1981.

Bourdon, C., and R. E. Levitt, *Union and Open Shop Construction*, Lexington Books, Lexington, Mass., 1980.

Business Roundtable, *Construction Industry Cost Effectiveness Project, Reports on Industrial Relations:* A-1 Construction Productivity Measurement: A-2 Construction Labor Motivation; C-1 Exclusive Jurisdiction in Construction; C-2 Scheduled Overtime Effect on Construction Projects; C-3 Contractor Supervision in Unionized Construction; C-4 Constraints Imposed by Collective Bargaining Agreements; C-5 Local Labor Practices; C-6 Absenteeism and Turnover; C-7 Impact of Local Union Politics; D-1 Use of Subjourneymen in the Union Sector; D-2 Government Limitations on Training Innovations; Utilization of Vocational Education in Construction Training; D-4 Training Problems in Open-Shop Construction; D-5 Labor Supply Information; E-1 Administration and Enforcement of Building Codes and Regulations, 1982, Park Avenue, New York, N.Y.

Christesen, R.J., and C. B. Tatum, "Labor Relations Considerations on PCM Projects," *Journal of the Construction Division*, ASCE, vol. 106, no. CO4, December 1980, pp. 535–549.

Christie, R. A. *Empire in Wood: A History of the Carpenters Union*, Cornell University Press, New York, 1956.

Construction Labor Report, Bureau of National Affairs, Washington, D.C. (weekly subscription periodical providing extensive reporting on industrial relations in construction.)

Fondahl, John and Boyd Paulson, *The Impact of Exclusive Craft Jurisdiction in the Construction Industry*, Technical Report 263, The Construction Institute, Department of Civil Engineering, Stanford University, Stanford, Calif., October, 1981.

Koehn, Enno, and Michael W. Jones, "Benefits and Costs of EEO Rules in Construction," *Journal of Construction Engineering and Management*, ASCE, vol. 109, no. 4, December 1983, pp. 435–446.

—— and Cesar A. Espaillat, "Costs and Benefits of MBE Rules in Construction," *Journal of Construction Engineering and Management*, ASCE, vol. 110, no. 2, June 1984, pp. 235–247.

Levitt, Raymond E., "Union vs. Non-Union Construction in the U.S.," *Journal of the Construction Division*, ASCE, vol. 105, no. CO4, December, 1979, pp. 289–303.

—— and Clinton C. Bourdon, "Cost Impacts of Prevailing Wage Laws in Construction," *Journal of the Construction Division*, ASCE, vol. 105, no. CO4, December, 1979, pp. 281–288.

—— and Donald S. Barrie, "Open Shop Movement," Chap. 12 in: *Directions in Managing Construction*, by Donald S. Barrie, ed., Wiley, New York, 1981.

—— and Joel B. Leighton, Employer and Owner Associations, Chap. 11 in: *Directions in Managing Construction*, by Donald S. Barrie, ed., Wiley, New York, 1981.

Mangum, G. L., *The Operating Engineers: Economic History of a Trade Union*, Harvard University Press, Cambridge, Mass., 1964.

Mills, Daniel Quinn, *Industrial Relations and Manpower in Construction*, MIT Press, Cambridge, Mass., 1972.

Monthly Labor Review, U.S. Dept. of Labor, Bureau of Labor Statistics.

Northrup, Herbert R., and Howard G. Foster, *Open Shop Construction*, University of Pennsylvania Press, Philadelphia, 1975.

O'Brian, J. J. and R. G. Zilly, eds., *Contractor's Management Handbook*, Chap. 9, D. Q. Mills, "The Labor Force and Industrial Relations," McGraw-Hill, Inc., New York, 1971.

Segal, M., *The Rise of the United Association*, Harvard University Press, Cambridge, Mass., 1969.

Claims, Liability, and Dispute Resolution

Avoiding and Resolving Disputes in Underground Construction, American Society of Civil Engineers, New York, June 1989.

Caspe, Marc, John Igoe, and S. R. McDonald, "The Disputes Resolution Clause," *Proceedings of the Annual Seminar/Symposium of the Project Management Institute*, San Francisco, CA, September 1988.

Collins, Carroll J., "Impact–The Real Effect of Change Orders," *Transactions of the American Association of Cost Engineers*, Morgantown, W. Va., June 1970, pp. 188–191.

Diekmann, James E., and Mark C. Nelson, "Construction Claims: Frequency and Severity," *Journal of Construction Engineering and Management*, ASCE, vol. 111, no. 1, March 1985, pp. 74–81.

Dunham, Clarence W., Robert D. Young, and Joseph T. Bockrath, *Contracts Specifications and Law for Engineers*, 3d ed., McGraw-Hill, Inc., New York, 1979.

Hester, Weston T., John A. Kuprenas, and H. Randolph Thomas, "Arbitration: A Look at Its Form and Performance," *Journal of Construction Engineering and Management*, ASCE, vol. 113, no. 3, September 1987, pp. 353–368.

Ibbs, C. William, and David B. Ashley, "Impact of Various Construction Contract Clauses," *Journal of Construction Engineering and Management*, ASCE, vol. 113, no. 3, September 1987, pp. 501–524.

Shanley, E. M., "A Better Way," *Civil Engineering*, December 1988, pp. 58–60.

Sweet, Justin, *Legal Aspects of Architecture, Engineering and the Construction Process*, 2d ed., West Publishing Co., St. Paul, Min., 1977.

Vlatas, D. A., "Owner and Contractor Review to Reduce Claims," *Journal of Construction Engineering and Management*, ASCE, vol. 112, no. 1, March 1986, pp. 104–111.

Wilson, Roy L., "Prevention and Resolution of Construction Claims," *Journal of The Construction Division*, ASCE, vol. 108, no. CO3, September 1982, pp. 390–405.

INDEX